Biofilm Associated Antimicrobial Resistance and Its Recovery

A biofilm is a collection of microbial cells that have adhered to biotic surfaces such as plant cuticles or animal epithelia, as well as abiotic surfaces such as rocky substratum or catheter exteriors. The life cycle of a microorganism includes the fundamental process of biofilm formation for survival in diverse and harsh environments since it is a protected mode of growth allowing colonisation of new habitats by dispersal of microbes from the microbial clusters. The biofilm bound microorganisms remain embedded in an extracellular polymeric matrix that protects the indwelling cells from surfactants, biocides, several invaders like protozoans, and defences offered by the hosts like phagocytic cells. The biofilm bound recalcitrant microbes induce chronic and nosocomial diseases, posing a serious threat to public health. It has been observed that various antimicrobial drugs are able to successfully remove the planktonic (freely suspended) states of microbes as compared to the sessile (substrate-bound) forms, thus resulting in the development of antimicrobial resistance.

Modern pharmacological strategies targeting the biofilm matrix differ from the conventional methods of antibiotic usage. This includes the use of natural compounds such as plant bioactive molecules, antimicrobial peptides, green synthesised nanoparticles, or secondary metabolites from other organisms that not only prevent the rise of antimicrobial resistance but are also safe for the host tissues.

Biofilm Associated Antimicrobial Resistance and Its Recovery provides a detailed and systematic review of alternative pharmacological developments in the field of biofilm research.

Features:

- A narrative overview of the mechanism of biofilm formation and its role in the emergence of antimicrobial resistance.
- Recent research in the development of antibiofilm remedies involving biogenic compounds.
- Advancements in biofilm detection methodologies with cutting-edge technologies.

This book serves as a resource for researchers who need to understand and analyze the progression of events during microbial biofilm formation, as well as design safer methodologies for its successful eradication. It may also be used as a textbook for a graduate level course in microbiology or microbial biotechnology.

Biofilm Associated Antimicrobial Resistance and Its Recovery

Edited by
Dibyajit Lahiri, Moupriya Nag, Rina Rani Ray and Sougata Ghosh

CRC Press
Taylor & Francis Group
Boca Raton London New York

CRC Press is an imprint of the
Taylor & Francis Group, an **informa** business

First edition published 2024
by CRC Press
6000 Broken Sound Parkway NW, Suite 300, Boca Raton, FL 33487-2742

and by CRC Press
4 Park Square, Milton Park, Abingdon, Oxon, OX14 4RN

CRC Press is an imprint of Taylor & Francis Group, LLC

ISBN: 9781032250335 (hbk)
ISBN: 9781032250366 (pbk)
ISBN: 9781003281238 (ebk)

DOI: 10.1201/9781003281238

Typeset in Times
by Newgen Publishing UK

Contents

Preface

Antimicrobial resistance (AMR) is one of the greatest emerging threats for global health sector. Days are coming when most of the pathogenic microbes will stop responding against the antimicrobials and a simple bacterial infection may invite death. A large number of antibiotic drugs, now being practiced to combat microbial pathogenesis will become gradually useless causing a huge economic loss worldwide. Unchecked AMR will take the lives of millions of livestock and other animals also.

Although development of AMR is a multifaceted and complex phenomenon, biofilm plays a pivotal role behind the acquisition of such resistance. Most of the microbes showing antibiotic resistance produce biofilm and within the architecturally complex matrices remain protected from external disturbances including antimicrobials, surfactants and immunological reactions. Biofilm bound cells are actually the bunch of sessile cells that adhere irretrievably on the abiotic or biotic surface and remain there secured within a biofilm matrix formed by extracellular polymeric substances.

Due to the recalcitrant nature, the biofilm bound cells are very difficult to approach and eradicate completely. This gives rise to persistent cells resulting in chronic diseases. The exopolymeric substances synthesised around the microbial cells do not allow the drug molecules to approach and thereby make the removal of biofilm formed on implants and medical devices extremely difficult.

Moreover, the extracellular matrix bound enzymes often show some antibiotic degrading effects. Such antibiotic resistance increases with horizontal gene transfer (HGT), hypermutation, and quorum sensing (QS), and extrusion of antibiotics through efflux mechanism.

Every day, microbes causing common or serious infections are found to acquire resistance against each new commercially available antibiotic. Moreover, the growth of microbial biofilms developed on indwelling medical devices like mechanical cardiac valves and pacemakers, various catheters, contact lenses, endotracheal tubes, artificial joints, and dental implants induce numerous bacterial infections that are difficult to manage with common antimicrobials.

Challenged with this reality, there came an increasing interest to find out some alternative therapeutic measures to manage such biofilm mediated AMR.

Various bioactive compounds, namely polyphenols, terpenoids, alkaloids, acetylenes, essential oils, and the like, antimicrobial peptides and biosurfactants with high antibiofilm efficacy, extracted from various natural sources are being used to disrupt the various events of biofilm formation, like interference of quorum-sensing mechanism, disintegration of biofilm matrix, downregulation of biofilm forming and quorum sensing genes. Antimicrobial peptides, for their potential broad-spectrum antimicrobial and antibiofilm efficacies and low environmental effect, are now considered to be a promising candidate to deal with AMR.

With the advent of nanotechnology, nanomaterials are now being increasingly used as effective therapeutic agents, or as nano carriers for previously known antimicrobials and there strategical application is expected to change the scenario of biofilm management.

Biofilm Associated Antimicrobial Resistance and Its Recovery would provide a comprehensive concept of every possible factor that are accountable for the expansion of biofilm mediated AMR. It will also focus on the alternate strategies that would help in reducing the robust trend in development of AMR depicting the mode of action of all such agents with antimicrobial and antibiofilm efficacies with a note of their relative advantages and shortcomings.

Preface

Editor Biographies

Dibyajit Lahiri is Associate Professor in the Department of Biotechnology, University of Engineering & Management, Kolkata. His research is precisely on biofilm isolated from human prosthesis and its inhibition by various novel phyto and nano compounds. He is keen to explore the molecular mechanism behind the removal of biofilm by natural compounds. His research area also encompasses computational drug development by using bioinformatics. His work is regularly presented before and appreciated by a number of experts of national and international repute. He is the editor of various international journals. He is a life member of the Indian Science Congress Association and Institution of Engineers (India). He has also edited books for Springer Nature and Elsevier.

Moupriya Nag is Associate Professor in the Department of Biotechnology, University of Engineering & Management, Kolkata. Her research interests include single molecule Biophysics of protein unfolding/refolding and protein aggregation kinetics and its mechanism of action especially the amyloid proteins in neurodegenerative disorders. Her recent research focuses on functional amyloids in bacteria and their relation in forming biofilms. Her work deals with developing novel antimicrobial and anti-biofilm compounds from natural sources including plants, microbes and green synthesized nanoparticles. She has about six years of teaching experience in Biophysics, Bioinformatics, Bioprocess and Biotechnology, Enzyme Technology, Computing, Molecular Modeling and Drug Design. She has also published many research articles in peer-reviewed international journals and authored or co-authored numerous book chapters. She is also a member of many scientific societies such as the Indian Biophysical Society and Indian Science Congress Association.

Rina Rani Ray is Associate Professor in the Department of Biotechnology, Maulana Abul Kalam Azad University of Technology, West Bengal. Her research focusses on the diverse field of microbial biotechnology. She has worked in academia and research for more than 25 years in several institutes of repute. She started her research career in the field of microbial enzyme technology. She successfully completed a number of research projects funded by different sponsoring agencies such as UGC, DST and DRDO and is working on a project sponsored by DBT, WB. She has published around 80 research papers in various peer-reviewed national and international journals and is on the editorial boards and a regular reviewer of many journals. At present her research group is working on the positive and negative aspects of bacterial biofilm. She is a member of various scientific associations and was nominated fellow of the Zoological Society, India and Academy for Environment and Life Sciences, India. She has also edited books for Springer Nature.

Sougata Ghosh is Associate Professor and Research Co-ordinator in the Department of Microbiology, School of Science, RK University, India. He obtained his BSc., MSc. and Ph.D. in Microbiology from Savitribai Phule Pune University, India. He is also Visiting Professor at the Department of Physics, Faculty of Science, Kasetsart University, Bangkok, Thailand and Department of Chemical Engineering, Northeastern University, Boston, USA. He was principle investigator of the Department of Biotechnology (DBT), Government of India funded Foldscope Research Project. He has filed three patents on novel nanodrugs and published more than 70 highly cited research articles in various international journals such as the *International Journal of Nanomedicine*, *Journal of Nanobiotechnology*, *Journal of Nanomaterials*, *Journal of Nanoscience and Nanotechnology*. He has 4077 citations, h-index 33 and i10 index 70. He has written one book entitled *Dioscorea bulbifera: A model system in nanobiotechnology and medicine*, which is a frequently consulted, voluminous and comprehensive report on traditional medicinal plants. He has edited eight books and contributed to 150 book chapters. As speaker he has addressed numerous international conferences, the most notable being IJAS-2019 and 2017 at Harvard University; ITNANO, Northeastern University, Boston; ASM Biodefense, Washington D.C.; Nano-Bio-Med 2013, Trieste, Italy; and BioMicroWorld2013, Madrid, Spain. He is a reviewer for various reputed journals such as *Nanoscale*, *Journal of Photochemistry & Photobiology, B: Biology*, *Journal of Diabetes Research*, *International Journal of Advances in Engineering Research*, *International Journal of Research in Science and Technology*, *Chemotherapy Research and Practice*, *Journal of Nanomedicine & Biotherapeutic Discovery*, *Journal of Nanomedicine and Nanotechnology*, *Applied Nanoscience and Arabian Journal of Chemistry (Elsevier)*. He is a life member of the Association of Microbiologists of India (AMI). His area of research consists of but is not limited to nanomedicine, microbial metabolism, bioprospecting, bioremediation, biofilms, cancer and diabetes.

Contributors

Saroj Ballav
Department of Human Physiology, Vidyasagar University,
Midnapore, West Bengal, India

Ankita Banerjee
Department of Biotechnology,
University of Engineering & Management, Kolkata, India

Debarati Basu
Department of Biotechnology,
Maulana Abul Kalam Azad University of Technology, West Bengal, India

Debasmita Bhattacharya
Department of Basic Science and Humanities, Institute of Engineering and Management, Salt Lake,
University of Engineering and Management
Kolkata, West Bengal, India

Mustansir Bhori
School of Biotechnology and Bioinformatics,
D. Y. Patil Deemed to be University,
CBD Belapur, Navi Mumbai, India

Ankurita Bhowmik
Department of Microbiology,
Tripura University, Suryamaninagar,
Tripura West, India

Madhurima Bhowmik
Department of Microbiology,
Tripura University, Suryamaninagar,
Tripura West, India

Archisman Bhunia
Department of Biotechnology,
University of Engineering & Management, Kolkata, India

Sambuddha Chakraborty
Department of Microbiology,
Tripura University, Suryamaninagar,
Tripura West, India

Anushree R. Chaphalkar
Docplexus Online Services Pvt. Ltd.,
Pune, India

Ashwini Chauhan
Department of Microbiology,
Tripura University, Suryamaninagar,
Tripura West, India

Ashish Cyril
School of Biotechnology and Bioinformatics,
D. Y. Patil Deemed to be University,
CBD Belapur, Navi Mumbai, India

Sharanya D.
Amity Institute of Biotechnology,
Amity University Maharashtra–Pune Expressway,
Bhatan, Panvel, Mumbai, Maharashtra, India

Arindam Das
Surface and Nanoscience Division, Material Science Group,
IGCAR, Kalpakkam, Tamil Nadu, India

Biswajit Das
Department of Human Physiology, Vidyasagar University,
Midnapore, West Bengal, India

Sanchari Das
Department of Biotechnology,
University of Engineering & Management, Kolkata, India

Tathagata Deb
Department of Biotechnology,
University of Engineering & Management, Kolkata, India

Rikthai Debbarma
Department of Microbiology,
Tripura University, Suryamaninagar,
Tripura West, India

Snigdha Debnath
Department of Microbiology,
Tripura University, Suryamaninagar,
Tripura West, India

Sergey Dobretsov
Centre of Excellence in Marine Biotechnology,
Sultan Qaboos University,
Department of Marine Science and Fisheries,
Sultan Qaboos University Al Khoud,
Muscat, Oman

Bandita Dutta
Department of Biotechnology,
Maulana Abul Kalam Azad University of Technology,
Kalyani, West Bengal, India

Anamika Dutta
Department of Microbiology,
Tripura University, Suryamaninagar,
Tripura West, India

Sayantani Garai
Department of Biotechnology,
University of Engineering & Management,
Kolkata, India

Chandradipa Ghosh
Department of Human Physiology, Vidyasagar University,
Midnapore, West Bengal, India

Sreejita Ghosh
Department of Biotechnology,
Maulana Abul Kalam Azad University of Technology,
Kalyani, West Bengal, India

Kartik Chandra Guchhait
Department of Human Physiology, Vidyasagar University,
Midnapore, West Bengal, India

Subrata Hazra
Department of Human Physiology, Vidyasagar University,
Midnapore, West Bengal, India

Debarati Jana
Department of Human Physiology, Vidyasagar University,
Midnapore, West Bengal, India

Santhosh Jathar
Amity Institute of Biotechnology,
Amity University Maharashtra–Pune Expressway,
Bhatan, Panvel, Mumbai, Maharashtra, India

Nidhi Jha
Department of Microbiology,
Tripura University, Suryamaninagar,
Tripura West, India

Renitta Jobby
Amity Institute of Biotechnology,
and
Centre of Excellence in Astrobiology,
Amity University Maharashtra–Pune Expressway,
Bhatan, Panvel, Mumbai, Maharashtra, India

Jun-ichi Kadokawa
Graduate School of Science and Engineering,
Kagoshima University,
1-21-24 Korimoto, Kagoshima, Japan

Bhavtosh A. Kikani
P. D. Patel Institute of Applied Sciences,
Charotar University of Science and Technology,
CHARUSAT Campus, Changa, Gujarat, India

Dibyajit Lahiri
Department of Biotechnology,
University of Engineering & Management
Kolkata, India

Suhasini Mallick
Department of Biotechnology,
Maulana Abul Kalam Azad University of Technology,
Kalyani, West Bengal, India

Mousumi Manna
Department of Human Physiology, Vidyasagar University,
Midnapore, West Bengal, India

Thankamani Marar
School of Biotechnology and Bioinformatics,
D. Y. Patil Deemed to be University, Plot no 50, Sector-15,
CBD Belapur, Navi Mumbai, India

Dipro Mukherjee
Department of Biotechnology,
University of Engineering & Management
Kolkata, India

P. Sriyutha Murthy
Biofouling & Biofilm Processes Section, Water & Steam Chemistry Division,
Bhabha Atomic Research Center, Kalpakkam, Tamil Nadu, India
and
Homi Bhabha National Institute, Anushaktinagar, Mumbai, India

Moupriya Nag
Department of Biotechnology,
University of Engineering & Management
Kolkata, India

Sneha P. Nair
Amity Institute of Biotechnology,
Amity University Maharashtra–Pune Expressway,
Bhatan, Panvel, Mumbai, Maharashtra, India

Aditi Pal
Department of Biotechnology,
University of Engineering & Management
Kolkata, India

Beula Subashini Panchatcharam
Department of Clinical Microbiology
Christian Medical College
Tamilnadu, Vellore, India

Amiya Kumar Panda
Department of Chemistry, Vidyasagar University,
Midnapore, West Bengal, India
and
Sadhu Ram Chand Murmu University of Jhargram,
Jhargram, West Bengal, India

Samriddha Paul
Department of Biotechnology,
University of Engineering & Management
Kolkata, India

Bharathi Ramalingam
Suzhou Xishan Biotech Inc., VRL Asia,
Suzhou, China

Rina Rani Ray
Department of Biotechnology,
Maulana Abul Kalam Azad University of Technology,
Kalyani, West Bengal, India

Katharina Richter
Department of Surgery, University of Adelaide,
Adelaide, Australia

Tanmay Sarkar
Malda Polytechnic, West Bengal State Council of Technical Education,
Government of West Bengal, Malda, India

Raju Sekar
Department of Biological Sciences
Xián Jiaotong-Liverpool University, Suzhou, China

Indu Sharma
Department of Microbiology,
Assam University, Silchar, Assam, India

Ankita Sil
Department of Microbiology,
Tripura University, Suryamaninagar,
Tripura West, India

Sobita Singha
Department of Microbiology,
Assam University, Silchar, Assam, India

Kanchanlata Tungare
School of Biotechnology and Bioinformatics,
D. Y. Patil Deemed to be University,
CBD Belapur, Navi Mumbai, India

Hanna Yumnam
Department of Microbiology,
Assam University, Silchar, Assam, India

1 Introduction

Ankurita Bhowmik, Sambuddha Chakraborty, Rikthai Debbarma, Snigdha Debnath, Nidhi Jha, Madhurima Bhowmik, Ankita Sil, Anamika Dutta and Ashwini Chauhan

ABBREVIATIONS

AHL: Acyl homoserine lactone
MATE: Multidrug and toxic compound extrusion
ABC: ATP-binding cassette
MFS: Major facilitator superfamily
SMR: Small multidrug resistance
DMT: Drug metabolite transporter
RND: Resistance-nodulation division
SAAT: Self-associating autotransporter proteins
MSCRAMMs: Microbial surface component recognizing adhesive matrix molecules
ROS: Reactive Oxygen Species
ARGs: Antibiotic resistance genes
HGT: Horizontal Gene Transfer
EPS: Extracellular polysaccharide
AMP: Antimicrobial peptide

1.1 BIOFILMS: AN INTRODUCTION

Antonie van Leeuwenhoek in the 17th century observed opaque deposits of "animalcules" on his teeth that remained even after diligent cleaning. These deposits of "animalcules" on tooth surfaces are what now we know as dental plaque. This was the first observation that documented the existence of bacteria in encased form (1). After Leeuwenhoek, in 1936, Zobel and Anderson observed increased bacterial growth in stored sea water above that in the natural environment. They concluded that the nutrients become concentrated on the solid surface of the container forming a film and provides bacteria a resting surface. They also assumed that the bacteria might adhere to the container surface and proliferate (2). In 1940, Heukelekian and Heller demonstrated that bacteria can grow and multiply on diluted media, in the presence of glass beads, when the medium was otherwise too diluted for sustaining bacterial growth. They also mentioned that the rate of biological reaction is greatly enhanced on substrate surfaces once the slime surface is active (3). Despite these aforementioned studies, the chemical and physical analysis of biofilm were not documented before the late 1960s (4, 5). In 1969, Jones et al, examined wastewater plant biofilms on trickling filters using electron microscopy and showed that polysaccharides primarily constitute the biofilm matrix and multiple species of bacteria were residing inside the biofilm. Characklis in 1973 revealed that industrial water biofilms were highly resistant to chlorine (5). It was in 1978 when Costerton's

DOI: 10.1201/9781003281238-1

analysis truly described biofilm per se. He stated that bacteria binds tenaciously and specifically to different surfaces such as cow intestine, human tooth and lungs, rock bed submerged in high-speed streams. This behavior is the crucial determinant of bacterial disease initiation and development. He further stated that bacteria do so by juxtaposing self-fabricated branched chain polysaccharides or glycocalyx to any surfaces to which they wish to adhere (6). In subsequent studies, Costerton stated that the glycocalyx acts as a nutrient trapping matrix by means of ion exchange and further transports this nutrient to the bacterial cells. Moreover, glycocalyx was also characterized as a physical barrier that enables limited protection to biofilm bacteria against antimicrobial agents. Biofilms developed under different settings possess diverse and distinct physiochemical features. Yet the development of biofilm design by Costerton et al in 1995 has laid down the first stone in understanding the basic concepts of the structural, chemical, and biophysical aspects of biofilm. This model indicates that in a biofilm, bacteria forms microcolonies on a surface surrounded by thick exopolysaccharides. Inside the biofilm, water channels are present that promote nutrient intake and outpour the toxic metabolites (7). Biofilm development facilitates bacteria in adapting to changing environments. Although, in general, scientists normally investigate pure mono bacterial cultures, most of the ecosystems are populated with a complex polymicrobial community that attaches to surfaces and is not free floating (1).

1.2 THE FORMATION OF BIOFILM

Biofilms are defined as surface adhered microbial communities. They aggregate, embed, and immobilize themselves on the substratum surface within an autogenerated extracellular polymeric (EPS) matrix. The process of biofilm formation is governed by multiple factors including type of bacteria, composition of the bacterial cell surface, hydrodynamics, cell signaling, gene regulation, availability of nutrients, and the like. Biofilm formation can be illustrated in three main stages (Figure 1.1): the attachment of microbial cells to a substratum, the maturation of the biofilm and the detachment or dispersal of cells (8, 9).

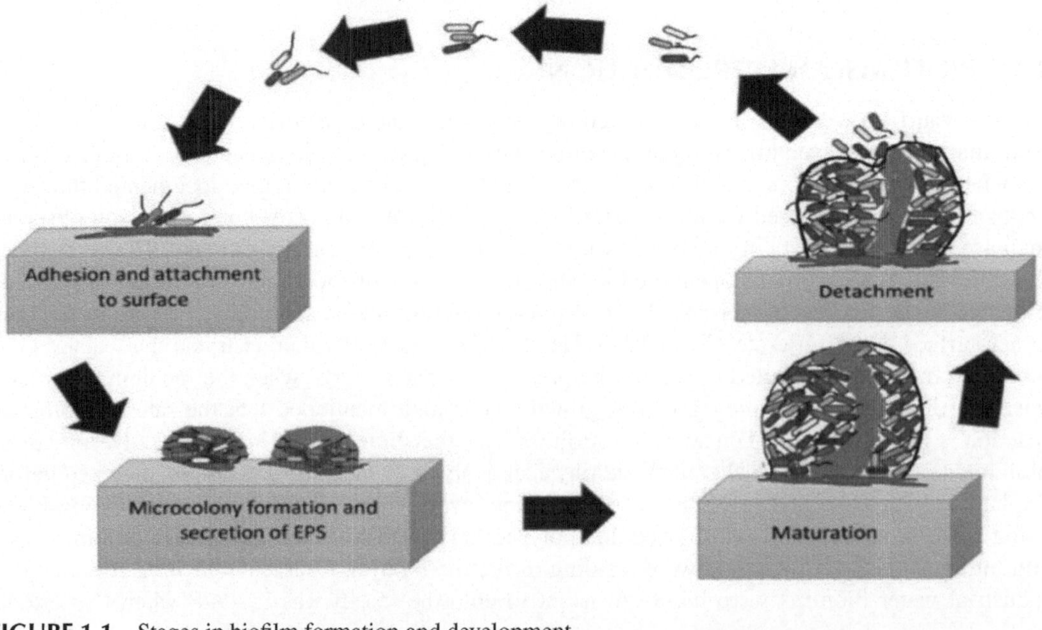

FIGURE 1.1 Stages in biofilm formation and development.

1.2.1 Attachment

Attachment of microorganisms to a substratum is an initial step in biofilm formation. Free floating bacteria approach any conditioned surface, such as blood factor coated vascular catheters, by flagellar movement or random collision driven by Brownian motion (10, 11).

Depending on environmental factors, the attachment can be reversible or irreversible. In reversible attachment the substratum-bacterial binding is fragile and may get disrupted very easily. Between abiotic substrate and bacterial cell surface, the primary attachment is mediated by electrostatic, van der Waals and hydrophobic interactions. However, specific receptor mediated interactions are important for the adhesion of bacteria to biotic surfaces (12).

For example, in *E.coli,* autotransporter protein antigen 43, self-associating autotransporter proteins (SAAT) such as AIDA-I and TibA impair bacteria motility and thus mediate bacterial cohesiveness to the surfaces (13).

In irreversible attachment bacteria commits for permanent attachment to the substratum and biofilm development. Bacteria usually influence this process by producing adhesins that bind to the substratum. The microbial cell surface hydrophobicity, and appendages such as fimbriae and EPS production have a major effect on adhesion. For instance, type 1 fimbriae in *Escherichia coli* adhere to eukaryotic cell surface receptors via tip adhesin. In *Staphylococcus epidermidis* and *Staphylococcus aureus* biofilms, microbial surface component recognizing adhesive matrix molecules (MSCRAMMs) are expressed which bind to host proteins such as fibronectin and fibrinogen and enable irreversible attachment (14, 15).

1.2.2 Maturation

The second stage in biofilm formation is the maturation stage. In this stage, the aggregated bacterial cells proliferate first, produce EPS and form microcolonies and then finally transform into an organised 3D mushroom-like structure.

1.2.2.1 EPS Secretion

Once the bacteria adhere to a surface, they start producing a sticky extracellular polymeric substance (EPS). EPS matrix is highly hydrated and is mainly composed of water (about 97%), with the remainder including exopolysaccharide polymer, proteins, lipids, nucleic acids and nutrients. EPS matrix acts as a source of nutrients, protects the embedded cells from external environmental factors, host immunity and antimicrobials (15).

1.2.2.2 Microcolony Formation

As the bacterial cells become attached to the surface, they aggregate, proliferate, and produce extracellular polymeric matrix. The bacterial cells become embedded within this hydrated EPS matrix and form a microcolony. Gradually, the base of the biofilm turns anaerobic when the biofilm reaches a thickness of 10–25 μm, which indicates that the maturation state is imminent in the biofilm. Within this microcolony the embedded bacterial cells coordinate with each other by quorum sensing. Bacterial cells use chemical molecules, called autoinducers (AI) for quorum sensing. By means of quorum sensing, the bacterial cells control population density, alter gene expression to carry out certain function such as the synthesis of virulence factor, biofilm colonization, antimicrobial resistance, and the like (15). Quorum sensing can be classified into two broad types:

1.2.2.2.1 Intraspecies Signaling

Gram negative bacteria utilize acyl homoserine lactone (AHL) for this type of communication. AHL produced by LuxI protein that diffuses out of the cells, become accumulated into the environment and then upon reaching a threshold level, AHLs are transported back into the cell. Inside the cell,

AHLs bind to LuxR protein and form a LuxR-AHL complex which further binds to a promoter gene for transcription (8, 16). In *P.aeruginosa,* mutation of the *lasI-rshiI* system, results in the formation of flat, thin and undifferentiated biofilms in comparison to the wild type (17).

Gram positive bacteria utilize a peptide-based signaling system for interspecies communication. They synthesize oligopeptide chains that are exported into their surrounding environment by an ATP-binding cassette (ABC) transporter. The signaling peptide molecules become accumulated and upon reaching a certain level, are transported back into the cells that bind to the target promoter to activate transcription. In *S.aureus,* the *agr* system plays a crucial role in biofilm formation. AgrD synthesizes the oligopeptide containing 46 amino acids that are further modified to a final autoinducing peptide (AIP) by AgrB. An AIP binding domain, along with AgrA and AgrC forms a two-component system and activates the RNAII and subsequently the RNAIII transcripts (18, 19).

1.2.2.2.2 Interspecies Signaling

This type of quorum sensing is used by both Gram positive and Gram negative bacteria containing a *luxS* gene. The autoinducer molecule (AI-2) or (2*S*,4*S*)-2-methyl-2,3,3,4-tetrahydroxytetrahydrofuran borate is produced by LuxS and is detected by a dedicated LuxPQ receptor (20).

1.2.3 Detachment

A variety of environmental factors like pH, temperature, low nutrient conditions, stress, and so forth, are known to affect biofilm detachment from the initial attachment (21). Enzymatic degradation of biofilm EPS matrix also leads to biofilm dispersal (22). Biofilm dispersals are regulated in different patterns in both Gram types of bacteria. *P.aeruginosa* produces hydrolases and rhamnolipids that degrade EPS matrix and facilitate dispersal. C-di-GMP is also known to regulate biofilm dispersal. Gram-positive bacteria, mainly *Staphylococcus aureus* produce certain proteases, nucleases, and detergents to facilitate dispersal (23). There are certain other agents that can cause dispersion, for example, compounds such as oligomer oligoG, nitric acids, unsaturated fatty acids, and so forth. The oligomer oligoG targets and degrades exopolysaccharide and thus leads to biofilm dispersal.

1.3 BIOFILM–ASSOCIATED FACTORS LEADING TO ANTIMICROBIAL RESISTANCE

Initial biofilm studies mostly focused on biofilm development in natural environments such as rivers, submerged rock beds, in sea water, and so forth. However, colonization of bacteria on medical devices and tissues has emerged as a human health hazard lately (24). Implanted devices are predominantly colonized by *Staphylococcus aureus, Enterococcus faecalis Staphylococcus epidermidis, Klebsiella pneumonia, Streptococcus spp., Escherichia coli, Pseudomonas aeruginosa, Acinetobacter spp,* and so forth. The biofilm developed by these bacteria contributes to more than 80% of the nosocomial infections (25). Moreover, inherent antimicrobial resistance in biofilm is a generic phenomenon. Several factors contribute to biofilm associated antimicrobial resistance, including impermeability of antibiotics through the EPS matrix, hypermutation and transfer of resistant genes, heterogeneity, slow growing persistent population, increased expression of efflux pumps, and the like (26).

1.3.1 Role of EPS in the Development of Antimicrobial Resistance (AMR)

EPS matrix primarily includes branched chain polysaccharides, eDNA, proteins and water that surrounds the biofilm cells. The polysaccharide chains tangle into a matt like structure that firmly holds the bacteria on the substrate surface, entraps nutrients for the biofilm members, and enables gradient formation that helps in the development of a heterogenic population. EPS acts as a sustainable barrier for the slow diffusion of antibiotics. Additionally, EPS interacts with the charged and hydrophilic antibiotic groups and sequesters them from penetrating the deeper layers of biofilm. In *P.aeruginosa*, the interaction of the alginate component of EPS with aminoglycosides results in their slow diffusion. The eDNA present in the biofilm matrix facilitates transfer of genetic materials amongst the bacteria and contributes to the structural integrity of the biofilm. Moreover, the development of a heterogenic population due to the gradient of nutrients, pH, oxygen, and so forth, in the biofilm contributes to antibiotic tolerance (11, 27).

1.3.2 The Role of Efflux Pumps in the Development of AMR

Efflux pumps play a key role in resistance development in bacteria. These systems regulate the intracellular levels of solutes including quorum sensing molecules, toxic metabolites, antibiotics, and so forth, by pumping them out of the cells (28–30). Based on the type of molecule exported by the pumps and the number of transmembrane spanning regions, these pumps are categorised into six families (29, 31–35),

1. Multidrug and toxic compound extrusion (MATE).
2. ATP-binding cassette (ABC).
3. Major facilitator superfamily (MFS).
4. Small multidrug resistance (SMR).
5. Drug metabolite transporter (DMT).
6. Resistance-nodulation division (RND).

In Gram negative bacteria, most of the resistance associated efflux pumps belong to the RND family, whereas, in Gram positive bacteria, the MFS and ABC type of efflux systems are predominant. Reports suggest that efflux pumps partially control QS signalling and the expression of virulence factors crucial for biofilm formation. In *B.pseudomallei*, a BpeAB-OprB pump is observed to be implicated with biofilm formation (36). In *P.aeruginosa*, the MexAB-OprM, MexEF-OprN, and MexXY systems classified under the RND family are crucial for antibiotic resistance (37). Particularly, MexAB-OprM is important in resistance towards gentamicin, tetracycline, aztreonam and tobramycin in *P.aeruginosa* biofilm. Other efflux systems expressed in *P.aeruginosa* biofilm, under ciprofloxacin and colistin exposure are PA1874-1877 and MesxCD-OprJ respectively (38). In clinical isolates of *E.coli*, AcrAB-TolC pumps are overexpressed and are reported to export chloramphenicol, fusidic acid, rifampicin, bile salts, fluoroquinolones, and so forth, in biofilm (39).

S.aureus is a clinically relevant Gram positive bacterium that contributes to major nosocomial infections. In *S.aureus,* six efflux systems have been identified that confer resistance to multiple antibiotics. For example, NorA, NorB and NorC pumps export quinolones, Tet38 exports tetracycline and MsrA exports macrolides and facilitate resistance towards these antibiotics (29) (40).

1.3.3 Horizontal Gene Transfer Facilitates AMR in Biofilm

Horizontal gene transfer (HGT) is a renowned driving phenomenon which enhances transfer of antibiotic resistance genes (ARGs) in multi species biofilm communities (41). It allows the bacteria to acquire new genes from the co-resistant species of bacteria that convey antimicrobial resistance to the overall multi species biofilm community. Back in 1993, Goodman and his coworkers revealed that the transfer rate of AGRs was higher in bacterial cells adhered on glass beads than in an aqueous phase. Since then, many studies have demonstrated biofilm as a reservoir of multidrug resistant bacteria. HGT can be acquired by three mechanisms in biofilm bacteria: transformation, conjugation, and bacteriophage mediated transduction. The type of genetic material transferred mostly depends on the above mechanisms. For example, conjugation transfers conjugative plasmids and elements while transformation transfers non-conjugative DNA and plasmids (42).

Conjugative plasmids and eDNA plays a crucial role in biofilm development. For instance, in *E.faecalis,* pCF10 encodes three cell wall proteins, namely, PrgA, PrgB and PrgC which promote initial adherence in biofilm. In *E.coli* and *K.pneumonia*, F plasmid and pOLA52 influence cell attachment and assist in biofilm formation respectively (42, 43). Moreover, one plasmid can carry multiple resistance genes. pAQU1is an unique example of a plasmid that carries *tet(B), tet(M), sul2, mph(G), mef(C),* $bla_{CARB\text{-}9}$ genes that confer resistance towards beta lactams and macrolides (42). Along with plasmids, eDNA also contributes to structural stability and antibiotic resistance in biofilm. eDNA is released inside the biofilm during DNA competence and cell death. Bacteria such as *S.pneumonia* releases bacteriocins and enzymes that degrade the cell wall of competent bacteria which leads to cell lysis and the release of eDNA (44).

1.3.4 The Role of Slow Growth Rate and Heterogeneity in AMR Development in Biofilm

The development of a resistant and persistent bacterial population is a characteristic feature of population heterogeneity in biofilm. Antibiotic resistance is generally defined as when a higher MIC concentration of antibiotics is required to inhibit bacterial growth (45). This phenomenon can be broadly termed as heteroresistance. Heteroresistance can be of two broad categories: polyclonal and monoclonal heteroresistance. Polyclonal heteroresistance is the classical form of antibiotic resistance that is observed in mixed species biofilm and displays only in a section of the population (45, 46). Colistin resistance in *Enterobacter cloacae* is a recent example of polyclonal heteroresistance. In a mouse model, *E.cloacae* isolates were rapidly treated with colistin which killed the susceptible population, leaving the heteroresistant population that had survived and multiplied in the presence of colistin (47). The surviving population displayed PhoQ histidine kinase dependent lipopolysaccharide modification of the outer membrane that conferred resistance towards colistin (48).

Monoclonal heteroresistance, however, develops from diversity in a clonal population. The changes in monoclonal heteroresistance are mostly acquired by gene amplification, which increases the expression of resistant genes and hence results in reduced sensitivity towards the antibiotic. But gene amplification is expensive and unstable in bacteria and hence is lost as the selective pressure disappears (45, 49, 50)

Growth rate is also a pioneer factor that impacts on the sensitivity of bacteria towards antibiotics. In general, antibiotics target metabolically active cells and hence dormant cells with reduced or no metabolism can render most of the antibiotics less sensitive. Classes of antibiotics that target the cell wall synthesis in bacteria such as penicillin and ampicillin are incapable of killing nondividing cells. A similar scenario can be observed inside the biofilm community. Gradients of oxygen, nutrients, and metabolic products, in the interior of the biofilm, result in development of slow growing phenotypes of bacteria that impart distinct tolerance and persistence patterns towards antibiotics (1). Persister cells are a subpopulation of biofilms that show higher tolerance to antibiotics. They can survive a lethal dose of antibiotics but cannot grow in the presence of antibiotics. They contribute to 0.1–10% of the biofilm population. It was in 1942, when Hobby et al first observed that penicillin was incapable of killing 1% of *S.aureus* cells. Later Bigger confirmed that this hyper tolerant surviving population of *S.aureus* does not undergo any genetic mutation and coined them as peristers (51). Persisters can develop spontaneously in the absence of stress when the level of persister proteins exceeds threshold concentration.

Toxin-antitoxin (TA) is a primary example of such proteins responsible for dormancy induction and persister cell formation. The toxin part of a TA system disrupts cellular processes, and the antitoxin part prevents toxicity. For example, in a TisB/lstR-1 system, the reduction of proton motive force and ATP levels by the TisB toxin induces persister formation. TisB producing cells show persistence to different antibiotics such as ciprofloxacin, ampicillin, streptomycin, and so forth (52). A different pattern of persistence was coined by Balaban et al in 2019, defined as triggered persistence, where a variety of signals including nutrient transition, pH, thermal and oxidative stress, starvation, sub-MIC antibiotic concentration, DNA damage, and the like, can trigger the development of persister cells (53, 54). In clinical settings, persister cells are estimated to confer higher tolerance to antibiotics and to cause recalcitrant infections (1).

1.4 EMERGING TRENDS IN COMBATING BIOFILM INFECTIONS

Biofilms have long been established as the preferred mode of bacterial lifestyle and the primary cause of disease chronicity, persistency, and infection recurrence. Antibiotic recalcitrance plays the most critical role behind this obstinacy. Intrinsic antibiotic resistance and biofilm structures both ensure protection to biofilm residing bacteria from antibiotics. The futility of antibiotics against bacterial pathogens points out an urgent need for an alternative to antibiotics. Several strategies have been reported earlier to deal with the biofilms that are assessed *in vitro* as well *in vivo* (55–57). Currently, biofilm remediation strategies aim to develop biofilm-dispersal agents to prevent early-stage bacterial surface colonization or disrupt the formed biofilm. Below, we will outline the emerging trends in antibiofilm agents.

1.4.1 Nano Particles against Biofilm

Globally, antibiotics have been rendered ineffective due to the rapid emergence of multi-drug resistant microbes. Failure of conventional antibiotic treatment against tissue-specific or device-specific biofilm infection led to the recruitment of alternatives to antibiotics. The application of nano particles (NPs) against bacterial biofilm has been considered as one of the promising approaches of the field (58–60). NPs, as the name indicates, are ultramicroscopic particles ranging from 1 to 100 nm in size (61), can be generally classified into; organic NPs, inorganic NPs and carbon-based NPs (Figure 1.2).

The high surface area to volume ratio granted NPs unique physio-chemical properties (62). Because of these properties, NPs are applied in energy-based research (63), as catalysts (64), in

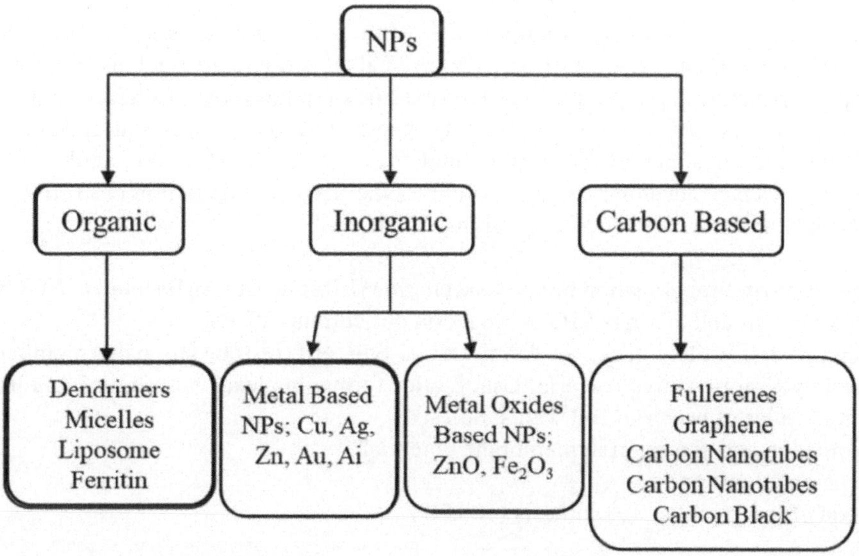

FIGURE 1.2 Broad Classification of NPs.

imaging applications (65), for drug delivery (66), and in gas, biological agents and chemical detection (67, 68). Moreover, these properties make some NPs bactericidal and a trending topic of microbiology (69–71). The killing mechanism of NPs are attributed to generation of oxidative stress by the formation of reactive oxygen species (ROS), the release of metal ions by the metal oxide based NPs and the non-oxidative mechanism inducing cell wall damage, pore formation, electrostatic attraction, and the like (72–74). The sizes of nanoparticles also has a determining effect on the antibiofilm efficacy of the particles. It is reported that smaller-sized (≤ 20nm) NPs show better bactericidal activity than relatively larger ones as it is easier to penetrate the bacterial cells (69, 70, 75). NPs can prevent surfaces from bacterial colonization or disrupt and disperse already formed biofilms. The coating of medical implantable device surfaces with NPs to reduce the risk of biofilm infections is receiving close attention from several research groups (76–79). The use of nanoparticles to eliminate biofilm communities of bacteria has been documented in a variety of reports. Silver NPs are widely used because of their antiseptic effect. Further, silver NPs do not show cytotoxicity (80–82). Chitosan is a biocompatible and biodegradable natural polymer deacetylated from chitin. Chitosan nanoparticles are positively charged and compel negatively charged bacterial cell walls and EPS. Chitosan coating used along with silver nanoparticles has been reported to to bring a more effective bactericidal effect (83–85). Gold nanoparticles are widely reported as antibiofilm agents(86, 87). Iron NPs are cost-effective, synthesized through coprecipitation or thermal decomposition. Iron NPs can act as magnetic nanoparticles and be used against bacterial biofilm (88, 89). SPIONs (super magnetic iron oxide nanoparticles), synthesized in a two-step process by chelation and metal conjugation, have been documented to use against biofilms of drug resistant bacteria (90). Copper (Cu) nanoparticles have EPS biomass degradation properties (91, 92). Cu NPs are antimicrobial as well as antifungal in nature and used as coating agents in implantable devices to prevent nosocomial infections (93, 94). ZnO has been considered as one of the potent bactericidal NPs and described to have a detrimental effect on bacterial biofilms (95). Green synthesis of nanoparticles is receiving widespread appreciation because of its cost-effectiveness, energy-efficiency, sustainability, and potent anti-biofilm properties (96–98). Needless to say, in this alarming situation of antibiotic resistance, NPs will play an indispensable role in preventing and eradicating bacterial biofilm in future.

1.4.2 Antimicrobial Peptides

Antimicrobial peptides (AMPs) are composed of from five to over a hundred amino acids, with molecular weight between 1–5 KDa (99). AMPs are called cationic host defense peptides as these are generally cationic peptides produced as the first line of defense by all multicellular organisms. On a structural basis, AMPs are classified into four types: alpha, beta, alpha-beta, and non-alpha-beta (99). AMPs have a broad bactericidal activity and can act as an alternative to antibiotic treatment, which lowers resistance development and eliminates the persister population (100–102). There are several modes of antibacterial action exerted by AMPs:

1. Neutralization or disaggregation lipopolysaccharide (LPS), where AMPs bind to LPS by electrostatic interactions and disrupts LPS aggregation or neutralizes LPS.
2. Inhibition of cell wall synthesis by lantibiotic, a type of post-translationally modified peptide, produced by Gram-positive bacteria. Lantibiotics form complexes with lipid I and lipid II precursor molecules of bacterial cell wall synthesis.
3. Pore formation and alteration in membrane potential.
4. Cell division inhibition.
5. Inhibition of macromolecule synthesis (103,104).

These peptides have been used as antimicrobial agents in biofilm infection in medical implants. The coating of antimicrobial peptides on the implant surfaces prevents the attachment of bacteria by exerting biocidal action (105–107). The antibacterial efficacy, stability, diffusion of peptides, and toxicity of AMP coated medical implants are highly dependent on the surface immobilization techniques (108). Recent developments focus on covalent immobilization approaches where acid or amine is assembled on the implant surface, and then AMPs are attached with the help of a cross-linking agent (108).

In a biofilm, microbial communities are dynamic and rapidly changing in different phenotypic forms. AMPs have first-killing kinetics compared to most conventional drugs, so they are suitable for the fast lane of the biofilm niche. AMPs have high potential to act on slow-growing and nongrowing bacteria (109). AMPs are reported to act through diverse antibiofilm mechanisms: disruption of biofilm structure, downregulation of quorum sensing, adhesion inhibition, alteration in biofilm pH, reduction of biofilm metabolic activity, downregulation of biofilm associated genes, and so forth (104, 110, 111). Various *in-vitro and in-vivo* studies have been reported to prevent or eradicate bacterial biofilm with several AMP-mediated anti-biofilm strategies (Figure 1.3).

A number of semi synthetic, synthetic, and naturally occurring AMPs have been used against bacterial biofilms. LL-37, a human antimicrobial peptide, has been used in a study against P. aeruginosa biofilms to decrease the attachment of the bacteria on plastic surfaces (112). Human saliva isolated iron-binding glycoprotein lactoferrin (LF) was found to have inhibitory effects against two periodontopathic bacteria *Porphyromonas gingivalis* and *Prevotella intermedia* biofilms (113). To minimize the destruction of naturally occurring flora by AMPs, an approach such as STAMPs, namely, specifically targeted antimicrobial peptides, have been developed. STAMPs conjugate a targeting and antimicrobial peptide domain and can be guided against target bacteria (114). Antimicrobial peptides with a combination of antibiotics have been reported as one efficient antibiofilm strategy. A study has reported a combination of tobramycin and STAMPs substantially reduced *P. aeruginosa* biofilm (115). A cationic peptide, citropin, has reportedly been used in the central venus catheter (CVC) rat model against *S. aureus* biofilm infection along with the antibiotics, minocycline and rifampicin (116). There are several more reports on the antibiofilm

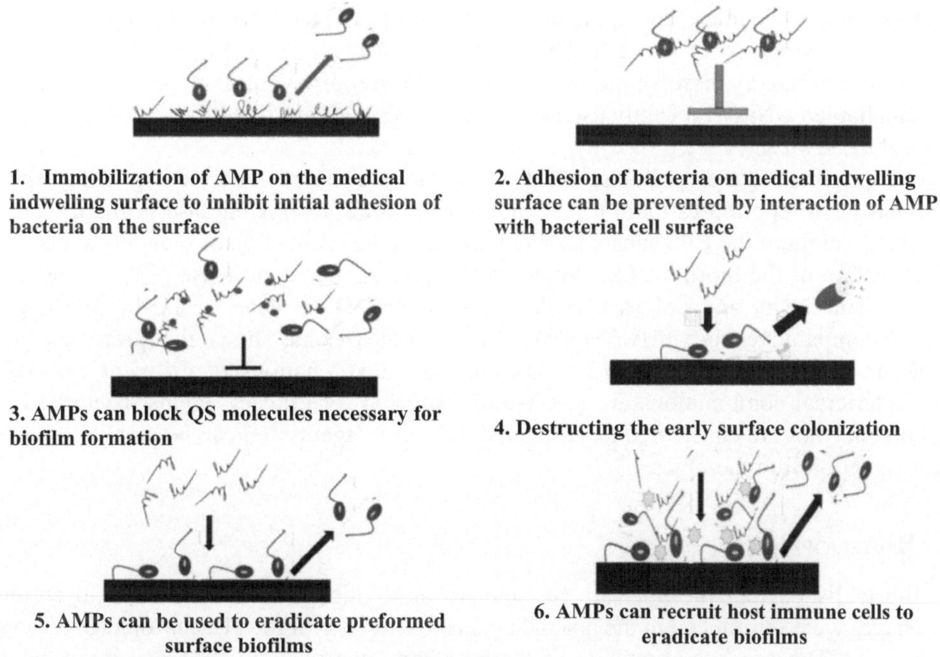

FIGURE 1.3 AMP-mediated anti-biofilm strategies.

activity of AMPs which we will be discussing further in this book. Nevertheless, it is obvious that in the conundrum of antibiotic resistance and biofilm-associated disease chronicity, finding an alternative to antibiotics with a broad spectrum anti-microbial activity is a critical assignment, and AMPs are representing a promising strategy against bacterial biofilm infection.

1.4.3 Enzymes

Some enzymes help to destroy the matrix component of biofilm and help stop biofilm formation (117, 118). These enzymes are taken into consideration as a therapeutic strategy against bacterial biofilm (119). Any enzyme that disperses a biofilm colony makes the bacterial cells accessible to various antimicrobial agents such as detergents, disinfectants, bacteriophages, and antibiotics. Another application of these enzymes is that they are extensively used as research tools that help identify the structural components of the biofilm matrix (120). Biofilm degrading enzymes primarily target the extracellular polymeric substances (EPSs), one of the most crucial components of bacterial biofilm. Biofilm eruption mechanisms vary with the types of enzyme. Some can directly target the biofilm components and destroy them. Some cause lysis of the cells, and a few interfere with the QS system. There are mainly four types of enzymes for the removal of biofilm: anti quorum sensing (Anti-QS) enzymes, oxidative enzymes, polysaccharide degrading enzymes, and lastly proteolytic enzymes (117, 121).

Among Anti-QS enzymes, lactonase is one of the efficient anti QS enzymes> Studies reported a 69–77% *Pseudomonas aeruginosa* biofilm reduction when treated with the enzyme (122). Lactonase was also able to downregulate the production of virulence factors in *Pseudomonas aeruginosa* (123). Oxidative enzymes primarily target the extracellular DNA of the biofilm matrix. Extracellular DNAs or eDNAs are reported to be secreted by the biofilm bacteria and act as one of the binding materials of biofilm architecture. Studies have reported the reduction of biofilm by deoxyribonuclease (DNase) (124–126). It is found that DNase-l has antibiofilm activity against many Gram-negative and Gram-positive species. DNase can inhibit as well as disrupt biofilms. In addition to inhibition and disruption, DNase was reported to make biofilm bacteria sensitive to killing by various biocides. The application of DNase to reduce *Streptococcus pneumoniae* has been reported by Moscoso et al (127). The addition of *Streptococcus*-derived deoxyribonuclease streptodornase can rapidly dissolve biofilms of clinical *P. aeruginosa* strains (128). Pulmozyme is a recombinant human DNase-l recently used in cystic fibrosis patients to reduce viscosity in the lungs, caused by *P. aeruginosa* (129).

Amylase, alginate lyase, cellulase, and lysozyme are polysaccharide-degrading enzymes. Cellulase has been reported to inhibit *P. aeruginosa* biofilm (130). Cellulase primarily decreases the molecular weight of the EPS substances and increases the production of reducing sugar, leading to the destruction of the biofilm EPS. Proteolytic enzymes such as protease (131), phage-derived enzymes such as lysin, and polysaccharide depolymerase (132) have strong biofilm degrading activities. A comparative study between amylases and proteases has shown that proteases are more efficacious in biofilm degradation (133). Nevertheless, combinations of different enzymes and enzyme-bactericidal combinations are also being evaluated for efficacy against bacterial biofilm. Antibiofilm enzymes are considered as efficient anti-biofilm agents and can be thought of as one of the effective antibiofilm therapies.

1.4.4 Bacteriophage

It is well-established that biofilms are resistant to antibiotics. As an alternative to antibiotics, bacteriophages were introduced in the arsenal against bacterial biofilms. Bacteriophages are viruses that kill bacteria. The term bacteriophage was derived from "bacteria" and the Greek word "phagein," meaning "to devour." Bacteriophages are omnipresent, from soil to air and sewage to oceans.

The basic theory of phage therapy is through the use of infectious bacteriophages against bacterial infection. In the classical approach, bacteriophages are isolated and purified from natural sources and then are applied against the targeted bacteria, but with the exciting advancement in biotechnology, it is possible to reinforce genetically modified or engineered bacteriophages (134). Interestingly phage therapy existed before the antibiotic era. It started in a village hospital in France in 1919 when d'Hérelle treated children suffering from severe *Shigella dysenteriae* infection with a bacteriophage cocktail (135). Phages are decorated with enzymes such as virion-associated peptidoglycans, hydrolases, polysaccharide depolymerizes, endolysins, and holins, capable of damaging and killing bacterial cells (136–138). Theoretically, bacteriophages can infect biofilm-forming bacteria faster than the planktonic form as, inside the biofilm, bacterial cells are in close proximity, providing a phage concentrated host population.

Biofilms have specific hallmark properties which make the residing biofilm bacteria invincible. Firstly, the EPS matrix works as a protective shield against antibiotics or the host's innate and adaptive immune system. Secondly, there is the bacterial cellular heterogeneity, which is recalcitrant towards antibiotics. Thirdly, there is the increased numbers of drug-resistant mutants. Besides these three significant factors, other associated factors are high functioning bacterial drug efflux pumps and high bacterial cell density. These multifaceted obstacles of biofilm demand a versatile antagonist which can take on all the mentioned impediments. Bacteriophages can diffuse through the EPS by water channels present on them (139, 140). Phages can readily infect persister cells and remain inside them until they become metabolically active. After activation, phages can initiate the lytic cycle and subsequently destroy the cell (141). An evolutionary trade-off phenomenon is studied where bacteria become increasingly sensitive to traditional antibiotics due to the process of acquiring phage resistance (142). Phage bacteriolytic enzymes such as Lysin and depolymerase helps in the lysis of the targeted host, and depolymerase leads to degradation of the capsule (CPS), exopolysaccharides (EPS), or lipopolysaccharides (LPS), decreasing bacterial virulence, followed by enhancement of immune system clearance (136). Hence, the primary rationale behind bacteriophage therapy against bacterial biofilms is based on: a) Phages can penetrate through the biofilm matrix. b) Phages can kill persister cells. c) Phages alone or with a combination of antibiotics can kill drug-resistant bacteria. Phage therapy as an alternative to antibiotics is receiving renewed attention. Using phages against drug resistant bacteria has already started in clinics (143). It is certain that lytic phages or virulent phages are more efficient bactericidals but advocation of lysis deficient phages cites evidence of a safer therapy in account of a moderate release of phage endotoxin and phage endotoxin associated immune response in the recipient (144).

Bacteriophages can be used to prevent biofilm formation or eradicate formed biofilms. Studies have shown that coating the surfaces with bacteriophages prevents biofilm formation. Single phage strain/phage cocktail/phage with combination of antibiotics/phage with a combination of antimicrobial agents has been experimentally applied to eradicate biofilms (145–147) (Figure 1.4).

Doolittle et al. (1994), was the first group to use the T4 bacteriophage against E Coli biofilm, the study contradicted the belief that biofilms due to the EPS layer are protected from phages (148). Hibma et al. (1997) presented an in-situ demonstration of bacteriophages preventing biofilm formation in stainless steel (149). Although antibiotic resistance is spiking up, antibiotics are still the drugs of choice against clinical infections. Bedi et al. (2009) isolated a *Klebsiella pneumonia* B5055 strain specific bacteriophage from a sewage sample which exerted approximately 5 log reduction of biofilms over eight days with a phage and antibiotic combination (150). Other than antibiotics, xylitol (151), hydroxyapatite (152), eicosanoic acid (153), chlorine (154), nanoparticles (155), and chestnut honey (156) has also been used with a combination of bacteriophages. Several clinical translations of phage therapy have been documented with promising results, promising a new horizon of antibacterial treatment.

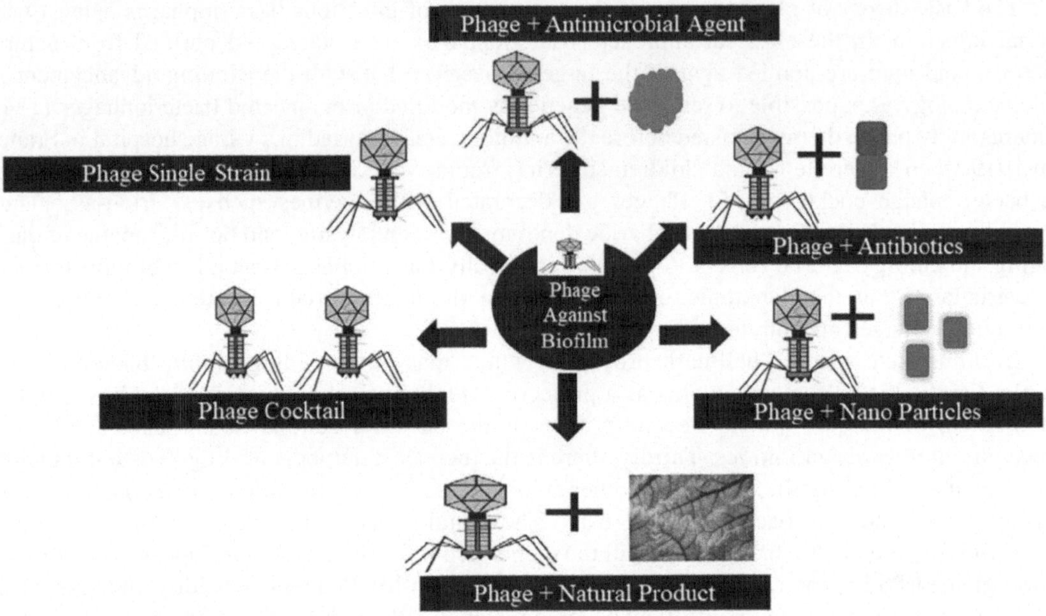

FIGURE 1.4 Phage therapy against bacterial biofilms.

1.4.5 CRISPR/Cas

Antibiotic resistance has become a global threat and is primarily associated with bacteria biofilm formation. To cease this life-threatening spread of drug resistance, microbiologists from all over the planet are trying to develop and implement efficient bacteriocidic approaches. One of the latest developments of this field is using CRISPR Cas to combat antimicrobial resistance (157). CRISPR Cas stands for clustered regularly interspaced short palindromic repeats linked with CRISPR-associated (Cas) genes. This system acts as an adaptive immune system of bacteria against viral infections and mobile genetic elements (MGEs). The CRISPR system works in three stages; primarily invaded foreign spacer sequences are recognized and integrated into a CRISPR array, in the subsequent step, synthesis/expression of CRISPR RNA (crRNA) occurs, and in the final stage, crRNA is recognized by the foreign genetic material and becomes annealed to it, and the base pair formation leads to cleavage of the crRNA–foreign gene complex by the associated Cas9 enzyme (158, 159). Based on genetic components, the CRISPR Cas system is classified into two main classes, and the classes are further divided into several types.

The slower pace of new antibiotic development and the rapid emergence of drug-resistant bacteria are leading society towards a significant disease burden. CRISPR Cas mediated antibacterial approach can be considered an outstanding approach against these drug-resistant bacterial pathogens. Unlike antibiotics, CRISPR can be targeted against specific bacteria of interest, saving the natural flora from collateral damage (157, 160). CRISPR Cas as an antibacterial strategy comes under the category of gene-editing tools such as zinc fingers, transcription activation-like effector nucleases (TALENs) and RNA interference (RNAi). However, CRISPR Cas is regarded as one of the efficient and affordable approaches as it functions on the basis of RNA-DNA interaction, while the other tools act on the more complex and expensive to apply Protein-DNA interaction mediated mechanisms (161) (162). The bactericidal strategy by CRISPR Cas can be as follows: (a) CRISPR can be used directly to cleave the genes present on the chromosome of pathogenic bacteria. Several studies on *S. aureus*, *S. pneumoniae* and *Salmonella* backed the strategy. The *S. aureus*-specific nuc gene was targeted by CRISPR, and the pathogen was successfully eliminated (163). (b) CRISPR can target AMR genes. The re-sensitization of *S. aureus* to kanamycin and methicillin is shown by a study reported by Yosef et al. (2005) (164). CRISPR mediated elimination of clinically significant, virulent, and highly drug resistant ESKAPE (*Enterococcus spp.*, *Staphylococcus aureus*, *Klebsiella pneumoniae*, *Acinetobacter baumannii*, *Pseudomonas aeruginosa*, and *Enterobacter spp.*) pathogens have been reported (165). In another in-vivo study on tetracycline and erythromycin-resistant *E. faecalis*, Rodrigues et al. successfully reduced antibiotic sensitivity by targeting the tet(M) and erm(B) genes by CRISPR (166). Despite being one of the most advanced tools to date, the technology is limited as it needs to minimize off-target effects in targeted cells.

1.5 CONCLUSION

The recalcitrance towards traditional antibiotics is one of the major concerns in treating persistent bacterial biofilm infection. The deteriorating efficacy of antibiotics against biofilm is multi-factorial such as altered microenvironment, higher rate of horizontal gene transfer, cellular heterogeneity including persister population, highly activated drug efflux pumps and deposition of drug degrading enzymes in the matrix. An effective alternative to antibiotics should be multi-factorial in order to tackle biofilms and associated infections. Several anti-biofilm strategies are in the research pipeline, including bacteriophages, nanoparticles, enzymes, anti-microbial peptides, gene-editing tools, phytochemicals, and the like. Research has revealed that combination therapy can also restore the effectiveness of existing antibiotics. Despite all the exclusive findings more and more research is warranted as pathogenic bacteria are ever evolving and biofilms are worsening medical outcomes. The future trend of anti-biofilm therapy will depend on the safety, efficacy, affordability, and longevity of the treatment.

REFERENCES

1. Bjarnsholt, T. (2011) Introduction to biofilms. *Biofilm Infect.* (May 2014):1–9.
2. Zobell, C.E., Anderson, D.Q. (1936) Observations on the multiplication of bacteria in different volumes of stored sea water and The influence of oxygen tension and solid surfaces. *Biol Bull.* 71(2):324–42.
3. Heukelekian, A. (1940) *Relation Between Food Concentration and Surface for Bacterial Growth.* (1933):547–58.
4. Characklis, W. G. (1973). *Attached Microbial Growths.*
5. Jones H.C., Roth I.L., Sanders, W.M. (1969) Electron microscopic study of a slime layer. *J Bacteriol.* 99(1):316–25.
6. Costerton, J.W., Geesey, G.G., Cheng, K.J. (1978) How bacteria stick. *Sci Am.* 238(1):86–95.
7. Lewandowski, J.W.C., Z.D.E.C., and DRKHML-S. (1995) Microbial biofilms. *Nanoscale Technol Biol Syst.* 341–57.
8. Bhowmik, A., Malhotra, A., Jana, S., and Chauhan, A. (2021) *Biofilm Formation and Pathogenesis.* 3–37 p.
9. Shirtliff, M.E., Mader, J.T., and Camper, A.K. (2002) Molecular interactions in biofilms. *Chem Biol.* 9(8):859–71.
10. O'Toole, G.A., and Kolter, R. (1998) Flagellar and twitching motility are necessary for.pdf. *Mol Microbiol.* 30(2):295–304.
11. Finch, R.G., Pace, J.L., and Rupp, M.E. (2005) *Biofilms, Infection, and Antimicrobial Therapy*. Taylor and Francis.
12. Krasowska, A., and Sigler, K. (2014) How microorganisms use hydrophobicity and what does this mean for human needs?. *Front Cell Infect Microbiol.* 4(AUG):1–7.
13. Nwoko, E.S.Q.A., and Okeke, I.N. (2021) Bacteria autoaggregation: How and why bacteria stick together. *Biochem Soc Trans.* 49(3):1147–57.
14. Foster, T.J. (2020) Surface proteins of Staphylococcus epidermidis. *Front Microbiol.* 11(July).

15. Rabin, Z., Opoku-Temeng, Du B and S.K. (2017) Biofilm formation mechanisms and targets for developing antibiofilm agents. *Future Med Chem*. 71(10):643.
16. Zhou, L., Zhang, Y., Ge, Y., Zhu, X., and Pan, J. (2020) Regulatory mechanisms and promising applications of quorum sensing-inhibiting agents in control of bacterial biofilm formation. *Front Microbiol*. 11(October):1–11.
17. Walters III, M.C., Roe, F., Bugnicourt, A., Franklin, M.J., and Stewart, P.S. (2003) Contributions of antibiotic penetration, Oxygen Limitation. *Antimicrob Agents Chemother*. 47(1):317–23.
18. Prazdnova, E.V., Gorovtsov, A.V., Vasilchenko, N.G., Kulikov, M.P., Statsenko, V.N., Bogdanova, A.A., et al. (2022) *Quorum-Sensing Inhibition by Gram-Positive Bacteria*. 1–24.
19. Ziemichód, A., and Skotarczak, B. (2017) QS–systems communication of Gram-positive bacterial cells. *Acta Biol*. 24:51–6.
20. Bridges, A.A., and Bassler, B.L. (2019) The intragenus and interspecies quorum-sensing autoinducers exert distinct control over vibrio cholerae biofilm formation and dispersal. *PLoS Biol*. 17(11):1–28.
21. McEldowney, S., and Fletcher, M. (1988) Effect of pH, temperature, and growth conditions on the adhesion of a gliding bacterium and three nongliding bacteria to polystyrene. *Microb Ecol*. 16(2):183–95.
22. Powell. M.S., and Slater, N.K.H. (1982) Removal rates of bacterial cells from glass surfaces by fluid shear. *Biotechnol Bioeng*. 24(11):2527–37.
23. Lister, J.L., and Horswill, A.R. (2014) Staphylococcus aureus biofilms: Recent developments in biofilm dispersal. *Front Cell Infect Microbiol*. 4(DEC):1–9.
24. Landry, K.S., Morey, J.M., Bharat, B., Haney, N.M., and Panesar, S.S. (2020) Biofilms—impacts on human health and its relevance to space travel. *Microorganisms*. 8(7):1–15.
25. Dhar, Y., and Han, Y. (2020) Current developments in biofilm treatments: Wound and implant infections. *Eng Regen*. 1(June):64–75.
26. Goel, N., Fatima, S.W., Kumar, S., Sinha, R., and Khare, S.K. (2021) Antimicrobial resistance in biofilms: Exploring marine actinobacteria as a potential source of antibiotics and biofilm inhibitors. *Biotechnol Reports*. 30:e00613.
27. Pinto, R.M., Soares, F.A., Reis, S., Nunes, C., and Van Dijck, P. (2020) Innovative strategies toward the disassembly of the EPS matrix in bacterial Biofilms. *Front Microbiol*. 11(May):1–20.
28. Alav, I., Sutton, J.M., and Rahman, K.M. (2018) Role of bacterial efflux pumps in biofilm formation. *J Antimicrob Chemother*. 73(8):2003–20.
29. Soto, S.M. (2013) Role of efflux pumps in the antibiotic resistance of bacteria embedded in a biofilm. 223–9.
30. Ebbensgaard, A.E., Løbner-Olesen, A., and Frimodt-Møller, J. (2020) The role of efflux pumps in the transition from low-level to clinical antibiotic resistance. *Antibiotics*. 9(12):1–7.
31. Saier, M.H., Beatty, J.T., Goffeau, A., Harley, K.T., Heijne, W.H.M., Huang, S.C., et al. (1999) The major facilitator superfamily. *J Mol Microbiol Biotechnol*. 1(2):257–79.
32. Lubelski, J., Konings, W.N., and Driessen, A.J.M. (2007) Distribution and physiology of ABC-Type transporters contributing to multidrug resistance in bacteria. *Microbiol Mol Biol Rev*. 71(3):463–76.
33. Nikaido, H., and Takatsuka, Y. (2009) Mechanisms of RND multidrug efflux pumps. *Biochim Biophys Acta–Proteins Proteomics*. 1794(5):769–81.
34. Kuroda, T., and Tsuchiya, T. (2009) Multidrug efflux transporters in the MATE family. *Biochim Biophys Acta–Proteins Proteomics*. 1794(5):763–8.
35. Jack, D.L., Yang, N.M., and Saier, M.H. (2001) The drug/metabolite transporter superfamily. *Eur J Biochem*. 268(13):3620–39.
36. Chan, Y.Y., and Chua, K.L. (2005) The Burkholderia pseudomallei BpeAB-OprB efflux pump: Expression and impact on quorum sensing and virulence. *J Bacteriol*. 187(14):4707–19.
37. De Kievit, T.R., Parkins, M.D., Gillis, R.J., Srikumar, R., Ceri, H., Poole, K., et al. (2001) Multidrug efflux pumps: Expression patterns and contribution to antibiotic resistance in Pseudomonas aeruginosa biofilms. *Antimicrob Agents Chemother*. 45(6):1761–70.
38. Pamp, S.J., Gjermansen, M., Johansen, H.K., and Tolker-Nielsen, T. (2008) Tolerance to the antimicrobial peptide colistin in Pseudomonas aeruginosa biofilms is linked to metabolically active cells, and depends on the pmr and mexAB-oprM genes. *Mol Microbiol*. 68(1):223–40.
39. Fernandes, P., Sommer Ferreira, B., and Sampaio Cabral, J.M. (2003) Solvent tolerance in bacteria: Role of efflux pumps and cross-resistance with antibiotics. *Int J Antimicrob Agents*. 22(3):211–6.

40. Ding, Y., Onodera, Y., Lee, J.C., and Hooper, D.C. (2008) NorB, an efflux pump in Staphylococcus aureus strain MW2, contributes to bacterial fitness in abscesses. *J Bacteriol.* 190(21):7123–9.
41. Wiedenbeck, J., and Cohan, F.M. (2011) Origins of bacterial diversity through horizontal genetic transfer and adaptation to new ecological niches. *FEMS Microbiol Rev.* 35(5):957–76.
42. Abe, K., Nomura, N., and Suzuki, S. (2021) Biofilms: Hot spots of horizontal gene transfer (HGT) in aquatic environments, with a focus on a new HGT mechanism. *FEMS Microbiol Ecol.* 96(5):1–12.
43. Ghigo, J-M. (2001) Natural conjugative plasmids induce bacterial bioflm development horizontal gene transfer is a principal source of evolution leading to change in the ecological character of bacterial species. *Nature.* 412(July):442–5.
44. Oggioni, M.R., Trappetti, C., Kadioglu, A., Cassone, M., Iannelli, F., Ricci, S., et al. (2006) Switch from planktonic to sessile life: A major event in pneumococcal pathogenesis. *Mol Microbiol.* 61(5):1196–210.
45. Dewachter, L., Fauvart, M., and Michiels, J. (2019) Bacterial Heterogeneity and Antibiotic Survival: Understanding and Combatting Persistence and Heteroresistance. *Mol Cell.* 76(2):255–67.
46. Nicoloff, H., Hjort, K., Levin, B.R., and Andersson, D.I. (2019) The high prevalence of antibiotic heteroresistance in pathogenic bacteria is mainly caused by gene amplification. *Nat Microbiol.* 4(3):504–14.
47. Band, V.I., Crispell, E.K., Napier, B.A., Herrera, C.M., Greg, K., Vavikolanu, K., et al. (2017) *Antibiotic Failure Mediated by a Resistant Subpopulation in Enterobacter cloacae.* 1(6):1–20.
48. Kang, K.N., Klein, D.R., Kazi, M.I., Guérin, F., Cattor, V., Brodbelt, J.S., et al. (2019) Colistin heteroresistance in rnterobacter cloacae is regulated by PhoPQ-dependent 4-amino-4-deoxy-l-arabinose addition to lipid A. *Mol Microbiol.* 111(6):1604–16.
49. McGann, P., Courvalin, P., Snesrud, E., Clifford, R.J., Yoon, E.J., Onmus-Leone, F., et al. (2014) Amplification of aminoglycoside resistance gene aphA1 in acinetobacter baumannii results in tobramycin therapy failure. *MBio.* 5(2):1–8.
50. Andersson, D.I., Nicoloff, H., Hjort, K. (2019) Mechanisms and clinical relevance of bacterial heteroresistance. *Nat Rev Microbiol.* 17(8):479–96.
51. Bigger, J.W. (1944) Treatment of Staphylococcal infections with Penicillin by intermittent sterilisation. *Lancet.* 244(6320):497–500.
52. Wood, T.K., Knabel, S.J., Kwan, B.W. (2013) Bacterial persister cell formation and dormancy. *Appl Environ Microbiol.* 79(23):7116–21.
53. Balaban, N.Q., Helaine, S., Lewis, K., Ackermann, M., Aldridge, B., Andersson, D.I., et al. (2019) Definitions and guidelines for research on antibiotic persistence. *Nat Rev Microbiol.* 17(7):441–8.
54. van den Bergh, B., Fauvart, M., and Michiels, J. (2017) Formation, physiology, ecology, evolution and clinical importance of bacterial persisters. *FEMS Microbiol Rev.* 41(3):219–51.
55. Chauhan, A., Lebeaux, D., Ghigo, J.M., and Beloin, C. (2012) Full and broad-spectrum in vivo eradication of catheter-associated biofilms using gentamicin-EDTA antibiotic lock therapy. *Antimicrob Agents Chemother.* 56(12):6310–8.
56. Chauhan, A., Ghigo, J.M., and Beloin, C. (2016) Study of in vivo catheter biofilm infections using pediatric central venous catheter implanted in rat. *Nat Protoc.* 11(3):525–41.
57. Singh, B.P., Ghosh, S., and Chauhan, A. (2021) Development, dynamics and control of antimicrobial-resistant bacterial biofilms: A review. *Environ Chem Lett.* 19(3):1983–93.
58. Qayyum, S., and Khan, A.U. (2016) Nanoparticles: Vs. biofilms: A battle against another paradigm of antibiotic resistance. Vol. 7, *MedChemComm.* 1479–1498 p.
59. Benoit, D.S.W., Sims, K.R., and Fraser, D. (2019) Nanoparticles for Oral Biofilm Treatments. *ACS Nano.* 13(5):4869–75.
60. Shkodenko, L., Kassirov, I., and Koshel, E. (2020) Metal oxide nanoparticles against bacterial biofilms: Perspectives and limitations. *Microorganisms.* 8(10):1–21.
61. Khan, I., Saeed, K., and Khan, I. (2019) Nanoparticles: Properties, applications and toxicities. *Arab J Chem.* 12(7):908–31.
62. Afolalu, S.A., Soetan, S.B., Ongbali, S.O., Abioye, A.A., and Oni, A.S. (2019) Morphological characterization and physio-chemical properties of nanoparticle–Review. *IOP Conf Ser Mater Sci Eng.* 640(1).

63. Li, Q., Luo, T.Y., Zhou, M., Abroshan, H., Huang, J., Kim, H.J., et al. (2016) Silicon nanoparticles with surface nitrogen: 90% quantum yield with narrow luminescence bandwidth and the ligand structure based energy law. *ACS Nano*. 10(9):8385–93.
64. Shaker Ardakani, L., Surendar, A., Thangavelu, L., and Mandal, T. (2021) Silver nanoparticles (Ag NPs) as catalyst in chemical reactions. *Synth Commun*. 51(10):1516–36.
65. Padmanabhan, P., Kumar, A., Kumar, S., Chaudhary, R.K., and Gulyás, B. (2016) Nanoparticles in practice for molecular-imaging applications: An overview. *Acta Biomater*. 41:1–16.
66. Parveen, S., Misra, R., and Sahoo, S.K. (2012) Nanoparticles: A boon to drug delivery, therapeutics, diagnostics and imaging. *Nanomedicine Nanotechnology, Biol Med*. 8(2):147–66.
67. Dilonardo, E., Penza, M., Alvisi, M., Di Franco, C., Rossi, R., Palmisano, F., et al. (2016) Electrophoretic deposition of Au NPs on MWCNT-based gas sensor for tailored gas detection with enhanced sensing properties. *Sensors Actuators, B Chem*. 223:417–28.
68. Jans, H., and Huo, Q. (2012) Gold nanoparticle-enabled biological and chemical detection and analysis. *Chem Soc Rev*. 41(7):2849–66.
69. Han, C., Romero, N., Fischer, S., Dookran, J., Berger, A., and Doiron, A.L. (2017) Recent developments in the use of nanoparticles for treatment of biofilms. *Nanotechnol Rev*. 6(5):383–404.
70. Tran, H.M., Tran, H., Booth, M.A., Fox, K.E., Nguyen, T.H., Tran, N., et al. (2020) Nanomaterials for treating bacterial biofilms on implantable medical devices. *Nanomaterials*. 10(11):1–19.
71. Das, P., and Karankar, V.S. (2019) New avenues of controlling microbial infections through antimicrobial and anti-biofilm potentials of green mono-and multi-metallic nanoparticles: A review. *J Microbiol Methods*. 167:105766.
72. Shaikh, S., Nazam, N., Rizvi, S.M.D., Ahmad, K., Baig, M.H., Lee, E.J., et al. (2019) Mechanistic insights into the antimicrobial actions of metallic nanoparticles and their implications for multidrug resistance. *Int J Mol Sci*. 20(10):1–15.
73. Vallet-Regí, M., González, B., and Izquierdo-Barba, I. (2019) Nanomaterials as promising alternative in the infection treatment. *Int J Mol Sci*. 20(15).
74. Wang L, Hu C, Shao L. The-antimicrobial-activity-of-nanoparticles—present-situati. *Int J Nanomedicine*. 2017;12:1227–49.
75. Singh, A., Gautam, P.K., Verma, A., Singh, V., Shivapriya, P.M., Shivalkar, S., et al. (2020) Green synthesis of metallic nanoparticles as effective alternatives to treat antibiotics resistant bacterial infections: A review. *Biotechnol Reports*. 25:e00427.
76. Arsiwala, A.M., Raval, A.J., and Patravale, V.B. (2013) Nanocoatings on implantable medical devices. *Pharm Pat Anal*. 2(4):499–512.
77. Knetsch, M.L.W., and Koole, L.H. (2011) New strategies in the development of antimicrobial coatings: The example of increasing usage of silver and silver nanoparticles. *Polymers (Basel)*. 3(1):340–66.
78. McGuffie, M.J., Hong, J., Bahng, J.H., Glynos, E., Green, P.F., Kotov, N.A., et al. (2016) Zinc oxide nanoparticle suspensions and layer-by-layer coatings inhibit staphylococcal growth. *Nanomedicine Nanotechnology, Biol Med*. 12(1):33–42.
79. Basova, T.V., Vikulova, E.S., Dorovskikh, S.I., Hassan, A., and Morozova, N.B. (2021) The use of noble metal coatings and nanoparticles for the modification of medical implant materials. *Mater Des*. 204:109672.
80. Ansari, M.A., Khan, H.M., Khan, A.A., Cameotra, S.S., and Pal, R. (2014) Antibiofilm efficacy of silver nanoparticles against biofilm of extended spectrum β-lactamase isolates of Escherichia coli and Klebsiella pneumoniae. *Appl Nanosci*. 4(7):859–68.
81. Franci, G., Falanga, A., Galdiero, S., Palomba, L., Rai, M., Morelli, G., et al. (2015) Silver nanoparticles as potential antibacterial agents. *Molecules*. 20(5):8856–74.
82. Markowska, K., Grudniak, A.M., and Wolska, K.I. (2013) Silver nanoparticles as an alternative strategy against bacterial biofilms. *Acta Biochim Pol*. 60(4):523–30.
83. Tan, Y., Ma, S., Leonhard, M., Moser, D., Haselmann, G.M., Wang, J., et al. (2018) Enhancing antibiofilm activity with functional chitosan nanoparticles targeting biofilm cells and biofilm matrix. *Carbohydr Polym*. 200:35–42.
84. de Paz, L.E.C., Resin, A., Howard, K.A., Sutherland, D.S., and Wejse, P.L. (2011) Antimicrobial effect of chitosan nanoparticles on Streptococcus mutans biofilms. *Appl Environ Microbiol*. 77(11):3892–5.

85. Karthik, C.S., Chethana, M.H., Manukumar, H.M., Ananda, A.P., Sandeep, S., Nagashree, S., et al. (2021;) Synthesis and characterization of chitosan silver nanoparticle decorated with benzodioxane coupled piperazine as an effective anti-biofilm agent against MRSA: A validation of molecular docking and dynamics. *Int J Biol Macromol*. 181:540–51.
86. Khan, F., Manivasagan, P., Lee, J.W., Pham, D.T.N., Oh, J., and Kim, Y.M. (2019) Fucoidan-stabilized gold nanoparticle-mediated biofilm inhibition, attenuation of virulence and motility properties in pseudomonas aeruginosa pao1. *Mar Drugs*. 17(4):1–19.
87. Ahmed, A., Khan, A.K., Anwar, A., Ali, S.A., and Shah, M.R. (2016) Biofilm inhibitory effect of chlorhexidine conjugated gold nanoparticles against Klebsiella pneumoniae. *Microb Pathog*. 98:50–6.
88. Armijo, L.M., Wawrzyniec, S.J., Kopciuch, M., Brandt, Y.I., Rivera, A.C., Withers, N.J., et al. (2020) Antibacterial activity of iron oxide, iron nitride, and tobramycin conjugated nanoparticles against Pseudomonas aeruginosa biofilms. *J Nanobiotechnology*. 18(1):1–27.
89. Prodan, A.M., Iconaru, S.L., Chifiriuc, C.M., Bleotu, C., Ciobanu, C.S., Motelica-Heino, M., et al. (2013) Magnetic properties and biological activity evaluation of iron oxide nanoparticles. *J Nanomater*. 2013.
90. Taylor, E.N., Kummer, K.M., Durmus, N.G., Leuba, K., Tarquinio, K.M., and Webster, T.J. (2012) Superparamagnetic iron oxide nanoparticles (SPION) for the treatment of antibiotic-resistant biofilms. *Small*. 8(19):3016–27.
91. LewisOscar, F., MubarakAli, D., Nithya, C., Priyanka, R., Gopinath, V., Alharbi, N.S., et al. (2015) One pot synthesis and anti-biofilm potential of copper nanoparticles (CuNPs) against clinical strains of Pseudomonas aeruginosa. *Biofouling*. 31(4):379–91.
92. Mao, M., Zhang, W., Huang, Z., Huang, J., Wang, J., Li, W., et al. (2021) Graphene oxide-copper nanocomposites suppress cariogenic Streptococcus mutans biofilm formation. *Int J Nanomedicine*. 16(January):7727–39.
93. Bayat, M., Zargar, M., Chudinova, E., Astarkhanova, T., and Pakina, E. (2021) In vitro evaluation of antibacterial and antifungal activity of biogenic silver and copper nanoparticles: The first report of applying biogenic nanoparticles against pilidium concavum and pestalotia sp. fungi. *Molecules*. 26(17).
94. Li, M., Gao, L., Schlaich, C., Zhang, J., Donskyi, I.S., Yu, G., et al. (2017) Construction of functional coatings with durable and broad-spectrum antibacterial potential based on mussel-inspired Dendritic Polyglycerol and in Situ-Formed copper nanoparticles. *ACS Appl Mater Interfaces*. 9(40):35411–8.
95. Lee, J.H., Kim, Y.G., Cho, M.H., and Lee, J. (2014) ZnO nanoparticles inhibit Pseudomonas aeruginosa biofilm formation and virulence factor production. Vol. 169, *Microbiological Research*. Elsevier GmbH. 888–896 p.
96. Ishwarya, R., Vaseeharan, B., Kalyani, S., Banumathi, B., Govindarajan, M., Alharbi, N.S., et al. (2018) Facile green synthesis of zinc oxide nanoparticles using Ulva lactuca seaweed extract and evaluation of their photocatalytic, antibiofilm and insecticidal activity. *J Photochem Photobiol B Biol*. 178:249–58.
97. Gopinath, K., Kumaraguru, S., Bhakyaraj, K., Mohan, S., Venkatesh, K.S., Esakkirajan, M., et al. (2016) Green synthesis of silver, gold and silver/gold bimetallic nanoparticles using the Gloriosa superba leaf extract and their antibacterial and antibiofilm activities. *Microb Pathog*. 101:1–11.
98. Punniyakotti, P., Panneerselvam, P., Perumal, D., Aruliah, R., and Angaiah, S. (2020) Anti-bacterial and anti-biofilm properties of green synthesized copper nanoparticles from Cardiospermum halicacabum leaf extract. *Bioprocess Biosyst Eng*. 43(9):1649–57.
99. Bahar, A.A., and Ren, D. (2013) Antimicrobial peptides. *Pharmaceuticals*. 6(12):1543–75.
100. Mahlapuu, M., Håkansson, J., Ringstad, L., and Björn, C. (2016) Antimicrobial peptides: An emerging category of therapeutic agents. *Front Cell Infect Microbiol*. 6(DEC):1–12.
101. Lei, J., Sun, L.C., Huang, S., Zhu, C., Li, P., He, J., et al. (2019) The antimicrobial peptides and their potential clinical applications. *Am J Transl Res*. 11(7):3919–31.
102. Zhang, Q.Y., Yan, Z.B,, Meng, Y.M., Hong, X.Y., Shao, G., Ma, J.J., et al. (2021) Antimicrobial peptides: Mechanism of action, activity and clinical potential. *Mil Med Res*. 8(1):1–25.
103. Raheem, N., and Straus, S.K. (2019) Mechanisms of Action for antimicrobial peptides with antibacterial and antibiofilm functions. *Front Microbiol*. 10(December):1–14.
104. Park, S.C., Park, Y., and Hahm, K.S. (2011) The role of antimicrobial peptides in preventing multidrug-resistant bacterial infections and biofilm formation. *Int J Mol Sci*. 12(9):5971–92.

105. Kazemzadeh-Narbat, M., Lai, B.F.L., Ding, C., Kizhakkedathu, J.N., Hancock, R.E.W., and Wang, R. (2013) Multilayered coating on titanium for controlled release of antimicrobial peptides for the prevention of implant-associated infections. *Biomaterials*. 34(24):5969–77.
106. Kazemzadeh-Narbat, M., Kindrachuk, J., Duan, K., Jenssen, H., Hancock, R.E.W., and Wang, R. (2010) Antimicrobial peptides on calcium phosphate-coated titanium for the prevention of implant-associated infections. *Biomaterials*. 31(36):9519–26.
107. Acosta, S., Ibañez-Fonseca, A., Aparicio, C., and Rodríguez-Cabello, J.C. (2020) Antibiofilm coatings based on protein-engineered polymers and antimicrobial peptides for preventing implant-associated infections. *Biomater Sci*. 8(10):2866–77.
108. Nicolas, M., Beito, B., Oliveira, M., Martins, M.T., Gallas, B., Salmain, M., et al. (2022) Strategies for antimicrobial peptides immobilization on surfaces to prevent biofilm growth on biomedical devices. *Antibiotics*. 11(1).
109. Batoni, G., Maisetta, G., Lisa Brancatisano, F., Esin, S., and Campa, M. (2011) Use of Antimicrobial peptides against microbial biofilms: Advantages and limits. *Curr Med Chem*. 18(2):256–79.
110. Yasir, M., Willcox, M.D.P., and Dutta, D. (2018) Action of antimicrobial peptides against bacterial biofilms. *Materials (Basel)*. 11(12).
111. Chung, P.Y., and Khanum, R. (2017) Antimicrobial peptides as potential anti-biofilm agents against multidrug-resistant bacteria. *J Microbiol Immunol Infect*. 50(4):405–10.
112. Overhage, J., Campisano, A., Bains, M., Torfs, E.C.W., Rehm, B.H.A., and Hancock, R.E.W. (2008) Human host defense peptide LL-37 prevents bacterial biofilm formation. *Infect Immun*. 76(9):4176–82.
113. Wakabayashi, H., Yamauchi, K., Kobayashi, T., Yaeshima, T., Iwatsuki, K., and Yoshie, H. (2009) Inhibitory effects of lactoferrin on growth and biofilm formation of Porphyromonas gingivalis and Prevotella intermedia. *Antimicrob Agents Chemother*. 53(8):3308–16.
114. Eckert, R., He, J., Yarbrough, D.K., Qi, F., Anderson, M.H., and Shi, W. (2006) Targeted killing of Streptococcus mutans by a pheromone-guided "smart" antimicrobial peptide. *Antimicrob Agents Chemother*. 50(11):3651–7.
115. Kang, H.K., Kim, C., Seo, C.H., and Park, Y. (2017) The therapeutic applications of antimicrobial peptides (AMPs): a patent review. *J Microbiol*. 55(1):1–12.
116. Cirioni, O., Giacometti, A., Ghiselli, R., Kamysz, W., Orlando, F., Mocchegiani, F., et al. (2006) Citropin 1.1-treated central venous catheters improve the efficacy of hydrophobic antibiotics in the treatment of experimental staphylococcal catheter-related infection. *Peptides*. 27(6):1210–6.
117. Kaplan, J.B. (2009) Therapeutic potential of biofilm-dispersing enzymes. 32(9):545–54.
118. Kaplan, J.B. (2014) Biofi lm Matrix-Degrading Enzymes. *Springer*. 1147:203–13.
119. Thallinger, B., Prasetyo, E.N., Nyanhongo, G.S., and Guebitz, G.M. (2013) Antimicrobial enzymes: An emerging strategy to fight microbes and microbial biofilms. *Biotechnol J*. 8(1):97–109.
120. Allison, D.G. (2003) The Biofilm Matrix. *Biofouling*. 19(SUPPL.):139–49.
121. Meireles, A., Borges, A., Giaouris, E., and Simões, M. (2016) The current knowledge on the application of anti-biofilm enzymes in the food industry. *Food Res Int*. 86:140–6.
122. Hraiech, S., Hiblot, J., Lafleur, J., Lepidi, H., Papazian, L., Rolain, J.M., et al. (2014) Inhaled lactonase reduces pseudomonas aeruginosa quorum sensing and mortality in rat pneumonia. *PLoS One*. 9(10):1–8.
123. Guendouze, A., Plener, L., Bzdrenga, J., Jacquet, P., Rémy, B., Elias, M., et al. (2017) Effect of quorum quenching lactonase in clinical isolates of pseudomonas aeruginosa and comparison with quorum sensing inhibitors. *Front Microbiol*. 8(FEB):1–10.
124. Nguyen, U.T., and Burrows, L.L. (2014) DNase I and proteinase K impair listeria monocytogenes biofilm formation and induce dispersal of pre-existing biofilms. *Int J Food Microbiol*. 187:26–32.
125. Kaplan, J.B., Lovetri, K., Cardona, S.T., Madhyastha, S., Sadovskaya, I., Jabbouri, S., et al. (2012) Recombinant human DNase i decreases biofilm and increases antimicrobial susceptibility in staphylococci. *J Antibiot (Tokyo)*. 65(2):73–7.
126. Tetz, V.V., and Tetz, G.V. (2010) Effect of extracellular DNA destruction by DNase I on characteristics of forming biofilms. *DNA Cell Biol*. 29(8):399–405.
127. Moscoso, M., García, E., and López, R. (2009) Pneumococcal biofilms. *Int Microbiol*. 12(2):77–85.
128. Nemoto, K., Hirota, K., Murakami, K., Taniguti, K., Murata, H., Viducic, D., et al. (2003) Effect of Varidase (streptodornase) on biofilm formed by Pseudomonas aeruginosa. *Chemotherapy*. 49(3):121–5.

129. Frederiksen, B., Pressler, T., Hansen, A., Koch, C., and Høiby, N. (2006) Effect of aerosolized rhDNase (Pulmozyme®) on pulmonary colonization in patients with cystic fibrosis. *Acta Paediatr Int J Paediatr*. 95(9):1070–4.
130. Loiselle, M., and Anderson, K.W. The use of cellulase in inhibiting biofilm formation from organisms commonly found on medical implants. *Biofouling*. 2003;19(2):77–85.
131. Selan, L., Berlutti, F., Passariello, C., Comodi-Ballanti, M.R., Thaller, M.C. (1993) Proteolytic enzymes: A new treatment strategy for prosthetic infections? *Antimicrob Agents Chemother*. 37(12):2618–21.
132. Topka-Bielecka, G., Dydecka, A., Necel, A., Bloch, S., Nejman-Faleńczyk, B., Węgrzyn, G., et al. (2021) Bacteriophage-derived depolymerases against bacterial biofilm. *Antibiotics*. 10(2):1–21.
133. Molobela, I.P., Cloete, T.E., and Beukes, M. (2010) Protease and amylase enzymes for biofilm removal and degradation of extracellular polymeric substances (EPS) produced by Pseudomonas fluorescens bacteria. *African J Microbiol Res*. 4(14):1515–24.
134. Pires, D.P., Cleto, S., Sillankorva, S., Azeredo, J., and Lu, T.K. (2016) Genetically engineered phages: A review of advances over the last decade. *Microbiol Mol Biol Rev*. 80(3):523–43.
135. Chanishvili, N. (2012) *Phage Therapy-History from Twort and d'Herelle Through Soviet Experience to Current Approaches*. 1st ed. Vol. 83, Advances in Virus Research. Elsevier Inc. 3–40 p.
136. Maciejewska, B., Olszak, T., and Drulis-Kawa, Z. (2018) Applications of bacteriophages versus phage enzymes to combat and cure bacterial infections: An ambitious and also a realistic application? *Appl Microbiol Biotechnol*. 102(6):2563–81.
137. Jado, I., López, R., García, E., Fenoll, A., Casal, J., García, P., et al. (2003) Phage lytic enzymes as therapy for antibiotic-resistant Streptococcus pneumoniae infection in a murine sepsis model. *J Antimicrob Chemother*. 52(6):967–73.
138. Fischetti, V.A. (2018) Development of phage lysins as novel therapeutics: A historical perspective. *Viruses*. 10(6).
139. Lacroix-Gueu, P., Briandet, R., Lévêque-Fort, S., Bellon-Fontaine, M.N., and Fontaine-Aupart, M.P. (2005) In situ measurements of viral particles diffusion inside mucoid biofilms. *Comptes Rendus–Biol*. 328(12):1065–72.
140. Ferriol-González, C., and Domingo-Calap, P. (2020) Phages for biofilm removal. *Antibiotics*. 9(5):1–16.
141. Harper, D.R., Parracho, H.M.R.T., Walker, J., Sharp, R., Hughes, G., Werthén, M., et al. (2014) Bacteriophages and biofilms. *Antibiotics*. 3(3):270–84.
142. Burmeister, A.R., and Turner, P.E. (2020) Trading-off and trading-up in the world of bacteria–phage evolution. *Curr Biol*. 30(19):R1120–4.
143. Dedrick, R.M., Guerrero-Bustamante, C.A., Garlena, R.A., Russell, D.A., Ford, K., Harris, K., et al. (2019) Engineered bacteriophages for treatment of a patient with a disseminated drug-resistant mycobacterium abscessus. *Nat Med*. 25(5):730–3.
144. Ali, Q., and Wahl, L.M. (2017) Mathematical modelling of CRISPR-Cas system effects on biofilm formation. *J Biol Dyn*. 11(0):264–84.
145. Lin, D.M., Koskella, B., and Lin, H.C. (2017) Phage therapy: An alternative to antibiotics in the age of multi-drug resistance. *World J Gastrointest Pharmacol Ther*. 8(3):162.
146. Santiago, A.J., and Donlan, R.M. (2020) Bacteriophage infections of biofilms of health care-associated pathogens: Klebsiella pneumoniae. *EcoSal Plus*. 9(1):1–12.
147. Khalifa, L., Shlezinger, M., Beyth, S., Houri-Haddad, Y., Coppenhagen-Glazer, S., Beyth, N., et al. (2016) Phage therapy against Enterococcus faecalis in dental root canals. *J Oral Microbiol*. 8(1):1–11.
148. Doolittle, M.M., Cooney, J.J., and Caldwell, D.E. (1995) Lytic infection of Escherichia coli biofilms by bacteriophage T4. *Can J Microbiol*. 41(1):12–8.
149. Hibma, A.M., Jassim, S.A.A., and Griffiths, M.W. (1997) Infection and removal of L-forms of listeria monocytogenes with bred bacteriophage. *Int J Food Microbiol*. 34(3):197–207.
150. Bedi, M.S., Verma, V., and Chhibber, S. (2009) Amoxicillin and specific bacteriophage can be used together for eradication of biofilm of Klebsiella pneumoniae B5055. *World J Microbiol Biotechnol*. 25(7):1145–51.
151. Chhibber, S., Bansal, S., and Kaur, S. (2015) Disrupting the mixed-species biofilm of klebsiella pneumoniae B5055 and pseudomonas aeruginosa PAO using bacteriophages alone or in combination with xylitol. *Microbiol (United Kingdom)*. 161(7):1369–77.

152. Papaianni, M., Ricciardelli, A., Fulgione, A., D'errico, G., Zoina, A., Lorito, M., et al. (2020) Antibiofilm activity of a trichoderma metabolite against xanthomonas campestris pv. Campestris, alone and in association with a phage. *Microorganisms*. 8(5).
153. Papaianni, M., Ricciardelli, A., Casillo, A., Corsaro, M.M., Borbone, F., Ventura, B. Della., et al. (2021) The union is strength: The synergic action of long fatty acids and a bacteriophage against xanthomonas campestris biofilm. *Microorganisms*. 9(1):1–18.
154. Zhang, Y., and Hu, Z. (2013) Combined treatment of Pseudomonas aeruginosa biofilms with bacteriophages and chlorine. *Biotechnol Bioeng*. 110(1):286–95.
155. Abdelsattar, A.S., Nofal, R., Makky, S., Safwat, A., Taha, A., and El-Shibiny, A. (2021) The synergistic effect of biosynthesized silver nanoparticles and phage zcse2 as a novel approach to combat multidrug-resistant salmonella enterica. *Antibiotics*. 10(6).
156. Oliveira, A., Sousa, J.C., Silva, A.C., Melo, L.D.R., and Sillankorva, S. (2018) Chestnut honey and bacteriophage application to control pseudomonas aeruginosa and escherichia coli biofilms: Evaluation in an ex vivo wound model. *Front Microbiol*. 9(July):1–13.
157. Gholizadeh, P., Köse, Ş., Dao, S., Ganbarov, K., Tanomand, A., Dal, T., et al. (2020) How CRISPR-Cas system could be used to combat antimicrobial resistance. *Infect Drug Resist*. 13:1111–21.
158. Makarova, K.S., Haft, D.H., Barrangou, R., Brouns, S.J.J., Charpentier, E., Horvath, P., et al. (2011) Evolution and classification of the CRISPR-Cas systems. *Nat Rev Microbiol*. 9(6):467–77.
159. Knott, G.J., and Doudna, J.A. (2018) CRISPR-Cas guides the future.pdf. *Science (80-)*. 361(August):866–9.
160. Shabbir, M.A.B., Shabbir, M.Z., Wu, Q., Mahmood, S., Sajid, A., Maan, M.K., et al. (2019) CRISPR-cas system: Biological function in microbes and its use to treat antimicrobial resistant pathogens. *Ann Clin Microbiol Antimicrob*. 18(1):1–9.
161. Ma, Y., Zhang, L., and Huang, X. (2014) Genome modification by CRISPR/Cas9. *FEBS J*. 281(23):5186–93.
162. Greene, A.C. (2018) CRISPR-Based Antibacterials: Transforming Bacterial Defense into Offense. *Trends Biotechnol*. 36(2):127–30.
163. Park, J.Y., Moon, B.Y., Park, J.W., Thornton, J.A., Park, Y.H., and Seo, K.S. (2017) Genetic engineering of a temperate phage-based delivery system for CRISPR/Cas9 antimicrobials against Staphylococcus aureus. *Sci Rep*. 7(August 2016):1–13.
164. Ates, A., Tastan, C., and Ermertcan, V. (2020) Alternative therapies to antibiotics: CRISPR-Cas antimicrobials. *Gene Ed*. 1(1):30–5.
165. de Aledo, M.G., González-Bardanca, M., Blasco, L., Pacios, O., Bleriot, I., Fernández-García, L., et al. (2021) Crispr-cas, a revolution in the treatment and study of eskape infections: Pre-clinical studies. *Antibiotics*. 10(7):1–23.
166. Enterococci, D.A., Rodrigues, M., Mcbride, S.W., Hullahalli, K., and Palmer, K.L. (2019) Conjugative delivery of CRISPR-Cas9 for the selective depletion of antibiotic-resistant Enterococci. *Antimicrob Agents Chemother*. 63(August):1–14.

2 Biofilm Formation and Metabolomic Regulation

Bhavtosh A. Kikani

2.1 INTRODUCTION

Microbial biofilms can be defined as a cooperative association of various microorganisms either belonging to the same species or to different species, which are embedded in a complex extracellular matrix adhered on a solid surface. On the contrary, same microbial species grows unicellularly with evident motility in the liquid media (Azeredo et al. 2017). Different microorganisms may produce biofilms on multitudes of objects to survive under adverse conditions (Landini et al. 2010). The process of biofilm formation initiates with attachment onto a solid substratum, shading of flagella, followed by secretion of exopolymeric substances, including proteins, polysaccharides, humic substances, and extracellular DNA. Although it seems very easy, microbial biofilm formation is precisely regulated by various physicochemical factors, mainly availability of nutrients and various environmental stress conditions (Assaidi et al. 2018; Favre et al. 2018; Jin et al. 2018; Landini et al. 2010). Biofilm formation may sometimes be beneficial, whereas in many instances it yields negative impacts, such as a prevalence of multidrug resistance with enhanced production of virulence factors by a range of pathogens. Microbial biofilms may block industrial pipes or may form over medical devices and inanimate objects in the hospital, which is one of the reasons for increased cases of nosocomial infections (Azeredo et al. 2017; Jin et al. 2018; Landini et al. 2010) (Figure 2.1).

Considering these facts, microbial biofilms are increasingly studied using metabolomics to determine metabolic and/or biochemical alterations happening when a cell decides to shift from the planktonic growth to the biofilm state. These approaches can also be explored to evaluate the composition of biofilms and also to monitor *in vitro* and *in vivo* biofilm generation processes. These approaches have helped the research fraternity to characterize several unique biochemical reactions and metabolites. Principally, the microbial metabolomics can be sub categorized into two domains: untargeted metabolomics and targeted metabolomics. The untargeted approach, also commonly known as global metabolomics, involves analysis and interpretation of an entire metabolome without any knowledge of the metabolites to be studied. On the other hand, targeted metabolomics exclusively deals with specific metabolites. Thus, both of the branches emphasize metabolite fingerprinting or metabolite foot printing (Baptista et al. 2018; Götz et al. 2018; López-Gresa et al. 2017; Sander et al. 2017).

The metabolomics generally starts with sampling, followed by an extraction process, which largely affects accurate data acquisition and interpretation (Azzollini et al. 2018; Chatzimitakos and Stalikas 2016; Maansson et al. 2016). The very first step ensures separation of intracellular and secretary metabolites produced by the microorganisms. The extraction procedure, which should be optimized each time, includes physical or chemical disruption of the microbial cells in the presence of various buffers or solvents to isolate different metabolites. An equally significant step

DOI: 10.1201/9781003281238-2

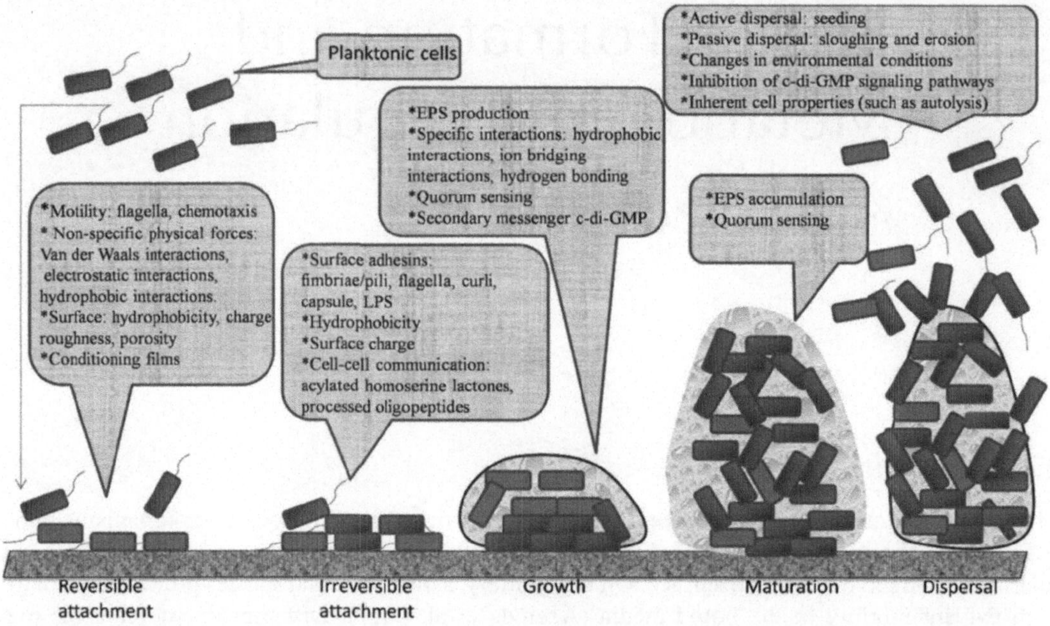

FIGURE 2.1 The five main phases leading to the development and dispersal of biofilm. Reprinted from Muhammad, M.H., Idris, A.L., Fan, X., Guo, Y., Yu, Y., Jin, X., Qiu, J., Guan, X. and Huang, T. (2020) Beyond risk: bacterial biofilms and their regulating approaches. *Frontiers in microbiology*, *11*, p.928. The article is an open access article distributed under the terms and conditions of the Creative Commons Attribution license (http://creativecommons.org/licenses/by/3.0/).

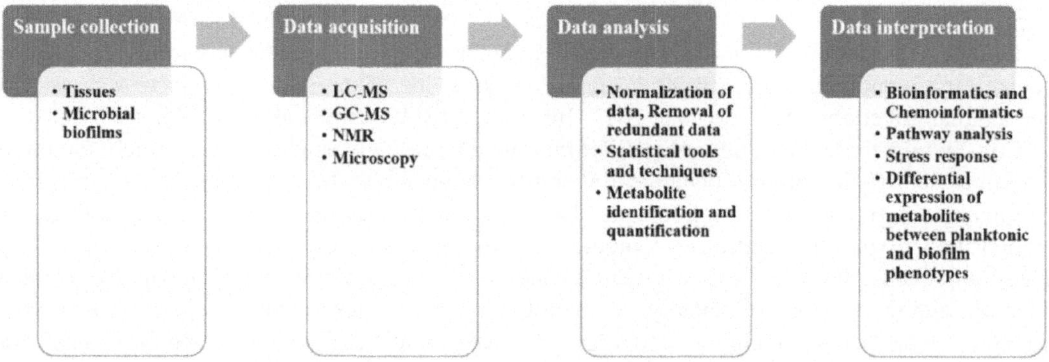

FIGURE 2.2 Work flow while studying metabolomics of microbial biofilms.

is data recording, where advanced tools and technologies are used to identify and quantify various metabolites present in the sample. Interpretation of the recorded data is truly a time and effort consuming task, where various software and algorithms are being used to reach an error free conclusion. The present chapter aims to discuss these aspects in depth (Figure 2.2).

2.2 METABOLOMICS: AN APPROACH TO LEARNING MICROBIAL BIOFILMS

Due to genetic simplicity and ease of engineering, the microorganisms are used as model systems to conduct metabolomics studies. The qualitative and quantitative determination of various metabolites may vary from organism to organism. Therefore, generalized approaches may not be useful and the experimental strategies may require extensive optimization as per the organism being explored (Fiehn, 2002; Bino et al. 2004). It is very obvious that the metabolic pathways are greatly inter-linked in the living systems and are quite sensitive to physicochemical changes. Hence, slight variation may lead to significant alterations in metabolite profiling. Moreover, the overall metabolome may consist of a very diverse range of compounds, mainly volatile compounds, alcohols, carbohydrates, lipids, organic acids and others types of metabolites (Nalbantoglu 2019). This further indicates that a single technological approach may not be sufficient to determine and quantify these compounds, each having its unique attributes. Therefore, it requires the use of multiple and very sensitive analytical tools and techniques to analyse these metabolites together. MS and NMR based analytical techniques are extensively used (Wang et al. 2010; Segers et al. 2019).

Primarily, the metabolite profiling can be performed by either untargeted or targeted approaches (Patti et al. 2012; Rai et al. 2013). The choice of approach depends upon the objectives of the research and numbers of metabolites to be analysed.

2.2.1 Untargeted Metabolomics

The untargeted or global metabolomics approach, as the name itself suggests, do not aim to identify or quantify all the metabolites of the sample. Instead, it determines the metabolite profile emerging at a given time point or under a specified physicochemical conditions. Thus, such approaches are extensively selected while performing drug discovery or other discovery based studies (Parfieniuk et al. 2018). In other words, the untargeted metabolomics emphasizes the determination of the maximum possible metabolites in a sample without any technical or strategy based bias and therefore it is used during a hypothesis-generating approach. An example is, untargeted metabolomics studies while treating any microbial infections under different physiological status (planktonic versus biofilm) (Stipetic et al. 2016; Sadiq et al. 2020) or using various drugs (Favre et al. 2019) would yield some significant information about overall life cycle, transition between various forms, other relevant physiological variations/adaptations, metabolic regulations, stress response and drug resistance. Although it is a very good approach to estimate metabolite profiling, it is associated with certain limitations, leading to experimental bias. Importantly, the extraction techniques should be optimized every time due to differential properties of the metabolome. Sensitivity of these compounds to certain trace detection methods may further cause issues during its analysis. Chromatography methods, columns, buffers, composition of the mobile phase and apt separation of metabolites are also crucially needed. For instance, MS may detect metabolites easily at higher concentration compared to those present in trace concentrations. Further, missing data and statistical analysis will make data analysis and interpretation very complicated (Schrimpe-Rutledge et al. 2016). Therefore, a range of free and commercially available data analysis softwares are available to ensure precise interpretation. Such freeware includes MathDAMP (Baran et al. 2006), XCMS (Smith et al. 2006), MetDAT (Biswas et al. 2010), MetExplore (Cottret et al. 2010), Mzmine 2 (Pluskal et al. 2010), datPAV (Biswas et al. 2011), MetAlign 3 (Lommen and Kools 2012) and MetaboAnalyst (Xia et al. 2012). On the contrary, Mass Profiler Professional (Chen et al. 2016; Vizcaino et al. 2016), Mass Hunter (Chen et al. 2016), Progenesis (Zhang et al. 2016), and Sieve (Hao et al. 2018) are commercial software. However, validation of the outcomes needs to be performed for accurate data interpretation (Longnecker et al. 2015). The untargeted metabolomics have proved to be an excellent tool in the drug discovery approach with a special focus to identify newer biofilm pathways (Griffiths et al. 2010; Patti et al. 2012) (Figure 2.3).

2.2.2 Targeted Metabolomics

The targeted metabolomics approaches involve evaluation of an exclusive hypothesis, or a particular biology based research problem. As these approaches are very specific and aim to target

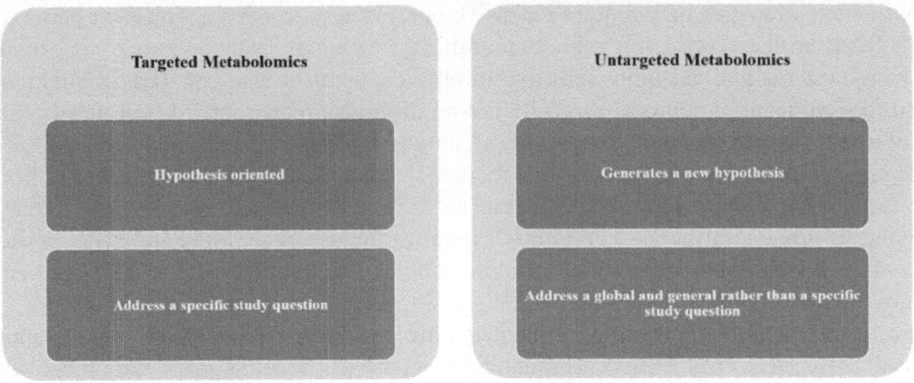

FIGURE 2.3 A comparison between the targeted and global metabolomics.

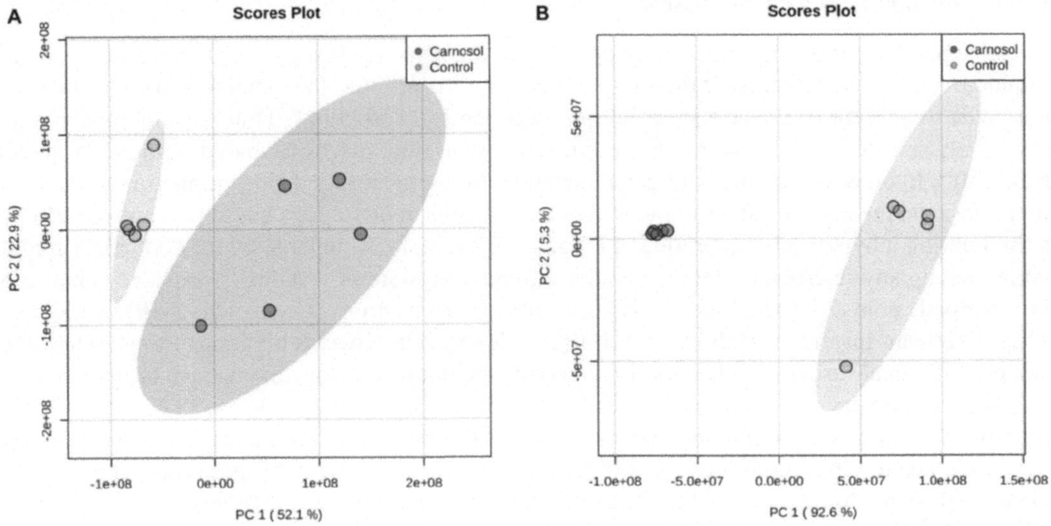

FIGURE 2.4 PCA score plot of metabolites in *Staphylococcus aureus* 29213 planktonic cells and biofilms treated with carnosol. (A) The S. aureus 29213 planktonic cells were treated with 48 µg/mL (1.5×MIC) carnosol (n = 5). The control groups were treated with the same volume of DMSO (n = 5). (B) The S. aureus 29213 biofilms were treated with 128 µg/mL (4×MIC) carnosol (n = 5). The control groups were treated with the same volume of DMSO (n = 5). Reprinted from Shen, F., Ge, C. and Yuan, P. (2020) Metabolomics Study Reveals Inhibition and Metabolic Dysregulation in Staphylococcus aureus Planktonic Cells and Biofilms Induced by Carnosol. *Frontiers in Microbiology*, p.2315. The article is an open access article distributed under the terms and conditions of the Creative Commons Attribution license (http://creativecommons.org/licenses/by/3.0/).

a specific pathway or metabolite, they are employed to address the biological question under study or to identify and quantify the metabolites of interest directly. In contrast to the untargeted metabolomics, these approaches are highly sensitive, permitting detection of metabolites at very trace concentration too (Roberts et al. 2012). Practically, the targeted metabolomics approaches are conducted after the untargeted metabolomics to get an interpretation with apt precision (Lu et al. 2019). These approaches inevitably need pre-existing knowledge of the metabolite of interest, which can be acquired using global approaches (Pinnu et al. 2019). Thus, they also require the metabolite in the purified form to perform the control experiments, which may help with data analysis and interpretation. Unlike untargeted metabolomics, the targeted approaches cannot be explored to discover and quantify any novel compounds (Cambiaghi et al. 2017). Even then, the researchers cannot correlate the role of the known metabolite in some novel metabolic pathways using targeted metabolomics. These are the reasons ffor the emergence of studies involving the use of untargeted and targeted metabolomics altogether to generate error free data analysis and interpretation (Figure 2.4).

2.3 METABOLOMICS FOR THE LEARNING OF METABOLIC PATHWAYS

Studies involving metabolomics aim to cater for metabolite profiling, whereas the metabolic flux analysis assists in determination of the metabolome over a particular time period. Stable or radioisotopes of the metabolites of microbial cultures are utilized to produce the next intermediates of any biochemical pathway in metabolic flux analysis. Stable isotopes are preferred over the radioisotopes as they are not associated with any hazards. For this reason, this approach overcomes classical conventional metabolomics by evaluating the biochemical pathways and their interrelatedness using labelled metabolites. In addition, the metabolic flux analysis determines the turnover of the metabolites rather than evaluating the alterations in their concentration, as in the case of classical metabolomics.

A study was conducted on *P. aeruginosa* biofilms by Wan et al. (2018), where some novel pathways were discovered in its planktonic and biofilm state using the ^{13}C tracing technique. The study revealed that glucose 6-phosphate depleted slowly in the biofilms, which highlighted slower metabolism in biofilms compared to the planktonic stage. On a similar note, Kromer et al. (2004) conducted a study on *Corynebacterium glutamicum*, where the glucose isotopes exhibited an increased flux via a TCA cycle and amino acid anabolism pathway. However, such trends were not evident at transcriptomics level, which further necessitates use of the multi-omics approach to learn bacterial metabolic variations.

2.4 EVALUATION OF THE IMPACT OF ENVIRONMENTAL FACTORS ON MICROBIAL BIOFILMS

The microbial biomarkers may furnish some important clue about microbial metabolism and may address various questions related to microbial stress response against multitudes of biotic and abiotic stress conditions. For instance, availability of oxygen, nutrients, water and incubation temperature are considered as important abiotic factors which may affect microbial metabolism remarkably and thereby microbial biofilm formation too.

A study was conducted on *Aspergillus nidulans*, where growth pattern was studied in the presence of light and dark conditions. During light exposure, it preferred asexual reproduction with enhanced expression of antitumor agents, terrequinone A and Emericellamide. Under dark conditions, the fungi preferred sexual reproduction with production of a mycotoxin sterigmatocystin (Bayram et al. 2016).

Similarly, oxygen availability also has a profound impact over metabolome composition of different microorganisms. Such an impact over microbolome fingerprint and foot print was studied in *Streptococcus spp.* using hydrophilic interaction chromatography (Fei et al. 2016). Oxygen scarcity increased the rate of metabolism and prompted the re-modelling of cell membrane composition. These increased plasticity and stability of the membranes against oxygen deprived conditions.

Interestingly, stress responses by some halophilic microbial strains against increasing salt concentration was evaluated using NMR. It was observed that the metabolites associated with some biochemical energy generating pathways, such as glycolysis, TCA and PP shunt increased upon increased salt stress (Joghee and Jayaraman 2014).

2.5 TECHNICAL BOTTLENECKS AND PROBABLE SOLUTIONS

The microorganisms exhibited a wide range of metabolic diversity under different environmental conditions and at different growth stages. This is the reason behind variations and fluctuations in the qualitative and quantitative estimation of various metabolites. Sometimes, the metabolites may be novel and difficult to identify and quantify. Thus, such observational facts make the analysis and interpretation of the metabolite profile under a given experimental conditions very difficult. During biofilm formation, composition and concentrations of the exopolysaccharides greatly vary in the matrix. Moreover, they interact with other components to different extents. These add further complexities during the studies, making the treatments inhibiting exopolysaccharide formation utterly cumbersome. Besides, extraction protocols should be precisely optimized so that they should represent all classes of metabolites present in the samples (Vuckovic, 2012; Putri et al. 2013). This can be achieved by isolating the metabolites under organic and hydrophobic phases, followed by a separate analysis and interpretation. While studying the antimicrobial resistance pattern in the biofilms, altered composition of the exopolysaccharides often complicates it (Lembre et al. 2012), as some of them are homopolysaccharides, whereas mostly heteropolysaccharides dominate in the same. In general, alginate, cellulose and poly-N-acetyl glucosamine are frequently present in the bacterial biofilms (Lembre et al. 2012; Flemming et al. 2016). It is believed that alginate ensures mechanical stability and strength in the biofilms. Moreover, it also promotes the development of microcolonies during various stages of microbial biofilm formation (Flemming et al. 2016). Cellulose is chemically a homopolysachharide consisting of glucose monomers linked by the β-1-4–linkages. It is vitally needed during biofilm formation by *S. typhi* and *E. coli* (Lembre et al. 2012). Similarly, poly-N-acetyl glucosamine is inevitably required during the adherence and biofilm formation of many bacterial species (Lembre et al. 2012). In the case of fungal biofilm matrix, β-1, 3 glucans and galactosaminogalactan are commonly present, which assist in impounding antifungal antibiotics and also prevent phagocytosis, being anti-phagocytic agents. On the other hand, the galactosaminogalactan directs adhesion towards the host surfaces and also promotes protection against host immune responses during *A. fumigatus* infections (Sheppard and Howell, 2016).

The major experimental problem includes the solubility of exopolysaccharides during the extraction protocols (Flemming et al. 2016). A majority of the exopolysaccharides consist of a large amount of carbohydrate moieties, which are frequently insoluble and tough to isolate. Moreover, these exopolysaccharides are either of the ordered or disordered forms, where the disordered forms are encouraged at higher temperatures and/or lower ionic strength (Nguyen et al. 2012). The biofilm matrix also composed of proteins, DNA, and lipids (Lembre et al. 2012).

Similarly, bacterial cell membranes are constituted by a very wide and diversified range of amphiphilic phospholipids. These may include glycolipids, and sphingolipids along with phosphatidylcholine, phosphatidylinositol, phosphatidylglycerol and phosphatidylethanolamine. Moreover, the same study also reported that the membrane composition may differ upon fluctuations in the environmental conditions (Sohlenkamp and Geiger 2015). Looking at the cell membrane of *Pseudomonas aeruginosa* before and after construction of biofilms over glass wool by Benamara et al. (2014) using GC-MS, the group tried to identify changes in the lipid profile across various stages of biofilm generation. Interestingly, the study reported that concentrations of phosphatidylethanolamines and phosphatidylglycerols were very high in the studied bacterial membrane. Moreover, there was no significant change in the lipidome that was observed during planktonic growth, whereas it altered remarkably during the biofilm generation process. For instance, the outer membrane was rich in long and branched phospholipids (Benamara et al. 2014). In another study, *Pseudoalteromonas lipolytica* membrane was reported to express phosphatidylethanolamine during biofilm generation, whereas during planktonic growth it was rich in ornithine lipids (Favre et al. 2018). Moreover, the metabolome of *Desulfovibrio vulgaris* was studied during planktonic and biofilm phenotypes using GC-MS and LC-MS. The study interestingly revealed that the metabolites which were involved in the fatty acid biosynthesis were up-regulated during the biofilm formation (Zhang et al. 2016).

Such observations established the fact that the cell membrane associated metabolites are crucially significant for the structural and functional integrity of the microbial biofilms. Exceptionally, *S. intermedius* exhibited adaptation in anaerobic conditions by altering membrane composition. Such changes might have altered the selective permeability of the membrane and so the membrane transport might have changed. This might be a reason for growth of the bacteria as in a biofilm phenotype within a healthy host (Fei et al. 2016).

In a study, the metabolite fingerprints and metabolite footprints of *Salmonella spp.* was studied under a different time period (Wong et al. 2015). The study involved the use of GC-MS to determine such alterations during planktonic and biofilm phenotypes of *Salmonella spp.*, where concentrations of secretary metabolites such as alanine, glutamic acid, glycine and ornithine fluctuated remarkably under both the phenotypes. Along the same lines, the intracellular metabolites, such as succinic acid, putrescine, pyroglutamic acid, and N-acetylglutamic acid altered during physiological transition between biofilm and planktonic growth.

On a similar note, the metabolism of carbon and nitrogen sources may also play a vital role during the physiological transformations. *Staphylococcus aureus* exhibited an altered arginine anabolism pathway while growing as a planktonic growth or biofilm state (Stipetic et al. 2016). Some of the extracellular amino acid concentrations and expressions were up regulated in *Aspegillus pallidofulvus* during biofilm generations. This was evaluated using a quantitative NMR technique (Ząbek et al. 2017).

Practically, the researchers need to manage isolation of all the metabolites with the highest degree of achievable purity whilst maintaining the stability of each metabolite, as sometimes the conditions favouring isolation of one metabolite may decrease the stability of other metabolites. This in turn necessitates the use of multiple methods to isolate and purify such a wide range of molecules from the sample.

Considering these points, the majority of the extraction strategies advocate employment of matrix-assisted laser desorption/ionization time-of-flight mass spectrometry (MALDI-TOF MS), permitting the study of intact bacterial cells (Schumann and Maier, 2014). Similarly, high resolution LC-MS was employed to evaluate sequential variations in types and relative concentrations of the metbolites during different stages of biofilm generation by *P. aeruginosa*. The study selected a global metabolomics approach and a novel metabolite, pyocyanine was identified (Borgos et al. 2015). It is one of the major virulence factors of the said bacterial pathogen, having good antibacterial activities against Gram positive bacteria. (Gharieb et al. 2013). The intracellular and extracellular metabolites of another human pathogen, *Staphylococcus aureus* was also studied by a quantitative NMR technique. Both methicillin resistance and methicillin sensitive *S. aureus* strains were studied to observe such dynamic variations during the shifts between biofilm and planktonic growth. Specifically, amino acid uptake, lipid catabolism, a significant metabolic shift from energy generation to cell wall formation and increased deposition of exopolymeric substances and butanediol fermentation were the main distinct phenotypes while studying biofilm generation (Ammons et al. 2014).

2.6 CURRENT APPLICATIONS OF BIOFILM METABOLOMICS

Microbial biofilms have extensively been studied with reference to the use of various strategies employing the concept of metabolomics very recently. Such studies either employed targeted metabolomics, global metabolomics or an amalgamation of both of the approaches. The choice of approach depends upon factors such as the particular research problem, the type of biofilms to be studied, the biofilm model employed and the varieties of metabolites to be analysed. Such approaches may sometimes also address a very wide range of phenotypic issues associated during biofilm generation. It sometimes also evaluates the marker phenotypes involved during the physiological shift between planktonic and biofilm growth patterns. Such studies are performed with *Pseudomonas aeruginosa* (Gjersing et al. 2007), *Acinetobacter baumanii* (Yeom et al. 2013), *Vibrio fischeri* (Chavez-Dozal et al. 2015), *Staphylococcus aureus* (Stipetic et al. 2016), *Desulfovibrio vulgaris* (Zhang et al. 2016), *Pseudoalteromonas lipolytica* (Favre et al. 2018), and *Aspergillus spp.* (Ząbek et al. 2017), respectively. Most of the studies employed an untargeted approach, whereas the studies involving *Vibrio fischeri* (Chavez-Dozal et al. 2015) and *Desulfovibrio vulgaris* (Zhang et al. 2016) used a targeted metabolomics approach.

The microbial biofilms produced by a mixed microbial population have been quite interesting and challenging to study. Metabolomics approaches may play a handy role for conducting studies involving polymicrobial biofilms. Biofilms produced by mixed populations of halophilic microorganisms were studied by Adamiak et al. (2017), whereas a biofilm produced by an opportunistic yeast, *Candida albicans* and a pathogenic bacteria *Staphylococcus aureus* was studied by Weidt et al. (2016). Oral cavities are the best example of where one can find biofilm generation using various microbial species. Similar studies were conducted *in vitro* from saliva samples before and after treating the dental cavities with sodium hypochloride (Agnello et al. 2017; Califf et al. 2017; Edlund et al. 2017). In another study, the biofilm produced by *Pseudomonas aeruginosa* near to a physical wound to the skin was studied using the targeted metabolomics approach.

Microbial biofilms and the metabolite profiling may greatly fluctuate during stress response. For instance, *Pseudomonas fluorescence* exhibited variations in the metabolite profiles in its planktonic growth and biofilm state when exposed to copper stress (Booth et al. 2011). On the contrary, some precise research problems can also be addressed using metabolomics. The entire mechanism of biofilm production by a human pathogen, *Helicobacter pylori* was studied by the global metabolomics approach. Interestingly, the study explained possible reasons behind the biofilm generation, physiological shifts and also various biomarkers which may influence biofilm formation (Wong et al. 2018). The microbial biofilms generated in teeth and dental cavities were studied using global metabolomics (Zandona et al. 2015).

Thus, the metabolomics approach proves quite promising with the ongoing technological advancements to overcome current technical bottlenecks during the analysis and interpretation of different physicochemical factors affecting various stages of biofilm generation.

2.7 CONCLUSION

Microorganisms form biofilms to survive under adverse conditions as a dynamic collaboration resulting in either inter-species or intra-species communication via attachment to a solid substratum. The microbial biofilm process is a precisely regulated process involving multiple stages. Further, various environmental factors contribute significantly during the process of microbial biofilm formation. It would be interesting to learn the biochemical and metabolic changes incurred by these microbes to survive the alterations to their environmental factors. Further, microbes express different metabolites during the biofilm phenotype and planktonic growth. Metabolomics aims to study qualitative and quantitative metabolic shifts during microbial biofilm formation. The studies can broadly be divided into two major strategies: untargeted or global metabolomics, which generally involve hypothesis generation and targeted metabolomics, which also generally involves hypothesis validation. Several studies use an amalgamation of both of the approaches to reach an accurate and error free interpretation. The use of statistical methods and advanced bioinformatics tools may help analyse data correctly. However, some recent studies advocate exploration of multi-omics approaches to understand the process of microbial biofilm better. The information obtained can be used to treat multidrug resistant microbial infections well.

ACKNOWLEDGEMENT

BAK acknowledges P. D. Patel Institute of Applied Sciences, Charotar University of Science and Technology, Changa for the required support.

REFERENCES

Adamiak, J., Bonifay, V., Otlewska, A., Sunner, J.A., Beech, I.B., Stryszewska, T., Kańka, S., Oracz, J., Żyżelewicz, D. and Gutarowska, B. (2017). Untargeted metabolomics approach in halophiles: Understanding the biodeterioration process of building materials. *Frontiers in Microbiology*, *8*, p.2448.

Agnello, M., Cen, L., Tran, N.C., Shi, W., McLean, J.S. and He, X. (2017) Arginine improves pH homeostasis via metabolism and microbiome modulation. *Journal of Dental Research*, *96*(8), pp.924–930.

Ammons, M.C.B., Tripet, B.P., Carlson, R.P., Kirker, K.R., Gross, M.A., Stanisich, J.J. and Copié, V. (2014) Quantitative NMR metabolite profiling of methicillin-resistant and methicillin-susceptible *Staphylococcus aureus* discriminates between biofilm and planktonic phenotypes. *Journal of Proteome Research*, *13*(6), pp.2973–2985.

Assaidi, A., Ellouali, M., Latrache, H., Mabrouki, M., Hamadi, F., Timinouni, M., Zahir, H., El Mdaghri, N., Barguigua, A. and Mliji, E.M. (2018) Effect of temperature and plumbing materials on biofilm formation by *Legionella pneumophila* serogroup 1 and 2–15. *Journal of Adhesion Science and Technology*, *32*(13), pp.1471–1484.

Azeredo, J., Azevedo, N.F., Briandet, R., Cerca, N., Coenye, T., Costa, A.R., Desvaux, M., Di Bonaventura, G., Hébraud, M., Jaglic, Z. and Kačániová, M. (2017) Critical review on biofilm methods. *Critical Reviews in Microbiology*, *43*(3), pp.313–351.

Azzollini, A., Boggia, L., Boccard, J., Sgorbini, B., Lecoultre, N., Allard, P.M., Rubiolo, P., Rudaz, S., Gindro, K., Bicchi, C. and Wolfender, J.L. (2018) Dynamics of metabolite induction in fungal co-cultures by metabolomics at both volatile and non-volatile levels. *Frontiers in Microbiology*, *9*, p.72.

Baptista, R., Fazakerley, D.M., Beckmann, M., Baillie, L. and Mur, L.A. (2018) Untargeted metabolomics reveals a new mode of action of pretomanid (PA-824). *Scientific Reports*, *8*(1), pp.1–7.

Baran, R., Kochi, H., Saito, N., Suematsu, M., Soga, T., Nishioka, T., Robert, M. and Tomita, M. (2006) MathDAMP: a package for differential analysis of metabolite profiles. *BMC Bioinformatics*, *7*(1), pp.1–9.

Bayram, Ö., Feussner, K., Dumkow, M., Herrfurth, C., Feussner, I. and Braus, G.H. (2016) Changes of global gene expression and secondary metabolite accumulation during light-dependent *Aspergillus nidulans* development. *Fungal Genetics and Biology*, *87*, pp.30–53.

Benamara, H., Rihouey, C., Abbes, I., Ben Mlouka, M.A., Hardouin, J., Jouenne, T. and Alexandre, S. (2014) Characterization of membrane lipidome changes in *Pseudomonas aeruginosa* during biofilm growth on glass wool. *PLoS One*, *9*(9), p.e108478.

Bino, R.J., Hall, R.D., Fiehn, O., Kopka, J., Saito, K., Draper, J., Nikolau, B.J., Mendes, P., Roessner-Tunali, U., Beale, M.H. and Trethewey, R.N. (2004) Potential of metabolomics as a functional genomics tool. *Trends in Plant Science*, *9*(9), pp.418–425.

Biswas, A., Mynampati, K.C., Umashankar, S., Reuben, S., Parab, G., Rao, R., Kannan, V.S. and Swarup, S. (2010) MetDAT: a modular and workflow-based free online pipeline for mass spectrometry data processing, analysis and interpretation. *Bioinformatics*, *26*(20), pp.2639–2640.

Biswas, A., Rao, R., Umashankar, S., Mynampati, K.C., Reuben, S., Parab, G. and Swarup, S. (2011) datPAV—an online processing, analysis and visualization tool for exploratory investigation of experimental data. *Bioinformatics*, *27*(11), pp.1585–1586.

Booth, S.C., Workentine, M.L., Wen, J., Shaykhutdinov, R., Vogel, H.J., Ceri, H., Turner, R.J. and Weljie, A.M. (2011) Differences in metabolism between the biofilm and planktonic response to metal stress. *Journal of Proteome Research*, *10*(7), pp.3190–3199.

Borgos, S.E.F., Skjåstad, R., Tøndervik, A., Aas, M., Aasen, I.M., Brunsvik, A., Holten, T., Iversen, O.J., Ahlen, C. and Zahlsen, K. (2015) Rapid metabolic profiling of developing *Pseudomonas aeruginosa* biofilms by high-resolution mass spectrometry fingerprinting. *Annals of Microbiology*, *65*(2), pp.891–898.

Califf, K.J., Schwarzberg-Lipson, K., Garg, N., Gibbons, S.M., Caporaso, J.G., Slots, J., Cohen, C., Dorrestein, P.C. and Kelley, S.T. (2017) Multi-omics analysis of periodontal pocket microbial communities' pre-and posttreatment. *Msystems*, *2*(3), pp.e00016–17.

Cambiaghi, A., Ferrario, M. and Masseroli, M. (2017) Analysis of metabolomic data: tools, current strategies and future challenges for omics data integration. *Briefings in Bioinformatics*, *18*(3), pp.498–510.

Chatzimitakos, T.G. and Stalikas, C.D. (2016) Qualitative alterations of bacterial metabolome after exposure to metal nanoparticles with bactericidal properties: a comprehensive workflow based on ^{1}H NMR, UHPLC-HRMS, and metabolic databases. *Journal of Proteome Research*, *15*(9), pp.3322–3330.

Chavez-Dozal, A., Gorman, C. and Nishiguchi, M.K. (2015) Proteomic and metabolomic profiles demonstrate variation among free-living and symbiotic *Vibrio fischeri* biofilms. *BMC Microbiology*, *15*(1), pp.1–11.

Chen, J., Deng, J., Zhou, J., Xie, Z., Gan, J. and Wen, C. (2016) Metabonomic study of traditional Chinese herb pair, Qinghao-Biejia in treating systemic lupus erythematosus mice. *Chinese Pharmacological Bulletin*, pp.727–731.

Cottret, L., Wildridge, D., Vinson, F., Barrett, M.P., Charles, H., Sagot, M.F. and Jourdan, F. (2010) MetExplore: a web server to link metabolomic experiments and genome-scale metabolic networks. *Nucleic Acids Research*, *38* (suppl_2), pp.W132–W137.

Edlund, A., Garg, N., Mohimani, H., Gurevich, A., He, X., Shi, W., Dorrestein, P.C. and McLean, J.S. (2017) Metabolic fingerprints from the human oral microbiome reveal a vast knowledge gap of secreted small peptidic molecules. *Msystems*, *2*(4), pp.e00058–17.

Favre, L., Ortalo-Magné, A., Kerloch, L., Pichereaux, C., Misson, B., Briand, J.F., Garnier, C. and Culioli, G. (2019) Metabolomic and proteomic changes induced by growth inhibitory concentrations of copper in the biofilm-forming marine bacterium *Pseudoalteromonas lipolytica*. *Metallomics*, *11*(11), pp.1887–1899.

Favre, L., Ortalo-Magné, A., Pichereaux, C., Gargaros, A., Burlet-Schiltz, O., Cotelle, V. and Culioli, G. (2018) Metabolome and proteome changes between biofilm and planktonic phenotypes of the marine bacterium *Pseudoalteromonas lipolytica* TC8. *Biofouling*, *34*(2), pp.132–148.

Fei, F., Mendonca, M.L., McCarry, B.E., Bowdish, D.M. and Surette, M.G. (2016) Metabolic and transcriptomic profiling of Streptococcus intermedius during aerobic and anaerobic growth. *Metabolomics*, *12*(3), pp.1–13.

Fiehn, O. (2002) Metabolomics—the link between genotypes and phenotypes. *Functional Genomics*, pp.155–171.

Flemming, H.C., Wingender, J., Szewzyk, U., Steinberg, P., Rice, S.A. and Kjelleberg, S. (2016) Biofilms: an emergent form of bacterial life. *Nature Reviews Microbiology*, *14*(9), pp.563–575.

Gharieb, M.M., El-Sheekh, M.M., El-Sabbagh, S.M. and Hamza, W.T. (2013) Efficacy of pyocyanin produced by *Pseudomonas aeruginosa* as a topical treatment of infected skin of rabbits. *BioTechnology: An Indian Journal*, *7*, pp.184–193.

Gjersing, E.L., Herberg, J.L., Horn, J., Schaldach, C.M. and Maxwell, R.S. (2007) NMR metabolomics of planktonic and biofilm modes of growth in *Pseudomonas aeruginosa*. *Analytical Chemistry*, *79*(21), pp.8037–8045.

Götz, F., Longnecker, K., Kido Soule, M.C., Becker, K.W., McNichol, J., Kujawinski, E.B. and Sievert, S.M. (2018) Targeted metabolomics reveals proline as a major osmolyte in the chemolithoautotroph *Sulfurimonas denitrificans*. *MicrobiologyOpen*, *7*(4), p.e00586.

Griffiths, W.J., Koal, T., Wang, Y., Kohl, M., Enot, D.P. and Deigner, H.P. (2010) Targeted metabolomics for biomarker discovery. *Angewandte Chemie International Edition*, *49*(32), pp.5426–5445.

Hao, L., Wang, J., Page, D., Asthana, S., Zetterberg, H., Carlsson, C., Okonkwo, O.C. and Li, L. (2018) Comparative evaluation of MS-based metabolomics software and its application to preclinical Alzheimer's disease. *Scientific Reports*, *8*(1), pp.1–10.

Jin, C., Yu, Z., Peng, S., Feng, K.E., Zhang, L. and Zhou, X. (2018) The characterization and comparison of exopolysaccharides from two benthic diatoms with different biofilm formation abilities. *Anais da Academia Brasileira de Ciências*, *90*, pp.1503–1519.

Joghee, N.N. and Jayaraman, G. (2014) Metabolomic characterization of halophilic bacterial isolates reveals strains synthesizing rare diaminoacids under salt stress. *Biochimie*, *102*, pp.102–111.

Kromer, J.O., Sorgenfrei, O., Klopprogge, K., Heinzle, E. and Wittmann, C. (2004) In-depth profiling of lysine-producing Corynebacterium glutamicum by combined analysis of the transcriptome, metabolome, and fluxome. *Journal of Bacteriology*, *186*(6), pp.1769–1784.

Landini, P., Antoniani, D., Burgess, J.G. and Nijland, R. (2010) Molecular mechanisms of compounds affecting bacterial biofilm formation and dispersal. *Applied Microbiology and Biotechnology*, *86*(3), pp.813–823.

Lembre, P., Lorentz, C. and Di Martino, P. (2012) Exopolysaccharides of the biofilm matrix: a complex biophysical world. *Complex world of Polysaccharides*, pp.371–392.

Lommen, A. and Kools, H.J. (2012) MetAlign 3.0: performance enhancement by efficient use of advances in computer hardware. *Metabolomics*, *8*(4), pp.719–726.

Longnecker, K., Futrelle, J., Coburn, E., Soule, M.C.K. and Kujawinski, E.B. (2015) Environmental metabolomics: Databases and tools for data analysis. *Marine Chemistry*, *177*, pp.366–373.

López-Gresa, M.P., Lisón, P., Campos, L., Rodrigo, I., Rambla, J.L., Granell, A., Conejero, V. and Bellés, J.M. (2017) A non-targeted metabolomics approach unravels the VOCs associated with the tomato immune response against *Pseudomonas syringae*. *Frontiers in Plant Science*, *8*, p.1188.

Lu, H., Que, Y., Wu, X., Guan, T. and Guo, H., (2019) Metabolomics deciphered metabolic reprogramming required for biofilm formation. *Scientific Reports*, *9*(1), pp.1–7.

Maansson, M., Vynne, N.G., Klitgaard, A., Nybo, J.L., Melchiorsen, J., Nguyen, D.D., Sanchez, L.M., Ziemert, N., Dorrestein, P.C., Andersen, M.R. and Gram, L. (2016) An integrated metabolomic and genomic mining workflow to uncover the biosynthetic potential of bacteria. *Msystems*, *1*(3), pp.e00028–15.

Muhammad, M.H., Idris, A.L., Fan, X., Guo, Y., Yu, Y., Jin, X., Qiu, J., Guan, X. and Huang, T. (2020) Beyond risk: bacterial biofilms and their regulating approaches. *Frontiers in Microbiology*, *11*, p.928.

Nalbantoglu, S. (2019) Metabolomics: basic principles and strategies. *Molecular Medicine*, *10*.

Nguyen, T., Roddick, F.A. and Fan, L. (2012) Biofouling of water treatment membranes: a review of the underlying causes, monitoring techniques and control measures. *Membranes*, *2*(4), pp.804–840.

Parfieniuk, E., Zbucka-Kretowska, M., Ciborowski, M., Kretowski, A. and Barbas, C. (2018) Untargeted metabolomics: an overview of its usefulness and future potential in prenatal diagnosis. *Expert Review of Proteomics*, *15*(10), pp.809–816.

Patti, G.J., Yanes, O. and Siuzdak, G. (2012) Metabolomics: the apogee of the omics trilogy. *Nature Reviews: Molecular Cell Biology*, *13*(4), pp.263–269.

Pinu, F.R., Goldansaz, S.A. and Jaine, J. (2019) Translational metabolomics: Current challenges and future opportunities. *Metabolites*, *9*(6), p.108.

Pluskal, T., Castillo, S., Villar-Briones, A. and Orešič, M. (2010) MZmine 2: modular framework for processing, visualizing, and analyzing mass spectrometry-based molecular profile data. *BMC Bioinformatics*, *11*(1), pp.1–11.

Putri, S.P., Nakayama, Y., Matsuda, F., Uchikata, T., Kobayashi, S., Matsubara, A. and Fukusaki, E. (2013) Current metabolomics: practical applications. *Journal of Bioscience and Bioengineering*, *115*(6), pp.579–589.

Rai, A., Umashankar, S. and Swarup, S. (2013) Plant metabolomics: from experimental design to knowledge extraction. In *Legume Genomics* (pp. 279–312). Humana Press, Totowa, NJ.

Roberts, L.D., Souza, A.L., Gerszten, R.E. and Clish, C.B. (2012) Targeted metabolomics. *Current protocols in Molecular Biology*, *98*(1), pp.30–2.

Sadiq, F.A., Yan, B., Zhao, J., Zhang, H. and Chen, W. (2020) Untargeted metabolomics reveals metabolic state of *Bifidobacterium bifidum* in the biofilm and planktonic states. *LWT*, *118*, p.108772.

Sander, K., Asano, K.G., Bhandari, D., Van Berkel, G.J., Brown, S.D., Davison, B. and Tschaplinski, T.J. (2017) Targeted redox and energy cofactor metabolomics in *Clostridium thermocellum* and *Thermoanaerobacterium saccharolyticum*. *Biotechnology for Biofuels*, *10*(1), pp.1–18.

Schrimpe-Rutledge, A.C., Codreanu, S.G., Sherrod, S.D. and McLean, J.A. (2016) Untargeted metabolomics strategies—challenges and emerging directions. *Journal of the American Society for Mass Spectrometry*, *27*(12), pp.1897–1905.

Schumann, P. and Maier, T. (2014) MALDI-TOF mass spectrometry applied to classification and identification of bacteria. In *Methods in Microbiology* (Vol. 41, pp. 275–306). Academic Press.

Segers, K., Declerck, S., Mangelings, D., Heyden, Y.V. and Eeckhaut, A.V. (2019) Analytical techniques for metabolomic studies: a review. *Bioanalysis*, *11*(24), pp.2297–2318.

Shen, F., Ge, C. and Yuan, P. (2020) Metabolomics study reveals inhibition and metabolic dysregulation in Staphylococcus aureus Planktonic cells and biofilms induced by Carnosol. *Frontiers in Microbiology*, p.2315.

Sheppard, D.C. and Howell, P.L. (2016) Biofilm exopolysaccharides of pathogenic fungi: lessons from bacteria. *Journal of Biological Chemistry*, *291*(24), pp.12529–12537.

Smith, C.A., Want, E.J., O'Maille, G., Abagyan, R. and Siuzdak, G. (2006) XCMS: processing mass spectrometry data for metabolite profiling using nonlinear peak alignment, matching, and identification. *Analytical chemistry*, *78*(3), pp.779–787.

Sohlenkamp, C. and Geiger, O. (2016) Bacterial membrane lipids: diversity in structures and pathways. *FEMS Microbiology Reviews*, *40*(1), pp.133–159.

Stipetic, L.H., Dalby, M.J., Davies, R.L., Morton, F.R., Ramage, G. and Burgess, K.E. (2016) A novel metabolomic approach used for the comparison of *Staphylococcus aureus* planktonic cells and biofilm samples. *Metabolomics*, *12*(4), pp.1–11.

Vizcaino, M.I. and Crawford, J.M. (2016) Secondary metabolic pathway-targeted metabolomics. In *Nonribosomal Peptide and Polyketide Biosynthesis* (pp. 175–195). Humana Press, New York, NY.

Vuckovic, D. (2012) Current trends and challenges in sample preparation for global metabolomics using liquid chromatography–mass spectrometry. *Analytical and Bioanalytical Chemistry*, *403*(6), pp.1523–1548.

Wan, N., Wang, H., Ng, C.K., Mukherjee, M., Ren, D., Cao, B. and Tang, Y.J. (2018) Bacterial metabolism during biofilm growth investigated by ^{13}C tracing. *Frontiers in Microbiology*, p.2657.

Wang, J.H., Byun, J. and Pennathur, S. (2010) September. Analytical approaches to metabolomics and applications to systems biology. In *Seminars in Nephrology* (Vol. 30, No. 5, pp. 500–511). WB Saunders.

Weidt, S., Haggarty, J., Kean, R., Cojocariu, C.I., Silcock, P.J., Rajendran, R., Ramage, G. and Burgess, K.E. (2016) A novel targeted/untargeted GC-Orbitrap metabolomics methodology applied to *Candida albicans* and *Staphylococcus aureus* biofilms. *Metabolomics*, *12*(12), pp.1–10.

Wong, E.H.J., Ng, C.G., Goh, K.L., Vadivelu, J., Ho, B. and Loke, M.F. (2018) Metabolomic analysis of low and high biofilm-forming *Helicobacter pylori* strains. *Scientific reports*, *8*(1), pp.1–9.

Wong, H.S., Maker, G.L., Trengove, R.D. and O'Handley, R.M. (2015) Gas chromatography-mass spectrometry-based metabolite profiling of *Salmonella enterica* serovar *typhimurium* differentiates between biofilm and planktonic phenotypes. *Applied Environmental Microbiology*, 81, pp.2660–2666.

Xia, J., Mandal, R., Sinelnikov, I.V., Broadhurst, D. and Wishart, D.S.(2012) MetaboAnalyst 2.0—a comprehensive server for metabolomic data analysis. *Nucleic Acids Research*, *40*(W1), pp.W127–W133.

Yeom, J., Shin, J.H., Yang, J.Y., Kim, J. and Hwang, G.S. (2013) ^{1}H NMR-based metabolite profiling of planktonic and biofilm cells in *Acinetobacter baumannii* 1656–2. *PloS One*, *8*(3), p.e57730.

Ząbek, A., Junka, A., Szymczyk, P., Wojtowicz, W., Klimek-Ochab, M. and Młynarz, P. (2017) Metabolomics analysis of fungal biofilm development and of arachidonic acid-based quorum sensing mechanism. *Journal of Basic Microbiology*, *57*(5), pp.428–439.

Zandona, F., Soini, H.A., Novotny, M.V., Santiago, E., Eckert, G.J., Preisser, J.S., Benecha, H.K., Arthur, R.A. and Zero, D.T. (2015) A potential biofilm metabolite signature for caries activity-A pilot clinical study. *Metabolomics: Open Access*, *5*(1).

Zhang, J., Yang, W., Li, S., Yao, S., Qi, P., Yang, Z., Feng, Z., Hou, J., Cai, L., Yang, M. and Wu, W. (2016) An intelligentized strategy for endogenous small molecules characterization and quality evaluation of earthworm from two geographic origins by ultra-high performance HILIC/QTOF MSE and Progenesis QI. *Analytical and Bioanalytical Chemistry*, *408*(14), pp.3881–3890.

Zhang, Y., Pei, G., Chen, L. and Zhang, W. (2016) Metabolic dynamics of *Desulfovibrio vulgaris* biofilm grown on a steel surface. *Biofouling*, *32*(7), pp.725–736.

3 Susceptibility Testing for the Determination of Biofilm Associated Resistance

Beula Subashini Panchatcharam and Katharina Richter

3.1 INTRODUCTION

Antimicrobial resistance (AMR) among pathogens is a global health threat predicted to cause over ten million deaths per year by 2050 (1) (2). Among various factors, the development of intrinsic resistance within biofilms has been increasingly acknowledged to contribute to AMR.

John William (Bill) Costerton (1934–2012) (3), pioneer and founding father of the field of biofilms, introduced the concept of biofilms in the 1970s and passionately studied the implications of these complex microbial populations in health and disease. The current understanding defines biofilms as communities of microorganisms attached to each other and frequently also to inert or living surfaces, embedded in a protective extracellular matrix. Switching from a free-living planktonic state to a sessile (4) form exhibits their specific identity (5). Compared to planktonic cells, biofilms are known to require up to 1000-fold (6) higher concentrations of antimicrobial agents to be eradicated.

The Clinical and Laboratory Standards Institute (CLSI) and the European Committee on Antimicrobial Susceptibility Testing (EUCAST) have laid guidelines for the susceptibility testing of planktonic cells that influence clinical treatment regimens of infectious diseases. However, as around 80% of microbial infections in the human body are associated with biofilms (7), there is the need to implement the susceptibility testing of biofilms in standard diagnostic procedures to improve patient outcomes and prevent sub-therapeutic treatment that could contribute to AMR.

This chapter presents a summary of current models for antibiofilm susceptibility testing in vitro and in vivo, exploring their use and limitations in the management of biofilm related infections.

3.2 LABORATORY MODELS FOR SUSCEPTIBILITY TESTING IN BIOFILMS

Biofilm growth models are classified as closed or open systems, depending on the type of nutrient delivery (Table 3.1). The model selection is mainly based on the needs and requirements of the clinical setting.

3.2.1 Closed ("Batch Culture" based) Models

In these models the supply of nutrients is finite within a closed system limiting the growth of biofilms over time due to the static nature of the media (8). The simplicity, rapidity and cost-effective nature make these models valuable for researchers (9).

DOI: 10.1201/9781003281238-3

3.2.1.1 Agar Plate Method

Agar plates are suitable for both disc diffusion and dilution methods, they can be enriched with substrates for growing bacteria for a defined period (10). The colony morphology and bacterial viability are better appreciated with this technique; however, extracellular matrix (ECM) formation is minimal and cannot be replicated using this model (11).

3.2.1.2 Microtiter Plates Method

Polystyrene well plates incorporating growth medium can be used to test susceptibility to antimicrobial treatments. Tissue culture plates and round bottomed plates are preferred to flat bottomed plates as this allows homogenous distribution of the nutritional media and uniform exposure to calorimetric reagents (12). Detachment of biofilms can occur during vigorous rinsing procedures hence caution should be taken not to disturb the biofilm layers during the early stages of formation. Advantages of this method is that biofilm growth and subsequent treatment with antibiotics to determine the susceptibility can be done in the same microtiter plate (13).

Quantification and analysis of biofilms can be performed in multiple ways, for example, by measuring the optical density and biofilm biomass after staining with crystal violet, or by analysing microbial viability using metabolic staining, such as Alamar blue, tetrazolium salts and florescent chromophores (14).

Microbial viability can also be assessed by confocal laser scanning microscopy using microtiter plates or chamber slides for optical analysis. The benefit of these assays is that multiple replicates in small volumes can be performed in high throughput to test antibiotic agents that inhibit and/or eradicate biofilms. For biofilm biomass measurements, care should be taken to discern if results are due to the biofilm forming process or due to sedimentation of the cells.

3.2.1.3 Calgary Biofilm Device

This is a modification of the microtiter plate where removable pegs (Figure 3.1) are attached to the lid that growths can be transferred to a new plate. The growth of biofilms on pegs provides an undisturbed environment for biofilm development over prolonged periods both in static and dynamic settings (15) (16). Following biofilm growth, the pegs on the lid can be transferred to a new plate

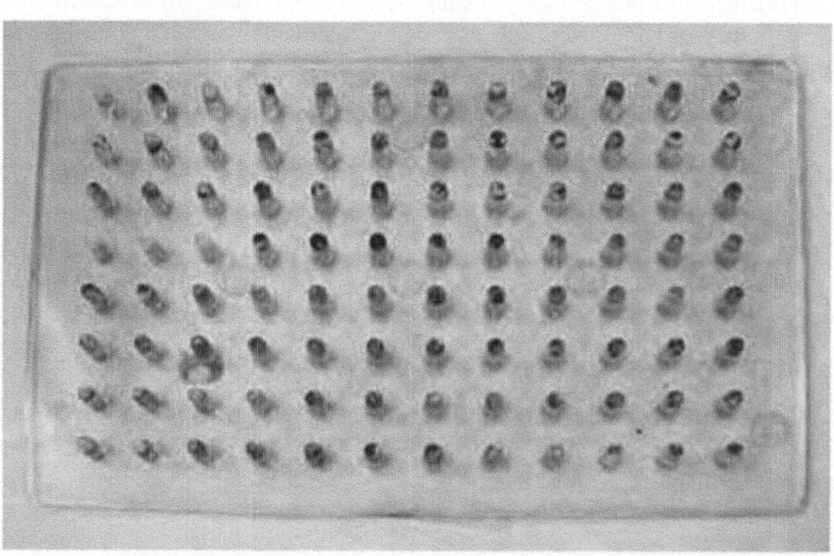

FIGURE 3.1 Calgary biofilm device (18).

TABLE 3.1
Salient Features of Closed and Open Systems to Test Biofilm Susceptibility

	Open system	Closed system
Flow of nutrients	Over the surface of biofilms	No flow
Supply of nutrients	Continuous supply	Limited (can be replenished)
Equipment	Specialized	Common laboratory equipment
Uses	Study of mature biofilms	Study early stages of biofilm development
Duration of experiments	Days to weeks	Days
Cost	Medium to high	Low
Sheer stress	Adjustable	Mild

with antibiotic treatments. The viability of biofilms exposed to antibiotics is evaluated by reading the optical density at 600 nm or by counting colony forming units.

The CBD has been approved by the American Society for Testing and Materials International as a standardized method for testing biofilm disinfectant efficacy (17).

3.2.1.4 Colony Biofilm Model

In this model, colonies of biofilms are grown on a permeable membrane placed on agar, which mimics the conditions of an infection site (19). Using a filter membrane with a pore size of 0.22 µm allows nutrient diffusion from the agar into the biofilm without contaminating the agar. Biofilm colonies can be exposed to treatments, followed by biofilm extraction and serial dilution to determine susceptibility by counting colony forming units. Various bacterial species have been tested using this model, including *S. aureus, MRSA* and *P. aeruginosa* (20) (21).

3.2.1.5 Biofilm Ring Test (BRT)

Immobilization of magnetic microbeads by bacterial biofilms in a microtiter plate (96-well polystyrene) is the principle behind the BRT (22). The adherence of bacteria is proportionately associated with the scattering of the magnetic field (23) (24). Biofilm producing capacity among non-typeable *H. influenzae* strains, *P. aeruginosa* and *S. aureus* were assessed by the this technique (25). The rapidity of obtaining results is one of the advantages of this method, allowing the evaluation of the extent of attachment of the biofilm layer to its surface. A newer modification of this method is the Antibiofilmogram®, which has been validated on isolates of *S. aureus* and *P. aeruginosa* from bone and joint prosthetic infections (26).

3.2.2 Dynamic Flow Systems: Open Continuous Culture Models

In dynamic flow systems the nutrients are continuously replenished by fresh medium (27). The growth of biofilms occurs on membranes inside chambers with laminar flow of media adjacent to the air flow interface (28). The drip flow reactor (DFR) and the Robbins biofilm sampler, Bioflux devices and Sorbarods methods are examples of these flow cell systems.

These models mimic in vivo conditions of certain diseases better than closed systems, are reproduceable and cost effective, but require trained personnel and specialized equipment for performance. The continuous flow of nutrients on the surface of biofilms could lead to decreased biomass which should be kept in mind during analysis and quantification of biofilms.

3.2.2.1 The Sorbarods Method

The Sorbarods method (29) uses cylinders that allow slower growth of biofilms and for a longer incubation period (30). This consists of paper sleeves containing cellulose fibres, bacterial cultures in the mid-logarithmic phase are inoculated using a syringe dropwise (Figure 3.2).

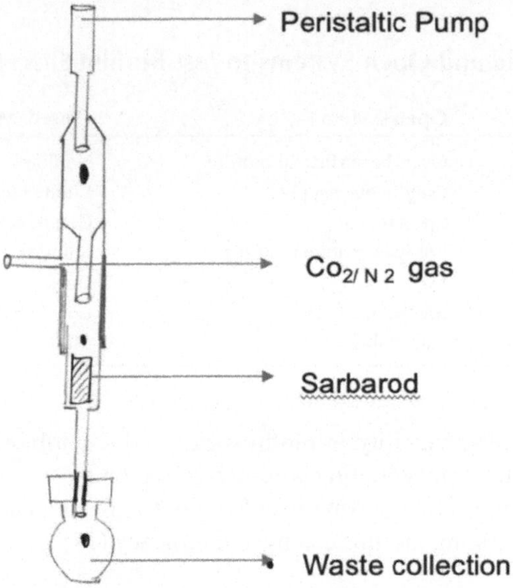

FIGURE 3.2 Sorbarod method of biofilm culture (32).

The entire system is inserted into a PVC tubing enclosing the Sorbarod along with a disposable needle through the seal made of rubber. Many of these units can be placed together and incubated at 37°C. Sterilization of this entire system can be performed by autoclave. This simple method has been used to perform biofilm cultures of *Pseudomonas aeruginosa* and *Staphylococcus aureus* (31).

3.2.2.2 Bioflux Devices

Bioflux devices (33) allow the growth of biofilms on glass slides that are usable for microscopic evaluation. This technique combines the well plate assay with a microfluidic system for 24 chambers for suitability with plate readers (34). Biofilm growth can be observed in real-time via light microscopy and biofilm viability, and matrix composition and dynamics can be visualised with microscopy dyes. Structural details such as the biomass and thickness after addition of antibiotics can be assessed by using software such as COMSTAT or Image J (35). Continuous flow systems are new approaches that depict Gram negative shift that can be demonstrated only in continuous flow conditions (36).

Small amounts of fluidic devices enable the detection of antimicrobial susceptibility testing. These models are robust however they are expensive and require expertise to monitor the system (37). Fluorescence with live/dead bacteria is performed to quantify the viability of the biofilm cultures.

3.2.2.2 The Drip Flow Biofilm Reactor (DFBR)

This system consists of separate channels with lids to maintain aseptic conditions, the nutrients enter a coupon incorporated inside each chamber. DFBR enables the study of disinfection measures for the removal and the eradication of biofilms (38). This model is a standardised technique to reproduce the air liquid interface medium in low shear conditions such as wounds, teeth models and cystic fibrosis lungs. Limitations of DFBR include the requirement of large volumes of nutritional media and only a few samples at a specific point of time can be performed.

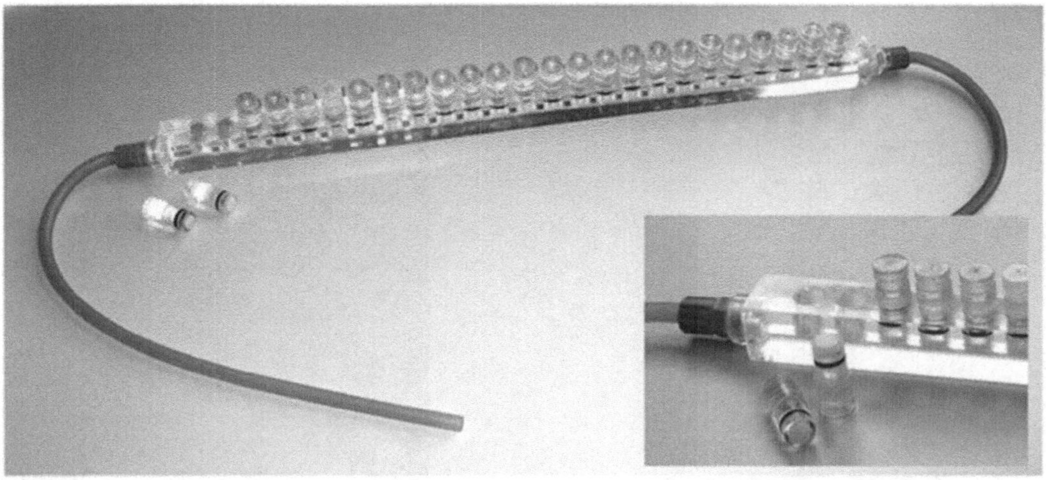

FIGURE 3.3 The modified Robbins device (39).

3.2.2.3 The Robbins Device

This device is made of a pipe with multiple holes attached to coupons at the ends of screws within a liquid stream. The coupons are arranged in parallel fashion to allow the fluid flow which can be removed independently (Figure 3.3). The growth of biofilms can be maintained for several weeks, however, with the higher rates of contamination regular sterilization is necessary (40). This device was mainly used for the study of the antimicrobial susceptibility of venous catheters and biofilms of the oral cavity (41).

3.2.2.4 Biofilm Reactors

Biofilm reactors can contain coupons that are immersed in growth medium to allow biofilm formation (42). For susceptibility testing, the biofilms are exposed to antimicrobial agents added to the fluid phase, following which the coupons are sonicated to release the cells adherent to the coupons and susceptibility performed by either counting colony forming units or by biofilm visualisation using confocal microscopy (43). Various systems such as the Centre for Disease Control (CDC) biofilm reactor (44), the rotating disk reactor and the constant depth film fermenter are provided with the coupons (Figure 3.4) which are positioned at a certain depth above static scraper blades (45). This is a standardised method, which has been used to determine antimicrobial susceptibility, for example, in biofilms on orthopaedic and oral devices (46). This model has been tested to assess the efficacy of antibiotics on polymicrobial biofilms such as *P. aeruginosa* (47), *S. aureus and Enterococcus spp.* Biofilm reactors are also used to assess the pharmacokinetics/pharmacodynamics of betalactam infections among *P. aeruginosa* isolates (48).

3.2.3 Ex vivo Biofilm Models

These models are a combination of in vitro and in vivo models, in which tissue extracted from porcine or murine models is applied to artificial media for experimental processes (50). These models provide controlled conditions alternative to in vivo models, and are valuable for the analysis of bacterial colonization on tissue surfaces (51). The times required for individual treatment of biofilms are easy to assess following application of antibiofilm treatments (52).

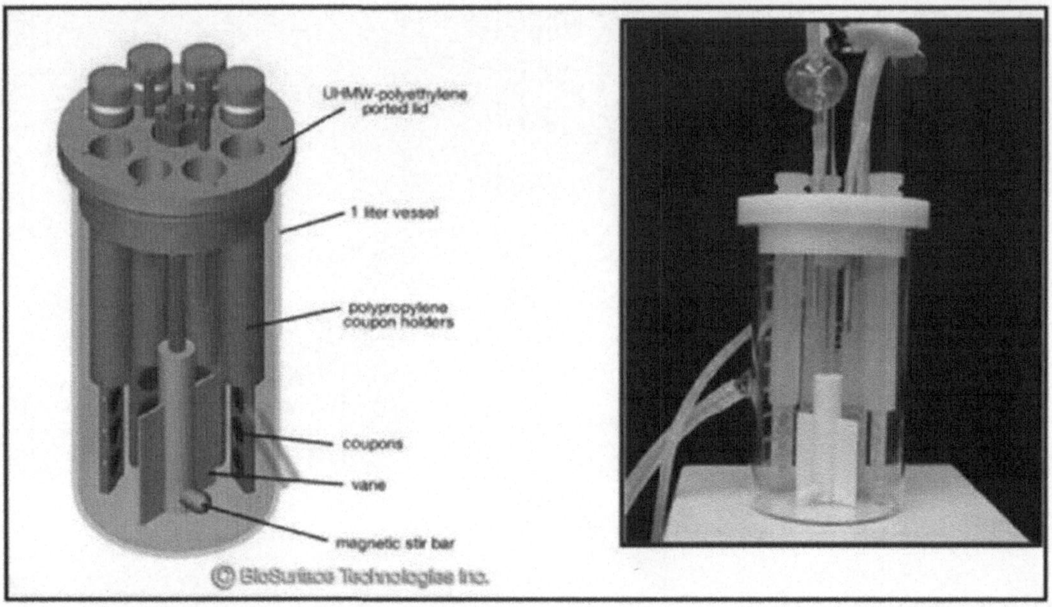

FIGURE 3.4 The Center for Disease Control (CDC) biofilm reactor (49).

3.2.4 Non-mammalian In vivo Models

Non-mammalian models such as the *Drosophila melanogaster, Danio rerio (zebra fish)* can be used to study gut colonization by biofilms (53), while invertebrates such as *Caenorhabditis elegans* worms and *Galleria mellonella* larvae (54) can be used to study the antibiofilm activity of treatments (21) and the virulence effects of bacteria, like *S. aureus* and *P. aeruginosa* (55) (56). The advantages of non-mammalian models include a short generation time, so animals can be grown rapidly and in large numbers to accelerate research outputs, affordable costs, easy handling, and no need for ethics approval. The small size of animals enables the use of microtiter plates for high throughput (57). However, complex immune responses cannot be studied using these models, as well as chronic infections due to the short animal life span.

3.2.5 In vivo Models of Tissue-associated Biofilm Infections

Tissue models closely mimic biofilm related infections and were created to study the pathogenesis of disease in detail.

- Lungs:
  Cystic Fibrosis disease that occurs due to mutations of single gene (CFTR), knockout mice are commonly used to study lung infection and the inflammation that occurs in the airway. These models were used to study the development of chronic *P. aeruginosa* lung infections and the molecular mechanisms involved while on treatment with azithromycin (58).

  Based on the route of infection produced in animal models, they are classified as intravenous and intratracheal models.

  Rat models were established to demonstrate *P. aeruginosa* chronic lung infections (54), later adapted to monkeys and guinea pigs to describe the pathogenicity of *Haemophilus influenzae*, Burkholderia *cenocepacia* and *S. aureus*. This model was also used to describe the role of Psl and Pel polysaccharides produced by biofilms of *P. aeruginosa* (59). Intravenous models were developed to assess the efficacy of antimicrobial agents such as carbapenems by injecting pathogenic organisms into agar beads that are inoculated intravenously.

- Urinary tract: Uropathogenic *E. coli* (UPEC) are the most commonly used bacteria to study the pathogenesis of chronic cystic infections (60), mouse cystitis models are used to establish *E. coli* and *K. pneumoniae* as aggregates in the epithelial cells of the bladder while elucidating the reasons for recurrent cystitis. Murine models were used to confirm the role of immune response in relation to outcome measures and emphasize the role of TLR4 signalling pathways (61). The involvement of biofilms in the pathogenesis of urinary stones was demonstrated using the scanning electron microscope, seen as bacteria enclosed within a glycocalyx matrix (62).
- Uropathogenic *E. coli* (UPEC) is the most studied organism causing cystitis, mouse models have been used for most of the studies, while rats have been used to demonstrate the effect of fosfomycin to prevent chronic cystitis (63).
- Gut: Biofilms in the gut are seen to colonize mucosal layers in animal models and play an important role in the dysbiosis theory, which is linked to the evolution of colorectal cancer (64).
- Wounds: The most studied microorganisms associated with wound infections are *S. aureus* and *P. aeruginosa* using animal tissue models (65).
- Heart: *S. aureus* is the most common organism causing endocarditis. Rabbit models have been used to study the modulation of virulence factors and their regulator systems (66). Rat model also facilitated the study of bacteraemia that mimics human infections to emphasize the role of fibrinogen and fibronectin in the induction of infective endocarditis.
- Sinuses: Immense evidence is available in literature on the presence of biofilms in chronic rhinosinusitis (CRS) as evidenced by scanning electron microscopy (67) and confocal laser scanning microscopy (68). In vivo models using sheep were found to be suitable, having similar CRS disease spectrum and histological features (69).
- Mouth: Different models were developed to study cariogenic biofilms. Of these, the rodent models to study streptococci biofilms are most used to evaluate antimicrobial susceptibility testing for dental plaques (70).
- Bones: Scanning electron microscopic studies have confirmed bacterial biofilms in bone tissue. Rabbit and rat models have been used to study bone infections to validate the advantage of fosfomycin over daptomycin in the treatment of osteomyelitis (71).

3.2.6 In vivo Models of Device-related Infections

The first proof of device-related biofilm *S. aureus* infections was identified by electron microscopy from a patient with a pacemaker who had recurrent episodes of bacteraemia (72). In vivo models are mainly site-specific, such as intravascular or intrafemoral models. In the subcutaneous models the foreign bodies are inserted in pockets that are not in contact with the specific organs. Site-specific models include vascular catheters, urinary catheters, orthopaedic implants, endotracheal tubes, vascular grafts, tissue filters and contact lenses. Subcutaneous foreign-body models mainly comprise of tissue cage, vascular catheters, cardiac alves, surgical meshes and pacing devices (73).

3.3 ANTIBIOTIC ACTIVITY AGAINST BIOFILMS: PHARMACODYNAMIC PARAMETERS

The end points using both the open and closed models can be used to analyse the effect of antimicrobials on biofilms (74), which includes the calculation of the minimum biofilm inhibitory concentration (MBIC) or the biofilm prevention concentration (BPC). MBIC is defined as the lowest amount of antibiotic needed to demonstrate a decrease in OD 600 by 10% in comparison to control cells (unexposed) (75). BPC is the lowest concentration that prevents the formation of biofilms by decreasing the cell density (76).

Biofilm bactericidal concentration (BBC) and minimum biofilm eradication concentration (MBEC) are used to measure effectiveness against preformed biofilms. BBC is defined as the lowest concentration of an antibiotic to exhibit a 99.9% decrease in the viability of a biofilm culture compared to the growth control. MBEC is lowest amount of antibiotic needed to eradicate biofilms that prevent growth visibly. A target value is necessary to provide information on the effective concentration of the antimicrobial agent needed to treat infections and serve as a guide to the treating physicians. Treatment decisions are made based on MBIC and MBEC cut-offs. Unfortunately, to date no consensus is available on target reference breakpoints. This leads to the opinion that complete eradication is essential for resolution of clinical symptoms and this mainly depends on the type of infection, the bacterial strain, and the immune status of the host.

3.4 THE NEED FOR ANTIMICROBIAL SUSCEPTIBILITY TESTING OF BIOFILMS

The antimicrobial susceptibility testing (AST) of planktonic growth differs significantly from those based on biofilm forms. The MBEC values are 100 to 1000 times higher compared to the MIC of planktonic bacteria, however, studies done with CF patients revealed that there was no difference in selected outcomes measured, for example, forced expiratory volume, quality of life and sputum density after comparing the biofilm AST and conventional AST levels (77).

In vivo biofilms are typically polymicrobial in nature, while the in vitro models are appropriate for the growth of monomicrobial isolates only. The effects of host immune systems, both innate and adaptive immune responses, are absent in the in vitro models, which could be the reasons for the poor predictability of in vitro models (64, 65). Clinical trials and research have proved that a longer duration of therapy could increase better outcome measures, hence, appropriate clinical and technical break points are required to determine treatment efficacy (78).

3.4.2 Bridging the Gap Between In vitro and In vivo Biofilm Models

3.4.2.1 Microfluidic-Electrochemical Coupled Systems

The release of electro active molecules are markers of cell virulence; the response of these markers can be measured using electrochemical techniques. Pyocyanin (PYO) released by *P. aeruginosa* is one such example, the advantage being that the in vivo status of infection can be monitored while the antibiotics are being used for the treatment.

3.4.2.2 The Alginate Bead System

In the alginate bead system, biofilms are grown on beads to form spatially structured aggregates like those seen in CF lungs and chronic wound infections (79). The alginate model was used, for example, for the anaerobic growth of *P. aeruginosa* to study resistance to disinfectants and antibiotics by incorporating NO_3 into the bead model (80). In the sodium alginate model, for analysis of biofilms the substratum can be dissolved to study the effect of treatment with specific antibiotics (81).

3.4.2.3 The Duckworth Biofilm Device

This device consists of a single instrument with separate individual channels for assessment, with biofilms cultured on cellulose acetate or nitrate discs or independent glass coverslips for microscopic examination (82). No technical expertise is required to operate the device and it can be 3D printed using different materials. The device is suited for wound modules grown at air-liquid interface and allows for application of wound dressings.

3.4.2.4 The 3D Wound Model

In the 3D collagen-containing matrix wound model, biofilms form cell aggregates along with matrix with serum proteins (83), closely resembling those seen in wounds. The wound model is commonly used to test the susceptibility of organisms commonly causing wound infections such as *S. aureus* and *P. aeruginosa*.

3.4.2.5 Isothermal Microcalorimetry (IMC)

The in vitro and in vivo effects of bacterial biofilms can be studied using the IMC method. IMC allows for the monitoring of the viability of biofilms in real time as biofilm formation occurs on glass ampoules. Analysis and measurements are performed using a microcalorimeter (84).

This technique is also used for quantifying the antimicrobial capacity of implants and biomaterials (85). Although this method is not a standardised technique, it is highly sensitive and reliable for biofilm susceptibility testing.

3.4.2.6 Poloxamer Thermo-Reversible Matrix

The use of improved biofilm poloxamers allows easy growth of biofilms on co-polymeric surfaces. These poloxamer matrices are inert and non-toxic, and biofilms on these surfaces form microcolonies as demonstrated by fluorescence microscopy after staining with live/dead stains (86).

3.4.2.7 Chip Calorimetry

The silicon chip contains thermocouples that translate the heat produced by bacteria into electrical signals. This method is validated for *Pseudomonas putida* biofilm susceptibility testing and for their cell viability. A calorimetric method is used for the assessment of biofilms containing multichannel wells for separate samples. The flexibility, real time monitoring and small size makes this model very convenient for research (87).

3.5 SELECTING AN APPROPRIATE MODEL FOR BIOFILM AST

Various AST models are available, but to choose the appropriate model that suits based on simplicity, easy reproducibility and available resources remains a challenge. The utmost care should be taken to select the right model that mimics clinical settings to identify and inform on treatment advice.

An ideal in vitro model should resemble in vivo conditions especially the O_2 levels, the pH and temperature levels. Understanding the host immune interactions between the biofilm and immune system is crucial to address effective treatment. Dynamic models mimic the flow of biological fluids better than static models. There is no gold standard method to give the best answer, researchers need to select an appropriate model based on their environment and the resources available. Most of the available in vitro models focus only on single species, however, biofilms can be polymicrobial in origin. Some of the major limitations (Table 3.2) of an in vivo model includes high variability of results between laboratories due to different strains among the same species, variations in the nutritional contents, time taken for the growth and method chosen to evaluate the biofilms (88).

In vivo models also have their own limitations, as immune mechanisms in small animals do not recapitulate human immune responses. Ethical issues of using animal models seem to be one of the main concerns. It is important to weigh the pros and cons of both in vivo and in vitro models to standardise a model appropriate for specific settings.

3.6 QUANTIFICATION OF BIOFILMS

3.6.1 Colony Forming Units

The biofilms are detached from the surface by sonication and vortexing. After homogenization and serial dilution, the liquid is transferred to agar plates (using the spread plate or the drop plate method), incubated and colony forming units are counted. Although colony counting is a standard method to check for viability, it is labour-intensive and the cost of resources and the restriction to a

TABLE 3.2
Advantages and Limitations of In vitro and In vivo Biofilm Models

In vitro models		In vivo models	
Advantages	Limitations	Advantages	Limitations
Low to high throughput	Real time monitoring not possible	Mimics human biological conditions	Subject to variability
Simple and easy to interpret	Experimental conditions influence biofilm variability	Allows study of complex interactions	Complex interpretation
Minimal ethical concerns	Nutrient variability differs from clinical biofilms	Translation of results to clinical conditions	Must adhere to ethical regulations
Non-invasive	Missing host interactions		Invasive
Low cost			Expensive
Rapid results			Time consuming

single microbial species is a limitation. Calorimetric methods are preferred in comparison to colony forming units.

3.6.2 SYTO 9/Propidium Iodide (Live/Dead BacLight Staining)

The fluorescent dye SYTO 9 binds to nucleic acids of both live and dead cells, which gives an estimate of cells within the biofilm matrix. Photo bleaching is one of the major limitations of SYTO 9 while staining intensity is an issue with other dyes such as SYBR green and acridine orange (73). The live/dead stains are usually a combination of SYTO 9 and propidium iodide, which can be used for different techniques in spectrophotometers and flow cytometers.

Acridine orange can be combined with povidone iodine to assess both live and dead cells in both *S. aureus* and *P. aeruginosa* biofilms while the combination of propidium iodine and SYTO 9 displays good staining intensities with *S. aureus* living cells.

3.6.3 Tetrazolium Salts

The metabolic activity of biofilm bacteria can be measured using tetrazolium salts. The reduction of these salts to formazan can be read calorimetrically. The rapidity of results using these techniques is an advantage, however, comparing results among different species and polymicrobial biofilms is a challenge. Their inhibitory nature on bacteria limits them to be used as end point assays. The sensitivity of these assays is relatively poor to detect bacteria in their dormant states (89).

3.6.4 Resazurin Dyes

These dyes are advantageous compared to tetrazolium salts due to less inhibitory effects on bacteria and the assays require less time. Good correlation is observed between colony counts and quantification methods (39). They are non-toxic, less time consuming and less labour intensive compared to colony counting methodologies (90).

3.6.5 Bioluminescence

Bioluminescence-based assays can estimate biofilm forms using adenosine triphosphate (ATP) which is required as an energy resource for life (91). These assays can be performed on microtiter plates as batch assays to decrease cost (92), however, they are expensive compared to other metabolic assays.

3.6.6 Total Biomass: Crystal Violet and Congo Red

One of the most common methods used for biofilm quantification by calorimetry is the crystal violet (CV) staining technique. Crystal violet stains the negatively charged polymers and molecules, thereby estimating the total biomass. It is mainly used for staining live and dead cells to assess the growth of biofilms and evaluate treatment response. Although it is a popular method, it has technical difficulties due to multiple washing steps making it less suitable for the quantification of *P.aeruginosa* biofilms (93). Hence, this method is opted for estimating large amount of biofilm biomass (78). Congo red staining is also used for the quantification of biofilms, however, because of the high background signal it is suitable only for biofilm models grown on a constant depth film fermenter. Other staining methods such as safranin and calcofluor white can also be used to stain cell components while fluorescein isothiocyanate (FITC) stains target the proteins of the biofilm structure (94).

3.6.7 Fluorescence Microscopy

Many fluorescent probes have been used to study the 3D biofilm structure by confocal laser scanning microscopy. A combination of probes with filters can be used to study different structures of the biofilms, namely the polysaccharides, proteins and live or dead bacteria (95). Confocal microscopy can be further used to assess the structural measurements, such as the biofilm surface area, thickness, and volume. The population of microorganisms can be estimated using fluorescence in situ hybridization (FISH), which is mainly used for permeabilized biofilms. Killing the microorganisms is one of the main disadvantages of FISH. FISH helps in the identification and detection of both metabolically active and inactive bacteria. Peptide nucleic acid FISH is one of the newer methods used to assess polymicrobial biofilms in chronic wound infections (96).

3.6.8 Electron Microscopy

The qualitative assessment of cell density is determined using scanning electron microscopy. The processing of specimens for electron microscopy includes dehydration and fixation, which affects the structure of biofilms. The environmental SEM (ESEM) does not require the dehydration of the samples and hence, is more advantageous to maintain the biofilm morphology (97).

3.7 CONCLUSIONS

It is of paramount importance to develop appropriate in vitro and in vivo techniques for biofilm susceptibility, keeping in mind the pros and cons of each system based on the availability of resources. Research related to biofilms has advanced since the work by Bill Costerton. The current focus has now shifted to advanced computing tools to pinpoint specific characteristics in biological systems to finely replicate in vivo models. Future biofilm research can enable the understanding of the physiological anatomy of biofilm communities and their interactions.

To improve clinical outcomes of infectious diseases, it is instrumental to include biofilm susceptibility testing into standard diagnostics in hospitals. This could provide guidance and inform on the selection of appropriate treatment regimens for faster patient recovery, optimised therapies, and less risk for the development of antibiotic resistance.

REFERENCES

1. www.who.int/news/item/29-04-2019-new-report-calls-for-urgent-action-to-avert-antimicrobial-resistance-crisis#

2. Goel, N., Fatima, S.W., Kumar, S., Sinha, R., and Khare, S.K. Antimicrobial resistance in biofilms: Exploring marine actinobacteria as a potential source of antibiotics and biofilm inhibitors. *Biotechnol Rep Amst Neth.* 30:e00613, 2021 Jun.
3. Lappin-Scott, H., Burton, S., and Stoodley, P. Revealing a world of biofilms — the pioneering research of Bill Costerton. *Nat Rev Microbiol.* 12(11):781–7, 2014 Nov.
4. Geesey, G.G., Mutch, R., Costerton, J.W., and Green, R.B. Sessile bacteria: An important component of the microbial population in small mountain streams. *Limnol Oceanogr.* 23(6):1214–23, 1978.
5. Bueno, J. Anti-biofilm drug susceptibility testing methods: Looking for new strategies against resistance mechanism. *J Microb Biochem Technol.* S3:1–9, 2014 May 16.
6. Sharma, D., Misba, L., and Khan A.U. Antibiotics versus biofilm: an emerging battleground in microbial communities. *Antimicrob Resist Infect Control.* 8(1):76, 2019 May 16.
7. Davies, D. Understanding biofilm resistance to antibacterial agents. *Nat Rev Drug Discov.* 2(2):114–22, 2003 Feb.
8. Christensen, G.D., Simpson, W.A., Younger, J.J., Baddour, L.M., Barrett, F.F., Melton, D.M., et al. Adherence of coagulase-negative staphylococci to plastic tissue culture plates: a quantitative model for the adherence of staphylococci to medical devices. *J Clin Microbiol.* 22(6):996–1006, 1985 Dec.
9. Stepanović, S., Vuković, D., Hola, V., Bonaventura, G.D., Djukić, S., Ćirković, I., et al. Quantification of biofilm in microtiter plates: overview of testing conditions and practical recommendations for assessment of biofilm production by staphylococci. *APMIS.* 115(8):891–9, 2007.
10. McBain, A.J. Chapter 4: In vitro biofilm models: an overview. *Adv Appl Microbiol.* 69:99–132, 2009.
11. Mah, T-F., Pitts, B., Pellock, B., Walker, G.C., Stewart, P.S., and O'Toole, G.A. A genetic basis for Pseudomonas aeruginosa biofilm antibiotic resistance. *Nature.* 426(6964):306–10, 2003 Nov 20.
12. Pitts, B., Hamilton, M.A., Zelver, N., and Stewart, P.S. A microtiter-plate screening method for biofilm disinfection and removal. *J Microbiol Methods.* 54(2):269–76, 2003 Aug.
13. Merritt, J.H., Kadouri, D.E., O'Toole, G.A. Growing and analyzing static biofilms. *Curr Protoc Microbiol.* 2005 Jul. doi: 10.1002/9780471729259.mc01b01s00. PMID: 18770545; PMCID: PMC4568995.
14. Wilson, C., Lukowicz, R., Merchant, S., et al. Quantitative and qualitative assessment methods for biofilm growth: A mini-review. *Res Rev J Eng Technol.* 6(4), 2017. www.rroij.com/open-access/quantitative-and-qualitative-assessment-methods-for-biofilm-growth-a-minireview
15. Ceri, H., Olson, M.E., Stremick, C., Read, R.R., Morck, D., and Buret, A. The Calgary biofilm device: New technology for rapid determination of antibiotic susceptibilities of bacterial biofilms. *J Clin Microbiol.* 37(6):1771–6, 1999 Jun.
16. Tré-Hardy, M., Traore, H., El Manssouri, N., Vanderbist, F., Vaneechoutte, M., and Devleeschouwer, M.J. Evaluation of long-term co-administration of tobramycin and clarithromycin in a mature biofilm model of cystic fibrosis clinical isolates of Pseudomonas aeruginosa. *Int J Antimicrob Agents.* 34(4):370–4, 2009 Oct.
17. Parker, A.E., Walker, D.K., Goeres, D.M., Allan, N., Olson, M.E., and Omar, A. Ruggedness and reproducibility of the MBEC biofilm disinfectant efficacy test. *J Microbiol Methods.* 102:55–64, 2014 Jul.
18. Holá, V., and Růžička, F. The Formation of Poly-Microbial Biofilms on Urinary Catheters. In: *Urinary tract infections.* 2011.
19. James, G.A., Ge Zhao, A., Usui, M., Underwood, R.A., Nguyen, H., Beyenal, H., et al. Microsensor and transcriptomic signatures of oxygen depletion in biofilms associated with chronic wounds. *Wound Repair Regen Off Publ Wound Heal Soc Eur Tissue Repair Soc.* 24(2):373–83, 2016 Mar.
20. Richter, K., Thomas, N., Claeys, J., McGuane, J., Prestidge, C.A., Coenye, T., et al. A topical hydrogel with deferiprone and gallium-protoporphyrin targets bacterial iron metabolism and has antibiofilm activity. *Antimicrob Agents Chemother.* 61(6):e00481–17, 2017.
21. Richter, K., Thomas, N., Zhang, G., Prestidge, C.A., Coenye, T., Wormald, P-J., et al. Deferiprone and gallium-protoporphyrin have the capacity to potentiate the activity of antibiotics in Staphylococcus aureus small colony variants. *Front Cell Infect Microbiol* [Internet]. 2017 [cited 2022 Mar 16];7. Available from: www.frontiersin.org/article/10.3389/fcimb.2017.00280
22. Chavant, P., Gaillard-Martinie, B., Talon, R., Hébraud, M., and Bernardi, T. A new device for rapid evaluation of biofilm formation potential by bacteria. *J Microbiol Methods.* 68(3):605–12, 2007 Mar.
23. Liesse Iyamba, J.M., Seil, M., Devleeschouwer, M., Takaisi Kikuni, N.B., and Dehaye, J.P. Study of the formation of a biofilm by clinical strains of Staphylococcus aureus. *Biofouling.* 27(8):811–21, 2011 Sep.

24. Puig, C., Domenech, A., Garmendia, J., Langereis, J.D., Mayer, P., Calatayud, L., et al. Increased biofilm formation by nontypeable Haemophilus influenzae isolates from patients with invasive disease or otitis media versus strains recovered from cases of respiratory infections. *Appl Environ Microbiol.* 80(22):7088–95, 2014 Nov.
25. Valour, F., Rasigade, J-P., Trouillet-Assant, S., Gagnaire, J., Bouaziz, A., Karsenty, J,. et al. Delta-toxin production deficiency in Staphylococcus aureus: a diagnostic marker of bone and joint infection chronicity linked with osteoblast invasion and biofilm formation. *Clin Microbiol Infect Off Publ Eur Soc Clin Microbiol Infect Dis.* 21(6):568.e1–11, 2015 Jun.
26. Tasse, J., Croisier, D., Badel-Berchoux, S., Chavanet, P., Bernardi, T., and Provot, C., et al. Preliminary results of a new antibiotic susceptibility test against biofilm installation in device-associated infections: the Antibiofilmogram®. *Pathog Dis.* 74(6):ftw057, 2016 Aug.
27. Coenye, T., Nelis, H.J. In vitro and in vivo model systems to study microbial biofilm formation. *J Microbiol Methods.* 83(2):89–105, 2010 Nov.
28. Pereira, M.O., Kuehn, M., Wuertz, S., Neu, T., and Melo, L.F. Effect of flow regime on the architecture of a Pseudomonas fluorescens biofilm. *Biotechnol Bioeng.* 78(2):164–71, 2002 Apr 20.
29. Hodgson, A.E., Nelson, S.M., Brown, M.R.W., and Gilbert, P. A simple in-vitro model for growth control of bacterial biofilms. *Journal of Applied Bacteriology.* 79, 87–93, 1995.
30. Hodgson, A.E., Nelson, S.M., Brown, M.R., and Gilbert, P. A simple in vitro model for growth control of bacterial biofilms. *J Appl Bacteriol.* 79(1):87–93, 1995 Jul.
31. Budhani, R.K., and Struthers, J.K. The use of Sorbarod biofilms to study the antimicrobial susceptibility of a strain of Streptococcus pneumoniae. *J Antimicrob Chemother.* 40(4):601–2, 1997 Oct.
32. Spencer, P., Greenman, J., McKenzie, C., Gafan, G., Spratt, D., and Flanagan, A. In vitro biofilm model for studying tongue flora and malodour. *J Appl Microbiol.* 103(4):985–92, 2007.
33. Volgenant, C.M.C., Hoogenkamp, M.A., Krom, B.P., Janus, M.M., Cate, J.M.T., de Soet, J.J., et al. Red and Green Fluorescence from Oral Biofilms. *PLoS ONE.* 11(12):e0168428, 2016 Dec 20.
34. Benoit, M.R., Conant, C.G., Ionescu-Zanetti, C., Schwartz, M., and Matin, A. New device for high-throughput viability screening of flow biofilms. *Appl Environ Microbiol.* 76(13):4136–42, 2010 Jul.
35. Heydorn, A., Nielsen, A.T., Hentzer, M., Sternberg, C., Givskov, M., Ersbøll, B.K., et al. Quantification of biofilm structures by the novel computer program COMSTAT. *Microbiol Read Engl.* 146 (Pt 10):2395–407, 2000 Oct.
36. Alves, D.R., Booth, S.P., Scavone, P., Schellenberger, P., Salvage, J., Dedi, C., et al. Development of a high-throughput ex-vivo burn wound model using porcine skin, and its application to evaluate new approaches to control wound infection. *Front Cell Infect Microbiol.* 8:196, 2018.
37. Richter, L., Stepper, C., Mak, A., Reinthaler, A., Heer, R., Kast, M., et al. Development of a microfluidic biochip for online monitoring of fungal biofilm dynamics. *Lab Chip.* 7(12):1723–31, 2007 Dec.
38. Curtin, J.J., and Donlan, R.M. Using bacteriophages to reduce formation of catheter-associated biofilms by Staphylococcus epidermidis. *Antimicrob Agents Chemother.* 50(4):1268–75, 2006 Apr.
39. Azeredo, J., Azevedo, N.F., Briandet, R., Cerca, N., Coenye, T., Costa, A.R., et al. Critical review on biofilm methods. *Crit Rev Microbiol.* 43(3):313–51, 2017 May.
40. Teodósio, J.S., Simões, M., Melo, L.F., and Mergulhão, F.J. Flow cell hydrodynamics and their effects on E. coli biofilm formation under different nutrient conditions and turbulent flow. *Biofouling.* 27(1):1–11, 2011 Jan.
41. Lasserre, J., Toma, S., Dos Santos-Gonçalvez, A-M., Leprince, J., Leloup, G., and Brecx, M. Evaluation of Emdogain® antimicrobial effectiveness against biofilms containing the keystone pathogen Porphyromonas gingivalis. *New Microbiol.* 41(1):73–6, 2018 Jan.
42. Ombaka, E.A., Cozens, R.M., and Brown, M.R. Influence of nutrient limitation of growth on stability and production of virulence factors of mucoid and nonmucoid strains of Pseudomonas aeruginosa. *Rev Infect Dis.* 5 Suppl 5:S880–888, 1983 Dec.
43. Kim, J., Pitts, B., Stewart, P.S., Camper, A., and Yoon, J. Comparison of the antimicrobial effects of chlorine, silver ion, and tobramycin on biofilm. *Antimicrob Agents Chemother.* 52(4):1446–53, 2008 Apr.
44. McNeer, N.A., Anandalingam, K., Fields, R.J., Caputo, C., Kopic, S., Gupta, A., et al. Nanoparticles that deliver triplex-forming peptide nucleic acid molecules correct F508del CFTR in airway epithelium. *Nat Commun.* 6:6952, 2015 Apr 27.

45. Peters, A.C., and Wimpenny, J.W. A constant-depth laboratory model film fermentor. *Biotechnol Bioeng*. 32(3):263–70, 1988 Jul 20.
46. Khatoon, Z., McTiernan, C..D, Suuronen, E.J., Mah, T-F., and Alarcon, E.I. Bacterial biofilm formation on implantable devices and approaches to its treatment and prevention. *Heliyon*. 4(12):e01067, 2018 Dec 28.
47. Cho, H.H., Kwon, K.C., Kim, S., Park, Y., Koo, S.H. Association between biofilm formation and antimicrobial resistance in carbapenem-resistant Pseudomonas aeruginosa. *Ann Clin Lab Sci*. 48(3):363–8, 2018 May.
48. Jahanbakhsh, S., Singh, N.B., Yim, J., Kebriaei, R., Smith, J.R., Lev, K.L., et al. Impact of daptomycin dose exposure alone or in combination with β-lactams or rifampin against vancomycin-resistant Enterococci in an in vitro biofilm model. *Antimicrob Agents Chemother*. 2020
49. Shelobolina, E.S., Walker, D.K., Parker, A.E., Lust, D.V., Schultz, J.M., and Dickerman, G.E. Inactivation of Pseudomonas aeruginosa biofilms formed under high shear stress on various hydrophilic and hydrophobic surfaces by a continuous flow of ozonated water. *Biofouling*. 34(7):826–34, 2018 Aug 9.
50. Lenton, P., Rudney, J., Chen, R., Fok, A., Aparicio, C., and Jones, R.S. Imaging in vivo secondary caries and ex vivo dental biofilms using cross-polarization optical coherence tomography. *Dent Mater Off Publ Acad Dent Mater*. 28(7):792–800, 2012 Jul.
51. Wolcott, R.D., Rumbaugh, K.P., James, G., Schultz, G., Phillips, P., Yang, Q., et al. Biofilm maturity studies indicate sharp debridement opens a time-dependent therapeutic window. *J Wound Care*. 19(8):320–8, 2010 Aug.
52. Harriott, M.M., Lilly, E.A., Rodriguez, T.E., Fidel, P.L., and Noverr, M.C. Candida albicans forms biofilms on the vaginal mucosa. *Microbiol Read Engl*. 156(Pt 12):3635–44, 2010 Dec.
53. Lemaitre, B., and Ausubel, F.M. Animal models for host-pathogen interactions. *Curr Opin Microbiol*. 11(3):249–50, 2008 Jun.
54. Wille, J., Teirlinck, E., Sass, A., Van Nieuwerburgh, F., Kaever, V., Braeckmans, K., et al. Does the mode of dispersion determine the properties of dispersed Pseudomonas aeruginosa biofilm cells? *Int J Antimicrob Agents*. 56(6):106194, 2020 Dec 1.
55. Zhang, Y., Hu, Y., Yang, B., Ma, F., Lu, P., and Li, L., et al. Duckweed (Lemna minor) as a Model Plant System for the Study of Human Microbial Pathogenesis. *PLoS ONE*. 5(10):e13527, 2010 Oct 25.
56. Richter, K., Facal, P., Thomas, N., Vandecandelaere, I., Ramezanpour, M., Cooksley, C., et al. Taking the silver bullet colloidal silver particles for the topical treatment of biofilm-related infections. *ACS Appl Mater Interfaces*. 9(26):21631–8, 2017.
57. Neely, M.N., Pfeifer, J.D., and Caparon, M. Streptococcus-zebrafish model of bacterial pathogenesis. *Infect Immun*. 70(7):3904–14, 2002 Jul.
58. Keiser, N.W., Engelhardt, J.F. New animal models of cystic fibrosis: what are they teaching us? *Curr Opin Pulm Med*. 17(6):478–83, 2011 Nov.
59. Yang, L., Hengzhuang, W., Wu, H., Damkiaer, S., Jochumsen, N., Song, Z., Givskov, M., Hoiby, N., and Molin, S. Polysaccharides serve as scaffold of biofilms formed by mucoid Pseudomonas aeruginosa. *FEMS Immunol. Med. Microbiol*. 65, 366–376, 2012.
60. Hannan, T.J., Totsika, M., Mansfield, K.J., Moore, K.H., Schembri, M.A., and Hultgren, S.J. Host-pathogen checkpoints and population bottlenecks in persistent and intracellular uropathogenic Escherichia coli bladder infection. *FEMS Microbiol Rev*. 36(3):616–48, 2012 May.
61. Fischer, H., Lutay, N., Ragnarsdóttir, B., Yadav, M., Jönsson, K., Urbano, A., et al. Pathogen specific, IRF3-dependent signaling and innate resistance to human kidney infection. *PLoS Pathog*. 6(9):e1001109, 2010 Sep 23.
62. Nemoy, N.J., and Staney, T.A. Surgical, bacteriological, and biochemical management of "infection stones." *JAMA*. 215(9):1470–6, 1971 Mar 1.
63. Ozok, H.U., Ekim, O., Saltas, H., Arikok, A.T., Babacan, O., Sagnak, L., Topaloglu, H., and Ersoy, H. The preventive role of transurethral antibiotic delivery in a rat model. *Drug Des Dev Ther*. 6, 187–194, 2012.
64. Xavier, R.J., and Podolsky, D.K. Unravelling the pathogenesis of inflammatory bowel disease. *Nature*. 448(7152):427–34, 2007 Jul 26.

65. Cambridge Media Journals. Fighting chronic wound infection – One model at a time. [Internet]. [cited 2022 Mar 21]. Available from: https://journals.cambridgemedia.com.au/wpr/volume-25-number-1/fighting-chronic-wound-infection-one-model-time
66. Akiyama, H., Kanzaki, H., Tada, J., and Arata, J. Staphylococcus aureus infection on cut wounds in the mouse skin: experimental staphylococcal botryomycosis. *J Dermatol Sci*. 11(3):234–8, 1996 Mar.
67. Cryer, J., Schipor, I., Perloff, J.R., Palmer, J.N. Evidence of bacterial biofilms in human chronic sinusitis. *ORL J Oto-Rhino-Laryngol Its Relat Spec*. 66(3):155–8, 2004.
68. Psaltis, A.J., Ha, K.R., Beule, A.G., Tan, L.W., and Wormald, P-J. Confocal scanning laser microscopy evidence of biofilms in patients with chronic rhinosinusitis. *The Laryngoscope*. 117(7):1302–6, 2007.
69. Ha, K.R., Psaltis, A.J., Tan, L., and Wormald, P-J. A sheep model for the study of biofilms in rhinosinusitis. *Am J Rhinol*. 21(3):339–45, 2007 Jun.
70. Fitzgerald, R.J., and Keyes, P.H. Demonstration of the etiologic role of streptococci in experimental caries in the hamster. *J Am Dent Assoc 1939*. 61:9–19, 1960 Jul.
71. Poeppl, W., Tobudic, S., Lingscheid, T., Plasenzotti, R., Kozakowski, N., Lagler, H., et al. Daptomycin, fosfomycin, or both for treatment of methicillin-resistant Staphylococcus aureus osteomyelitis in an experimental rat model. *Antimicrob Agents Chemother*. 55(11):4999–5003, 2011 Nov.
72. Marrie, T.J., Nelligan, J., and Costerton, J.W. A scanning and transmission electron microscopic study of an infected endocardial pacemaker lead. *Circulation*. 66(6):1339–41, 1982 Dec.
73. Kokai-Kun, J.F., Chanturiya, T., Mond, J.J. Lysostaphin eradicates established Staphylococcus aureus biofilms in jugular vein catheterized mice. *J Antimicrob Chemother*. 64(1):94–100, 2009 Jul.
74. Macià, M.D., Rojo-Molinero, E., and Oliver, A. Antimicrobial susceptibility testing in biofilm-growing bacteria. *Clin Microbiol Infect Off Publ Eur Soc Clin Microbiol Infect Dis*. 20(10):981–90, 2014 Oct.
75. Moskowitz, S.M., Foster, J.M., Emerson, J., and Burns, J.L. Clinically feasible biofilm susceptibility assay for isolates of Pseudomonas aeruginosa from patients with cystic fibrosis. *J Clin Microbiol*. 42(5):1915–22, 2004 May.
76. Kolpen, M., Hansen, C.R., Bjarnsholt, T., Moser, C., Christensen, L.D., and van Gennip, M., et al. Polymorphonuclear leucocytes consume oxygen in sputum from chronic Pseudomonas aeruginosa pneumonia in cystic fibrosis. *Thorax*. 65(1):57–62, 2010 Jan.
77. Waters, V., and Ratjen, F. Standard versus biofilm antimicrobial susceptibility testing to guide antibiotic therapy in cystic fibrosis. *Cochrane Database Syst Rev*. 10:CD009528, 2017 Oct 5.
78. Castaneda, P., McLaren, A., Tavaziva, G., and Overstreet, D. Biofilm antimicrobial susceptibility increases with antimicrobial exposure time. *Clin Orthop*. 474(7):1659–64, 2016 Jul.
79. Bjarnsholt, T., Alhede, M., Alhede, M., Eickhardt-Sørensen, S.R., Moser, C., Kühl, M., et al. The in vivo biofilm. *Trends Microbiol*. 21(9):466–74, 2013 Sep.
80. Cao, B., Christophersen, L., Thomsen, K., Sønderholm, M., Bjarnsholt, T., Jensen, P.Ø., et al. Antibiotic penetration and bacterial killing in a Pseudomonas aeruginosa biofilm model. *J Antimicrob Chemother*. 70(7):2057–63, 2015 Jul.
81. Dall, G.F., Tsang, S.T.J., Gwynne, P.J., Wilkinson, A.J., Simpson, A.H.R.W., Breusch, S.J.B., et al. The dissolvable bead: A novel in vitro biofilm model for evaluating antimicrobial resistance. *J Microbiol Methods*. 142:46–51, 2017 Nov.
82. Duckworth, P.F., Rowlands, R.S., Barbour, M.E., and Maddocks, S.E. A novel flow-system to establish experimental biofilms for modelling chronic wound infection and testing the efficacy of wound dressings. *Microbiol Res*. 215:141–7, 2018 Oct.
83. Werthén, M., Henriksson, L., Jensen, P.Ø., Sternberg, C., Givskov, M., and Bjarnsholt, T. An in vitro model of bacterial infections in wounds and other soft tissues. *APMIS Acta Pathol Microbiol Immunol Scand*. 118(2):156–64, 2010 Feb.
84. Braissant, O., Wirz, D., Göpfert, B., Daniels, A.U. Use of isothermal microcalorimetry to monitor microbial activities. *FEMS Microbiol Lett*. 303(1):1–8, 2010 Feb;.
85. Di Luca, M., Navari, E., Esin, S., Menichini, M., Barnini, S., Trampuz, A., et al. Detection of Biofilms in Biopsies from Chronic Rhinosinusitis Patients: In Vitro Biofilm Forming Ability and Antimicrobial Susceptibility Testing in Biofilm Mode of Growth of Isolated Bacteria. *Adv Exp Med Biol*. 1057:1–27, 2018.
86. Carlson, R.P., Taffs, R., Davison, W.M., Stewart, P.S. Anti-biofilm properties of chitosan-coated surfaces. *J Biomater Sci Polym Ed*. 19(8):1035–46, 2008.

87. Buchholz, F., Wolf, A., Lerchner, J., Mertens, F., Harms, H., and Maskow, T. Chip calorimetry for fast and reliable evaluation of bactericidal and bacteriostatic treatments of biofilms. *Antimicrob Agents Chemother*. 54(1):312–9, 2010 Jan.
88. Guzmán-Soto, I., McTiernan, C., Gonzalez-Gomez, M., Ross, A., Gupta, K., Suuronen, E.J., et al. Mimicking biofilm formation and development: Recent progress in in vitro and in vivo biofilm models. *iScience*. 24(5):102443, 2021 May.
89. Braissant, O., Astasov-Frauenhoffer, M., Waltimo, T., and Bonkat, G. A review of methods to determine viability, vitality, and metabolic rates in microbiology. *Front Microbiol* [Internet]. 11. Available from: www.frontiersin.org/article/10.3389/fmicb.2020.547458, 2020 [cited 2022 Feb 15].
90. Van den Driessche, F., Rigole, P., Brackman, G., and Coenye, T. Optimization of resazurin-based viability staining for quantification of microbial biofilms. *J Microbiol Methods*. 98:31–4, 2014 Mar.
91. Balouiri, M., Sadiki, M., and Ibnsouda, S.K. Methods for in vitro evaluating antimicrobial activity: A review. *J Pharm Anal*. 6(2):71–9, 2016 Apr.
92. Bahamondez-Canas, T.F., Zhang, H., Tewes, F., Leal, J., Smyth, H.D.C. PEGylation of Tobramycin Improves Mucus Penetration and Antimicrobial Activity against Pseudomonas aeruginosa Biofilms in Vitro. *Mol Pharm*. 15(4):1643–52, 2018 Apr 2.
93. Peeters, E., Nelis, H.J., and Coenye, T. Comparison of multiple methods for quantification of microbial biofilms grown in microtiter plates. *J Microbiol Methods*. 72(2):157–65, 2008 Feb.
94. Baird, F.J., Wadsworth, M.P., and Hill, J.E. Evaluation and optimization of multiple fluorophore analysis of a Pseudomonas aeruginosa biofilm. J *Microbiol Methods*. 90(3):192–6, 2012 Sep.
95. Lewandowski, Z., and Beyenal, H. Fundamentals of Biofilm Research [Internet]. 0 ed. CRC Press; Available from: www.taylorfrancis.com/books/9781466559608 2013 [cited 2022 Feb 15].
96. Kirketerp-Møller, K., Jensen, P.Ø., Fazlim M,. Madsenm K,G,, Pedersenm J,. Moserm C., et al. Distribution, organization, and ecology of bacteria in chronic wounds. *J Clin Microbiol*. 46(8):2717–22, 2008 Aug.
97. Abed, S.E., Ibnsouda, S.K., Latrache, H., and Hamadi, F. Scanning electron microscopy (SEM) and environmental SEM: Suitable tools for study of adhesion stage and biofilm formation. In: Kazmiruk V, editor, *Scanning Elec/tron Microscopy* [Internet]. London: IntechOpen; Available from: www.intechopen.com/chapters/30960 2012 [cited 2022 Feb 16].

4 Nutritional and Stress Responses Resulting in the Development of AMR

Indu Sharma, Hanna Yumnam and Sobita Singha

4.1 INTRODUCTION

Antimicrobial resistance (AMR) has been recognized as a current global threat to human health (Hakanen, A., et al. 2017). We have major challenges in treating bacterial infections as a result of the extensive use of antibiotics, which is leading to a range of resistance mechanisms (Gullberg, E., et al. 2011). Millions of human diseases are caused worldwide by pathogens that have evolved antimicrobial resistance (AMR) in response to stressful environmental conditions. Fortunately, the NCBI Pathogen Detection Isolates Browser (NPDIB) has compiled the identified stress response and AMR genes from a variety of pathogen isolates sampled globally (Pei, R., et al. 2021). Antibiotics that are used therapeutically have the potential to select for resistant microorganisms at high concentrations. However, it is still unclear how selection may be impacted by the significantly lower antibiotic concentrations present in many habitats (Gullberg, E., et al. 2011).

According to Scallan, E., et al. (2011), pathogens are to blame for 1351 fatalities, 55961 hospitalizations, and 9.4 million cases of foodborne disease in the US annually. Before they come into contact with humans and spread disease, these pathogens managed to withstand the strain that adverse settings placed on them. Therefore, the pathogens must endure any decontaminating procedures intended to get rid of them, especially if the infections are spread by water or food (Boor, K. J., 2006). As for example *E. coli* has been identified as a common foodborne pathogen that can survive in temperatures below freezing (Gao, W., et al. 2006; Weiser, R. S., and Hargiss, C. O., 1946). In a study, it was discovered that when the temperature of the environment rapidly lowers, *E. coli* strains acquire genes that cause thermal stress reactions. The pathogen controls this stress response by producing cold shock proteins. These proteins were discovered, that control vital cellular processes such as, translation, transcription, and recombination and aided in the microorganism's survival (Chung, H. J., et al. 2006). Pathogens must endure a variety of pressures in addition to the cold stress in order to survive the harsh environments and multiply more quickly before infecting humans. In order to survive against the body's inherent defences, pathogens must take precautions. Examples of stresses are: gastric acid which is, pH-stress, fevers which are thermal stress and nutrient restriction which is starvation stress (Dühring, S., et al. 2015). Therefore, study of the mechanisms of microbial stress response has become very important where pathogens quickly perceive and then adapt themselves according to alterations in the environment that could otherwise cause them harm or destroy them (Boor, K. J. 2006). Pathogens can be further treated with antimicrobials even if stress response mechanisms may help them survive in a hostile environment such as high or low temperature, low pH, and restricted nutrition, and so forth. Antimicrobial resistance comes as a result of the pathogens that endure antimicrobial treatments (AMR). AMR gene-encoded proteins in particular have the potential to modify the targets of antibiotics and destroy them. They can also use efflux pumps to remove the antimicrobials from the system or lessen

DOI: 10.1201/9781003281238-4

the permeability of the antimicrobials' cell membranes (Reygaert, W. C. 2018; O'Brien, T. F. 2002; Harbottle, H., et al. 2006). Unfortunately, it has been observed that AMR genes may be spread horizontally through conjugation, transformation, or transduction or by the use of portable genetic carriers like plasmids (O'Brien, T. F. 2002).

Numerous adverse situations in the environment as well as in its hosts have led to the evolution of several survival tactics by bacteria. Responses to nutritional limitation, nitrosative or oxidative stress, exposure to antibiotics, envelope stress and many other growth-limiting conditions have been discovered to be linked to the emergence of AMR in Gram negative bacteria. All of this came about as a result of the activation of protective changes to cellular physiology, such as decreased permeability of the outer membrane, resistance transporters, or drug efflux pumps, resistant lifestyles such as biofilms and resistance mutations such as target mutations, production of antibiotic modification, or enzyme degradation. Antibiotics are sometimes seen as stressors, to which protective reactions have developed by focusing on and obstructing vital physiological functions (Dam, S., et al. 2018). This chapter focuses on the effects of stress and nutrition on the development of antibiotic resistance.

4.2 NUTRITIONAL STRESS RESPONSES RESULT IN THE DEVELOPMENT OF ANTIMICROBIAL RESISTANCE (AMR)

The rapidly replicating bacteria are preferentially generally killed by the antibiotics (Eng, R. H., et al. 1991). It has been noted that resistance in *Pseudomonas aeruginosa* that has been attributed to the biofilm mode of growth emerged from established biofilm cells that were present in a very oxygen-starved environment and and which happened in an anaerobic stationary phase rendered them resistant to antimicrobials. When compared to the normal condition it was found that young biofilm cells which are more aerobic and in a growing stage and are thus antimicrobial susceptible (Borriello, G., et al. 2004). Furthermore, it was discovered through the screening of both active and dormant cells in a *P. aeruginosa* biofilm that the latter were less sensitive to the aminoglycoside tobramycin (Kim, J., et al. 2009). It is considered that the lower metabolic activity of anoxic *P. aeruginosa* biofilm cells is one of the mechanisms that contributes to their tolerance to antibiotics (Borriello, G., et al. 2006). In *Mycobacterium tuberculosis*, oxygen limitation also led to stopage of the process of replication followed by adoption of an antibiotic-tolerant state (Wayne, L. G., and Hayes, L. G. 1996). The *in vivo* antibiotic tolerance of this organism in the presence of effective antibiotic activity *in vitro* is caused by the lowered growth and metabolic activity of quiescent *M. tuberculosis* infections (Mitchison, D. A., and Coates, A. R. 2004).

4.2.1 Nutritional Constraints and the Adverse Response

One of the case examples of nutritional stress is amino acid deprivation which activates the adverse response (Chatterji, D., and Ojha, A. K., et al. 2001). The alarmone guanosine 5′-(tri) diphosphate 3′-diphosphate [(p)ppGpp] is produced more frequently in adverse reactions (Potrykus, K., and Cashel, M. 2008) and is characterised by transcriptional switching, which results in increased expression of survival genes that efficiently use limiting resources while decreasing expression of genes generally linked with growth (Sharma, U. K., and Chatterji, D. 2010).The physiology of the bacterial cell is impacted by the ppGpp rise caused by the stringent response. and demonstrates its effect on susceptibility to antibiotics (Wu, J., et al. 2010). For instance, lower penicillin susceptibility in *Escherichia coli* has been linked to amino acid deficiency and the stringent response. Due to the inability of the mutant *E. coli* strains to synthesise ppGpp and increased susceptibility to penicillin-dependent lysis under amino acid restriction, their penicillin resistance was lost in the absence of relA (Kusser, W., and Ishiguro, E. E. 1985, Kusser, W., and Ishiguro, E. E. 1987). Resistance to penicillin and mecillin is likewise connected to elevations in ppGpp levels brought on by the cloned relA gene in a wild-type *E. coli* strain (Rodionov, D. G., and Ishiguro, E. E. (1995) and mecillinam resistance (Vinella,

D., et al. 1992), and mecillinam resistance of in vitro-selected mecillinam-resistant mutants. Additionally, it has been noted that *E. coli* exposed to a small amount of Fe exhibited an increase in ppGpp levels and mecillinam resistance (Vinella, D., et al. 2005). AMR has not been associated with an increase in ppGpp accumulation, with the exception of *Enterococcus faecalis*, where a mutant that was unable to synthesise ppGpp showed increased susceptibility to the antibiotic vancomycin (Abranches, J., et al. 2009), mupirocin (Abranches, J., et al. 2009), and penicillin (Rodionov, D. G., Abranches, J., et al. 2009). Additionally, *Streptomyces coelicolor* ppGpp synthesis has been connected to vancomycin and bacitracin resistance (Hesketh, A., et al. 2011).

4.3 OXIDATIVE AND NITROSATIVE STRESSES RESULTING IN ANTIMICROBIAL RESISTANCE (AMR)

Oxidative or nitrosative stress has an impact on antimicrobial resistance determinants and their regulatory proteins come in a wide range (Poole, K. 2012). *E. coli* and other intestinal organisms have been shown to respond to oxidative stress in a critical way that is regulated by SoxRS. A two-component regulatory system (TCS) called SoxRS was first described as being superoxide (O_2) responsive and controlling an adaptive O_2–stress response (Touati, D. 2000), but it is now recognized that it interacts immediately with these redox-cycling agents (Gu, M., and Imlay, J. A. 2011). The protective SoxRS-mediated response to redox-cycling medicines relies heavily on the resistance-nodulation-division (RND) family AcrAB-TolC multidrug efflux system, in which acrAB expression is favourably regulated by soxS. Mutations that elevated acrAB expression and AMR boosted constitutive soxS expression (Poole, K. 2012). The MexR repressor of the *P. aeruginosa* mex AB-oprM multidrug efflux operon is another recognised regulator of the multidrug efflux system. It is also recognised to react to OS. It has been discovered that the protein separates from its target DNA, the mexAB-oprM promoter, when it is in vitro oxidised by peroxide or cumene hydroperoxide. Thus, mexAB-oprM expression was satisfactorily increased after *P. aeruginosa* was exposed to these reactive oxygen species (ROS) in vivo (Chen, H., et al. 2008). The principal determinants of innate and acquired antimicrobial resistance seen in Gram-negative bacteria are multidrug efflux systems of the RND family, which are a part of bacterial stress responses, including those of oxidative and nitrosative stress responses (Poole, K. 2008).

It is also observed that stress can induce changes in the mexEF-oprN multidrug efflux locus in *P. aeruginosa*. Response to nitrosative stress induces the expression of this operon, which is normally inactive (Fetar, H., et al. 2011). A number of MexT targets discovered in an array study of MexT-dependent genes (Tian, Z. X., et al. 2009) were also found to be induced in response to nitrosative stress (Fetar, H., et al. 2011), suggesting that MexT regulates the expression of a regulon with some role in a response to nitrosative stress. It was also noted that, intriguingly, MexT was required for chloramphenicol's induction of mexEF-oprN expression (Fetar, H., et al. 2011). The significance of the nitro moiety of chloramphenicol for this induction was thus underscored by this event. It is intriguing to note that nitrated amino acids are occasionally produced by nitrosative stress in bacteria since chloramphenicol may mimic a nitrated nitrosative stress product that is meant to signal MexT and act as a substrate for MexEF-OprN (Poole, K. 2012).

The end result of antimicrobial exposure is the development of ROS in both bacteria and OS. Typically, ROS are blamed for the fatal side effects of bactericidal antibiotics (Kolodkin-Gal, I., et al. 2008; Kohanski, M. A., et al. 2008). Antioxidant and oxidative stress protective responses were discovered in bacteria as a result, which is likely to increase AMR. It is undeniable that ROS defense genes were removed from *E. coli*, increasing their vulnerability to antibiotics (Goswami, M., et al. 2006). Indole, an extracellular signaling molecule whose production is stimulated by antibiotic exposure, may have had a role in the OS defensive mechanism that increased *E. coli's* susceptibility to antibiotics (Han, T. H., et al. 2011). The tnaA tryptophanase gene, whose product catalysed the production of indole, may have been stimulated by oxidative stress–ROS (Kuczyńska-Wiśnik, D., et al. 2010). Most importantly it was found that endogenous indole production or exogenous indole addition promoted AMR (Han, T. H., et al. 2011). It has also been demonstrated that a heterogeneous population's otherwise vulnerable members express more OS protection genes as a result of highly antibiotic-resistant *E. coli* strains' synthesis of indole (Lee, H. H., et al. 2010). Additionally, it was shown that polyamines act as antioxidants and also stimulate the development of genes that defend against ROS in *E. coli* (Tkachenko, A. G., and Fedotova, M. V. 2007). Antibiotics have recently been found to have a favorable effect on polyamine formation in *E. coli*, which seems to somewhat reduce ROS production that is dependent on antimicrobials and makes cells more resistant to antimicrobials (Tkachenko, A. G. et al. 2012). Recently, it has been discovered that *P. aeruginosa* and *E. coli* H_2S gas production, which was previously not connected to any specific physiological activity of bacteria, is related to antibiotic resistance. Due in part to its reduction of oxidative damage caused by antibiotic dependence (Shatalin, K., et al. 2011). The expression of the genes and proteins that produce H_2S was not examined, nor was the regulation of H_2S synthesis. Because of this, it is unclear how environmental stress circumstances would affect this antimicrobial resistance pathway. Oxidative stress, which has a close connection to antimicrobial resistance, causes the expression of antimicrobial resistance determinants and anti-oxidant responses that protect cells from ROS produced by antimicrobials (Poole, K. 2012). It has been revealed that *A. oleivorans* evolved resistance to redox-active substances and antibiotics in response to oxidative stress and QS signals (Shin, B., et al. 2020). The promoter regions of the genes encoding AOLE 03905 (SapA is a potential surface adhesion protein) and AOLE_ 11355 are directly bound by AqsR. (L-asparaginase). In the presence of 1 mM indole, the expression of these genes has been seen to be suppressed. The regulation of genes involved in ROS defence is discovered to be significantly influenced by two key global regulators, namely OxyR, which senses H_2O_2, and SoxR, which senses superoxide. Catalase (katG), a minor catalase gene (katAc), thioredoxin-disulfide reductase (trxB), and a potential outer membrane receptor protein are all directly regulated by the oxygen receptor (oxyR) (oprC). SinE, which encodes an endoribonuclease, is regulated by SoxR, and both SoxR and SinE are involved in the emergence of antibiotic and redox-active chemical resistance (Shin, B., et al. 2020).

4.4 ENVELOPE STRESS REACTIONS RESULTING IN THE EMERGENCE OF AMR

Bacterial adaptive responses to environmental insults to the cell envelope, such as membranes and cell walls, structure, and function, are intended to allow survival in the face of membrane or cell wall disruption. Alternative sigma factors, also known as extra-cytoplasmic function (ECF) sigmas, are often highly regulated and are responsible for these so-called envelope stress responses. For instance, RpoE (sE) is found in many different organisms, including *E. coli* (Ruiz, N., and Silhavy, T. J. 2005). TCSs, such as AmgRS were discovered in *P. aeruginosa* (Lee, S., et al., 2009) and CpxRA discovered in *E. coli* organisms (Ruiz, N., and Silhavy, T. J. 2005). Numerous antimicrobials have been shown to either target the elements of the cell envelope or to get beyond the barrier that the cell envelope creates in order to reach their targets inside the cell. Therefore, it may not come as a surprise that modifications brought on by envelope stress responses may impact antimicrobial susceptibility. Gram negative bacteria have specialised reactions to stress that target the cell wall and membranes as well as other components of the cell envelope. The phage shock response, the stress-responsive alternative sigma factor E, EnvZ/OmpR, CpxAR (Cpx), BaeRS, and Rcs phosphorelays, as well as the envelope stress responses (ESRs) in *E. coli* and the closely related *Enterobacteriaceae* are some of these ESRs. (Dam, S., et al. 2018). Each of these ESRs is triggered by exposure to specific environmental stresses or by the disruption of specific envelope components. Stress proteins play important roles in maintaining the fundamental cellular physiology in addition to their roles in responding to harmful conditions (Hayden, J. D., and Ades, S. E. 2008; Delhaye, A., et al. 2016). This is especially true for the *E. coli* rpoE-dependent stress response because the rpoE gene, which encodes E, is crucial for the survival of the organism (De Las Penas, et al. 1997). According to reports, a lot of antimicrobials either target parts of the cell envelope or they have to get beyond the barrier that the cell membrane creates in order to reach their targets inside the cell. Therefore, it shouldn't come as a surprise that modifications brought on by envelope stress responses can impact the susceptibility to antimicrobials (Poole, K. 2012). Perhaps the MexCD-OprJ multidrug efflux system of *P. aeruginosa* is the most direct link between envelope stress, or E, and antimicrobial resistance. Detergents, organic solvents, biocides, and CAPs are among the membrane-damaging agents (MDAs) that have been found to induce this efflux system (Fraud, S., et al. 2008). This effect is mediated by the algU-encoded envelope stress response sigma factor (Fraud, S. et al. 2008). Although MexCD-OprJ would appear to be a component of an envelope stress response in *P. aeruginosa* but its role in this is unclear (Poole, K. 2012). The amgRS locus present *P. aeruginosa* has been observed to encode a homolog of *E. coli* OmpR–EnvZ TCS that has been observed in osmoregulation of porin gene expression in *E. coli* (Mizuno, T., and Mizushima, S. 1990). In contrast to OmpR-EnvZ, AmgRS, in response to aminoglycoside exposure, showed regulation of several membrane transporter and protease genes in a manner more similar to the *E. coli* CpxRA envelope stress response regulon (Lee, S. et al. 2009). According to Kohanski, M. A., et al. (2008), AmgRS responded to envelope stress caused by abnormal polypeptides that are thought to accumulate following aminoglycoside treatment. The AmgRS-controlled stress response may protect cells against abnormal peptide-induced membrane damage (Lee, S., et al. 2009). This is corroborated by the discovery that numerous AmgRS-regulated protease genes as well as numerous other genes contribute to aminoglycoside resistance, and that the absence of these genes in mutants reduced the resistance (Hinz, A., et al. 2011). An observation revealed that an amgRS mutant was more vulnerable to ß-lactams, alkali stress, and salt stress. This suggested that AmgRS may be a general envelope stress response TCS in *P. aeruginosa*, similar to CpxRA found in *E. coli* (Hinz, A., et al. 2011). Thus, it can be concluded that antimicrobials, whether directly or indirectly compromising the integrity of the cell envelope, activate stress responses that encourage antimicrobial resistance when membrane modification, antimicrobial efflux, and amelioration of antimicrobial-dependent membrane perturbation occur. The destruction of the outer membrane (OM) and/or cytoplasmic membrane (CM) by MDAs, including detergents, solvents, biocides, ethanol, and cationic antimicrobial peptides causes the release of these fatty acids, which must be replaced for the membrane to function properly. These released or damaged fatty acids send out signals to the CM-associated Muc proteins, which in turn regulate the activity of the RpoE-like envelope stress sigma factor AlgU, upregulating the expression of mexCD-oprJ. The damaged or released fatty acids are then eventually exported through the MexCD-OprJ pump (Poole, K. 2012).

4.5 HEAT SHOCK RESPONSES RESULTING IN THE DEVELOPMENT OF ANTIMICROBIAL RESISTANCE

The finest research on the heat shock response has been done on *P. aeruginosa* and *E. coli* (Guisbert, E., et al. 2008) (Allan, B., et al. 1988). Heat shock response is a brief reaction to a rise in temperature and the potentially harmful effects it can have on protein folding. The heat shock elements are frequently discovered to play a function in promoting the refolding of misfolded proteins or turning them over (Kindrachuk, K. N., et al. 2011). Since ß-lactams do not produce abnormal or misfolded proteins, it is possible that chaperone elements of the heat shock response, as found in *Streptococcus pneumonia*, contribute to ß-lactam resistance by stabilizing cell wall synthesis enzymes (Tran, T. D. H., et al. 2011). *E. coli* and *Acinetobacter baumannii* have both been found to exhibit aminoglycoside-induced activation of heat shock genes (Kindrachuk, K. N., et al. 2011). It is interesting to note that aminoglycosides (AGs), including tobramycin, have been found to stimulate the expression of heat shock genes in *P. aeruginosa* (Kindrachuk, K. N., et al. 2011). The alternative Lon protease AsrA, whose expression is increased in response to tobramycin or heat shock and is reliant upon the RpoH heat shock sigma factor, mediates this process. It was found that an engineered strain's overexpression of asrA had a modestly beneficial effect on AG resistance (Kindrachuk, K. N., et al. 2011). Therefore, it is proposed that the heat shock reaction may help partially protect against the threat of AG. One hypothesis for the association between AGs and heat shock is that the elements responsible for clearing out misfolded proteins may target the aberrant mistranslated polypeptides produced by the AG disrupted ribosomes and that they might insert into and disrupt the bacterial membranes (Davis, B. D., et al. 1986). They described the AG-mediated membrane damage, which is probably seen as a crucial stage in the deadly activity of these drugs (Kohanski, M. A., et al. 2008; Lee, S., et al. 2009). So, eliminating any aberrant polypeptides that may be at fault would surely reduce their toxicity to cells. In response to hyperosmolarity, heat shock, and prolonged peroxide exposure, the general stress response sigma factor RpoS was activated in *P. aeruginosa* (Suh, S. J., et al. 1999). Additionally connected to antibiotic resistance in this organism is the RpoS sigma factor. Its overexpression was partially shown to restore ofloxacin tolerance in an ofloxacin-sensitive LasIR– strain of *P. aeruginosa* that is lacking in acyl homoserine lactone production and quorum-sensing-dependent gene expression (Kayama, S., et al. 2009). RpoS has also been related to carbapenem tolerance in *P. aeruginosa* in addition to this. It has been noted that a rpoS mutant strain becomes more susceptible to carbapenems in the stationary phase (Murakami, K., et al. 2005). Similar to this, it has been observed that rpoS is required for a heat-shock-induced increase in carbapenem tolerance in *P. aeruginosa* (Murakami, K., et al. 2005). However, it is still uncertain which RpoS targets are responsible for *P. aeruginosa's* resistance to carbapenem and quinolone antibiotics.

4.6 ANTIBIOTIC STRESS RESULTING IN THE DEVELOPMENT OF AMR

Numerous harmful substances or situations that impede growth are produced by a wide variety of bacteria that are found in the environment. These include antibiotics that are frequently employed in both agricultural and medicinal settings. There has been a constant increase in multi drug-resistant (MDR) clinical isolates in the last few decades. Thus, antibiotics are found almost everywhere at different concentrations (Andersson, D. I., and Hughes, D. 2014). According to studies, bacteria were negatively impacted by low antibiotic concentrations on at least four separate levels:

1. As resistance selectors, that is, by enhancing populations of resistant bacteria and then choosing them for de novo resistance mutations (Gullberg, E., et al. 2011)
2. As contributors of genotypic and phenotypic heterogeneity (Gutierrez, A., et al. 2013)
3. As intercellular signals (Lee, H. H., et al. 2010)
4. As inducers of persistence (Pu, Y., et al. 2016).

In connection with this, Viveiros and his collaborators have shown how exposure to subinhibitory quantities of the antibiotic tetracycline (TET) can be used to induce high levels of resistance in susceptible *E. coli* K12 cells (Viveiros, M., et al. 2005). Increased expression of the AcrAB efflux pump, which can be inhibited by the efflux pump inhibitor phenylalanine-arginine-b-naphthylamide, led to the development of TET resistance (PAbN). It was interesting to see that the TET-resistant strain also exhibited MDR brought on by the suppression of OmpF and OmpC expression(Viveiros, M., et al. 2007).

4.7 STRESS INDUCED BIOFILM FORMATION RESULTING IN THE DEVELOPMENT OF AMR

Infected hosts and biofilms in the natural world frequently contain bacteria (López, D., et al. 2010). In addition to its numerous advantages, biofilm also confers resistance to antimicrobials (Hoiby, N., et al. 2010). Environmental factors that cause or contribute to the production of bacterial biofilms will therefore positively affect AMR. Numerous stresses have been identified to be related to the development of biofilms, and the production of biofilms itself may represent a stress response (Landini, P. 2009). Although Mg^{2+} limitation has been found to have a favourable impact on *P. aeruginosa* biofilm development, *S. aureus* and *Legionella pneumophila* have both been demonstrated to promote biofilm formation in response to nutritional restriction in the form of Fe limitation (Johnson, M., et al. 2008), (Mulcahy, H., and Lewenza, S. 2011). It has been discovered that the latter results from a PhoPQ-mediated low Mg^{2+}-dependent regulation of the retS "biofilm repressor" gene. (Mulcahy, H., and Lewenza, S. 2011). The periodontopathogen *Aggregatibacter actinomycetemcomitans* produced biofilm and enhanced transcription and translation of established biofilm determinants (Amarasinghe, J. J. et al. 2009). However, restricting Fe may have a detrimental effect on biofilm formation in some organisms, such as *P. aeruginosa* (Patriquin, G. M., et al. 2008), and high metal concentrations have been seen to have a beneficial effect on biofilm formation in a number of organisms (Harrison, J. J. 2007). As a result of a SpoT-dependent decrease in ppGpp and an increase in the biofilm-promoting second messenger c-di-GMP [bis-(3′-5′)-cyclic di-GMP], strong biofilm induction in *E. coli* is also triggered by ribosome disruption brought on by ribosome-targeting medicines, or ribosome stress (A. Boehm et al. 2009). The ppGpp alarmone connects ribosome status to the genes required for the start of biofilm development as a result. *E. faecalis* mutants unable to produce this alarmone shown a reduced capacity to sustain biofilm development, indicating ppGpp connection to biofilm formation (Chǻvez de Paz, L. E., et al. 2012). Oxidative stress is seen to play a role in biofilm development in *E. coli*. In reaction to oxidative stress, it has been discovered that the antitoxin MqsA is degraded (by the Lon protease), which results in a decrease in c-di-GMP levels and a reduction in the production of biofilms (Wang, X., et al. 2011). Although it hasn't been precisely researched, the increased rpoS expression caused by the oxidative stress (Wang, X., et al. 2011) is likely to increase the production of biofilms. However, RpoS has also been linked to the promotion of biofilm formation by low temperatures in *E. coli* (White-Ziegler, C. A. 2008).

4.8 GENERAL STRESS RESPONSE RESULTING IN THE DEVELOPMENT OF AMR

The general stress response sigma factor is encoded by the rpoS gene. In *E. coli*, it reacts to food deprivation, pH change, hyperosmolarity, and unfavourable low and high temperatures (Hengge-Aronis, R. 2002). Additionally, *P. aeruginosa*'s rpoS gene responds to heat shock, hyperosmolarity, and protracted peroxide exposure (Suh, S. J., et al. 1999). Both *P. aeruginosa* and *E. coli* have been shown to carry this gene, which has been connected to antibiotic resistance. Constitutive expression of the rpoS gene in *E. coli* partially reduces the mutant strain's drug hypersusceptibility since it lacks the acrAB multidrug efflux genes (Rami, A., et al. 2005). However, its overexpression resulted in a partial restoration of ofloxacin tolerance in a LasIR2 *P. aeruginosa* strain that was quorum-sensing reliant and susceptible to ofloxacin (Kayama, S., et al. 2009). Researchers found that a *P. aeruginosa* strain with the rpoS mutation had increased susceptibility to carbapenems, but only when it was in the stationary phase (Murakami, K., et al. 2005). However, intriguingly, it has been found that the RpoS gene is required for a heat shock-promoted increase in carbapenem resistance in *P. aeruginosa*, which is a blatant example of a stress response sigma factor mediating stress-promoted antimicrobial resistance (Murakami, K., et al. 2005). RpoS has recently been found to be associated with persisted development in *E. coli* and *P. aeruginosa*. According to Hong, S. H., et al. (2012), rpoS null mutants showed increased production of *P. aeruginosa* persisters resistant to ciprofloxacin and *E. coli* persisters resistant to ampicillin, providing yet another way that this sigma factor may contribute to antimicrobial resistance brought on by stress. The ClpXP protease is responsible for discarding damaged proteins as part of the cell's protein quality control mechanisms (Baker, T. A., and Sauer, R. T. 2012). In both Gram-negative (Kroh, H. E., and Simon, L. D. 1990) and Gram-positive (Gerth, U., et al. 1998; Chastanet, A., et al. 2001) bacteria, the heat shock and other stresses have been observed to induce the clpP gene, which encodes the proteolytic component of ClpXP, and is crucial for stress tolerance. Heat and other stressors are reported to make clpP mutants more vulnerable (Gerth, U., et al. 1998; Chastanet, A., et al. 2001; Frees, D., et al. 2003; Thomsen, L. E., et al. 2002). However, clpP null mutants of *P. aeruginosa* showed modest, 2-fold increased susceptibility to ciprofloxacin, (Fernández, L., et al. 2012), thus highlighting a link to antimicrobial resistance. Recently, clpP mutations in *S. aureus* also showed reduced susceptibility to vancomycin antibiotic in several VISA isolates.

4.9 SOS RESPONSE RESULTING IN THE DEVELOPMENT OF AMR

Numerous "stresses" such as DNA damage, pH extremes, food shortages, exposure to certain antimicrobials, and oxidative stress can cause bacteria to launch their SOS response. In light of this, SOS response has the potential to encourage antimicrobial resistance under a range of stress-related circumstances. Beyond its roles in stress-induced mutagenesis and chromosomal rearrangements, the SOS response has been found to contribute to resistance. For instance, the survival of a deadly beta-lactam exposure is facilitated by the stimulation of lactam caused by an SOS response-promoted transitory pause in cell division S in *E. coli* (Miller, C., et al. 2004). The pause in cell division is protective, as lactams can only have an effect on developing cells. It has been discovered that the expression of qnrB, one of numerous plasmid-borne quinolone-resistance (qnr) genes that are widely present in the *Enterobacteriaceae*, is ciprofloxacin-inducible and is reliant on the drug's stimulation of the SOS response (Wang, M., et al. 2009). Additionally, it has recently been discovered that the activation of the SOS response is crucial for the emergence of rifampicin and ciprofloxacin resistance in *E. coli*, and that suppressing this response prevents the development of resistance to these medications in vitro. This increases the effectiveness of the FQ killing of drug-resistant, biofilm, and persister cells of *E. coli* (Lu, T. K., and Collins, J. J. 2009). The activation of the SOS response in *V. cholerae* by ciprofloxacin and other stresses results in the transfer of an integrative conjugative element (ICE) known as STX, which carries an antimicrobial resistance gene. The SOS response has also been observed to promote the horizontal dissemination of drug resistance genes (Beaber, J. W., et al. 2004). Thus, conditions that trigger the SOS response also result in increased expression and activity of the integron-encoded recombinases known as integrases (Guerin, É., et al. 2009). By catalyzing recombination events that led in the capture of new resistance genes or the reorganization of integron resident resistance genes to increase their expression, integrons are mobile genetic components that could contain numerous antibiotic resistance genes with integrases (Cambray, G., et al. 2010). The SOS response can also encourage integron elements to express and acquire resistant genes, hence encouraging the development of resistance. According to Guerin, É., et al. (2009), antibiotics that are known to trigger the SOS response have the ability to activate integrase expression and activity. The fact that many of today's common integrons have genes for resistance to these agents is one of the fascinating observations (Erill, I.,et al. 2007). In a wide range of integrons and superintegrons (large chromosomally located integrons that carry many genes, including the antibiotic resistant genes; Cambray, G., et al. 2010; Cambray, G., et al. 2011), binding sites for the LexA regulator of the SOS response are frequently found. This suggests that SOS control of integrase-mediated integron capture and expression of antibiotic resistance is regulated.

4.10 STRESS-INDUCED MUTAGENESIS RESULTING IN THE DEVELOPMENT OF AMR

In response to 'growth-limiting' stress brought on by nutritional deprivation, low pH, hypoxia, increased osmotic pressure, abrupt temperature changes, or antimicrobial exposure, bacteria trigger a stress-induced rise in the mutation rate (reviewed in Galhardo et al., (Galhardo, R. S. et al. 2007). Cirz and Romesberg (Cirz, R. T., and Romesberg, F. E. 2007) and Foster (Foster, P. L. 2007)).

Adaptive mutagenesis, another name for stress-induced mutagenesis, results from the stress-dependent occurrence of double-stranded breaks in chromosomal DNA, which is followed by the activation of error-prone repair, which is controlled by the SOS response (Galhardo, R. S., et al. 2009) the sigma E (sE)-dependent envelope stress response (Frisch, R. L., et al. 2010) and the RpoS-mediated general stress response (Gibson, J. L., et al. 2010). The intensity of mutations is boosted by the higher mutation rate, allowing for increased stressor adaptation.When studied predominantly in *E. coli*, (Foster, P. L., 2007) evidence for stress induced mutagenesis is also observed in *B. subtilis* (Sung, H. M., et al. 2003), *Pseudomonas putida* (Tegova, R., et al. 2004), *Caulobacter crescentus* (Galhardo, R. S., et al. 2005) and *M. tuberculosis* (Boshoff, H. I., et al. 2003) and FQ-induced mutation typical of stress-induced mutagenesis has been demonstrated in bacteria such as *S. aureus* (Mesak, L. R., and Davies, J. et al. 2009), *S. pneumoniae* (Henderson-Begg, S. K. 2006), *Streptococcus uberis* (Varhimo, E., et al. 2008), *P. aeruginosa* (Cirz, R. T., et al. 2006), *Salmonella typhimurium* (Power, E. G., and Phillips, I. 1993). Numerous non-antibiotic stresses that affect mutational events have also been noted, such as osmotic stress and UV light exposure that cause chromosomal DNA rearrangements in *E. coli* and *S. aureus*, respectively, and promote the development of ciprofloxacin resistance in *S. aureus* in response to the SOS response (Cirz, R. T., et al. 2007). Surprisingly, the development of antibiotic resistance has been related to the stimulation of the SOS and RpoS responses by nutritional deprivation. When starved of lactose, Lac2 *E. coli* that was genetically modified to possess the ampD ampC determinants of lactam resistance produces ampD mutants that express ampC lactamase (Petrosino, J. F., et al. 2009).

4.11 PERSISTERS AS A STRESS RESPONSE RESULTING IN THE DEVELOPMENT OF AMR

Many researchers have long observed that some bacteria may resist medication exposure while belonging to a small, genetically distinct population (Gefen, O., and Balaban, N. Q. 2009; Lewis, K. 2010). Slow growth and dormancy, which these so-called persisters possess, are advantageous to them because they effectively shield them from the fatal effects of antibiotics that would otherwise work preferentially on fast growing bacteria (Lewis, K. 2010). Persisters form in reaction to an external trigger, such as the environment, or arise stochastically and continually during population increase. (Gefen, O., and Balaban, N. Q. 2009; Lewis, K. 2010), by ensuring that a portion of its members can withstand insults, such as antimicrobial exposure, that would otherwise kill or destroy a phenotypically homogeneous population. The persistence of chronic infections, such as *P. aeruginosa* lung infections in CF and *M. tuberculosis* lung infections, following antimicrobial therapy has been linked to the AMR of biofilms (Lewis, K. 2008; Kim, J. et al. 2009) (Mulcahy, L. R., et al. 2010; Fauvart, M., et al. 2011). According to research findings, ciprofloxacin treatment can encourage the development of ciprofloxacin-resistant persisters in *E. coli*, and this action was discovered to be dependent on the SOS response (Dörr, T., et al. 2009). Additionally, according to recent research, it has been hypothesised that toxin-antitoxin (TA) gene pairs or modules are crucial for persister development (Wang, X., and Wood, T. K. 2011). The major players in the macromolecular synthesis were shown to be the targets of the toxin components of these modules, which would shut them down and consequently impair growth but not viability (Lewis, K. 2010). As a result, dormancy was promoted by stress-mediated toxin level increases, which led to the emergence of persisters that are resistant to antibiotics. In contrast, expression of the relE gene, which encodes a translation inhibitor, has been seen to favour the growth of persisters in *E. coli*, which is consistent with this concept (Keren, I., et al. 2004). The SOS response-mediated ciprofloxacin-inducible generation of persisters that has been reported in *E. coli* is discovered dependent on the SOS response-inducible TA locus tisAB/istR, and the expression of TA genes has also been observed to occur in drug-resistant persisters (Dörr, T., et al. 2009). This locus' knockout significantly decreased the number of persisters that were ciprofloxacin-tolerant (Dörr, T., et al. 2010). According to research, excessive doses of ciprofloxacin cause the production of the tisB toxin gene, and cells that express TisB are multidrug resistant (Dörr, T., et al. 2010). An important aspect of persisters' antimicrobial resistance seems to be growth impairment brought on by the toxin component of stress-inducible TA modules.

4.12 LINKAGE BETWEEN STRESS RESPONSE GENES AND ANTIMICROBIAL RESISTANT (AMR) GENES

In order to better understand how these two categories of genes co-occur, Pei, R., and his colleagues combined analysis of stress response genes and AMR genes in 2021. Seven genes, namely, *arsR, asr, merC, merP, merR, merT,* and *qacdelta1*—were discovered to be present in all five countries (namely, US, the UK, China, Australia, and South Africa). The main pathogens reported to carry the stress response genes were *E. coli* and *Shigella, Salmonella enterica*, and *Klebsiella pneumoniae* (Pei, R., et al. 2021). The results of this study could make a significant contribution to the future development of drugs that target the genes involved in the stress response in order to weaken bacteria that are resistant to antibiotics.

4.13 VARIOUS STRATEGIES TO COMBAT ANTIMICROBIAL RESISTANCE (AMR)

AMR is a growing issue on a worldwide scale. AMR contributes to show an upsurge in morbidity and mortality in populations from infections that were once under control. As a result of the widespread use of antibiotics, the clinical setting has become more and more linked to the emergence of AMR. The extensive spread of antibiotic resistance to additional environmental reservoirs makes it more difficult to develop effective AMR prevention techniques. As the rate of AMR growth has surpassed the rate of antimicrobial drug development, it is now necessary to speed up the process of generating new antimicrobials. However, it is sad that finding new, potent medicines is also not an easy task. Therefore, in order to counteract AMR, researchers have adopted a variety of alternate tactics (Moo, C. L., et al. 2020).

4.14 CONCLUSION

Although a connection between stress and numerous well-known antimicrobial resistance determinants has frequently been demonstrated, it is frequently unknown what role the resistant genes play in the stress response itself. For instance, even though several *P. aeruginosa* RND family multidrug exporters have been identified as stress-inducible, it is still unclear how these exporters contribute to the related stress responses as well as the molecules that act as the inducing effector and efflux substrate molecules. Understanding the biology of these transporters in contexts like a hospital or an infection site, where their induction may jeopardize an antimicrobial treatment, requires knowledge of the environmental factors and the effector molecules responsible for enlisting multidrug efflux systems. Stressor activation of these efflux mechanisms has frequently been demonstrated not to increase resistance (Fraud, S., and Poole, K. 2011). Only instances of mutational overexpression result in resistance (Poole, K., et al. 2005). Numerous studies have shown a connection between stress and resistance, and it is possible that stress reactions contribute to the emergence of antibiotic resistance. So, it is possible to think of stress response pathways as good candidates for therapeutic intervention. This is supported by the findings that inactivation of the SR in *P. aeruginosa* increases susceptibility to various antimicrobials and inactivation of the AmgRS regulators of an envelope stress response increases the susceptibility to AGs (Krahn, T., et al. 2012) (Nguyen, D., et al. 2011). To fully comprehend the significance of stress responses as resistance determinants and their value as therapeutic targets, as well as how to target them, a better understanding of the relationship between stress and antimicrobial resistance, including the identification of stress-induced effectors that recruit resistance determinants and promote resistance, as well as the genes involved, is required.

Given that stress response was directly expressed to be coordinated with the pathogen's response to antibiotics (Brazas, M. D., and Hancock, 2005; Lee, S., et al. 2009), and given that more effective stress responses have been tentatively linked to an increased antimicrobial resistance, it is also reasonable to conclude that pathogens with stronger stress response capacity might have higher chances of surviving the antimicrobial treatments (Lee, S., et al. 2009; Ebinesh, A. 2016; Poole, K. 2014). Therefore, it is crucial to identify the genes in pathogens that coordinate the stress response and antibiotic resistance (Li, K., Zheng, et al. 2019; Yang, K., et al. 2010; Hua, M., et al. 2020).

REFERENCES

Abranches, J., Martinez, A. R., Kajfasz, J. K., Chávez, V., Garsin, D. A., and Lemos, J. A. (2009). The molecular alarmone (p) ppGpp mediates stress responses, vancomycin tolerance, and virulence in Enterococcus faecalis. *Journal of Bacteriology*, *191*(7), 2248–2256.

Allan, B., Linseman, M., MacDonald, L. A., Lam, J. S., and Kropinski, A. M. (1988). Heat shock response of Pseudomonas aeruginosa. *Journal of Bacteriology*, *170*(8), 3668–3674.

Amarasinghe, J. J., Scannapieco, F. A., and Haase, E. M. (2009). Transcriptional and translational analysis of biofilm determinants of Aggregatibacter actinomycetemcomitans in response to environmental perturbation. *Infection and Immunity*, *77*(7), 2896–2907.

Andersson, D. I., and Hughes, D. (2014). Microbiological effects of sublethal levels of antibiotics. *Nature Reviews Microbiology*, *12*(7), 465–478.

Baker, T. A., and Sauer, R. T. (2012). ClpXP, an ATP-powered unfolding and protein-degradation machine. *Biochimica et Biophysica Acta (BBA)-Molecular Cell Research*, *1823*(1), 15–28.

Beaber, J. W., Hochhut, B., and Waldor, M. K. (2004). SOS response promotes horizontal dissemination of antibiotic resistance genes. *Nature*, *427*(6969), 72–74.

Boehm, A., Steiner, S., Zaehringer, F., Casanova, A., Hamburger, F., Ritz, D., et al. (2009). Second messenger signalling governs Escherichia coli biofilm induction upon ribosomal stress. *Molecular Microbiology*, *72*(6), 1500–1516.

Boor, K. J. (2006). Bacterial stress responses: what doesn't kill them can make them stronger. *PLoS Biology*, *4*(1), e23.

Borriello, G., Richards, L., Ehrlich, G. D., and Stewart, P. S. (2006). Arginine or nitrate enhances antibiotic susceptibility of Pseudomonas aeruginosa in biofilms. *Antimicrobial Agents and Chemotherapy*, *50*(1), 382–384.

Borriello, G., Werner, E., Roe, F., Kim, A. M., Ehrlich, G. D., and Stewart, P. S. (2004). Oxygen limitation contributes to antibiotic tolerance of Pseudomonas aeruginosa in biofilms. *Antimicrobial Agents and Chemotherapy*, *48*(7), 2659–2664.

Boshoff, H. I., Reed, M. B., Barry III, C. E., and Mizrahi, V. (2003). DnaE2 polymerase contributes to in vivo survival and the emergence of drug resistance in Mycobacterium tuberculosis. *Cell*, *113*(2), 183–193.

Brazas, M. D., and Hancock, R. E. (2005). Using microarray gene signatures to elucidate mechanisms of antibiotic action and resistance. *Drug Discovery Today*, *10*(18), 1245–1252.

Cambray, G., Guerout, A. M., and Mazel, D. (2010). Integrons. *Annual Review of Genetics*, *44*, 141–166.

Cambray, G., Sanchez-Alberola, N., Campoy, S., Guerin, E., Da Re, S., González-Zorn, B. and Erill, I. (2011). Prevalence of SOS-mediated control of integron integrase expression as an adaptive trait of chromosomal and mobile integrons. *Mob DNA* 2011; 2: 6; PMID: 21529368.

Chen, H., Hu, J., Chen, P. R., Lan, L., Li, Z., Hicks, L. M., et al. C. (2008). The Pseudomonas aeruginosa multidrug efflux regulator MexR uses an oxidation-sensing mechanism. *Proceedings of the National Academy of Sciences*, *105*(36), 13586–13591.

Chastanet, A., Prudhomme, M., Claverys, J. P., and Msadek, T. (2001). Regulation of Streptococcus pneumoniae clp genes and their role in competence development and stress survival.

Chatterji, D., and Ojha, A. K. (2001). Revisiting the stringent response, ppGpp and starvation signaling. *Current Opinion in Microbiology*, *4*(2), 160–165.

Chávez de Paz, L. E., Lemos, J. A., Wickström, C., and Sedgley, C. M. (2012). Role of (p) ppGpp in biofilm formation by Enterococcus faecalis. *Applied and Environmental Microbiology*, *78*(5), 1627–1630.

Chung, H. J., Bang, W., and Drake, M. A. (2006). Stress response of Escherichia coli. *Comprehensive Reviews in Food Science and Food Safety*, *5*(3), 52–64.

Cirz, R. T., Chin, J. K., Andes, D. R., de Crécy-Lagard, V., Craig, W. A., and Romesberg, F. E. (2005). Inhibition of mutation and combating the evolution of antibiotic resistance. *PLoS Biology*, *3*(6), e176.

Cirz, R. T., Jones, M. B., Gingles, N. A., Minogue, T. D., Jarrahi, B., Peterson, S. N., and Romesberg, F. E. (2007). Complete and SOS-mediated response of Staphylococcus aureus to the antibiotic ciprofloxacin. *Journal of Bacteriology*, *189*(2), 531–539.

Cirz, R. T., O'Neill, B. M., Hammond, J. A., Head, S. R., and Romesberg, F. E. (2006). Defining the Pseudomonas aeruginosa SOS response and its role in the global response to the antibiotic ciprofloxacin. *Journal of Bacteriology*, *188*(20), 7101–7110.

Cirz, R. T., and Romesberg, F. E. (2007). Controlling mutation: intervening in evolution as a therapeutic strategy. *Critical Reviews in Biochemistry and Molecular Biology*, *42*(5), 341–354.

Dam, S., Pages, J. M., and Masi, M. (2018). Stress responses, outer membrane permeability control and antimicrobial resistance in Enterobacteriaceae. *Microbiology*, *164*(3), 260–267.

Davis, B. D., Chen, L. L., and Tai, P. C. (1986). Misread protein creates membrane channels: an essential step in the bactericidal action of aminoglycosides. *Proceedings of the National Academy of Sciences*, *83*(16), 6164–6168.

De Las Peñas, A., Connolly, L., and Gross, C. A. (1997). SigmaE is an essential sigma factor in Escherichia coli. *Journal of Bacteriology*, *179*(21), 6862–6864.

Delhaye, A., Collet, J. F., and Laloux, G. (2016). Fine-tuning of the Cpx envelope stress response is required for cell wall homeostasis in Escherichia coli. *MBio*, *7*(1), e00047–16.

Dörr, T., Lewis, K., and Vulić, M. (2009). SOS response induces persistence to fluoroquinolones in Escherichia coli. *PLoS Genetics*, *5*(12), e1000760.

Dörr, T., Vulić, M., and Lewis, K. (2010). Ciprofloxacin causes persister formation by inducing the TisB toxin in Escherichia coli. *PLoS Biology*, *8*(2), e1000317.

Dühring, S., Germerodt, S., Skerka, C., Zipfel, P. F., Dandekar, T., and Schuster, S. (2015). Host-pathogen interactions between the human innate immune system and Candida albicans—understanding and modeling defense and evasion strategies. *Frontiers in Microbiology*, *6*, 625.

Ebinesh, A. (2016). Bacterial stress response and cross resistance to antibiotics in the light of natural selection. *Journal of Infectious Disease and Immune Therapies*, *1*, 1.

Eng, R. H., Padberg, F. T., Smith, S. M., Tan, E. N., and Cherubin, C. E. (1991). Bactericidal effects of antibiotics on slowly growing and nongrowing bacteria. *Antimicrobial Agents and Chemotherapy*, *35*(9), 1824–1828.

Erill, I., Campoy, S., and Barbé, J. (2007). Aeons of distress: an evolutionary perspective on the bacterial SOS response. *FEMS Microbiology Reviews*, *31*(6), 637–656.

Fabrega, A., Martin, R. G., Rosner, J. L., Tavio, M. M., and Vila, J. (2010). Constitutive SoxS expression in a fluoroquinolone-resistant strain with a truncated SoxR protein and identification of a new member of the marA-soxS-rob regulon, mdtG. *Antimicrobial Agents and Chemotherapy*, *54*(3), 1218–1225.

Fauvart, M., De Groote, V. N., and Michiels, J. (2011). Role of persister cells in chronic infections: clinical relevance and perspectives on anti-persister therapies. *Journal of Medical Microbiology*, *60*(6), 699–709.

Fernández, L., Breidenstein, E. B., Song, D., and Hancock, R. E. (2012). Role of intracellular proteases in the antibiotic resistance, motility, and biofilm formation of Pseudomonas aeruginosa. *Antimicrobial Agents and Chemotherapy*, *56*(2), 1128–1132.

Fetar, H., Gilmour, C., Klinoski, R., Daigle, D. M., Dean, C. R., and Poole, K. (2011). mexEF-oprN multidrug efflux operon of Pseudomonas aeruginosa: regulation by the MexT activator in response to nitrosative stress and chloramphenicol. *Antimicrobial Agents and Chemotherapy*, *55*(2), 508–514.

Foster, P. L. (2007). Stress-induced mutagenesis in bacteria. *Critical Reviews in Biochemistry and Molecular Biology*, *42*(5), 373–397.

Fraud, S., Campigotto, A. J., Chen, Z., and Poole, K. (2008). MexCD-OprJ multidrug efflux system of Pseudomonas aeruginosa: involvement in chlorhexidine resistance and induction by membrane-damaging agents dependent upon the AlgU stress response sigma factor. *Antimicrobial Agents and Chemotherapy*, *52*(12), 4478–4482.

Fraud, S., and Poole, K. (2011). Oxidative stress induction of the MexXY multidrug efflux genes and promotion of aminoglycoside resistance development in Pseudomonas aeruginosa. *Antimicrobial Agents and Chemotherapy*, *55*(3), 1068–1074.

Frees, D., Qazi, S. N., Hill, P. J., and Ingmer, H. (2003). Alternative roles of ClpX and ClpP in Staphylococcus aureus stress tolerance and virulence. *Molecular Microbiology*, *48*(6), 1565–1578.

Frisch, R. L., Su, Y., Thornton, P. C., Gibson, J. L., Rosenberg, S. M., and Hastings, P. J. (2010). Separate DNA Pol II-and Pol IV-dependent pathways of stress-induced mutation during double-strand-break repair in Escherichia coli are controlled by RpoS. *Journal of Bacteriology*, *192*(18), 4694–4700.

Galhardo, R. S., Do, R., Yamada, M., Friedberg, E. C., Hastings, P. J., Nohmi, T., and Rosenberg, S. M. (2009). DinB upregulation is the sole role of the SOS response in stress-induced mutagenesis in Escherichia coli. *Genetics*, *182*(1), 55–68.

Galhardo, R. S., Hastings, P. J., and Rosenberg, S. M. (2007). Mutation as a stress response and the regulation of evolvability. *Critical Reviews in Biochemistry and Molecular Biology*, *42*(5), 399–435.

Galhardo, R. S., Rocha, R. P., Marques, M. V., and Menck, C. F. (2005). An SOS-regulated operon involved in damage-inducible mutagenesis in Caulobacter crescentus. *Nucleic Acids Research*, *33*(8), 2603–2614.

Gao, W., Smith, D. W., and Li, Y. (2006). Natural freezing as a wastewater treatment method: E. coli inactivation capacity. *Water Research*, *40*(12), 2321–2326.

Gefen, O., and Balaban, N. Q. (2009). The importance of being persistent: heterogeneity of bacterial populations under antibiotic stress. *FEMS Microbiology Reviews*, *33*(4), 704–717.

Gerth, U., Krüger, E., Derré, I., Msadek, T., and Hecker, M. (1998). Stress induction of the Bacillus subtilis clpP gene encoding a homologue of the proteolytic component of the Clp protease and the involvement of ClpP and ClpX in stress tolerance. *Molecular Microbiology*, *28*(4), 787–802.

Gibson, J. L., Lombardo, M. J., Thornton, P. C., Hu, K. H., Galhardo, R. S., Beadle, B., et al (2010). The σE stress response is required for stress-induced mutation and amplification in Escherichia coli. *Molecular Microbiology*, *77*(2), 415–430.

Goswami, M., Mangoli, S. H., and Jawali, N. (2006). Involvement of reactive oxygen species in the action of ciprofloxacin against Escherichia coli. *Antimicrobial Agents and Chemotherapy*, *50*(3), 949–954.

Gu, M., and Imlay, J. A. (2011). The SoxRS response of Escherichia coli is directly activated by redox-cycling drugs rather than by superoxide. *Molecular Microbiology*, *79*(5), 1136–1150.

Guerin, É., Cambray, G., Sanchez-Alberola, N., Campoy, S., Erill, I., Da Re, S., et al. (2009). The SOS response controls integron recombination. *Science*, *324*(5930), 1034–1034.

Guisbert, E., Yura, T., Rhodius, V. A., and Gross, C. A. (2008). Convergence of molecular, modeling, and systems approaches for an understanding of the Escherichia coli heat shock response. *Microbiology and Molecular Biology Reviews*, *72*(3), 545–554.

Gullberg, E., Cao, S., Berg, O. G., Ilbäck, C., Sandegren, L., Hughes, D., and Andersson, D. I. (2011). Selection of resistant bacteria at very low antibiotic concentrations. *PLoS Pathogens*, *7*(7), e1002158.

Gutierrez, A., Laureti, L., Crussard, S., Abida, H., Rodriguez-Rojas, A., Blázquez, J., et al. (2013). β-Lactam antibiotics promote bacterial mutagenesis via an RpoS-mediated reduction in replication fidelity. *Nature Communications*, *4*(1), 1–9.

Han, T. H., Lee, J. H., Cho, M. H., Wood, T. K., and Lee, J. (2011). Environmental factors affecting indole production in Escherichia coli. *Research in Microbiology*, *162*(2), 108–116.

Hakanen, A., Jalava, J., and Kaartinen, L. (2017). The National Action Plan on Antimicrobial Resistance 2017–2021.

Harbottle, H., Thakur, S., Zhao, S., and White, D. G. (2006). Genetics of antimicrobial resistance. *Animal Biotechnology*, *17*(2), 111–124.

Harrison, J. J., Ceri, H., and Turner, R. J. (2007). Multimetal resistance and tolerance in microbial biofilms. *Nature Reviews Microbiology*, *5*(12), 928–938.

Hayden, J. D., and Ades, S. E. (2008). The extracytoplasmic stress factor, σE, is required to maintain cell envelope integrity in Escherichia coli. *PloS ONE*, *3*(2), e1573.

Henderson-Begg, S. K., Livermore, D. M., and Hall, L. M. (2006). Effect of subinhibitory concentrations of antibiotics on mutation frequency in Streptococcus pneumoniae. *Journal of Antimicrobial Chemotherapy*, *57*(5), 849–854.

Hengge-Aronis, R. (2002). Signal transduction and regulatory mechanisms involved in control of the σS (RpoS) subunit of RNA polymerase. *Microbiology and Molecular Biology Reviews*, *66*(3), 373–395.

Hesketh, A., Hill, C., Mokhtar, J., Novotna, G., Tran, N., Bibb, M., and Hong, H. J. (2011). Genome-wide dynamics of a bacterial response to antibiotics that target the cell envelope. *BMC Genomics*, 12: 226.

Hindre, T., Brüggemann, H., Buchrieser, C., and Hechard, Y. (2008). Transcriptional profiling of Legionella pneumophila biofilm cells and the influence of iron on biofilm formation. *Microbiology*, *154*(1), 30–41.

Hinz, A., Lee, S., Jacoby, K., and Manoil, C. (2011). Membrane proteases and aminoglycoside antibiotic resistance. *Journal of Bacteriology*, *193*(18), 4790–4797.

Høiby, N., Bjarnsholt, T., Givskov, M., Molin, S., and Ciofu, O. (2010). Antibiotic resistance of bacterial biofilms. *International Journal of Antimicrobial Agents*, *35*(4), 322–332.

Hong, S. H., Wang, X., O'Connor, H. F., Benedik, M. J., and Wood, T. K. (2012). Bacterial persistence increases as environmental fitness decreases. *Microbial Biotechnology*, *5*(4), 509–522.

Hua, M., Huang, W., Chen, A., Rehmet, M., Jin, C., and Huang, Z. (2020). Comparison of antimicrobial resistance detected in environmental and clinical isolates from historical data for the US. *BioMed Research International*, *2020*.

Johnson, M., Cockayne, A., and Morrissey, J. A. (2008). Iron-regulated biofilm formation in Staphylococcus aureus Newman requires ica and the secreted protein Emp. *Infection and Immunity*, *76*(4), 1756–1765.

Jolivet-Gougeon, A., David-Jobert, S., Tamanai-Shacoori, Z., Ménard, C., and Cormier, M. (2000). Osmotic stress-induced genetic rearrangements in Escherichia coli H10407 detected by randomly amplified polymorphic DNA analysis. *Applied and Environmental Microbiology*, *66*(12), 5484–5487.

Kayama, S., Murakami, K., Ono, T., Ushimaru, M., Yamamoto, A., Hirota, K., and Miyake, Y. (2009). The role of rpoS gene and quorum-sensing system in ofloxacin tolerance in Pseudomonas aeruginosa. *FEMS Microbiology Letters*, *298*(2), 184–192.

Keren, I., Shah, D., Spoering, A., Kaldalu, N., and Lewis, K. (2004). Specialized persister cells and the mechanism of multidrug tolerance in Escherichia coli. *Journal of Bacteriology*, *186*(24), 8172–8180.

Kim, J., Hahn, J. S., Franklin, M. J., Stewart, P. S., and Yoon, J. (2009). Tolerance of dormant and active cells in Pseudomonas aeruginosa PA01 biofilm to antimicrobial agents. *Journal of Antimicrobial Chemotherapy*, *63*(1), 129–135.

Kindrachuk, K. N., Fernández, L., Bains, M., and Hancock, R. E. (2011). Involvement of an ATP-dependent protease, PA0779/AsrA, in inducing heat shock in response to tobramycin in Pseudomonas aeruginosa. *Antimicrobial Agents and Chemotherapy*, *55*(5), 1874–1882.

Kohanski, M. A., Dwyer, D. J., Wierzbowski, J., Cottarel, G., and Collins, J. J. (2008). Mistranslation of membrane proteins and two-component system activation trigger antibiotic-mediated cell death. *Cell*, *135*(4), 679–690.

Kolodkin-Gal, I., Sat, B., Keshet, A., and Kulka, H. E. (2008). The communication factor EDF and the toxin–antitoxin module mazEF determine the mode of action of antibiotics. *PLoS Biology*, *6*(12), e319.

Krahn, T., Gilmour, C., Tilak, J., Fraud, S., Kerr, N., Lau, C. H. F., and Poole, K. (2012). Determinants of intrinsic aminoglycoside resistance in Pseudomonas aeruginosa. *Antimicrobial Agents and Chemotherapy*, *56*(11), 5591–5602.

Kroh, H. E., and Simon, L. D. (1990). The ClpP component of Clp protease is the sigma 32-dependent heat shock protein F21. 5. *Journal of Bacteriology*, *172*(10), 6026–6034.

Kuczyńska-Wiśnik, D., Matuszewska, E., and Laskowska, E. (2010). Escherichia coli heat-shock proteins IbpA and IbpB affect biofilm formation by influencing the level of extracellular indole. *Microbiology*, *156*(1), 148–157.

Kusser, W., and Ishiguro, E. E. (1985). Involvement of the relA gene in the autolysis of Escherichia coli induced by inhibitors of peptidoglycan biosynthesis. *Journal of Bacteriology*, *164*(2), 861–865.

Kusser, W., and Ishiguro, E. E. (1987). Suppression of mutations conferring penicillin tolerance by interference with the stringent control mechanism of Escherichia coli. *Journal of Bacteriology*, *169*(9), 4396–4398.

Landini, P. (2009). Cross-talk mechanisms in biofilm formation and responses to environmental and physiological stress in Escherichia coli. *Research in Microbiology*, *160*(4), 259–266.

Lee, S., Hinz, A., Bauerle, E., Angermeyer, A., Juhaszova, K., Kaneko, Y., et al. (2009). Targeting a bacterial stress response to enhance antibiotic action. *Proceedings of the National Academy of Sciences*, *106*(34), 14570–14575.

Lee, H. H., Molla, M. N., Cantor, C. R., and Collins, J. J. (2010). Bacterial charity work leads to population-wide resistance. *Nature*, *467*(7311), 82–85.

Lewis, K. (2008). Multidrug tolerance of biofilms and persister cells. *Bacterial Biofilms*, 107–131.

Lewis, K. (2010). Persister cells. *Annual Review of Microbiology*, *64*, 357–372.

Li, K., Zheng, J., Deng, T., Peng, J., Daniel, D., Jia, Q., and Huang, Z. (2019). An analysis of antimicrobial resistance of clinical pathogens from historical samples for six countries. *Processes*, *7*(12), 964.

López, D., Vlamakis, H., and Kolter, R. (2010). Biofilms. *Cold Spring Harbor Perspectives in Biology*, *2*(7), a000398.

Lu, T. K., and Collins, J. J. (2009). Engineered bacteriophage targeting gene networks as adjuvants for antibiotic therapy. *Proceedings of the National Academy of Sciences*, *106*(12), 4629–4634.

Mesak, L. R., and Davies, J. (2009). Phenotypic changes in ciprofloxacin-resistant Staphylococcus aureus. *Research in Microbiology*, *160*(10), 785–791.

Miller, C., Thomsen, L. E., Gaggero, C., Mosseri, R., Ingmer, H., and Cohen, S. N. (2004). SOS response induction by ß-lactams and bacterial defense against antibiotic lethality. *Science*, *305*(5690), 1629–1631.

Mitchison, D. A., and Coates, A. R. (2004). Predictive in vitro models of the sterilizing activity of anti-tuberculosis drugs. *Current Pharmaceutical Design*, *10*(26), 3285–3295.

Mizuno, T., and Mizushima, S. (1990). Signal transduction and gene regulation through the phosphorylation of two regulatory components: the molecular basis for the osmotic regulation of the porin genes. *Molecular Microbiology*, *4*(7), 1077–1082.

Moo, C. L., Yang, S. K., Yusoff, K., Ajat, M., Thomas, W., Abushelaibi, A., et al. (2020). Mechanisms of antimicrobial resistance (AMR) and alternative approaches to overcome AMR. *Current Drug Discovery Technologies*, *17*(4), 430–447.

Mulcahy, H., and Lewenza, S. (2011). Magnesium limitation is an environmental trigger of the Pseudomonas aeruginosa biofilm lifestyle. *PLoS One*, *6*(8), e23307.

Mulcahy, L. R., Burns, J. L., Lory, S., and Lewis, K. (2010). Emergence of Pseudomonas aeruginosa strains producing high levels of persister cells in patients with cystic fibrosis. *Journal of Bacteriology*, *192*(23), 6191–6199.

Murakami, K., Ono, T., Viducic, D., Kayama, S., Mori, M., Hirota, K., et al. (2005). Role for rpoS gene of Pseudomonas aeruginosa in antibiotic tolerance. *FEMS Microbiology Letters*, *242*(1), 161–167.

Nguyen, D., Joshi-Datar, A., Lepine, F., Bauerle, E., Olakanmi, O., Beer, K., et al. (2011). Active starvation responses mediate antibiotic tolerance in biofilms and nutrient-limited bacteria. *Science*, *334*(6058), 982–986.

O'Brien, T. F. (2002). Emergence, spread, and environmental effect of antimicrobial resistance: how use of an antimicrobial anywhere can increase resistance to any antimicrobial anywhere else. *Clinical Infectious Diseases*, *34*(Supplement 3), S78–S84.

Patriquin, G. M., Banin, E., Gilmour, C., Tuchman, R., Greenberg, E. P., and Poole, K. (2008). Influence of quorum sensing and iron on twitching motility and biofilm formation in Pseudomonas aeruginosa. *Journal of Bacteriology*, *190*(2), 662–671.

Pei, R., Zhang, L., Duan, C., Gao, M., Feng, R., Jia, Q., and Huang, Z. J. (2021). Investigation of Stress Response Genes in Antimicrobial Resistant Pathogens Sampled from Five Countries. *Processes*, *9*(6), 927.

Petrosino, J. F., Galhardo, R. S., Morales, L. D., and Rosenberg, S. M. (2009). Stress-induced β-lactam antibiotic resistance mutation and sequences of stationary-phase mutations in the Escherichia coli chromosome. *Journal of Bacteriology*, *191*(19), 5881–5889.

Poole, K. (2005). Aminoglycoside resistance in Pseudomonas aeruginosa. *Antimicrobial Agents and Chemotherapy*, *49*(2), 479–487.

Poole, K. (2008). Bacterial multidrug efflux pumps serve other functions. *Microbe-American Society for Microbiology*, *3*(4), 179.

Poole, K. (2012). Stress responses as determinants of antimicrobial resistance in Gram-negative bacteria. *Trends in Microbiology*, *20*(5), 227–234.

Poole, K. (2014). Stress responses as determinants of antimicrobial resistance in Pseudomonas aeruginosa: multidrug efflux and more. *Canadian Journal of Microbiology*, *60*(12), 783–791.

Potrykus, K., and Cashel, M. (2008). (p)ppGpp: still magical?. *Annual Review of Microbiology*, *62*, 35–51.

Power, E. G., and Phillips, I. (1993). Correlation between umuC induction and Salmonella mutagenicity assay for quinolone antimicrobial agents. *FEMS Microbiology Letters*, *112*(3), 251–254.

Pu, Y., Zhao, Z., Li, Y., Zou, J., Ma, Q., Zhao, Y., et al. (2016). Enhanced efflux activity facilitates drug tolerance in dormant bacterial cells. *Molecular Cell*, *62*(2), 284–294.

Rami, A., Toutain, C. M., and Jacq, A. (2005). An increased level of alternative sigma factor RpoS partially suppresses drug hypersensitivity associated with inactivation of the multidrug resistance pump AcrAB in Escherichia coli. *Research in Microbiology*, *156*(3), 356–360.

Reygaert, W. C. (2018). An overview of the antimicrobial resistance mechanisms of bacteria. *AIMS Microbiology*, *4*(3), 482.

Rodionov, D. G., and Ishiguro, E. E. (1995). Direct correlation between overproduction of guanosine 3', 5'-bispyrophosphate (ppGpp) and penicillin tolerance in Escherichia coli. *Journal of Bacteriology*, *177*(15), 4224–4229.

Ruiz, N., and Silhavy, T. J. (2005). Sensing external stress: watchdogs of the Escherichia coli cell envelope. *Current Opinion in Microbiology*, *8*(2), 122–126.

Scallan, E., Hoekstra, R. M., Widdowson, M. A., Hall, A. J., and Griffin, P. M. (2011). Foodborne illness acquired in the United States. *Emerging Infectious Diseases*, *17*(7), 1339.

Sharma, U. K., and Chatterji, D. (2010). Transcriptional switching in Escherichia coli during stress and starvation by modulation of σ70 activity. *FEMS Microbiology Reviews*, *34*(5), 646–657.

Shatalin, K., Shatalina, E., Mironov, A., and Nudler, E. (2011). H_2S: a universal defense against antibiotics in bacteria. *Science*, *334*(6058), 986–990.

Shaw, K. J., Miller, N., Liu, X., Lerner, D., Wan, J., Bittner, A., and Morrow, B. J. (2003). Comparison of the changes in global gene expression of Escherichia coli induced by four bactericidal agents. *Journal of Molecular Microbiology and Biotechnology*, *5*(2), 105–122.

Shee, C., Gibson, J. L., Darrow, M. C., Gonzalez, C., and Rosenberg, S. M. (2011). Impact of a stress-inducible switch to mutagenic repair of DNA breaks on mutation in Escherichia coli. *Proceedings of the National Academy of Sciences*, *108*(33), 13659–13664.

Shin, B., Park, C., and Park, W. (2020). Stress responses linked to antimicrobial resistance in Acinetobacter species. *Applied Microbiology and Biotechnology*, *104*(4), 1423–1435.

Shoji, M., Cui, L., Iizuka, R., Komoto, A., Neoh, H. M., Watanabe, Y., et al. (2011). walK and clpP mutations confer reduced vancomycin susceptibility in Staphylococcus aureus. *Antimicrobial Agents and Chemotherapy*, *55*(8), 3870–3881.

Suh, S. J., Silo-Suh, L., Woods, D. E., Hassett, D. J., West, S. E., and Ohman, D. E. (1999). Effect of rpoS mutation on the stress response and expression of virulence factors in Pseudomonas aeruginosa. *Journal of Bacteriology*, *181*(13), 3890–3897.

Sung, H. M., Yeamans, G., Ross, C. A., and Yasbin, R. E. (2003). Roles of YqjH and YqjW, homologs of the Escherichia coli UmuC/DinB or Y superfamily of DNA polymerases, in stationary-phase mutagenesis and UV-induced mutagenesis of Bacillus subtilis. *Journal of Bacteriology*, *185*(7), 2153–2160.

Tegova, R., Tover, A., Tarassova, K., Tark, M., and Kivisaar, M. (2004). Involvement of error-prone DNA polymerase IV in stationary-phase mutagenesis in Pseudomonas putida. *Journal of Bacteriology*, *186*(9), 2735–2744.

Thomsen, L. E., Olsen, J. E., Foster, J. W., and Ingmer, H. (2002). ClpP is involved in the stress response and degradation of misfolded proteins in Salmonella enterica serovar Typhimurium. *Microbiology*, *148*(9), 2727–2733.

Tian, Z. X., Fargier, E., Mac Aogain, M., Adams, C., Wang, Y. P., and O'Gara, F. (2009). Transcriptome profiling defines a novel regulon modulated by the LysR-type transcriptional regulator MexT in Pseudomonas aeruginosa. *Nucleic Acids Research*, *37*(22), 7546–7559.

Tkachenko, A. G., and Fedotova, M. V. (2007). Dependence of protective functions of Escherichia coli polyamines on strength of stress caused by superoxide radicals. *Biochemistry (Moscow)*, *72*(1), 109–116.

Tkachenko, A. G., Akhova, A. V., Shumkov, M. S., and Nesterova, L. Y. (2012). Polyamines reduce oxidative stress in Escherichia coli cells exposed to bactericidal antibiotics. *Research in Microbiology*, *163*(2), 83–91.

Touati, D. (2000). Sensing and protecting against superoxide stress in Escherichia coli – how many ways are there to trigger soxRS response?. *Redox Report*, *5*(5), 287–293.

Tran, T. D. H., Kwon, H. Y., Kim, E. H., Kim, K. W., Briles, D. E., Pyo, S., and Rhee, D. K. (2011). Decrease in penicillin susceptibility due to heat shock protein ClpL in Streptococcus pneumoniae. *Antimicrobial Agents and Chemotherapy*, *55*(6), 2714–2728.

Varhimo, E., Savijoki, K., Jefremoff, H., Jalava, J., Sukura, A., and Varmanen, P. (2008). Ciprofloxacin induces mutagenesis to antibiotic resistance independent of UmuC in Streptococcus uberis. *Environmental Microbiology*, *10*(8), 2179–2183.

Vinella, D., Albrecht, C., Cashel, M., and d'Ari, R. (2005). Iron limitation induces SpoT-dependent accumulation of ppGpp in Escherichia coli. *Molecular Microbiology*, *56*(4), 958–970.

Vinella, D., D'Ari, R., Jaffe, A., and Bouloc, P. (1992). Penicillin binding protein 2 is dispensable in Escherichia coli when ppGpp synthesis is induced. *The EMBO Journal*, *11*(4), 1493–1501.

Viveiros, M., Dupont, M., Rodrigues, L., Couto, I., Davin-Regli, A., Martins, M., et al. (2007). Antibiotic stress, genetic response and altered permeability of E. coli. *PloS One*, *2*(4), e365.

Viveiros, M., Jesus, A., Brito, M., Leandro, C., Martins, M., Ordway, D., et al. (2005). Inducement and reversal of tetracycline resistance in Escherichia coli K-12 and expression of proton gradient-dependent multidrug efflux pump genes. *Antimicrobial Agents and Chemotherapy*, *49*(8), 3578–3582.

Wang, M., Jacoby, G. A., Mills, D. M., and Hooper, D. C. (2009). SOS regulation of qnrB expression. *Antimicrobial Agents and Chemotherapy*, *53*(2), 821–823.

Wang, X., Kim, Y., Hong, S. H., Ma, Q., Brown, B. L., Pu, M., et al. K. (2011). Antitoxin MqsA FRANCOS helps mediate the bacterial general stress response. *Nature Chemical Biology*, *7*(6), 359–366.

Wang, X., and Wood, T. K. (2011). Toxin-antitoxin systems influence biofilm and persister cell formation and the general stress response. *Applied and Environmental Microbiology*, *77*(16), 5577–5583.

Wayne, L. G., and Hayes, L. G. (1996). An in vitro model for sequential study of shiftdown of Mycobacterium tuberculosis through two stages of nonreplicating persistence. *Infection and Immunity*, *64*(6), 2062–2069.

Weiser, R. S., and Hargiss, C. O. (1946). Studies on the Death of Bacteria at Low Temperatures: II. The Comparative Effects of Crystallization, Vitromelting, and Devitrification on the Mortality of Escherichia coli. *Journal of Bacteriology*, *52*(1), 71–79.

Wu, J., Long, Q., and Xie, J. (2010). (p) ppGpp and drug resistance. *Journal of Cellular Physiology*, *224*(2), 300–304.

White-Ziegler, C. A., Um, S., Perez, N. M., Berns, A. L., Malhowski, A. J., and Young, S. (2008). Low temperature (23 C) increases expression of biofilm-, cold-shock-and RpoS-dependent genes in Escherichia coli K-12. *Microbiology*, *154*(1), 148–166.

Yang, K., Wang, A., Fu, M., Wang, A., Chen, K., Jia, Q., and Huang, Z. (2020). Investigation of incidents and trends of antimicrobial resistance in foodborne pathogens in eight countries from historical sample data. *International journal of environmental research and public health*, *17*(2), 472.

5 Efflux Pumps and Their Role in the Development of Multi-drug Resistant Bacterial Biofilm

Saroj Ballav, Debarati Jana, Kartik Chandra Guchhait, Biswajit Das, Mousumi Manna, Subrata Hazra, Jun-ichi Kadokawa, Amiya Kumar Panda and Chandradipa Ghosh

5.1 INTRODUCTION

Biofilms are aggregations of planktonic microorganisms that can grow on many different surfaces where cells are surrounded by self-produced extracellular polymeric substances (EPSs). Biofilm formation often becomes a defense for relief from antimicrobial treatment, for enduring a range of environmental stresses, and for the resistance to a host immune system. Biofilm resistance means the inability of antimicrobial agents to inhibit the growth or kill bacteria surviving within a biofilm to the same extent as living in a planktonic culture (Brauner et al., 2016). Biofilms are involved in a large spectrum of infections and survive in the presence of high concentrations of antimicrobials through a variety of mechanisms. Cell to cell communications through quorum sensing (QS) affect the expression of bacterial biofilm virulence factors (Fong et al., 2019). The formation of biofilm and the action of quorum sensing reflects the ineffectiveness of multiple antibiotics or antimicrobials and thereby inability to kill the microbes. Several multi-drug resistant microorganisms empirically known as "Superbugs" are resistant to the first-line and second-line antibiotics. There are so many examples of drug resistant bacteria that spread infection in uncontrolled way due to lack of effective antimicrobials. The World Health Organization (WHO) refers to these superbugs in several subdivisions according to priority (M Campos et al., 2020). Antimicrobial resistance (AMR) and drug efflux mechanisms are included in the ability to form biofilm. Quorum sensing participates in biofilm organization by various bacteria including *Staphylococcus aureus*, *Pseudomonas aeruginosa*, *Aeromonas hydrophilia*, *Streptococcus mutans* and many other bacteria (Aleksandra and Lucja, 2016). Biofilm regulatory genes and drug tolerance impose challenges for the clinical apprehension and abetment of conventional antimicrobials that impact on society.

Efflux pumps are membrane associated protein transporters found in microorganisms that remove drugs from cells causing drug resistance. Multi-drug resistant efflux pumps are a means by which bacteria can evade the effects of selected antimicrobials (Li et al., 2015), are age-old components

DOI: 10.1201/9781003281238-5

encoded in the chromosomes of microorganisms and which can bestow resistance to antimicrobials at different levels. Multi-drug resistant (MDR) efflux pumps not only come up with AMR but also participate in the removal of environmental toxins. Efflux pump gene expression makes cells more capable of expelling antimicrobials.

The evolution, global spread of antibiotic-resistant lineages of groups of pathogens such as *Acinetobacter baumannii, Enterococcus faecium, Klebsiella pneumoniae, Pseudomonas aeruginosa, Staphylococcus aureus* and the like, pose multiple emerging challenges on the future. These pathogens symbolize a remarkable capacity for both the development and acquisition mechanisms of AMR. Technically multidrug-resistance is the ability of any organism to carry out normal functioning against at least three or more antibiotic classes. The other two levels of antimicrobial resistances are extensive drug resistance means lack of susceptibility to one agent in all classes, and total drug resistance (TDR) which is defined as not being susceptible to all agents in each of the available antibiotic classes (Magiorakos et al., 2012; Shriram et al., 2018). AMR in the community has been increasing steadily day by day. In the communities of low and middle income countries, sub-Saharan Africa has the highest burden, followed by South Asia (Murray et al., 2022). The problem is exacerbated by general practitioners prescribing excessive antibiotics without appropriate indication. Easily available antimicrobial drugs can be readily purchased without prescription by a physician or other qualified health professional. Diagnostic uncertainty and self-medication are also underlying reasons for antibiotic resistance (Larson et al., 2007). Some drug-resistant infections can be transmitted between livestock and humans. In Southern India, 53% of isolates from patients of community acquired bacteremia with ESBL-production were found to be resistant to multiple groups of antibiotics caused by *E. coli* and *Klebsiella spp.* (Abhilash et al., 2010). The intensive and prolonged use of antimicrobial drugs may contribute to the emergence and spread of highly antibiotic-resistant nosocomial infections, especially in highly immuno-suppressed patients such as AIDS patients, transplant recipients and delicate aged patients. Moreover, invasive surgical procedures also play an important role in the spread from one patient to another (Vincent, 2003). In Mumbai, 18.5% of clinical isolates showed a dual carbapenemase gene from a molecular characterization study of carbapenem-resistant *Enterobacteriaceae* (Kazi et al., 2015). The rates of AMR among hospital and community settings have increased and the economic outcome of this remains largely unknown. Chris Dall reported in the file "Antimicrobial Stewardship", CIDRAP news that 1.2 million people died from drug-resistant infections in 2019 (Chris, 2022). The ECDC (European Centre for Disease Prevention and Control) reported around 6.7 million people affected by drug-resistant pathogens and among them nearly 33,000 died in only one year (namely, in 2020) in Europe (WHO, 2022). Lord Jim O'Neill (2016) in his review on antimicrobial resistance, in accordance with estimated data on AMR death in global scale, cited that six pathogens with AMR were responsible for at least 7,00,000 deaths a year and concluded that AMR could kill up to 10 million per year by 2050 if no action was taken. It is reported by the WHO that the number of drug-resistant bacterial superbugs are rising day by day. For more than seventy years antimicrobial agents have been used to treat patients with infectious diseases in a prospective battle between human and bacteria (Nadeem et al., 2020). The antimicrobials have been able to reduce fatal illnesses with great effect on the microbes and have been widely used for a long time, but unfortunately with time the microorganisms have developed adaptive features to withstand the antimicrobials' designs and have thus rendered these drugs less effective.

This chapter focuses on the current state of the expression of efflux pumps, efflux systems induced in biofilm development, the development of antibiotic resistance and bacterial persistence. A holistic strategy therefore needs to be implemented with a thorough understanding of biofilm formation and the role of efflux pumps in multi-drug resistant microbes.

5.2 MULTI-DRUG RESISTANCE

Drug resistance is the ability of a microorganism to resist or no longer respond to medicines that previously have been used to treat them. Antibiotics are substances that act to inhibit the growth (bacteriostat) of cells or cause cell death (bactericide), and are also responsible for establishing a strong selective pressure and/or adaptive advantage for those microbes that have acquired mechanisms to survive in the presence of antimicrobial agents. If we look back to the "antibiotic era", the abrupt use of antibiotics often resulted in a vertical transfer of an antibiotics resistant gene from the bacterial population. The resistance of bacteria to antibiotics is now a common phenomenon. Generally, antibiotic resistance takes place through natural resistance, genetic mutation and from one species acquiring resistance from another species. Several factors promote antibiotic resistance, such as exposure to sub-optimal level of antimicrobials, exposure to broad spectrum antibiotics, exposure to microbes carrying a resistant gene, lack of development of new antibacterial with a spectrum directed against Gram-negative bacteria and Gram-positive bacteria, lack of hygiene in the clinical environment and the use of antibiotics in food/agriculture (Nadeem et al., 2020). Antibiotic therapy can reverse disease symptoms but can't remove completely the bacterial colony from the infected organism. As a result, after the cycle of antibiotics the infection grows again and this time the previously applied antibiotics remain no longer effective on that bacterial population against the presence of unattached planktonic counterparts. In that situation bacterial cells are released from the biofilm and induce disease. These biofilms are not only resistant to antibiotics but also to bacteriophages and other chemical synthetic bactericidal elements due to failure of their full depth penetration into biofilms. In some cells, biofilms form slowly in which case antimicrobial activity is not effective.

5.3 THE ROLE OF EFFLUX PUMPS IN MULTI-DRUG RESISTANCE

A natural property of bacteria is to combat antibiotics due to the presence of warfare elements carried in plasmids, the mobile genetic elements. Bacteria can also acquire and develop antibiotic resistance through induced or spontaneous mutation in their chromosomal genes and by the incorporation of new genes via horizontal gene transfer (HGT). The antibiotic resistance mechanism includes, firstly, antibiotic inactivation through hydrolysis or modification, for example, phosphorylation by an enzyme, and, secondly, modification of the antibiotic target through genetic mutations or post translational modification by overproduction of the target of the antibiotic or its uptake by the cell or its post-acquisition of one or more molecular mechanisms for AMR, such as inactivation of antibiotics or increased antibiotic efflux out of the cell through efflux pumps and porins (Blair et al., 2015). Antibiotic resistance is further worsened by biofilms, due to higher genotypic and phenotypic resistance to antimicrobials than planktonic cells (Høiby et al., 2010). Efflux pumps might also play roles in QS and biofilm formation (Alav et al., 2018).

Efflux pumps in Gram-positive bacteria are encoded by both plasmid and chromosomally expressed genes. The transmissible drug-specific efflux pumps are encoded by plasmids, whereas those who are MDR are usually encoded by chromosome and not donated to other organisms. The plasmid-based efflux pumps are multi copy in nature without any additional need for any subsequent mutation. On the other hand, MDR occurs due to increased gene expression or to the occurrence of regulatory mutations. Bacterial efflux pumps diligently transport several antibiotics out of the cell and are contributors to the intrinsic resistance of Gram-negative bacteria to many of the drugs that are used to treat Gram-positive bacterial infections. Over expression of efflux pumps can also give high resistance to useful antibiotics which were previously effective. They can be specific for a single substrate (Tet pumps) or can transport a wide range of structurally diverse substrates and are known as MDR efflux pumps (Piddock, 2006). There are five main superfamilies of efflux pumps associated with MDR according to their composition, energy source, number of transmembrane spanning regions and substrates (Figure 5.1):

- The multi-drug and toxic compound extrusion (MATE) family: proteins that are predicted to have 12 alpha-helical transmembrane regions (Kuroda and Tsuchiya, 2009), such as MepA.
- The small multidrug resistance (SMR) family: composed of four transmembrane alpha-helices of approximately 100–140 amino acids in length, for example, Smr (Jack et al., 2001; Bay et al., 2017)
- The major facilitator superfamily (MFS): proteins that transport a broad spectrum of ions and solutes across membranes via facilitated diffusion, symport, or antiport, such as NorA, NorB, NorC, MdeA, SdrM, LmrS (Pao et al., 1998, Yan, 2015)
- The ATP-binding cassette (ABC) superfamily (Lubelski et al., 2007): transporters that are remarkably conserved in terms of the primary sequence and the organization of the domains or subunits. Characteristically, ABC transporters have a high conserved ATPase domain which bind and hydrolyze ATP to provide energy for the import and export of a wide variety of substrates, for example, MsrA
- The resistance-nodulation-division (RND) superfamily (Li and Nikaido, 2004; Seeger et al., 2008; Nikaido and Takatsuka, 2009): protein which dictates the substrate for the completed transport systems including metal ions, xenobiotics or drugs. Transport of hydrophobic and amphiphilic compounds are carried out by the HAE-RND subfamily. Moreover, the efflux of heavy metals is performed by HME-RND, for example, FarE. RND efflux pumps were found in GNBs and MFS efflux system while ABC transporters are the most common efflux system in case of GPBs. Only the ABC superfamily utilizes energy from ATP hydrolysis for mediating efflux of the molecule from inside the cell and all the other efflux superfamilies utilize energy from proton or sodium motive force (Van Bambeke et al., 2000).

In GNBs, the RND superfamily is the most clinically relevant of the multi-drug efflux pumps. The AcrAB-TolC protein complex found in *Enterobacteriaceae* such as *Escherichia coli*,

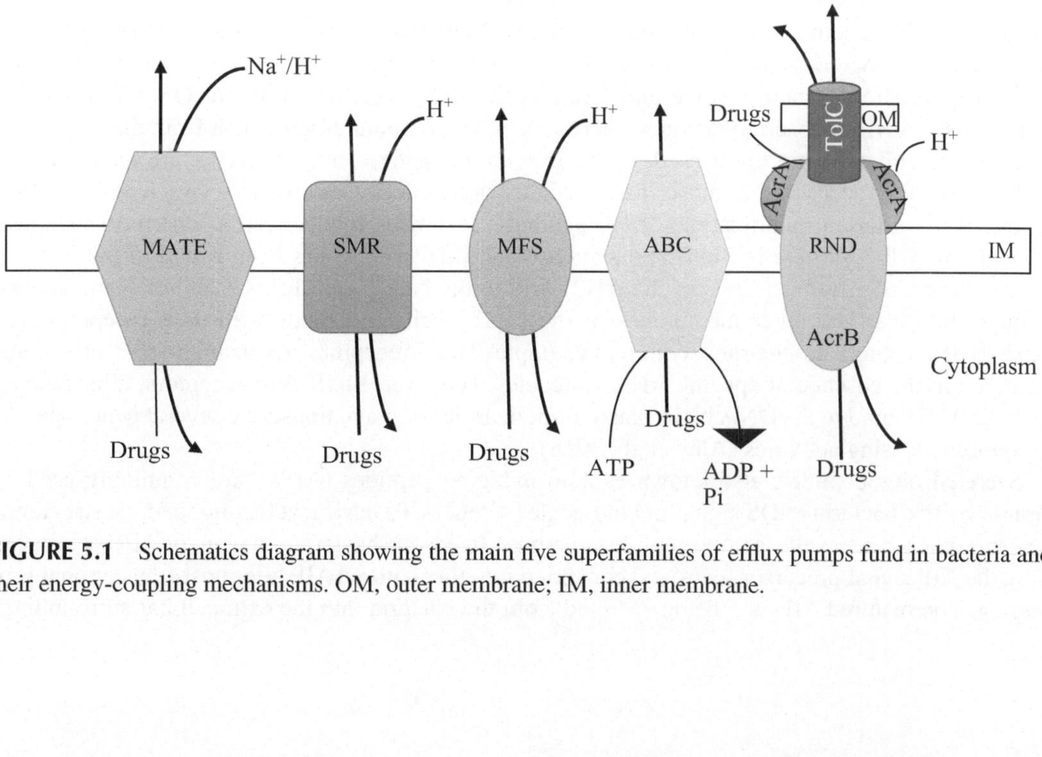

FIGURE 5.1 Schematics diagram showing the main five superfamilies of efflux pumps fund in bacteria and their energy-coupling mechanisms. OM, outer membrane; IM, inner membrane.

Salmonella species, Klebsiella pneumoniae. Staphylococcus aureus exports fluoroquinolones through NorA and the related efflux system of the MFS superfamilies (Poole, 2005). The homotrimeric inner membrane component acriflavine resistance B (AcrB) is found to be present in cells and it drops significantly when bacteria switch from a planktonic to a biofilm-related lifestyle, and it continues getting lower as the biofilms mature (Gardini et al., 2021). The RND superfamily of efflux pumps, for example, MexAB-OprM of *Pseudomonas aeruginosa* and AdeABC of *Acinetobacter baumannii* are also relevant. All species of bacteria can express single or multiple types of efflux system from the same or another superfamily (Piddock, 2006). Efflux pumps also express different substrate profiles, which differ within and between the superfamily(s) (Poole, 2005; Poole, 2007).

5.4 QUORUM SENSING AND ITS ASSOCIATION WITH THE EFFLUX PUMP

The regulated expression of certain genes in response to several extra-cellular signalling molecules which are produced by the bacterial population is known as quorum sensing (QS) (Pearson et al., 1999). The extra-cellular signalling molecules are termed autoinducers (AIs). Bacteria use the AIs concentration in their surrounding extracellular environment to monitor the cell density and to regulate the expression of several genes specific to QS (Li and Titan, 2012). These genes are responsible for antibiotic synthesis, biofilm development, bioluminescence, genetic competence, sporulation, and pathogenicity, and other bacterial behaviors (Li and Titan, 2012). Quorum sensing occurs inside and among the bacterial species, as well as in between the bacteria and other microorganisms, and is essential for the persistence of the bacteria in a variety of environmental cues. Different species of bacteria use different types of AIs, have different signal relay mechanisms, and have different identities for quorum-specific genes. However, all the QS pathways are based on three fundamental processes: synthesis of the AI molecules, AI detection by the receptors present within or on the surface of the bacteria, and the stimulation of specific transcription factors of quorum-specific genes (Alav et al., 2018). Under the condition of low bacterial population, the AIs concentration remains under the threshold concentration for detection by the receptors. Increased density of the bacterial population leads numerous cells to synthesize the AIs in the surroundings, resulting in an accumulation of AIs and thereby, reach in its threshold concentrations for detection by receptors, which can be intra-cellular, or membrane associated (Li and Titan, 2012). The expression of QS specific genes, and the genes associated with the synthesis of AIs, is triggered when AIs bind to their respective receptors. Gram-positive and Gram-negative bacteria use different QS systems (Alav et al., 2018). Acyl-homoserine lactones (AHLs) or other compounds generated from the compound S-adenosylmethionine are often used as AIs by Gram-negative bacteria. AHLs are made up of a lactone ring of N-acetylated homoserine molecules and an acyl chain with varying lengths that are produced by different modifications. AHL synthase enzymes produce AHLs. The most prevalent example of AHL synthase is LuxI-type synthase and this enzyme has been found in hundreds of bacterial species. After synthesis of the AHLs within the bacteria, it diffuses and/or is transported through the inner and outer membranes of the nearby cells and binds with their receptors. The LuxR-type receptors are the most common example. These receptors are unable to fold or become degraded in the absence of specific AHL molecules. However, LuxR-type receptors, after binding with the AHLs bind to the DNA by its stable dimerization form and transcribe several genes specific for quorum-sensing activities (Alav et al., 2018).

Secreted oligopeptides, also known as auto inducing peptides (AIPs), are commonly used by Gram-positive bacteria as QS signalling molecules. Some AIPs have acyclic oligopeptide structures, whereas others have cyclic lactone peptide structures. Inside the bacteria, a precursor AIP is produced from the AIP signal precursor locus and converted into the matured AIP after post-translational processing. The matured AIPs are being secreted from the bacteria into the extracellular surroundings

(Rutherford and Bassler, 2012). Once the cell population density reaches a particular degree, the AIPs attach to the histidine sensor kinase (a membrane-bound two-component system), activate the kinase activity of this enzyme and causes auto-phosphorylation. This phosphate group is transferred from the active sensor kinase to an intracellular response regulator, which is then phosphorylated. As a result, the response regulator (after the phosphorylation) is activated and binds to the DNA, causing quorum-specific genes to be transcribed. Although the QS signalling cascades for Gram-negative and Gram-positive bacteria employ different AIs, certain commonly found AIs (namely, AI-2) regulate the QS among the inter-species organisms (Federle and Bassler, 2003).

Quorum sensing regulation has been linked to efflux systems. It is well-known that QS regulates virulence expression and also the differentiation of biofilm. In *Burkholderia pseudomallei*, QS-mediated activities like biofilm formation are reliant on BpeAB-OprB efflux pump function (Chan and Chua, 2005). A mutation in a RND-like efflux pump transporter in *Pseudomonas aeruginosa* has been found to suppress the expression of QS-dependent *lec*A*::lux*. Previous studies also reported that the QS signal system is partly dependent on the efflux pump activity. In *P. aeruginosa*, the signal molecule N-3-oxododecanoyl homoserine lactone (3OC12-HSL) requires active trafficking through the bacterial membrane via an efflux pump (Pearson et al., 1999). As a result, an increased activity of the efflux pump could have a variety of consequences on biofilm formation by increasing QS molecule extrusion or incursion.

5.5 EFFLUX PUMPS IN DIFFERENT BACTERIA

5.5.1 The Role of Efflux Pumps in the Biofilm of *Acinetobacter baumannii*

Acinetobacter baumannii is an important opportunistic pathogen. It's an oxidase negative, non-fermentive, Gram-negative coccobacillus. It causes a broad range of infections, mainly hospital-related infections as well as meningitis, pneumonia, blood stream infections, urinary tract infections, skin and soft tissue infections (Nait Chabane et al., 2014). The Center for Disease Control and Prevention (CDC) in the USA acknowledges the MDR *A. baumannii* as a potent source of global outbreaks and epidemic. Reports are there that this bacterium increase its resistance rate annually to commercially available antibiotics, namely, beta-lactam antibiotics, tetracyclines, fluoroquinolones, and particularly the rate of infection increased with carbapenem-resistance (Handal et al., 2017).

Clinical isolates of *A. baumannii* have been shown to form biofilm which can accelerate their ability to survive in stress-related environmental conditions due to resistance towards antibiotics (Nait Chabane et al., 2014). If *A. baumannii* are treated with sub-inhibitory concentrations of antibiotics, then *A. baumannii* has been found to be even more extensive in biofilm formation by development of clinical antibiotics tolerance (Alav et al., 2018).

The report clearly reveals that an efflux system encoded by various genes including the RND (Resistance-Nodulation-Division) pump family efflux genes, namely, *A1S-0009, A1S-0116, A1S-0538* and the MFS pump (major facilitator superfamily) efflux gene *A1S-1316,* get up regulated in biofilm cells (Rumbo-Feal et al., 2013). In addition, the efflux genes *A1S-1117*, *A1S-1751* and *ade*T could be expressed only in biofilm cells, but not in planktonic cells. The *A1S-1117* belonging to the MFS family, *A1S-1751* encodes a membrane fusion protein and *ade*T, *ade*A are RND type efflux pump genes (Alav et al., 2018).

Further investigations have shown that an over-expression of the AdeFGH efflux gene remains associated with biofilm formation by clinical isolates of *Acinetobacter baumannii.* Moreover, the significantly increased biofilm could be seen through continuous up-regulation of genes *aba*I and *aba*G that encode AHL synthase, which is necessary for biofilm formation (He et al., 2015).

Another study has shown that mutant over-expression of three genes *ade*ABC, *ade*FGH and *ade*IJK are the RND family efflux pump genes that remarkably decrease biofilm formation (Yoon et al., 2015). The AdeABC efflux pump system is the main RND type efflux system in *A. baumannii*

(Coyne et al., 2011). The AdeABC efflux pump-mediated drug resistance gene in *A. baumannii* is the most studied pump gene among the three RND efflux pump genes. A study has shown that the AdeABC efflux pump gene encodes carbapenem resistance in the absence of mutation in *A. baumannii*. Clinical isolates over-expressing AdeABC efflux pump genes in strain ATCC17978 were resistant to different commercial antibiotics that include tigecycline, minocycline, gentamicin, tetracycline, chloramphenicol and erythromycin. The AdeFGH over-expression was found in chloramphenicol resistance acquired in mutant strains, and the AdeIJK efflux pump gene confers resistance to beta-lactam, chloramphenicol, tetracycline, erythromycin and fluoroquinolones (Xu et al., 2019).

The mutant over expressing AdeABC and AdeIJK efflux genes accomplished the under-expression of several genes that encoded proteins CsuA/B, CsuC and FimA. These proteins are associated with pilus systems that have important roles in the preliminary phase of biofilm formation when they initiate primary adhesion, colonization and micro colony formation (Yoon et al., 2015). Mutant strains over-expressing AdeABC and AdeIJK pumps reduced biofilm formation and by the down-regulation of genes considered with the pilus system they weaken the initial stages of biofilm formation.

It is apparent thus that clinical MDR isolates of *A. baumannii* and mutant over-expression of one of three RND pumps is the most important factor for developing resistance to several commercially available antibiotics.

5.5.2 The Role of Efflux Pumps in Biofilm Formation of *Pseudomonas aeruginosa*

Pseudomonas aeruginosa is a Gram-negative, opportunistic microorganism found in soil and water bodies. It can acquire adaptive and intensive resistance to survive and colonize in a variety of natural and artificial environments (Lyczak et al., 2000). This pathogen causes severe infectious diseases including cystic fibrosis, a chronic lung infection (Sing et al., 2000). The quorum sensing (QS) systems happen by the secretion of acylated homoserine lactones through gene expression (Davies et al., 1998; Withers et al., 2001).

After treatment with a cytoplasmic membrane proton gradient inhibitor, intracellular accumulation of N-3-oxododecanoyl-L-homoserine lactone (3OC12-HSL) occurs into different species of *Pseudomonas* that increase the quorum sensing process (Pearson et al., 1999). Intracellular accumulation of 3OC12-HSL is also found in mutant species to decrease biofilm formation where mexABoprM-encoded efflux pumps are not present. The MexAB-OprM and MexCD-OprJ efflux pumps are unable to form a biofilm in the presence of azithromycin, a potent antibiotic (Gillis et al., 2005).

The efflux pumps of the resistance nodulation cell division RND family are clinically the most important efflux systems in different Gram-negative bacteria. In *Burkholderia cenocepacia*, these systems are involved in various mechanisms including quorum sensing, biofilm formation, chemotaxis and motility and general physiological functions that increase virulence factor of the bacteria (Buroni et al., 2009). The extracellular secretion of acyl-homoserine lactones is dependent on the BpeAB-OprB efflux pump (Chan et al., 2007).

RND efflux pumps contain a cytoplasmic membrane pump, a periplasmic adaptor protein and an outer membrane protein channel through which they can efflux different types of drugs from the inside of the cell that increase biofilm formation and antibiotic resistance (Fahmy et al., 2016).

A reactive compound 5-methylphenazine-1-carboxylate (5-Me-PCA) induces the expression of efflux pump MexGHI-OpmD in different species of *Pseudomonas* (Sakhtah et al., 2016).

There are three processes of antibiotic resistance mechanisms: intrinsic, acquired and adaptive. Intrinsic resistance mechanisms are a genetic modification in the core genome of the organism, adaptive resistance mechanisms are those induced by environmental stimuli, and acquired resistances are rises from acquired resistance genes of other organisms or a consequence of selective advantageous mutations. Efflux pumps of the RND superfamily provide resistance to antibiotics and are responsible for multi-drug resistance (MDR) in Gram-negative bacteria. The RND drug anti-porter AcrB is a part of the AcrAB-TolC efflux pump that causes the efflux of noxious agents. (Tenover, 2006; Dreier and Ruggerone, 2015).

In *Pseudomonas*, the MexAB-OprM efflux pump offers intrinsic resistance to different antimicrobial agents including tetracycline, chloramphenicol, quinolones, β-lactams (De Kievit et al., 2001).

Antimicrobial peptides are a new potential antimicrobial therapeutic agent, which developed resistance towards these compounds. In biofilm, *P. aeruginosa* developed tolerance to the antimicrobial peptide colistin by a specific adaptation mechanism mediated through the pmr operon (Pamp et al., 2008).

Biofilm formation is facilitated by the efflux of exo-polysaccharide (EPS) and quorum sensing (QS) molecules from the interior of microbial cells. Due to mutation in efflux pump related genes, the biofilm-specific resistance is lower in aminoglycoside, fluoroquinolone and other anti-bacterial agents (Zhang and Mah, 2008) and for this mutation, the expression patterns of efflux pumps become changed which develops microbial colony formation that results chronic infection in the host (Baugh et al., 2012). The efflux pump inhibitors including chlorpromazine and phenyl-arginine-β-naphthylamide prevent biofilm formation in Gram-negative microbes. The anti-biofilm activity occurs through genetic inactivation and chemical inhibition (Baugh et al., 2012).

5.5.3 The Role of the Efflux Pump in the Biofilm of *Escherichia coli*

Escherichia coli is a gram-negative bacterium of *Enterobacteriaceae* and it can be found in the gastrointestinal tracts (GI tract) of vertebrates as well as a harmless or non-pathogenic commensal human gut flora by virtue of their ability to prevent colonization of pathogenic bacteria through the production of bacteriocins and other mechanisms (Alav et al., 2018; Tenaillon et al., 2010). Clinically, *E. coli* can also be found as an infectious or pathogenic agent in urinary tract infections (UTIs), sepsis, and so forth. Biofilms of *E. coli* exhibit higher antibiotic resistance than is to be found in planktonic cells, and also the abundance of several gene encoded efflux pumps in biofilms (Ito et al., 2009). Previous studies reported that in *E. coli*, several types of efflux pumps such as AcrB, AcrF, MdtB and MdtF are widely distributed and play a significant role in resistance against antibiotics (Bohnert et al., 2007).

Initial studies to evidence a link between efflux pumps and biofilm formation emphasised higher expression of efflux genes compared with planktonic growth within the biofilm. Global gene expression studies in *E. coli* biofilms indicate that the expression of several gene encoding putative efflux and transport proteins gets upregulated. The transport genes *mdt*F and *lsr*A, belonging to the RND and ABC superfamilies, were reported to be expressed at higher levels during biofilm growth compared with growth in exponential–and stationary-phases (Schembri et al., 2003). A study reported that under anaerobic conditions, the growth of *E. coli* cells displayed more expression of the MdtEF efflux pump compared with control, and under anaerobic respiration of nitrate, a mutant strain lacking the *mdt*EF gene had significantly lower survival rates (Zhang et al., 2011). Another study reported that in *E. coli* biofilm expression of several genes encoding efflux pumps were found to be increased that have higher antibiotic resistance (Webber and Piddock, 2001). The best characterised efflux pump in *E. coli* is AcrAB-TolC, which belongs to the RND family which exports the substrates including Fluoroquinolones, Rifampicin, Tetracycline, and so forth, and is also associated with bacterial colonization and persistence (Fernandes et al., 2003; Piddock et al., 2006). The AcrAB-TolC efflux pump has been found to be upregulated under growth in biofilm and exposure to several antibiotics (Baily et al., 2006; White et al., 1997). The TolC protein also plays an important role in the formation of biofilm as well as the export of antibiotics (Koronakis, 2003) but the mutant *tol*C gene showed a decrease in adhesion and biofilm formation that related to reduction in aggregative fimbriae expression (Imuta et al., 2008). In *E. coli*, the expression of AcrAB or

MarA did not correlate with ciprofloxacin resistance in biofilm, however, studies revealed that the expression of AcrAB only protects biofilm formation against low concentrations of ciprofloxacin (Maira-Litrán et al., 2000). Besides, in uropathogenic *E. coli,* the putative multi-drug resistant pump YhcQ could be observed to be involved in antibiotic resistance of bacteria in biofilm. The transcription of putative multi-drug resistance pump gene *ybc*Q was increased by RapA regulatory protein. Additionally, the protein was also found to contribute to the biofilm-specific penicillin G resistance (Lynch et al., 2007).

Efflux pumps also guide biofilm architecture in *E.coli* by impacting on QS regulation. QS in *E.coli* gets facilitated by the AI-2 signalling molecule. The LsrABCD complex, involving the *lsr*A gene as coding factor, interfere with the transport of the AI-2 signalling molecule (Xavier and Bassler, 2005). Efflux pumps have a significant role in the transport of AIs in *E.coli* biofilms, which may facilitate QS within the biofilm for promoting biofilm maturation (Alav et al., 2018).

A recent study reported that *E. coli* mutant strains lacking the efflux genes *acr*B, *acr*E and *tol*C displayed significant reductions in biofilm growth and expressed increasing antibiotic susceptibility compared with a control (Alav et al., 2018). Mutant strains lacking the efflux genes *acr*D, *emr*A, *emr*E, *mdt*K displayed enhanced biofilm growth when compared with the control (Bay et al., 2017). The efflux pump inhibitors (EPIs) have a significant role, which binds directly to the pump and blocks it either in a competitive or a non-competitive manner (Marquez, 2005).

5.5.4 The Role of the Efflux Pump in the Biofilm of *Staphylococcus aureus*

Staphylococcus aureus, a gram-positive coccal bacterium, has become a key concern for us due to its immense capability to cause both community-acquired and nosocomial diseases. It affects mainly skin and soft tissues and also causes haemorrhagic necrotizing pneumonia, one of the deadliest infection of *S. aureus* (Karmakar et al., 2018; Gillet et al., 2002). Especially noteworthy is that *S. aureus* is now resistant to most beta lactam antibiotics (Methicillin, Penicillin, *and so forth.*) and is also resistant to the recent antibiotic Vancomycin. Not only that, but *S. aureus* also develops biofilms, which promote resistance to multiple antibiotics, enhance longevity, and make treatment more challenging and complicated (Plata et al., 2009). The efflux pump is one of the most significant factors for the emergence of multi-drug resistant strains of *S. aureus*. Various investigations have shown that expression of numerous efflux pump genes was changed during biofilm formation. Exclusively, MFS (Major Facilitator Superfamily) type efflux pumps have been found to be significantly dynamic in *S. aureus* biofilm formation (Alav et al., 2018). During biofilm formation, the comparative array of expression of the efflux pump genes, mainly *nor*B, *nor*C and *mde*A, were studied and they were observed to be up-regulated during biofilm growth (He and Ahn, 2011). Osmolytes inhibit bacterial survival (Csonka, 1989). So, osmolyte transporters are up-regulated and play a crucial role in guarding bacterial cells from osmotic stress by enhancing the export of osmolytes across the membrane in the early phases of biofilm. The *nor*B and *nor*C export quinolones, ethidiumbromide, cetrimide and tetraphenylphosphonium, whereas the *mde*A can export a variety of antibiotics and quaternary ammonium compounds (Ding et al., 2008; Huang et al., 2004). *nor*B expression is enhanced in *S. aureus* in response to acid shock and decreased aeration, indicating that *nor*B is stimulated in response to hypoxic conditions and low pH inside biofilms (Costa et al., 2013). During biofilm formation, *S. aureus* ferments glucose, resulting in a deposition of different organic acids (acetic acid, lactic acid, and so forth) and thus resulting in a lowering of pH. Organic acids suppress bacterial growth and induce stress response pathways. Hence, *nor*B could shield biofilm from the harmful effects of those organic acids released from anaerobic respiration (Zhu et al., 2007; Rode et al., 2010).

In addition to these, two unidentified efflux pump genes, *SA2261*and *SA2131*, were shown to be down-regulated in the development of biofilm. It's very difficult to say how these two genes and their gene products participate in biofilm as very limited information is available about them

(Beenken et al., 2004). The expression of numerous efflux pump genes was greater in biofilm cells than in planktonic cells in accordance with a comparative transcriptomic study of *S. aureus*. The gene expressions inside the biofilms were assessed throughout, from 6 hrs to 48 hrs of biofilm formation. Genes that were expressed in 6 hrs and 8 hrs, were far more engaged in biofilm formation than those that were expressed in 48 hrs. At 8 hrs of biofilm growth, the *pro*P, *SA0589* and *SA2142* were predominantly expressed at about fivefold greater in biofilm than in the planktonic phase of bacterial life (Resch et al., 2005).

In an experiment using insertional mutant library, *bfd2* gene was identified from a high biofilm forming clinical strain of *S. aureus* that encodes a putative protein with MFS properties, but its substrate profile is still not clear (To Quoc et al., 2007). Not only that, a pleiotropic regulator, MgrA has been found to serve as a negative regulator of *nor*B and *nor*C and also inhibits biofilm development, implying a relationship amongst efflux pump and biofilm in *S. aureus* (Truong-Bolduc et al., 2006; Trotonda et al., 2008).

5.5.5 The Role of the Efflux Pump in the Biofilm of *Listeria monocytogene*

Listeria monocytogenes is a Gram-positive bacterium, which is the cause of listeriosis disease, that occurrs from contaminated food consumption (Hamon et al., 2006). Listeriosis diseases cause acute infections in fetuses, new born and immune competent individuals, in addition to having an approximately 28% mortality rate (Zhu et al., 2008). It is estimated in the United States that the most listeriosis cases in human have mainly originated from consuming contaminated RTF (Ready to Eat) meat products (FDA, 2003) and RTF dairy products (Ho et al., 2007). If it enters into the body, *L. monocytogene* can cross three biological obstacles: the blood brain barrier, the intestinal barrier, and the maternal-fetal barrier (Camejo et al., 2009) leading to developing complexity including gastroenteritis, meningitis, still-birth or spontaneous abortions (Allerberger and Wagner, 2010).

These bacteria are found to have some important properties such as, they form biofilms, they replicate in colder climates, and they have disinfectant tolerance and also resistance to phages. Biofilm formation ability is one of essential features in a bacterial community. The persistence of a static biofilm community is very challenging. There remains the chance that it could be cleared out by washing or scrubbing the adherent cells. Besides, cells surviving in a biofilm also acquire adaptive characteristics that result in increasing tolerance to several stresses and various antimicrobial agents as well as antibiotics (Bridier et al., 2011).

Studies have shown a relationship between multi-drug resistant efflux pumps and biofilm formation by *L. monocytogenes*. An ABC (ATP binding cassette) transporter is a large family of protein accountable for multi-drug resistance (Spengler et al., 2017) and it participates in negative regulation of biofilm production by *L. monocytogenes,* called Lm. G 1771 a novel ABC transporter. An ABC transporter generally consists of structural domains, namely, a nucleotide -binding domain, a transmembrane domain, or a solute-binding domain (SBD) (Tam et al., 1993). The ABC transporters act either as importers or exporters. The importers always characteristically have at least one SBD while the exporters do not have any SBD. Lm. G 1772 encodes an ATP-binding protein but does not encode any SBD. Therefore Lm. G 1772 transporter acts as an exporter (Zhu et al., 2008). This is basically an efflux protein exporting a signaling molecule that stimulates a pattern of genome expression. It could be seen that mutation in one of the components of this ABC transporter could significantly intensify a biofilm's formation capacity (Zhu et al., 2008; Soto, 2013). Lm G 1771 has been designated as a novel ABC transporter as it does not show any sequence similarity to other transporters of a signaling molecule (Alav et al., 2018).

Another study investigated the phenotypic, proteomic and genomic characterization by using a gene deletion mutation of the *Lm. G 1771* ABC transporter. Various cell surface proteins, cell anchor proteins and transcriptional regulators of biofilm formation were expressed distinctively in the *Lm. G 1771* deletion mutation as portrayed by DNA microarray and two-dimensional gel electrophoresis.

These findings could establish that efflux pumps might act as blockers of biofilm formation by changing the expression of biofilm forming genes (Zhu et al., 2008). Moreover, it was also found in *L. monocytogenes* that biofilm formation is regulated via exporting signaling molecules through the Lm. G 1771 transporter (Zhu et al., 2008; Zhu et al., 2011).

In *L .monocytogene,* two signalling systems are found; the Agr peptide and the AI-2 signals. Agr signals are not transported through the Lm. G1771 transporter (Le and Otto, 2015), whereas this transporter transports the AI-2 signal molecules (Herzberg et al., 2006). It has already been considered that AI-2 or its derivatives act as universal signals participating in interspecies communication (Basseler et al., 1997). The *luxS* gene encodes enzymes that are responsible for AI-2 biosynthesis that regulates biofilm formation (Zhu et al., 2008). Studies have shown that mature stage biofilm conformation increases antibiotic resistance. However, various mechanisms are imperative for the development of antibiotic resistance in *L. monocytogenes*. It is worth mentioning that efflux pumps in *L. monocytogenes* are associated with fluoroquinolone, macrolide and cefotaxime resistance (Olaimat et al., 2018).

5.6 CONCLUSION

Antimicrobial resistance among bacteria inside biofilm matrix has surged, contributing to the spread of chronic diseases. Efflux pumps have long been explored in the field of antibiotic resistance, but they are now known to have a significant role in biofilm development as well. The precise involvement of efflux pumps in various stages of biofilm development is unclear and needs to be researched further. It's questionable if efflux pumps are required for cell adhesion, biofilm development, or biofilm persistence. This varies greatly across various species and specific bacteria, including between diverse substrata. Biofilms provide these bacteria with the potential to tolerate traditional antibiotic therapy, leading to severe outbreaks. So, efflux pumps might be a great target for successful antimicrobial prophylaxis to combat lethal infections and also to reduce mortality, triggered by multidrug-resistant bacteria.

REFERENCES

Abhilash, K. P., Veeraraghavan, B., and Abraham, O. C. (2010). Epidemiology and outcome of bacteremia caused by extended spectrum beta-lactamase (ESBL)-producing Escherichia coli and Klebsiella spp. in a tertiary care teaching hospital in south India. *J Assoc Physicians India*, *58*(Suppl), 13–17.

Alav, I., Sutton, J. M., and Rahman, K. M. (2018). Role of bacterial efflux pumps in biofilm formation. Journal of Antimicrobial Chemotherapy, 73(8), 2003–2020.

Aleksandra, S., and Lucja, L. T. (2016). Biofilm formation by Pseudomonas aeruginosa and disinfectant susceptibility of planktonic and biofilm cells. *C J F Sci*. 34(3), 204–210.

Allerberger, F., and Wagner, M. (2010). Listeriosis: A resurgent foodborne infection. *Clinical Microbiology and Infection*, *16*(1), 16–23.

Bailey, A. M., Webber, M. A., and Piddock, L. J. (2006). Medium plays a role in determining expression of acrB, marA, and soxS in Escherichia coli. *Antimicrobial Agents and Chemotherapy*, *50*(3), 1071–1074.

Bassler, B. L., Greenberg, E. P., and Stevens, A. M. (1997). Cross-species induction of luminescence in the quorum-sensing bacterium Vibrio harveyi. *Journal of Bacteriology*, *179*(12), 4043–4045.

Baugh, S., Ekanayaka, A. S., Piddock, L. J., and Webber, M. A. (2012). Loss of or inhibition of all multidrug resistance efflux pumps of Salmonella enterica serovar typhimurium results in impaired ability to form a biofilm. *Journal of Antimicrobial Chemotherapy*, *67*(10), 2409–2417.

Bay, D. C., Stremick, C. A., Slipski, C. J., and Turner, R. J. (2017). Secondary multidrug efflux pump mutants alter Escherichia coli biofilm growth in the presence of cationic antimicrobial compounds. *Research in Microbiology*, *168*(3), 208–221.

Beenken, K. E., Dunman, P. M., McAleese, F., Macapagal, D., Murphy, E., Projan, S. J., Blevins, J. S., and Smeltzer, M. S. (2004). Global gene expression in Staphylococcus aureus biofilms. *Journal of Bacteriology*, *186*(14), 4665–4684.

Blair, J. M., Bavro, V. N., Ricci, V., Modi, N., Cacciotto, P., Kleinekathöfer, U., Ruggerone, P., Vargiu, A. V., Baylay, A. J., Smith, H. E., Brandon, Y., Galloway, D., and Piddock, L. J. 2015. AcrB drug-binding pocket substitution confers clinically relevant resistance and altered substrate specificity. *P N A Sci. 112*(11), 3511–3516.

Bohnert, J. A., Schuster, S., Fähnrich, E., Trittler, R., and Kern, W. V. (2007). Altered spectrum of multidrug resistance associated with a single point mutation in the Escherichia coli RND-type MDR efflux pump YhiV (MdtF). *Journal of antimicrobial Chemotherapy, 59*(6), 1216–1222.

Brauner, A., Fridman, O., Gefen, O., and Balaban, N. Q. (2016). Distinguishing between resistance, tolerance and persistence to antibiotic treatment. *N Rev. Micro. 14*(5), 320–330.

Bridier, A., Briandet, R., Thomas, V., and Dubois-Brissonnet, F. (2011). Resistance of bacterial biofilms to disinfectants: A review. *Biofouling, 27*(9), 1017–1032.

Buroni S, Pasca M. R., Flannagan R. S., et al. (2009) Assessment of three resistance nodulation–cell division drug efflux transporters of Burkholderia cenocepacia in intrinsic antibiotic resistance. *BMC Microbiol, 9*, 200.

Camejo, A., Buchrieser, C., Couvé, E., Carvalho, F., Reis, O., Ferreira, P., Sousa, S., Cossart, P., and Cabanes, D. (2009). In vivo transcriptional profiling of listeria monocytogenes and mutagenesis identify new virulence factors involved in infection. *PLoS pathogens, 5*(5), e1000449.

Chan, Y. Y., and Chua, K. L. (2005). The Burkholderia pseudomallei BpeAB-OprB efflux pump: Expression and impact on quorum sensing and virulence. *Journal of bacteriology, 187*(14), 4707–4719.

Chan, Y. Y., Bian, H. S., Tan, T. M. C., Mattmann, M. E., Geske, G. D., Igarashi, J., Hatano, T., Suga, H., Blackwell, H. E., and Chua, K. L. (2007). Control of quorum sensing by a Burkholderia pseudomallei multidrug efflux pump. *Journal of bacteriology, 189*(11), 4320–4324.

Chris, D. (2022). *Report on Antimicrobial Resistance far Deadlier than Thought.* Antimicrobial Stewardship. CIDRAP news.

Costa, S. S., Viveiros, M., Amaral, L., and Couto, I. (2013). Multidrug efflux pumps in Staphylococcus aureus: An update. *The Open Microbiology Journal, 7*, 59.

Coyne, S., Courvalin, P., and Périchon, B. (2011). Efflux-mediated antibiotic resistance in Acinetobacter spp. *Antimicrobial Agents and Chemotherapy, 55*(3), 947–953.

Csonka, L. N. (1989). Physiological and genetic responses of bacteria to osmotic stress. *Microbiological Reviews, 53*(1), 121–147.

Davies, D. G., Parsek, M. R., Pearson, J. P., Iglewski, B. H., Costerton, J. W. and Greenberg, E. P. (1998). The involvement of cell-to-cell signals in the development of a bacterial biofilm. *Science, 280*(5361), 295–298.

De Kievit, T. R., Parkins, M. D., Gillis, R. J., Srikumar, R., Ceri, H., Poole, K., Iglewski, B. H., and Storey, D. G. (2001). Multidrug efflux pumps: Expression patterns and contribution to antibiotic resistance in Pseudomonas aeruginosa biofilms. *Antimicrobial Agents and Chemotherapy, 45*(6), 1761–1770.

Ding, Y., Onodera, Y., Lee, J. C., and Hooper, D. C. (2008). NorB, an efflux pump in Staphylococcus aureus strain MW2, contributes to bacterial fitness in abscesses. *Journal of Bacteriology, 190*(21), 7123–7129.

Dreier, J., and Ruggerone, P. (2015). Interaction of antibacterial compounds with RND efflux pumps in Pseudomonas aeruginosa. *Frontiers in Microbiology, 6*, 660.

Fahmy, A., Srinivasan, A., and Webber, M. A. (2016). The relationship between bacterial multidrug efflux pumps and biofilm formation. In *Efflux-Mediated Antimicrobial Resistance in Bacteria* (pp. 651–663). Adis, Cham.

Federle, M. J., and Bassler, B. L. (2003). Interspecies communication in bacteria. *The Journal of Clinical Investigation, 112*(9), 1291–1299.

Fernandes, P., Ferreira, B. S., and Cabral, J. M. S. (2003). Solvent tolerance in bacteria: role of efflux pumps and cross-resistance with antibiotics. *International Journal of Antimicrobial Agents, 22*(3), 211–216.

Fong, J., Mortensen, K. T., Nørskov, A., Qvortrup, K., Yang, L., Tan, C. H., Nielsen, T. E., and Givskov, M. (2019). Itaconimides as novel quorum sensing inhibitors of Pseudomonas aeruginosa. *Frontiers in Cellular and Infection Microbiology, 8*, 443.

Food and Drug Administration, (2003). 'Bacteriologycal analytical manual online. Detection and enumeration of Listeria monocytogenes in foods', FDA, USA, Fecha de consulta: 27 de mayo de 2009.

Gardini, L., Vignolini, T., Capitanio, M., and Pavone, F. S. (2021). Quantitative superresolution imaging of Efflux pumps in Biofilm-Associated bacteria. *Biophysical Journal, 120*(3), 13a.

Gillet, Y., Issartel, B., Vanhems, P., Fournet, J. C., Lina, G., Bes, M., Vandenesch, F., Piemont, Y., Brousse, N., Floret, D., & Etienne, J. (2002). Association between Staphylococcus aureus strains carrying gene for Panton-valentine leukocidin and highly lethal necrotising pneumonia in young immunocompetent patients. *The Lancet*, *359*(9308), 753–759.

Gillis, R. J., White, K. G., Choi, K. H., Wagner, V. E., Schweizer, H. P., and Iglewski, B. H. (2005). Molecular basis of azithromycin-resistant Pseudomonas aeruginosa biofilms. *Antimicrobial Agents and Chemotherapy*, *49*(9), 3858–3867.

Hamon, M., Bierne, H., and Cossart, P. (2006). Listeria monocytogenes: A multifaceted model. *Nature Reviews Microbiology*, *4*(6), 423–434.

Handal, R., Qunibi, L., Sahouri, I., Juhari, M., Dawodi, R., Marzouqa, H., and Hindiyeh, M. (2017). Characterization of carbapenem-resistant acinetobacter baumannii strains isolated from hospitalized patients in palestine. *International Journal of Microbiology*, 2017.

He, X., and Ahn, J. (2011). Differential gene expression in planktonic and biofilm cells of multiple antibiotic-resistant Salmonella Typhimurium and Staphylococcus aureus. *FEMS Microbiology Letters*, *325*(2), 180–188.

He, X., Lu, F., Yuan, F., Jiang, D., Zhao, P., Zhu, J., Cheng, H., Cao, J., and Lu, G. (2015). Biofilm formation caused by clinical acinetobacter baumannii isolates is associated with overexpression of the AdeFGH efflux pump. *Antimicrobial Agents and Chemotherapy*, *59*(8), 4817–4825.

Herzberg, M., Kaye, I. K., Peti, W., and Wood, T. K. (2006). YdgG (TqsA) controls biofilm formation in Escherichia coli K-12 through autoinducer 2 transport. *Journal of Bacteriology*, *188*(2), 587–598.

Ho, A. J., Lappi, V. R., and Wiedmann, M. (2007). Longitudinal monitoring of listeria monocytogenes contamination patterns in a farmstead dairy processing facility. *Journal of Dairy Science*, *90*(5), 2517–2524.

Høiby, N., Bjarnsholt, T., Givskov, M., Molin, S., and Ciofu, O. (2010). Antibiotic resistance of bacterial biofilms. *International Journal of Antimicrobial Agents*, *35*(4), 322–332.

Huang, J., O'Toole, P. W., Shen, W., Amrine-Madsen, H., Jiang, X., Lobo, N., Palmer, L. M., Voelker, L., Fan, F., Gwynn, M. N., and McDevitt, D. (2004). Novel chromosomally encoded multidrug efflux transporter MdeA in Staphylococcus aureus. *Antimicrobial Agents and Chemotherapy*, *48*(3), 909–917.

Imuta, N., Nishi, J., Tokuda, K., Fujiyama, R., Manago, K., Iwashita, M., Sarantuya, J., and Kawano, Y. (2008). The Escherichia coli efflux pump TolC promotes aggregation of enteroaggregative E. coli 042. *Infection and Immunity*, *76*(3), 1247–1256.

Ito, A., Taniuchi, A., May, T., Kawata, K., and Okabe, S. (2009). Increased antibiotic resistance of Escherichia coli in mature biofilms. *Applied and Environmental Microbiology*, *75*(12), 4093–4100.

Jack, D. L., Yang, N. M., and H. Saier Jr, M. (2001). The drug/metabolite transporter superfamily. *European Journal of Biochemistry*, *268*(13), 3620–3639.

Karmakar, A., Jana, D., Dutta, K., Dua, P., and Ghosh, C. (2018). Prevalence of panton-valentine leukocidin gene among community acquired Staphylococcus aureus: A real-time PCR study. *Journal of Pathogens*, *2018*.

Kazi, M., Drego, L., Nikam, C., Ajbani, K., Soman, R., Shetty, A., and Rodrigues, C. (2015). Molecular characterization of carbapenem-resistant Enterobacteriaceae at a tertiary care laboratory in Mumbai. *European Journal of Clinical Microbiology & Infectious Diseases*, *34*(3), 467–472.

Koronakis, V. (2003). TolC–the bacterial exit duct for proteins and drugs. *FEBS letters*, *555*(1), 66–71.

Kuroda, T., and Tsuchiya, T. (2009). Multidrug efflux transporters in the MATE family. *Biochimica et Biophysica Acta (BBA)-Proteins and Proteomics*, *1794*(5), 763–768.

Larson, E. (2007). Community factors in the development of antibiotic resistance. *Annu. Rev. Public Health*, *28*, 435–447.

Le, K. Y., and Otto, M. (2015). Quorum-sensing regulation in staphylococci—an overview. *Frontiers in Microbiology*, *6*, 1174.

Li, X. Z., and Nikaido, H. (2004). Efflux-mediated drug resistance in bacteria. *Drugs*, *64*(2), 159–204.

Li, X. Z., Plésiat, P., and Nikaido, H. (2015). The challenge of efflux-mediated antibiotic resistance in gram-negative bacteria. *Clinical Microbiology Reviews*, *28*(2), 337–418.

Li, Y. H., and Tian, X. (2012). Quorum sensing and bacteriaLII social interactions in biofilms. *Sensors*, *12*(3), 2519–2538.

Lubelski, J., Konings, W. N., and Driessen, A. J. (2007). Distribution and physiology of ABC-type transporters contributing to multidrug resistance in bacteria. *Microbiology and Molecular Biology Reviews*, *71*(3), 463–476.

Lyczak, J. B., Cannon, C. L. and Pier, G. B. (2000). Establishment of Pseudomonas aeruginosa infection: Lessons from a versatile opportunist. *Microbes and infection*, *2*(9), 1051–1060.

Lynch, S. V., Dixon, L., Benoit, M. R., Brodie, E. L., Keyhan, M., Hu, P., and Matin, A. (2007). Role of the rapA gene in controlling antibiotic resistance of Escherichia coli biofilms. *Antimicrobial Agents and Cemotherapy*, *51*(10), 3650–3658.

M Campos, J. C. D., Antunes, L. C., & Ferreira, R. B. (2020). Global priority pathogens: Virulence, antimicrobial resistance and prospective treatment options. *Future microbiology*, *15*(8), 649–677.

Magiorakos, A. P., Srinivasan, A., Carey, R. B., Carmeli, Y., Falagas, M. E., Giske, C. G., Harbarth, S., Hindler, J. F., Kahlmeter, G., Olsson-Liljequist, B., Paterson, D. L., Rice, L. B., Stelling, J., Struelens, M. J., Vatopoulos, A., Weber, J. T., and Monnet, D. L. (2012). Multidrug-resistant, extensively drug-resistant and pandrug-resistant bacteria: An international expert proposal for interim standard definitions for acquired resistance. *Clinical Microbiology and Infection*, *18*(3), 268–281.

Maira-Litrán, T., Allison, D. G., and Gilbert, P. (2000). Expression of the multiple antibiotic resistance operon (mar) during growth of Escherichia coli as a biofilm. *Journal of Applied Microbiology*, *88*(2), 243–247.

Marquez, B. (2005). Bacterial efflux systems and efflux pumps inhibitors. *Biochimie*, *87*(12), 1137–1147.

Murray, C. J., Ikuta, K. S., Sharara, F., Swetschinski, L., Aguilar, G. R., Gray, A., Murray, C. J., Ikuta, K. S., Sharara, F., Swetschinski, L., Aguilar, G. R., Gray, A., Han, C., Bisignano, C., Rao, P., Wool, E., Johnson, S. C., Browne, A. J., Chipeta, M. G., Fell, F., Hackett, S., Haines-Woodhouse, G., Hamadani, B. H. K., Kumaran, E. A. P., McManigal, B., Agarwal, R., Akech, S., Albertson, S., Amuasi, J., Andrews, J., Aravkin, A., Ashley, E., Bailey, F., Baker, S., Basnyat, B., Bekker, A., Bender, R., Bethou, A., Bielicki, J., Boonkasidecha, S., Bukosia, J., Carvalheiro, C., Castañeda-Orjuela, C., Chansamouth, V., Chaurasia, S., Chiurchiù, S., Chowdhury, F., Cook, A. J., Cooper, B., Cressey, T. R., Criollo-Mora, E., Cunningham, M., Darboe, S., Day, N. P. J., Luca, M., De Dokova, K., Dramowski, A., Dunachie, S. J., Eckmanns, T., Eibach, D., Emami, A., Feasey, N., Fisher-Pearson, N., Forrest, K., Garrett, D., Gastmeier, P., Giref, A. Z., Greer, R. C., Gupta, V., Haller, S., Haselbeck, A., Hay, S. I., Holm, M., Hopkins, S., Iregbu, K. C., Jacobs, J., Jarovsky, D., Javanmardi, F., Khorana, M., Kissoon, N., Kobeissi, E., Kostyanev, T., Krapp, F., Krumkamp, R., Kumar, A., Kyu, H. H., Lim, C., Limmathurotsakul, D., Loftus, M. J., Lunn, M., Ma, J., Mturi, N., Munera-Huertas, T., Musicha, P., Mussi-Pinhata, M. M., Nakamura, T., Nanavati, R., Nangia, S., Newton, P., Ngoun, C., Novotney, A., Nwakanma, D., Obiero, C. W., Olivas-Martinez, A., Olliaro, P., Ooko, E., Ortiz-Brizuela, E., Peleg, A. Y., Perrone, C., Plakkal, N., Ponce-de-Leon, A., Raad, M., Ramdin, T., Riddell, A., Roberts, T., Robotham, J. V., Roca, A., Rudd, K. E., Russell, N., Schnall, J., Scott, J. A. G., Shivamallappa, M., Sifuentes-Osornio, J., Steenkeste, N., Stewardson, A. J., Stoeva, T., Tasak, N., Thaiprakong, A., Thwaites, G., Turner, C., Turner, P., Doorn, H. R., van Velaphi, S., Vongpradith, A., Vu, H., Walsh, T., Waner, S., Wangrangsimakul, T., Wozniak, T., Zheng, P., Sartorius, B., Lopez, A. D., Stergachis, A., Moore, C., Dolecek, C., Naghavi, M. (2022). Global burden of bacterial antimicrobial resistance in 2019: A systematic analysis. *Lancet*.

Nadeem, S. F., Gohar, U. F., Tahir, S. F., Mukhtar, H., Pornpukdeewattana, S., Nukthamna, P., Moula Ali, A. M., Bavisetty, C. B., and Massa, S. (2020). Antimicrobial resistance: More than 70 years of war between humans and bacteria. *Critical Reviews in Microbiology*, *46*(5), 578–599.

Nait Chabane, Y., Mlouka, M. B., Alexandre, S., Nicol, M., Marti, S., Pestel-Caron, M., Vila, J., Jouenne, T., and Dé, E. (2014). Virstatin inhibits biofilm formation and motility of Acinetobacter baumannii. *BMC microbiology*, *14*(1), 1–7.

Nikaido, H., and Takatsuka, Y. (2009). Mechanisms of RND multidrug efflux pumps. *Biochimica et Biophysica Acta (BBA)-Proteins and Proteomics*, *1794*(5), 769–781.

O'Neil, J. (2016). *Tackling Drug-Resistant Infections Globally: Final Report and Recommendations. Review on Antimicrobial Resistance*. Welcome Trust and UK Government.

Olaimat, A. N., Al-Holy, M. A., Shahbaz, H. M., Al-Nabulsi, A. A., Abu Ghoush, M. H., Osaili, T. M., et al. (2018). Emergence of antibiotic resistance in listeria monocytogenes isolated from food products: A comprehensive review. *Comprehensive Reviews in Food Science and Food Safety*, *17*(5), 1277–1292.

Pamp, S. J., Gjermansen, M., Johansen, H. K., and Tolker-Nielsen, T. (2008). Tolerance to the antimicrobial peptide colistin in Pseudomonas aeruginosa biofilms is linked to metabolically active cells, and depends on the pmr and mexAB-oprM genes. *Molecular microbiology*, *68*(1), 223–240.

Pao, S. S., Paulsen, I. T., and Saier Jr, M. H. (1998). Major facilitator superfamily. *Microbiology and Molecular Biology Reviews*, *62*(1), 1–34.

Pearson, J. P., Van Delden, C., and Iglewski, B. H. (1999). Active efflux and diffusion are involved in transport of Pseudomonas aeruginosa cell-to-cell signals. *Journal of Bacteriology*, *181*(4), 1203–1210.

Piddock, L. J. (2006). Multidrug-resistance efflux pumps? not just for resistance. *Nature Reviews Microbiology*, *4*(8), 629–636.

Plata, K., Rosato, A. E., and Wegrzyn, G. (2009). Staphylococcus aureus as an infectious agent: Overview of biochemistry and molecular genetics of its pathogenicity. *Acta Biochimica Polonica*, *56*(4).

Poole, K. (2005). Efflux-mediated antimicrobial resistance. *Journal of Antimicrobial Chemotherapy*, *56*(1), 20–51.

Poole, K. (2007). Efflux pumps as antimicrobial resistance mechanisms. *Annals of medicine*, *39*(3), 162–176.

Resch, A., Rosenstein, R., Nerz, C., and Götz, F. (2005). Differential gene expression profiling of Staphylococcus aureus cultivated under biofilm and planktonic conditions. *Applied and Environmental Microbiology*, *71*(5), 2663–2676.

Rode, T. M., Møretrø, T., Langsrud, S., Langsrud, Ø., Vogt, G., and Holck, A. (2010). Responses of Staphylococcus aureus exposed to HCl and organic acid stress. *Canadian journal of microbiology*, *56*(9), 777–792.

Rumbo-Feal, S., Gomez, M. J., Gayoso, C., Álvarez-Fraga, L., Cabral, M. P., Aransay, A. M., et al. (2013). Whole transcriptome analysis of acinetobacter baumannii assessed by RNA-sequencing reveals different mRNA expression profiles in biofilm compared to planktonic cells. *PloS one*, *8*(8), e72968.

Rutherford, S. T., and Bassler, B. L. (2012). Bacterial quorum sensing: Its role in virulence and possibilities for its control. *Cold Spring Harbor Perspectives in Medicine*, *2*(11), a012427.

Sakhtah, H., Koyama, L., Zhang, Y., Morales, D. K., Fields, B. L., Price-Whelan, A., Hogan, D. A., Shepard, K., and Dietrich, L. E. (2016). The Pseudomonas aeruginosa efflux pump MexGHI-OpmD transports a natural phenazine that controls gene expression and biofilm development. *Proceedings of the National Academy of Sciences*, *113*(25), E3538–E3547.

Schembri, M. A., Kjærgaard, K., and Klemm, P. (2003). Global gene expression in Escherichia coli biofilms. *Molecular Microbiology*, *48*(1), 253–267.

Seeger, M. A., Diederichs, K., Eicher, T., Brandstatter, L., Schiefner, A., Verrey, F., and Pos, K. M. (2008). The AcrB efflux pump: Conformational cycling and peristalsis lead to multidrug resistance. *Current Drug Targets*, *9*(9), 729–749.

Shriram, V., Khare, T., Bhagwat, R., Shukla, R., and Kumar, V. (2018). Inhibiting bacterial drug efflux pumps via phyto-therapeutics to combat threatening antimicrobial resistance. *Frontiers in Microbiology*, *2990*.

Singh, P. K., Schaefer, A. L., Parsek, M. R., Moninger, T. O., Welsh, M. J. and Greenberg, E. P. (2000). Quorum-sensing signals indicate that cystic fibrosis lungs are infected with bacterial biofilms. *Nature*, *407*(6805), 762–764.

Soto, S. M. (2013). Role of efflux pumps in the antibiotic resistance of bacteria embedded in a biofilm. *Virulence*, *4*(3), 223–229.

Spengler, G., Kincses, A., Gajdács, M., and Amaral, L. (2017). New roads leading to old destinations: Efflux pumps as targets to reverse multidrug resistance in bacteria. *Molecules*, *22*(3), 468.

Tam, R., and Saier Jr, M. H. (1993). Structural, functional, and evolutionary relationships among extracellular solute-binding receptors of bacteria. *Microbiological Reviews*, *57*(2), 320–346.

Tenaillon, O., Skurnik, D., Picard, B., and Denamur, E. (2010). The population genetics of commensal Escherichia coli. *Nature Reviews Microbiology*, *8*(3), 207–217.

Tenover, F. C. (2006). Mechanisms of antimicrobial resistance in bacteria. The *American Journal of Medicine*, *119*(6), S3–S10.

Trotonda, M. P., Tamber, S., Memmi, G., & Cheung, A. L. (2008). MgrA represses biofilm formation in Staphylococcus aureus. *Infection and immunity*, *76*(12), 5645–5654.

Truong-Bolduc, Q. C., Strahilevitz, J., & Hooper, D. C. (2006). NorC, a new efflux pump regulated by MgrA of Staphylococcus aureus. Antimicrobial *Agents and Chemotherapy*, *50*(3), 1104–1107.

Tu Quoc, P. H., Genevaux, P., Pajunen, M., Savilahti, H., Georgopoulos, C., Schrenzel, J., & Kelley, W. L. (2007). Isolation and characterization of biofilm formation-defective mutants of Staphylococcus aureus. *Infection and Immunity*, *75*(3), 1079–1088.

Van Bambeke, F., Balzi, E., and Tulkens, P. M. (2000). Antibiotic efflux pumps. *Biochemical Pharmacology*, *60*(4), 457–470.

Vincent, J. L. (2003). Nosocomial infections in adult intensive-care units. *The lancet*, *361*(9374), 2068–2077.

Webber, M. A., and Piddock, L. J. (2001). Absence of mutations in marRAB or soxRS in acrB-overexpressing fluoroquinolone-resistant clinical and veterinary isolates of Escherichia coli. *Antimicrobial Agents and Chemotherapy*, *45*(5), 1550–1552.

White, D. G., Maneewannakul, K., Von Hofe, E., Zillman, M., Eisenberg, W., Field, A. K., and Levy, S. B. (1997). Inhibition of the multiple antibiotic resistance (mar) operon in Escherichia coli by antisense DNA analogs. *Antimicrobial Agents and Chemotherapy*, *41*(12), 2699–2704.

Withers, H., Swift, S. and Williams, P. (2001). Quorum sensing as an integral component of gene regulatory networks in gram-negative bacteria. *Current Opinion in Microbiology*, *4*(2), 186–193.

World Health Organization. (2022). "Antimicrobial resistance surveillance in Europe 2022–2020 data". ISBN 978-92-9498-552-1.

Xavier, K. B., and Bassler, B. L. (2005). Regulation of uptake and processing of the quorum-sensing autoinducer AI-2 in Escherichia coli. *Journal of bacteriology*, *187*(1), 238–248.

Xu, C. F., Bilya, S. R., and Xu, W. (2019). AdeABC efflux gene in Acinetobacter baumannii. *New Microbes and New Infections*, *30*, 100549.

Yan, N. (2015). Structural biology of the major facilitator superfamily transporters. *Annual review of biophysics*, *44*, 257–283.

Yoon, E. J., Nait Chabane, Y., Goussard, S., Snesrud, E., Courvalin, P., Dé, E., and Grillot-Courvalin, C. (2015). Contribution of resistance-nodulation-cell division efflux systems to antibiotic resistance and biofilm formation in acinetobacter baumannii. *MBio*, *6*(2), e00309–15.

Zhang, L., and Mah, T. F. (2008) Involvement of a novel efflux system in biofilm-specific resistance to antibiotics. *J. Bacteriol. 190*, 4447–4452.

Zhang, Y., Xiao, M., Horiyama, T., Zhang, Y., Li, X., Nishino, K., and Yan, A. (2011). The multidrug efflux pump MdtEF protects against nitrosative damage during the anaerobic respiration in Escherichia coli. *Journal of Biological Chemistry*, *286*(30), 26576–26584.

Zhu, X., Liu, W., Lametsch, R., Aarestrup, F., Shi, C., She, Q., Shi, X., and Knøchel, S. (2011). Phenotypic, proteomic, and genomic characterization of a putative ABC-transporter permease involved in listeria monocytogenes biofilm formation. *Foodborne Pathogens and Disease*, *8*(4), 495–501.

Zhu, X., Long, F., Chen, Y., Knøchel, S., She, Q., and Shi, X. (2008). A putative ABC transporter is involved in negative regulation of biofilm formation by listeria monocytogenes. *Applied and Environmental Microbiology*, *74*(24), 7675–7683.

Zhu, Y., Weiss, E. C., Otto, M., Fey, P. D., Smeltzer, M. S., and Somerville, G. A. (2007). Staphylococcus aureus biofilm metabolism and the influence of arginine on polysaccharide intercellular adhesin synthesis, biofilm formation, and pathogenesis. *Infection and Immunity*, *75*(9), 4219–4226.

6 Genetic Diversity Resulting in the Development of AMR

Sreejita Ghosh, Bandita Dutta, Rina Rani Ray, Moupriya Nag, and Dibyajit Lahiri

6.1 INTRODUCTION

Presently, AMR in different bacteria has evolved as a serious threat against the treatment strategies available for infections caused by bacteria and has become a subject of interest for immediate research. AMR is not just a major complication but also one of the top ten global public health concerns dominating the entire world. Common antimicrobials include antivirals, antibiotics, antiparasitics and antifungals. The present condition of the adverse impacts of AMR is much more challenging than can be imagined. AMR has also become a major threat to the immunocompromised patients. In May, 2015, the World Health Organization (WHO) formulated a Global Action Plan on AMR (GAP) having five objectives for fighting AMR development. Fascinatingly, in November, 2020, World Antimicrobial Awareness week was observed along with a slogan saying "Antimicrobials Handle with Care" for overcoming AMR development in bacteria. a Few bacteria that exist naturally within the environment are able to acquire AMR determinants maybe because of genetic evolution, co–existence and/or a horizontal gene transfer (HGT) process (Das et al., 2020). Hence, all these factors and the increased use of antimicrobials have led to AMR development in microbial isolates. Threats imposed by both Gram–negative and Gram–positive bacterial species encouraged researchers to immediately explore new antimicrobials and novel therapeutic strategies for handling the problem of AMR development.

In addition to becoming a threat for the healthcare sectors, AMR is also leading to economic burdens in developing as well as developed countries. Different therapies such as chemotherapy, organ transplantation and surgery are prone to risks without proper antimicrobials. Antimicrobials are not just required for the treatment of minor injuries and infections but, they are also used for a prolonged period of treatment, which exposes the pathogens to those antimicrobials for a longer time period and consequently they will become resistant to those antimicrobials, leading to life threatening risks to the patients.

When antimicrobials were first discovered, it was presumed that AMR development was not very likely to happen. This was because of the assumption that genetic mutation frequency making bacteria resistant was unlikely. However, the opposite results appeared. It was not anticipated that microbes would ultimately react to the chemical actions of antimicrobials by becoming adapted to the altered environment AMR development. Furthermore, their capacity for gene interchanging or HGT was never expected to produce AMR. Then it was discovered that AMR had eemerged in *Escherichica coli* against the first β–lactamase antibiotic penicillin. Other antibiotics of classes aminoglycosides and macrolides also faced these resistance challenges. Some examples of bacteria developing AMR include methicillin–resistant *Staphylococcus aureus* (MRSA), vancomycin–resistant *Enterococci* (VRE) (Jevons et al., 1963).

DOI: 10.1201/9781003281238-6

Bacteria develop AMR via several mechanisms as follows (Wright, 2005):

1. Bacterial cells actively efflux the antimicrobials out of the respective cells.
2. Changes in permeability of the cell wall of bacteria prevent the entry of antimicrobials into the cells.
3. Antimicrobial degradation.
4. Antimicrobial modification through bacterial enzymes.
5. Antimicrobial target modification.
6. Target enzymes are overproduced.

Such AMR phenotypes in bacteria may develop due to mutations of chromosomal DNA that can modify the already present proteins of bacterial origin via transformation thereby creating mosaic proteins and also due to the transfer or the acquiring of new genetic substances amongst bacteria of the same or different genera or species (Ochman et al., 2000).

There are several examples of AMR developed due to mutation. For instance, resistance against macrolides may arise from substitution of nucleotide bases in the 23s rRNA gene. On the other hand, identical phenotypes may also arise from ribosomal protein (L22 and L4) mutations (Vester and Douthwaite, 2001). Single nucleotide polymorphisms (SNPs) may also lead to resistance against antimicrobials such as sulfonilamides, quinolones and trimethoprime and *rpsL*gene (encoding the ribosomal protein S12) mutation may cause high–level streptomycin resistance (Ruiz, 2003). Chromosomal frameshift mutation in the *ddl*gene, which encodes a cytoplasmic enzyme D–Ala–D–Alaligase may lead to resistance against glycopeptides (Casadewall and Courvalin, 1999).

6.2 ACQUIRED AMR

The genetic elements or genes giving rise to acquired resistance may be extrachromosomal or chromosomal. AMR occurs through diverse mechanisms associated with virulence factors, proteins present on the outer membrane, particular enzymatic factors present in cellular envelopes, formation of biofilm, QS and the secretory systems of proteins (Ndagi et al., 2020). Studies have demonstrated that bacteria are susceptible to genetic recombination and thereby exchanging different virulence factors and AMR–associated gene cassettes. Comparative transcriptomic and genetic studies of AMR strains revealed their pattern of gene expression and phylogenetic arrangement, which help researchers to identify the resistant determinants in various enteric pathogenic microbes and environmental reservoirs of genetic elements. HGT allows transmission of virulence and resistance factors between bacterial species and genera. Traits such as biofilms and mobile genetic elements aid in bacterial survival thereby overcoming the antimicrobial action of the therapeutics.

The genome island contains different clusters of genes, which give rise to virulence and other roles in enteric bacteria. The omnipresent mechanisms of DNA material exchange and the high frequency of spontaneous mutations in bacteria are the main controllers of AMR. Biofilm is one of the usual characteristics of different strains of bacteria that help them to survive in the presence of rigid conditions. The biofilms prevent antimicrobial penetration in the cells of bacteria thereby protecting them from antimicrobial activity and giving rise to the AMR mechanism (Verderosa et al., 2019). Wastewater and sewage treatment plants, effluents from hospitals, agriculture, aquaculture and waste from slaughterhouses are the main hotspots for genetic material exchange. Some external factors also lead to increased AMR levels such as improper methods of sanitization, unhygienic settings in hospitals and improper therapeutic and diagnostic standards (Laxminarayan et al., 2020).

Hospital–acquired community infections has remained as one of the major heath issues until now. The resistant gene reservoirs and hospital settings of the hospitals allow the emergence as well as transmission of new types of AMR. Some bacteria possess various mechanisms of resistance (acquired or intrinsic) giving rise to the resistant strains (Nikaido, 2009). Recombination or HGT allow the genetic transfer from resistant strains to other susceptible strains and genetic elements such as integrons and transposons and are responsible for providing AMR. Hence, acquired resistance is one of the most important traits that are further subdivided in the following sections.

6.2.1 Antimicrobial Efflux

Resistance mechanisms of different pathogens mainly include permeability restrictors for rapid efflux of the antimicrobial molecules (Ogawara, 2019). On analysing this mechanism, suitable targets for the design of novel drugs can be obtained. The synergistic activity of decreased uptake of drug and efflux is involved in multiple activities of the permeability barrier present on the outer membrane leading to the acquisition of increased levels of intrinsic or acquired resistance in numerous bacteria that are clinically important (Table 6.1). Trials ongoing are targeted towards studying the function, physical structure and efflux system regulation, and their functions will allow the exploitation of these efflux pumps as new targets for the novel drugs.

6.2.2 Degradation or Modification of the Antimicrobials

Degradation or modification of antimicrobials is one of the most frequent and significant mechanisms of acquired AMR. Bacteria contain gene clusters or genes that encode several antimicrobial–modifying enzymes. These enzymes work against different classes of antimicrobials like β–lactams, chloramphenicol and aminoglycosides (Table 6.1). A few of such enzymes are found in pathogenic microbes while others are found in non–pathogenic as well as pathogenic microbes (Baker et al., 2018).

Enzymes physically alter antimicrobials thereby actively decreasing antimicrobial concentration within the surrounding environment. The biofilm formed by bacteria has a matrix that contains enzymes, mainly β–lactamases, which has the capacity for destroying antimicrobials, thereby preventing them from acting upon the targeted cells. It has been reported that biofilms formed by *Klebsiella pneumoniae* secrete β–lactamases that can degrade ampicillin thereby preventing antimicrobials form destroying the sessile cells within the biofilm (Anderl et al., 2000). Various studies demonstrated that excessive secretion of chromosomally coded AmpC cephalosporinase has been held responsible for developing resistance in *P. aeruginosa* isolates inside cyctic fibrosis (CF) against β–lactam antibiotics. Studies indicated that unstable green fluorescent protein fusion with AmpC in imipenem and ceftazimide absence induced AmpC expression in *P. aeruginosa* (Bagge et al., 2004).

6.2.3 Reduced Membrane Permeability

Mutations in porins among the resistant strains lead to modifications in the permeability of the membranes thereby reducing uptake of drugs inside the cells. For instance, in strains of *P. aeruginosa*, a particular porin, called OprD causes carbapenem resistant gene mutation. Above all, decrease in permeability of the external membrane can play a significant role in providing resistance against aminoglycosides and quinolones (Pang et al., 2019). Porin alteration might be achieved by three general procedures: modification in porin expression levels, shift in types of porin expression and impaired porin functionality. Significantly, permeability changes via any of these mechanisms often giving rise to a low–level of resistance and are usually associated with other resistance mechanisms such enhanced efflux pump activations.

TABLE 6.1
Classes of Antibiotics and the Bacteria with Their Potential Resistant Mechanisms

Classes of antibiotics	Sub classes of antibiotics	Resistance mechanisms	Resistant bacteria	References
Aminoglycosides	Streptomycin, gentamicin, amikacin, spectinomycin, tobramycin, etc.	Enzymatic antimicrobial inactivation through phosphorylation (via phorylase enzyme), acetylation (via acetylase enzyme), efflux, nucleotidylation, modified target ribosomal binding site (RBS), reduced drug uptake through decreased permeability, other modifying enzymes like adenyltransferase, acetyl transferase and phosphotransferase	*Staphylococci*, Entercocci, Bacteriodes, *Streptococci*, *Vibrio cholera*, *P. aeruginosa*, *K. pneumoniae*, *E. coli*, *Acinetobacterbaumanii*, etc.	Das et al., 2020
β–lactams	Carbapenems, cephalosporin, etc.	Efflux, hydrolysis, decreased permeability, modified target, antibiotic inactivation through β–lactamases and β–lactamases hydrolyzingcarbapenems	*P. aeruginosa*, *S. aureus*, *Streptococcus penumoniae*, Enteric bacteria, *E. coli*, *V. cholera*, *A. baumanii*, *K. pneumoniae*	Peterson and Kaur, 2018
Tetracyclines	Minocycline, tigecycline, doxycycline	ABC efflux pumps, monooxygenation, enzyme inactivating tetracycline, modification of the ribosomes	*Streptococcii*, *Staphylococcii*, Enterobacteriaceae, Enterococci, *Listeria*, *Haemophilus*, *A. baumanii*, etc.	Sultan et al., 2018
Glycopeptides	Teicoplanin, vancomycin	Target modification	Lactobacilli, Enterococci, *Enetrococcusfaecium*, *Staphylococcus haemolyticus*, *Enterococcus faecalis*, etc.	Kapoor et al., 2017
Phenicols	Azidamphenicol, chloramphenicol, thiamphenicol	Cholramphenicol acetylation by acetyl transferase, target site modification, efflux pumps	*S. penumoniae*, *B. subtilis*, *Haemophilus influenza*, Enterobacteriaceae, *E. coli*, *V. cholera*, etc.	Raygaert, 2018
Macrolides	Azithromycin, erythromycin	Efflux, glycosylation, rRNA target methylation	Enterococci, Streptococci, *A. baumanii*, Staphylococci, etc.	Zaman et al., 2017
Rifamycins	Rifampin	Efflux, ribosylation of ADP, DNA–dependent modification of target RNA	Staphylococci, enteric bacteria, *Mycobacterium tuberculosis*, Streptococci, *E. coli*, *V. cholera*, *A. baumanii*, *K. penumoniae*, etc.	Das et al., 2020
Folic acid inhibitors	Sulfamethoxazole, trimethoprim	Target modification, efflux	Streptococci, Staphylococci, *Neisseria*, Enterobacteriaceae, *A. baumanii*, etc.	Peterson and Kaur, 2018
Cationic peptides	Polymyxin–B, colistin	Efflux, target modification	*Salmonella typhimurium*, *E. coli*, *A. baumanii*, etc.	Sultan et al., 2018
Quinolones	Ciprofloxacin, nalidixic acid, ofloxacin, levofloxacin, norfloxacin	Modified subunit A (parC) efflux pump of DNA topoisomerase IV or DNA gyrase, decreased permeability	*P. aeruginosa*, *S. aureus*, *S. penumoniae*, *S. epidermis*, *A. baumanii*, etc.	Kapoor et al., 2017

6.2.4 Using an Alternate Metabolic Path

Bacteria can also acquire AMR by use of alternate metabolic pathways. An identical mechanism resistance was observed in *S. aureus* small colony variants (SCVs), which involves activation of an alternate transcriptional program resulting in elevated levels of production of ATP while preserving AMR in SCVs. The rate of growth of the AMR SCVs also increased. Likewise, bacteria can acquire the characteristic of using folate available in the surrounding environment rather biosynthesizing folate (Fernandez–Villa et al., 2019).

6.3 INTRINSIC/NATURAL AMR

Natural resistance can be described as the capacity of a bacterium to resist the action of an antimicrobial because of its inbuilt characteristics. The best example of a natural resistance mechanism is the lack of a vulnerable target site for a particular antimicrobial in *Pseudomonas* (Zhu et al., 2010). Triclosan is a wide spectrum biocide particularly acting against Gram–positive bacteria and numerous Gram–negative bacteria however it cannot prevent *Pseudomonas* growth. At the initial stages, the resistance was thought to be because of active efflux, but in the later stages, the presence of an intensive allele (*fabI)* encoding an additional enoyl–acyl carrier protein called, the reductase enzyme, which is targeted by triclosan was found to be responsible for AMR. Likewise, a lipopeptide, called daptomycin, which was approved in 2003 for clinical trials, was an active antimicrobial against Gram–positive bacteria. Nonetheless, it is not active against Gram–negative bacteria. The membrane of the cytoplasm in Gram–negative bacteria is composed of a very small quantity of anionic phospholipids in comparison to Gram–positive bacteria (Saha and Sarkar, 2021). Hence, the efficiency of calcium ion mediated daptomycin penetration inside the membrane of cytoplasm is reduced (Randall et al., 2013). In addition to this, some antimicrobial compounds, which cannot penetrate the external membrane, are considered as a mechanism of natural resistance. In case of Gram–positive bacteria, vancomycin hinders peptidoglycan cross–linking via targeting D–Ala–D–Ala peptides, while in Gram–negative bacteria, vancomycin is unable to cross the external membrane. β–lactamse (carbapenemase, extended spectrum β–lactamses (ESBLs) and metallo–β–lactamses) biosynthesis has been reported to enhance the frequency of acquired AMR infections particularly leading to resistance against third generation antibiotics like carbapenem and cephalosporin (Lebeaux et al., 2014). Remarkably, different genes related to natural resistance against various classes of antibiotics like aminoglycosides, lactams and fluoroquinolones in *E. coli*, *S. aureus* and *P. aeruginosa* have been recognized using high–throughput techniques of high–density genome mutant libraries. Genes, which are responsible for AMR in *Salmonella* and *E. coli* are *bla- TEM-1*, *AmpC*, *VIM- 1*, *bla- CTX- M- 15*, *flo- R*, *NDM- 1*, *mcr- 1* and *tet- G* together encode resistance against colistin (Darwich et al., 2019).

For example, inactivation of the redundant genes in *E. coli* allowed putative target identification such as SapC, FabI, TrxA (thioredoxin), RecQ, TrxB (thioredoxinreductase), D- Ala- D-Alacarboxypeptidase and DacA for the phenotyping of susceptibility. Inhibition of all these genes can noticeably stimulate activities of antimicrobials like rifampicin, triclosan, nitrofurantoin, aminoglycosides and β-lactams. Bacteria can acquire AMR through diverse mechanisms such as improper antimicrobial penetration or antimicrobial efflux leading to reduced intracellular antimicrobial concentration, alteration of the antimicrobial target site by posttranslational modification or gene mutation of the target sites and antimicrobial inactivation through hydrolysis or modification (Alav et al., 2018). The plasmid, which encodes colistin resistant genes (dependent on *mcr- 1)* in *E. coli*, was first identified in animals, raw meat and people of China (Muktan et al., 2020). The *E. coli* harbouring *mcr- 1* gene was first reported in clinical isolates from humans in Pakistan and in migratory birds and AMR (quinolones and ESBLs) in *E. coli*, was first isolated from wild birds, poultry farms, dogs, pigeons and humans (Anyanwu et al., 2020). For example, *Neisseria gonorrhoea* acquired resistance against penicillins, sulphonilamides, macrolides, tetracyclines, early generation cephalosporins and fluoroquinolones (Sanchez–Buso and Harris, 2019). Besides rifampicin, resistant strains of *M. tuberculosis* also developed multi–drug resistance (MDR) posing a major health issue in treating patients suffering from tuberculosis (Prasad et al., 2018).

6.4 THE GENETIC BASIS OF AMR DEVELOPMENT

6.4.1 Mobile Genetic Elements Giving Rise to AMR

Acquired genes of AMR are often present inside mobile DNA material and can be roughly defined as any part or DNA segment, which can translocate from one portion of the genome to another portion or to some other genomes. This definition involves a diverse range of different mobile genetic elements. The main participants in HGT are the mobilizable and conjugative elements. The conjugative element is composed of all the genetic information needed to be transferred from one bacterium to the other whereas the mobilizable element utilizes the conjugative roles of the co–existing conjugative elements (conjugative transposons or plasmids) to aid in getting transferred to other hosts. Bacteriophages also take part in spreading DNA from one bacterium to the other through the process of transduction, in which the DNA of the bacteria and not the DNA of the phage, gets packaged inside the phage head followed by injection inside the recipient bacteria. In addition to this, there exist some of the elements that can be translocated to new genomic sites but themselves cannot be transferred within a new host (obviously if they get transposed to a conjugative element, they may get transferred inside new hosts). Such mobile genetic elements consist of mobile introns and transposons.

Bacteria may even acquire AMR genes through transformation. This process takes place in Gram–positive as well as Gram–negative bacteria. Bacteria, which can take up naked DNA from the environment, are called competent bacterial cells. Certain microbes such as most of the streptococci become competent at a particular phase during their growth, whereas other bacteria may not have any definite competence window. A few bacteria such as *Neisseria* have particular sequence requirement for successfully taking up DNA, whereas bacteria like *Bacillus subtilis* have no such requirements (Smith et al., 1999). In this mechanism, the recipient bacterium takes up naked DNA and incorporates that DNA within the genome of the host either through transposition or through homologous recombination. Conversely, a molecule of DNA may be capable of autonomous replication (such as plasmids). Mobile genetic elements are frequently acquired through conjugation and through transformation.

6.4.2 Plasmids or Conjugative Elements

Plasmids are typically the double stranded, circular, extra chromosomal DNA materials having the origins for self replication. They are present in nearly all genera of bacteria and the simplest of the conjugative elements only consist of a replication origin site and genes that encode for the replicative functions. Plasmids even sometimes comprise of an origin for transfer and genes, which encode functions allowing them to get transferred to new host bacteria through conjugation (Smillie et al., 2010). Plasmids containing genes for conjugation are termed as conjugative whilst the plasmids that just possess an origin for transfer (*oriT*) but no genes for conjugation are termed as mobilizable since they can utilize the conjugation functions of conjugative plasmid and become transferred to a new bacterial host.

Apart from taking part in replication and transfer, plasmids are also responsible for providing resistance against antimicrorbials. The relation between resistant genes and plasmids was first identified in *Pseudomonas aeruginosa* (Holloway, 1969). If a gene that is resistant is present on a mobilizable or a conjugative plasmid, then it possesses the capability to be transferred to new host bacteria. Certain plasmids have a wide range of hosts and may get transferred amongst diverse species, while others have a very narrow spectrum of hosts and thus remain confined within a single species or genus. Also, there are plasmids that are capable of being transferred to a specific host but do not replicate well or do not replicate at all within the new host. Under such circumstances, the plasmids may get lost, on the other hand, if the plasmid contains an AMR gene on the transposon, then this gentetic material may get translocated to the chromosome of the bacteria and hence can be maintained forever even if the plasmid is lost. Hence, a plasmid does not essentially need to be present in a specific host for spreading AMR among bacterial species.

From different studies, it has been found that R plasmids are capable of translocating pieces of AMR genes having DNA and this mechanism is not dependent on homologous recombination processes (Bennett and Richmond, 1976). Conjugative plasmids are composed of processes that are genetically complicated for horizontal transfer of plasmids thereby remarkably increasing the preserved backbone size. The transfer (tra) portion of the conjugative plasmids constitutes genes encoding for proteins, which are utilized for mating pair formation (MPF) leading to development of Type IV secretion system pores (T4SS). This also aids the DNA transfer replication (DTR) proteins, which aid in plasmid DNA processing. T4SS creates a filamentous surface appendage, a conjugative pilus in Gram–negative bacteria and hence aids in the establishment of cellular communications. The primitive type of plasmid that was observed to confer resistance against antimicrobials was F plasmid. FII plasmid R100 is composed of a class In2 within a Tn21 and this is again present within IS–mediated TN having chloramphenicol–resistant gene (*catA1*) (Liebert et al., 1999).

Various staphylococcal strains are composed of plasmids, which provide resistance against various antimicrobials, heavy metals and disinfectants, which are already known to be present in the plasmids. Three main types of resistant plasmids in *Staphylococcus* are multiresistant plasmids, small plasmids and conjugative multiresistant plasmids.

Circular plasmids are generally more extensively studied in comparison to the linear plasmids. This is maybe due to the easy extraction and separation of circular plasmids from the chromosomes of the bacteria. However, nowadays linear plasmids are well studied and have been demonstrated to confer beneficial phenotypes to the host bacteria. Just as in the case of circular plasmids, linear plasmids are also able to participate in conjugation (Chaconas and Kobryn, 2010).

Some types of resistant plasmids are not able to co–exist within the same bacterial cell and such plasmids fall in an incompatible group of plasmids (Couturier et al., 1988). Four major groups of plasmids have been defined on the basis of the structure of pilus and genetic relations. Those groups are: the IncI group (including IncB and IncK), the IncF group (including IncD, InnC, IncJ, IncF and IncS), the IncP group (including IncU, IncM and IncW) and the Ti plasmids.

6.4.3 Integrative Conjugated Elements (ICE)

ICE are also known as conjugative transposons just as the conjugative plasmids constitute an origin for transfer along with the genes involved in making the conjugation machinery (Roberts et al., 2008). Unlike plasmids, such elements are not composed of an origin of replication and need to be integrated within a replicon so that they can be maintained. This replicon may be a chromosome or a plasmid. This gives them an advantage over plasmids since they are not required to have replication apparatus, which has to be compatible with the host, hence these ICEs have a wider spectrum of hosts in comparison to plasmids.

ICEs are highly heterogeneous classes of genetic elements having varied ranges of hosts and properties. On the other hand, generally, they comprise of a modular organization, that is, recombination, conjugation, accessory and regulatory modules. The accessory modules are generally composed of genes encoding AMR.

Present also are integrative elements without any region of conjugation but which may be mobilized using conjugative plasmids or co–existent conjugative ICE. This in turn leads to the spreading of AMR.

6.4.4 Insertion Sequences (IS)

Generally, IS are small mobile genetic elements, which carry more than one transposase gene (*tnp*). They are the simplest types of mobile genetic elements. Initially, IS were not thought of as passenger genes but later on, it has been observed that they are capable of carrying resistance genes serving as composite transposon elements. These elements only contain the gene that is required for mobility of the element and inverted repeats present at the end of the elements. IS are as short as 1 kb.

6.4.5 Transposons

When such elements comprise of accessory genes, which do not take part in translocation of the elements, then those IS are known as transposons. Such composite transposons constitute of a portion, which is surrounded by two lookalike copies of the identical IS which altogether serve as a mono–unit. A strong promoter is present in several IS helping in capture gene expression and the intrinsic insertion of a chromosomal gene at the upstream even impacts AMR (Table 6.2) (Kamruzzaman et al., 2015). There exist several complicated types of transposons, which move with the help of various mechanisms involving class II transposons.

These transposons do not take part in conjugation, so for dissemination, they must be confined inside the conjugative elements. On the other hand, few of the ICE will get transferred to the new host, so they get transposed to new genomic regions. Their capacity for using various sites of integration is dependent on the recombinase type that they constitute. For instance, Tn*916* is capable of using a wide variety of sites for integration in the majority of the hosts (Roberts and Mullany, 2009) whilst few elements are extremely site–specific like Tn*916*. It is assumed that Tn*916* evolved to make use of various sites for integration and for increasing their range of hosts. Elements, which

TABLE 6.2
Bacterial genes and their potential AMR mechanisms against some of the antibiotics

AMR genes	Resistant antibiotics	AMR mechanism	References
SagS	Norfloxacin, tobramycin	*BrlR*activation via promotion of high levels of c–di–GMP	Webb et al., 2003
brlR	Norfloxacin, tobramycin, tetracycline, trimethoprim, chloramphenicol, kanamycin	Up–regulating the multidrug efflux pumps	Spoering and Lewis, 2001
PA1875–1877	Gentamicin, tobramycin, ciprofloxacin	Antibiotic efflux from the cell	Zheng and Stewart, 2004
ndvB	Gentamicin, tobramycin, ciprofloxacin	Antibiotic sequestration by up–regulating genes required for oxidation of ethanol	Thomas et al., 2009
rapA	Norfloxacin, penicillin G, gentamicin, chloramphenicol	Up–regulating YhcQfor biosynthesis of exopolysaccharides	
spot, relA	Meropenem, ofloxacin, gentamicin, colistin	Up–regulation of antioxidants for defence and down–regulation of pro–oxidants	Zheng and Stewart, 2004
epaOX	Gentamicin	Maintaining integrity of cell wall	Whiteley et al., 2001
yafQ	Cefazolin, tobramycin	Persistent formation of cell wall	Kohanski et al., 2010
dltABCD	Gentamicin	Reduces negativity of cell wall	Van Acker and Coenye, 2016

utilize only a specific number of sites for insertion, are confined within the hosts. They can only use other sites for insertion, if those sites are occupied or mutated.

Each transposon is larger in comparison to IS elements and is surrounded by an IR pair. It is also composed of an intrinsic passenger gene, which probably contains the encoding gene and even has an AMR site and transposase gene. The Tn3 transposon family is mainly linked with AMR genes and composed of a resolvase gene called *tnpR* including a site of resolution (res) containing the passenger genes. Tn3, Tn2 and Tn1 were the primary transposons first discovered inside Gram–negative bacteria. The res site is situated between *tnpR* and *tnpA* genes, which get transcribed in alternate directions. On either portion of the res site, homologous recombination occurs followed by recombination mediated by res. In hybrids, segments or variants of Tn3, Tn2 and Tn1, different inhibitor resistant (IRT) variants are being found. *strB*is a gene providing resistance against streptomycin and it is present on Tn5393 in aposition similar to that of bla_{TEM} in Tn3. Tn1546 is a significant member of this Tn3 group since it is related to the AMR of Gram–positive bacteria. It consists of a vanA gene cluster, which aids in coding for genes giving resistance against vancomycin. Expression of the vanA cluster of genes is mediated by a gene called a vanRS regulator (Lopez et al., 2010). Tn1721 contains Tn1722 having a tetracycline–resistant determinant at the IR_R end. Operon resistance against mercury (mer operon) has been found in the Tn21 group and it is very critical in moving AMR genes. Stpahylococcal β–lactamase genes have been observed to have elements such as Tn552 making the strains of *S. aureus* resistant against penicillin.

6.4.6 Transduction

There are current instances of AMR genes and also a whole set of mobile genetic elements, which get immobilized due to transduction. Transduction is a mechanism, where the phage particles get packed with the DNA of the bacteria and not with the phage itself. Transduction can be of two types: generalized transduction, where a DNA segment of the bacteria gets packaged within the phage head, and specialized transduction, where the DNA, which is flanking to the insertion site of the phage gets packaged.

6.4.7 Integrons

Integrons are also known as gene capture elements, which comprise of the components of the site–specific system of recombination, allowing them to mobilize and capture genes, particularly AMR determinants (Depardieu et al., 2007). They contain an *intI* gene, which encodes an integrase, which is site–specific belonging to the family of tyrosine recombinase carrying out recombination amongst two separate targeted sites, which are an attC site (59–base element) and an attI (recombination site), which is a targeted site for the promoter and the integration of the cassette (Mazel, 2006). However, integrons are not surrounded with the repeat sequences and they also do not contain any protein encoding genes for their movement catalyzation.Transfer of integrons to other bacteria via HGT is mainly facilitated by transposons or plasmids.

The *intI* genes are utilized as a foundation for classifying integrons into various groups. At present, four groups of integrons have been identified, those integrons having *intI* genes are called class I, integrons with *intI2* are called class II, those with *intI3* are class III and those with *intI4* are class IV (Partridge et al., 2009).

6.5 PARAMETERS CONTROLLING MOBILE GENETIC ELEMENT ACQUISITION

The capacity of mobile genetic elements constituing AMR genes for spreading is regulated by a set of parameters, such as selective environmental pressures, hosts and characteristics of the genetic elements itself. Each factor is discussed in detail in the sections below.

6.5.1 Environmental Selection Pressure

All factors mentioned in the previous section are significant for modulation of the AMR transmission but definitely, in the presence of antimicrobials within the environment, there exists a strong selection pressure for spreading the AMR and these parameters, which facilitate the spreading of resistance, will be selected for whilst those preventing the stretch of mobile genetic elements will be selected against (van Hoek et al., 2011).

Gene transfer is even more likely in conditions in which bacteria remain in close contact with each other and in comparatively higher density such as in the oral cavity and in the gut. For curbing the spread of AMR, it is important to have knowledge of the molecular biology of the various mobile genetic elements and of the ecology of the surrounding environment, where spread is likely to occur.

6.5.2 Non–Specific Host Encoded Factors

Certain non–specific parameters, which can serve as restrictors of HGT have been elucidated to have no target sites for a specific ICE or have an incompatible system of replication, which prevents plasmid replication within a host. Even, the cell surface architecture will not probably facilitate conjugation machinery to function productively for all the mobile genetic elements. In addition to this, one member from the mating pair is likely to secrete inhibitory compounds. Bacteria produce many antimicrobials, among which the most produced is the peptide antibiotic. The well known ones are colicins, which are produced from *E. coli.* Gram–positive bacteria also secrete a wide range of antimicrobial peptides (AMPs) (Riley and Wertz, 2002).

6.5.3 Specific Host Factors

Bacteria possess several mechanisms protecting them from any incoming foreign DNA and these systems are either CRISPR–Cas9 or restriction–modification (R–M) processes (Makarova et al., 2011). Although, these two mechanisms differ widely mechanically, they have identical end points for identification and subsequently destruction of foreign DNA. R–M systems function by recognizing specific sequences present in the arriving foreign DNA, which are unprotected through methylation, and digest them. On the contrary, CRISPR serves as an immunogenic memory of previous infection via a mobile genetic element and can annihilate that element when the bacteria are exposed to it for the second time. Both systems are efficient in preventing the spread of plasmids, ICE and phages.

A particular host factor attracting mobile genetic elements has been recognized in pheromone responsive systems, where a plasmid that is not much of a recipient produces a pheromone towards which strains containing plasmids respond to and transport their own plasmid to the recipient (Palmer et al., 2010).

6.5.4 Factors Encoded by Mobile Genetic Elements

Mobile genetic elements consist of a surplus of ways of overcoming defences posed by the bacterial systems. Numerous ICE and plasmids encode for anti–restriction proteins, which renders the restriction system of the host inactive so that the element can gain entry to a new host and keep on surviving. Several mobile genetic elements do not posses any recognition sites for the restriction systems so that they are not influenced by the presence of any restriction enzymes. Some of them, much like the common family of Tn–*916* conjugative transposons, code for anti–restriction proteins that have been demonstrated to imitate the DNA and become identified by the restriction enzymes. ArdA is a well characterized anti–restriction protein produced by Tn*916* (McMahon et al., 2009).

Nmeorus ICE and transposons may be transposed to necessary genes. If such a thing happens, the host dies, so to prevent host death, a few mobile genetic elements favourably target inter–genic portions or they are site specific (Cookson et al., 2011). In addition to this, many of the transposable elements such as ICE are strongly synchronized such that they can just be transposed with a low frequency or are transposed on subjection of the bacteria to stresses like the presence of antibiotics in the surroundings (Wozniak and Waldor, 2010). For instance, members of the CTndot class of ICE, are transferred with an increased frequency in the presence of tetracycline (an antibiotic, towards which they are resistant). This response is beneficial to both the host bacteria and the element (Moon et al., 2005).

6.6 ARRANGEMENT, MAINTENANCE AND RESISTANCE GENE TRANSFER

R plasmids usually constitute several resistance genes. These genes are stably maintained within the host bacterial strains and very effectively get transferred to the surrounding susceptible receptor cells. The underlying mechanisms of these processes are discussed below in detail.

6.6.1 Assembly of R Plasmids/Resistant Genes

We know that most drug resistant genes are efficient at being expressed in plasmids. Significantly, several of these genes usually exist over a single R plasmid such that the AMR gets transported to a vulnerable neighbouring bacterial cell during a single event of conjugation. When R plasmids were first found in Japan in 1950, several of them were found to be already containing resistant genes against tetracycline, aminoglycoside, sulfonilamides and chloramphenicol. Many plasmid sequencing activities have revealed the occurrence of these clusters.

In the sequences of early–generation R plasmids, it has been found that a majority of the resistant genes including constituents of transposons are capable of deliverying the genes to any piece of DNA (Nikaido, 2009).

Tn21 is an especially significant example of a complex, large and multiply composite transposon (Liebert and Hall, 1999). It also contains genes resistant to mercury. The discovery that several resistant genes in R plasmid comprise of a novel 59 base 3'–sequence tag resulted in the breakthrough of integron. Upon integration, the resistant genes are distinct because of the tag such that they can be effortlessly integrated in another integron, which contains a different resistant gene set. Furthermore, with increased mobility, the resistant genes on insertion within an integron become assembled into a single operon having identical transcription orientation in the presence of a strong promoter supplied by the integron structure (Hall and Stokes, 1993). As seen in the case of *Tn21*, the intergon already has the sulfonilamide resistant gene, called *sul1* along with a condensed version of a multidrug efflux gene, called *qacE*, which further integrates the aminoglycoside resistant gene, called *aadA1* at a particular site of integration, called *attI*. An integron has the capacity to contain a maximum of eight resistant genes (Rowe–Magnus and Mazel, 2002).

Additonally, most of these integrons possess an enzymatic apparatus to transpose the whole structure of the integron to other hosts.In addition to this, the integron is usually inserted within a large transposon like *Tn21* thereby allowing the total array of resistant genes to hop among various plasmids or in between chromosomes and plasmids. The evolution of modern integrons is not yet clearly understood but a similar underlying mechanism exists for assembling the genes of *V. cholerae* (Mazel, 2006).

Nowadays, numerous structures of integrons have been reported to remain associated with a downstream structure known as an IS*C* R element having a putative transposase gene (Toleman et al., 2006). Seemingly, it functions not in the usual way, but it acts in an open–ended transposition process recruiting different resistant genes and delivering them closely to the structure of the integron leading to the assembly of more such resistant genes.

6.6.2 R Plasmid Maintenance in Hosts

Recombinant plasmids obtained from cloning vectors are usually missing from the cells of the host even if they are present in a comparatively large number of copies. In contrast to this, the majority of the natural R plasmids are significantly stable and very seldom get lost throughout host cell multiplication even if the number of copies is relatively lower. This is due to the presence of genes ensuring proper partitioning of the copies to the daughter cells in the R plasmids (Nordstrom and Austin, 1989).

Furthermore, certain natural plasmids constitute of killer elements comprising of an mRNA or a stable killer protein and an unstable antisense RNA or inhibitor protein such that the plasmid loss leads to the death of host cells (Nordstrom and Austin, 1989).

Due to the existence of such regulatory networks, plasmids that share a common partition/replication apparatus, are not able to co–exist stably within the identical host cell and this process is used to classify plasmids as incompatible groups of plasmids. In enteric bacteria, this type of plasmid classification includes around two dozen of these plasmid groups (Novick et al., 1976).

6.6.3 R Plasmid Transfer

R plasmids are not just maintained stably but are generally transferred between cells of bacteria at a very high rate, in most cases at a rate of approximately 100%. The molecular mechanism of such cell-to-cell transport has been mainly observed in *E. coli* and its other strains and within the plasmids of incompatibility groups F1 (F for sex plasmid), W and P (RP4). Recently, the genes responsible for the transfer of plasmid were observed to be the homologues of *virB* genes in *Agrobacterium tumefaciens* that transfer a portion of the DNA of bacteria into the nucleus of a plant and numerous effector protein injection systems in animal and human pathogens (Christie et al., 2005). Now, such systems are known as Type IV secretion systems. A particular sequence in double stranded, circular DNA of plasmids is identified and a single strand cut is made by using the VirD2/TraI enzyme along with the enzymatic protein remaining attached covalently with the 5' end of the cut strand. This complex is known as a relaxosome since the nicking of one strand leads to the relaxation of the supercoiled plasmid DNA. The single–stranded DNA–protein complex is further exported by the formation of a mating pair complex containing almost a dozen of various proteins. Relaxosome export rather than the export of DNA free from protein, explains the phenotype that a few of the Type IV secretion systems export purified proteins. Prior to this, the leading 5' end of the exported DNA along with the VirD2/TraI protein attached need to be taken up by ATPase (TraG or VirD4). The ATPase is also known as a coupling protein since it takes part in exporting the substrate to the export machinery. Even though one strand of the DNA of R plasmid is completely transferred to the recipient cells, within the donor cell, this strand gets substituted with a mechanism of rolling circle replication such that whole R plasmids can now survive in recipient as well as in donor cells. It is known that foreign DNA entry is easily identified by the strain–specific restriction endonucleases present in the recipient cells. In contrast to this, the initial piece of DNA enters the single strand (although a complementary strand will soon get synthesized) and so it can escape the restriction modification mechanism. Furthermore, a few R plasmids are well equipped with functions antagonizing the restriction endonuclease attack (Lessl and Lanka, 1994).

Certain R plasmids are extremely small in size and they do not have the genes required to construct a mating pair formation complex. On the other hand, such plasmids are mobilized through the transfer of genes of the other plasmids and can efficiently get transferred to the recipient cells. Ultimately, it needs to be emphasized that plasmid transfer through conjugation is not a unique mechanism providing AMR, but the HGT by conjugation has been the major underlying mechanism for producing AMR bacteria (Ochman et al., 2000).

Gram–positive bacteria can also perform cell–cell plasmid transfer (Grohmann et al., 2003). Additionally, large transposons belonging to a special class known as conjugative transposons, play a major function in the bacterial cells (Rice, 1998).They make use of a λ–phage-like process of transposition, generating an intermediate circular DNA. This intermediate gets inserted in a predetermined position within the chromosomal or plasmid DNA but this can even act like a conjugative plasmid transferring its copy to another bacterium via conjugation. Significantly, inside the transposons having the resistant genes against tetracycline, the transfer function is robustly induced by tetracycline presence. Hence, in such transposons, the AMR genes are just not passive passengers but they become components of an integrated regulation mechanism facilitating conjugational transfer. This transfer even takes place readily among distantly related organisms, for instance, Gram–negative and Gram–positive bacteria (Courvalin, 1996).

6.7 CONCLUSION AND FUTURE PERSPECTIVES

Despite the measures taken by Centres for Disease Control and prevention (CDC) and WHO, a number of people in various parts of the world are still not especially aware of rising AMR and its consequences. This challenge is not just limited to the healthcare sectors but it is also a big issue for the world economy. With increased AMR globally, the available treatment strategies in medicine are facing a number of hurdles. The shortage of alternative and new antimicrobials is a major issue in the healthcare sectors. Hence, immediate enforcement of surveillance and epidemiological studies are needed to analyse and prevent AMR. Adequate information and knowledge are the key factors to control AMR. Stringent government regulations and policies all over the world are also necessary to monitor the antimicrobial use and various other aspects. Ultimately, multidisciplinary strategies must be considered for the elimination of serious threats posed by AMR. Collaborative studies of nanobiotechnology with computational biology can open new avenues for combating the problem of AMR worldwide. New information about mobile genetic elements can enrich our understanding of the origins of pathogens and evolution of their AMR variants. This may also aid in the development of potential therapeutic approaches for the spreading of pathogens and the inhibition of pathogenic strain growth. In addition to this, a model for reduced antimicrobial consumption may be fabricated by knowing the rate of transmission of AMR from animals to humans.

The goal of this chapter is to provide some fundamental ideas about the proper use of antimicrobials. It has focussed on various genetic aspects of AMR and the molecular mechanism through which AMR develops along with the evolution of AMR superbugs. There exists an ongoing need to prevent the further rise of AMR bacteria. Since the overuse of antimicrobials is the reason for selection in bacteria, so we need to minimize the excessive use of antimicrobials. There is currently a strong connection between the use of β–lactams and the rate at which penicillin–resistant pneumococci emerges in the member countries of the European Union (Bronzwaer et al., 2002).

REFERENCES

Alav, I., Sutton, J. M., and Rahman, K. M. (2018). Role of bacterial efflux pumps in biofilm formation. *The Journal of Antimicrobial Chemotherapy*, *73*(8), 2003–2020.

Anderl, J. N., Franklin, M. J., and Stewart, P. S. (2000). Role of antibiotic penetration limitation in Klebsiellapneumoniae biofilm resistance to ampicillin and ciprofloxacin. *Antimicrobial Agents and Chemotherapy*, *44*(7), 1818–1824.

Anyanwu, M. U., Jaja, I. F., and Nwobi, O. C. (2020). Occurrence and characteristics of mobile colistin resistance (*mcr*) gene-containing isolates from the environment: A review. *International Journal of Environmental Research and Public Health*, *17*(3), 1028.

Bagge, N., Hentzer, M., Andersen, J. B., Ciofu, O., Givskov, M., and Høiby, N. (2004). Dynamics and spatial distribution of beta-lactamase expression in pseudomonas aeruginosa biofilms. *Antimicrobial Agents and Chemotherapy*, *48*(4), 1168–1174.

Baker, S. J., Payne, D. J., Rappuoli, R., and De Gregorio, E. (2018). Technologies to address antimicrobial resistance. *Proceedings of the National Academy of Sciences of the United States of America, 115*(51), 12887–12895.

Bennett, P. M., and Richmond, M. H. (1976). Translocation of a discrete piece of deoxyribonucleic acid carrying an amp gene between replicons in eschericha coli. *Journal of Bacteriology, 126*(1), 1–6.

Bronzwaer, S. L., Cars, O., Buchholz, U., Mölstad, S., Goettsch, W., Veldhuijzen, I. K., Kool, J. L., Sprenger, M. J., Degener, J. E., and European Antimicrobial Resistance Surveillance System (2002). A European study on the relationship between antimicrobial use and antimicrobial resistance. *Emerging Infectious Diseases, 8*(3), 278–282.

Casadewall, B., and Courvalin, P. (1999). Characterization of the vanDglycopeptide resistance gene cluster from Enterococcus faecium BM4339. *Journal of Bacteriology, 181*(12), 3644–3648.

Chaconas, G., and Kobryn, K. (2010). Structure, function, and evolution of linear replicons in Borrelia. *Annual Review of Microbiology, 64*, 185–202.

Christie, P. J., Atmakuri, K., Krishnamoorthy, V., Jakubowski, S., and Cascales, E. (2005). Biogenesis, architecture, and function of bacterial type IV secretion systems. *Annual Review of Microbiology, 59*, 451–485.

Cookson, A. L., Noel, S., Hussein, H., Perry, R., Sang, C., Moon, C. D., Leahy, S. C., Altermann, E., Kelly, W. J., and Attwood, G. T. (2011). Transposition of Tn916 in the four replicons of the Butyrivibrioproteocla sticusB316(T) genome. *FEMS Microbiology Letters, 316*(2), 144–151.

Courvalin, P. (1996). The garrod lecture. Evasion of antibiotic action by bacteria. *The Journal of Antimicrobial Chemotherapy, 37*(5), 855–869.

Couturier, M., Bex, F., Bergquist, P. L., and Maas, W. K. (1988). Identification and classification of bacterial plasmids. *Microbiological Reviews, 52*(3), 375–395.

Darwich, L., Vidal, A., Seminati, C., Albamonte, A., Casado, A., López, F., Molina-López, R. A., and Migura-Garcia, L. (2019). High prevalence and diversity of extended-spectrum β-lactamase and emergence of OXA-48 producing enterobacterales in wildlife in catalonia. *PloS One, 14*(8), e0210686.

Das, B., Verma, J., Kumar, P., Ghosh, A., and Ramamurthy, T. (2020). Antibiotic resistance in vibrio cholerae: Understanding the ecology of resistance genes and mechanisms. Vaccine, 38 (1), A83–A92.

Depardieu, F., Podglajen, I., Leclercq, R., Collatz, E., and Courvalin, P. (2007). Modes and modulations of antibiotic resistance gene expression. *Clinical Microbiology Reviews, 20*(1), 79–114.

Fernández-Villa, D., Aguilar, M. R., and Rojo, L. (2019). Folic acid antagonists: Antimicrobial and immunomodulating mechanisms and applications. *International Journal of Molecular Sciences, 20*(20), 4996.

Grohmann, E., Muth, G., and Espinosa, M. (2003). Conjugative plasmid transfer in gram-positive bacteria. *Microbiology and Molecular Biology Reviews: MMBR, 67*(2), 277–301.

Hall, R. M., and Stokes, H. W. (1993). Integrons: Novel DNA elements which capture genes by site-specific recombination. *Genetica, 90*(2–3), 115–132.

Holloway, B. W. (1969). Genetics of pseudomonas. *Bacteriological Reviews, 33*(3), 419–443.

Jevons, M. P., Coe, A. W., and Parker, M. T. (1963). Methicillin resistance in staphylococci. *Lancet (London, England), 1*(7287), 904–907.

Kamruzzaman, M., Patterson, J. D., Shoma, S., Ginn, A. N., Partridge, S. R., and Iredell, J. R. (2015). Relative strengths of promoters provided by common mobile genetic elements associated with resistance gene expression in gram-negative bacteria. *Antimicrobial Agents and Chemotherapy, 59*(8), 5088–5091.

Kapoor, G., Saigal, S., and Elongavan, A. (2017). Action and resistance mechanisms of antibiotics: A guide for clinicians. *Journal of Anaesthesiology, Clinical Pharmacology, 33*(3), 300–305.

Kohanski, M. A., Dwyer, D. J., and Collins, J. J. (2010). How antibiotics kill bacteria: From targets to networks. *Nature Reviews. Microbiology, 8*(6), 423–435.

Laxminarayan, R., Van Boeckel, T., Frost, I., Kariuki, S., Khan, E. A., Limmathurotsakul, D., Larsson, D., Levy-Hara, G., Mendelson, M., Outterson, K., Peacock, S. J., and Zhu, Y. G. (2020). The Lancet Infectious Diseases Commission on antimicrobial resistance: 6 years later. *The Lancet. Infectious Diseases*, 20(4), e51–e60.

Lebeaux, D., Ghigo, J. M., and Beloin, C. (2014). Biofilm-related infections: Bridging the gap between clinical management and fundamental aspects of recalcitrance toward antibiotics. *Microbiology and Molecular Biology Reviews: MMBR, 78*(3), 510–543.

Lessl, M., and Lanka, E. (1994). Common mechanisms in bacterial conjugation and Ti-mediated T-DNA transfer to plant cells. *Cell, 77*(3), 321–324.

Liebert, C. A., Hall, R. M., and Summers, A. O. (1999). Transposon Tn21, flagship of the floating genome. *Microbiology and Molecular Biology Reviews: MMBR, 63*(3), 507–522.

López, M., Sáenz, Y., Alvarez-Martínez, M. J., Marco, F., Robredo, B., Rojo-Bezares, B., Ruiz-Larrea, F., Zarazaga, M., and Torres, C. (2010). Tn1546 structures and multilocus sequence typing of vanA-containing enterococci of animal, human and food origin. *The Journal of Antimicrobial Chemotherapy, 65*(8), 1570–1575.

Makarova, K. S., Haft, D. H., Barrangou, R., Brouns, S. J., Charpentier, E., Horvath, P., Moineau, S., Mojica, F. J., Wolf, Y. I., Yakunin, A. F., van der Oost, J., and Koonin, E. V. (2011). Evolution and classification of the CRISPR-Cas systems. *Nature reviews. Microbiology, 9*(6), 467–477.

Mazel D. (2006). Integrons: Agents of bacterial evolution. *Nature Reviews. Microbiology, 4*(8), 608–620.

McMahon, S. A., Roberts, G. A., Johnson, K. A., Cooper, L. P., Liu, H., White, J. H., Carter, L. G., Sanghvi, B., Oke, M., Walkinshaw, M. D., Blakely, G. W., Naismith, J. H., and Dryden, D. T. (2009). Extensive DNA mimicry by the ArdA anti-restriction protein and its role in the spread of antibiotic resistance. *Nucleic Acids Research, 37*(15), 4887–4897.

Moon, K., Shoemaker, N. B., Gardner, J. F., and Salyers, A. A. (2005). Regulation of excision genes of the Bacteroides conjugative transposon CTnDOT. *Journal of Bacteriology, 187*(16), 5732–5741.

Muktan, B., ThapaShrestha, U., Dhungel, B., Mishra, B. C., Shrestha, N., Adhikari, N., Banjara, M. R., Adhikari, B., Rijal, K. R., and Ghimire, P. (2020). Plasmid mediated colistin resistant mcr-1 and co-existence of OXA-48 among Escherichia coli from clinical and poultry isolates: First report from Nepal. *Gut pathogens, 12*, 44.

Ndagi, U., Falaki, A. A., Abdullahi, M., Lawal, M. M., and Soliman, M. E. (2020). Antibiotic resistance: Bioinformatics-based understanding as a functional strategy for drug design. *RSC Advances, 10*(31), 18451–18468.

Nikaido H. (2009). Multidrug resistance in bacteria. *Annual Review of Biochemistry, 78*, 119–146.

Nordström, K., and Austin, S. J. (1989). Mechanisms that contribute to the stable segregation of plasmids. *Annual review of genetics, 23*, 37–69.

Novick, R. P., Clowes, R. C., Cohen, S. N., Curtiss, R., 3rd, Datta, N., and Falkow, S. (1976). Uniform nomenclature for bacterial plasmids: A proposal. *Bacteriological Reviews, 40*(1), 168–189.

Ochman, H., Lawrence, J. G., and Groisman, E. A. (2000). Lateral gene transfer and the nature of bacterial innovation. *Nature, 405*(6784), 299–304

Ogawara H. (2019). Comparison of antibiotic resistance mechanisms in antibiotic-producing and pathogenic bacteria. *Molecules (Basel, Switzerland), 24*(19), 3430.

Palmer, K. L., Kos, V. N., and Gilmore, M. S. (2010). Horizontal gene transfer and the genomics of enterococcal antibiotic resistance. *Current opinion in microbiology, 13*(5), 632–639.

Pang, Z., Raudonis, R., Glick, B. R., Lin, T. J., and Cheng, Z. (2019). Antibiotic resistance in Pseudomonas aeruginosa: Mechanisms and alternative therapeutic strategies. *Biotechnology Advances, 37*(1), 177–192.

Partridge, S. R., Tsafnat, G., Coiera, E., and Iredell, J. R. (2009). Gene cassettes and cassette arrays in mobile resistance integrons. *FEMS Microbiology Reviews, 33*(4), 757–784.

Peterson, E., and Kaur, P. (2018). Antibiotic resistance mechanisms in bacteria: Relationships between resistance determinants of antibiotic producers, environmental bacteria, and clinical pathogens. *Frontiers in Microbiology, 9*, 2928.

Prasad, R., Gupta, N., and Banka, A. (2018). Multidrug-resistant tuberculosis/rifampicin-resistant tuberculosis: Principles of management. *Lung India: Official Organ of Indian Chest Society, 35*(1), 78–81.

Randall, C. P., Mariner, K. R., Chopra, I., and O'Neill, A. J. (2013). The target of daptomycin is absent from Escherichia coli and other gram-negative pathogens. *Antimicrobial Agents and Chemotherapy, 57*(1), 637–639.

Reygaert W. C. (2018). An overview of the antimicrobial resistance mechanisms of bacteria. *AIMS Microbiology, 4*(3), 482–501.

Rice L. B. (1998). Tn916 family conjugative transposons and dissemination of antimicrobial resistance determinants. *Antimicrobial Agents and Chemotherapy, 42*(8), 1871–1877.

Riley, M. A., and Wertz, J. E. (2002). Bacteriocins: Evolution, ecology, and application. *Annual Review of Microbiology, 56*, 117–137.

Roberts M. C. (2008). Update on macrolide-lincosamide-streptogramin, ketolide, and oxazolidinone resistance genes. *FEMS Microbiology Letters, 282*(2), 147–159.

Roberts, A. P., and Mullany, P. (2009). A modular master on the move: The Tn916 family of mobile genetic elements. *Trends in Microbiology*, *17*(6), 251–258.

Rowe-Magnus, D. A., and Mazel, D. (2002). The role of integrons in antibiotic resistance gene capture. *International Journal of Medical Microbiology: IJMM*, *292*(2), 115–125.

Ruiz J. (2003). Mechanisms of resistance to quinolones: Target alterations, decreased accumulation and DNA gyrase protection. *The Journal of Antimicrobial Chemotherapy*, *51*(5), 1109–1117.

Saha, M., and Sarkar, A. (2021). Review on multiple facets of drug resistance: A sising challenge in the 21st century. *Journal of Xenobiotics*, *11*(4), 197–214.

Sánchez-Busó, L., and Harris, S. R. (2019). Using genomics to understand antimicrobial resistance and transmission in Neisseria gonorrhoeae. *Microbial Genomics*, *5*(2), e000239.

Smith, H. O., Gwinn, M. L., and Salzberg, S. L. (1999). DNA uptake signal sequences in naturally transformable bacteria. *Research in Microbiology*, *150*(9–10), 603–616.

Spoering, A. L., and Lewis, K. (2001). Biofilms and planktonic cells of Pseudomonas aeruginosa have similar resistance to killing by antimicrobials. *Journal of Bacteriology*, *183*(23), 6746–6751.

Sultan, I., Rahman, S., Jan, A. T., Siddiqui, M. T., Mondal, A. H., and Haq, Q. (2018). Antibiotics, resistome and resistance mechanisms: A bacterial perspective. *Frontiers in Microbiology*, *9*, 2066.

Thomas, V. C., Hiromasa, Y., Harms, N., Thurlow, L., Tomich, J., and Hancock, L. E. (2009). A fratricidal mechanism is responsible for eDNA release and contributes to biofilm development of Enterococcus faecalis. *Molecular Microbiology*, *72*(4), 1022–1036.

Toleman, M. A., Bennett, P. M., and Walsh, T. R. (2006). ISCR elements: Novel gene-capturing systems of the 21st century?. *Microbiology and Molecular Biology Reviews: MMBR*, *70*(2), 296–316.

Van Acker, H., Sass, A., Bazzini, S., De Roy, K., Udine, C., Messiaen, T., Riccardi, G., Boon, N., Nelis, H. J., Mahenthiralingam, E., and Coenye, T. (2013). Biofilm-grown Burkholderiacepacia complex cells survive antibiotic treatment by avoiding production of reactive oxygen species. *PloS One*, *8*(3), e58943.

vanHoek, A. H., Mevius, D., Guerra, B., Mullany, P., Roberts, A. P., and Aarts, H. J. (2011). Acquired antibiotic esistance genes: An overview. *Frontiers in Microbiology*, *2*.

Verderosa, A. D., Totsika, M., and Fairfull-Smith, K. E. (2019). Bacterial biofilm eradication agents: A current review. *Frontiers in Chemistry*, *7*, 824.

Vester, B., and Douthwaite, S. (2001). Macrolide resistance conferred by base substitutions in 23S rRNA. *Antimicrobial Agents and Chemotherapy*, *45*(1), 1–12.

Webb, J. S., Thompson, L. S., James, S., Charlton, T., Tolker-Nielsen, T., Koch, B., Givskov, M., and Kjelleberg, S. (2003). Cell death in Pseudomonas aeruginosa biofilm development. *Journal of Bacteriology*, *185*(15), 4585–4592.

Whiteley, M., Bangera, M. G., Bumgarner, R. E., Parsek, M. R., Teitzel, G. M., Lory, S., and Greenberg, E. P. (2001). Gene expression in Pseudomonas aeruginosa biofilms. *Nature*, *413*(6858), 860–864.

Wozniak, R. A., and Waldor, M. K. (2010). Integrative and conjugative elements: Mosaic mobile genetic elements enabling dynamic lateral gene flow. *Nature Reviews. Microbiology*, *8*(8), 552–563.

Wright G. D. (2005). Bacterial resistance to antibiotics: Enzymatic degradation and modification. *Advanced Drug Delivery Reviews*, *57*(10), 1451–1470.

Zaman, S. B., Hussain, M. A., Nye, R., Mehta, V., Mamun, K. T., and Hossain, N. (2017). A review on antibiotic resistance: Alarm bells are ringing. *Cureus*, *9*(6), e1403.

Zheng, Z., and Stewart, P. S. (2004). Growth limitation of Staphylococcus epidermidis in biofilms contributes to rifampin tolerance. *Biofilms*, *1*(1), 31–35.

Zhu, L., Lin, J., Ma, J., Cronan, J. E., and Wang, H. (2010). Triclosan resistance of Pseudomonas aeruginosa PAO1 is due to FabV, a triclosan-resistant enoyl-acyl carrier protein reductase. *Antimicrobial Agents and Chemotherapy*, *54*(2), 689–698.

7 Multispecies Interactions Resulting in the Development of Antimicrobial Resistance

Dibyajit Lahiri and Moupriya Nag

7.1 INTRODUCTION

Recently, it has been observed that the behaviour of bacteria is strongly dependent on its social life (Burmølle et al. 2014). Various research activities at present are being performed in unravelling the intensity of interaction that takes place with respect to cooperation and competition among various species present in the spotlight. The mechanism of cooperation helps in benefiting either one organism or the entire population whereas the mechanism of competition results in the reduction of fitness among the group of opposing individuals (Mitri and Foster 2013). We often observe that the social interactions that take place amongst the microbes remain partitioned by intraspecies or interspecies interaction (Mitri and Foster 2013). The interactions between the same species are common due to their genetic resemblance but interspecies interaction is a rare phenomenon as they do not possess the same type of evolutionary interest. Thus the mechanism of interspecies interaction is based on the same type of benefits received by the organisms on the basis of the help received from the environment. As in most of the time the interspecies interaction is based on the phenomenon of competition thereby it shows a dominative aggression by secretion of toxins or VI secretion systems (Cianfanelli et al. 2016).

The mechanism that helps the biofilm to withstand higher concentration of antimicrobial agents is still not properly known. The mechanism of antibiotic resistance is the ability of the microbes to grow in the presence of the antibiotics by bringing about an the genetic expression (Mah and O'Toole 2001). The mechanism by which the biofilm protect themselves from the antibiotics are referred to as "recalcitrance" (Lewis 2008).

7.2 THE CAUSE OF POLYMICROBIAL INFECTIONS

The cause of persistent infections in a body is due to the development of multispecies biofilm within the oral cavity (Kolenbrander 2000), lungs (Boisvert et al. 2016), urinary tract (Ronald 2003) middle ear and both in chronic and surgical wounds (Percival et al. 2012). It has been observed that blood stream associated polymicrobial infections are associated with higher rates of mortality in comparison to those caused by monospecies (Weinstein et al. 1983). The patients suffering from polymicrobial lung infections are diagnosed with cystic fibrosis. These are co-infected by *Staphylococcus aureus* and *Pseudomonas aeruginosa*. This provides the evidence of the interaction among the microbes during the time of infection. Microbial physiology is greatly controlled by the interaction taking place amongst the species causing progression of disease and enhancement in the expression of the virulence factor (Duan et al. 2003). The interaction taking place between the

DOI: 10.1201/9781003281238-7

microbial species and the host helps in the modification of immune response to the co-infecting organism (Chekabab et al. 2015). Hence, the worst physiological conditions are observed in patients suffering from co-infections rather than monoinfections.

7.3 THE MECHANISM OF DEVELOPMENT OF RESISTANCE IN POLYMICROBIAL BIOFILM

Although the mechanism of mutation is responsible for alteration in gene expression, the process of horizontal gene transfer (HGT) is considered to be another important mechanism for the development of the resistance gene. The mechanism of HGT by the process of conjugation is observed to be enhanced in the biofilm (Sørensen et al. 2005). Studies have revealed that the cells dwelling within the biofilm possess the highest probability of undergoing HGT by the process of conjugation with respect to their planktonic counterpart. But the frequency of HGT is not uniform amongst the cells dwelling within the biofilm, that is, the highest rate of conjugation may be associated with certain sub-populations of cells within the biofilm during their initial stages of development (Christensen et al. 1996). The conjugative pili help in cellular adhesion within *Escherichia coli* and are responsible for the process of transferring the conjugative plasmids growing within the communities. Further it has been found that the rate of transformation has been observed to be enhanced within the biofilm with respect to their planktonic counterparts (Li et al. 2001). The development of competence within the cells is also influenced by various factors including antibiotic, pH, starvation, exposure and presence of specific type of carbon source (Cvitkovitch et al. 2003). The mechanism of quorum sensing is also responsible for the development of resistance among the dwelling sessile microcolonies existing within the biofilm (Li and Tian 2012). The process of HGT is observed to take place between the species existing within the polymicrobial biofilm. Studies have shown that carbapenemase resistance gene can be transferred from *E.coli* to either *P. aeruginosa* or *Acinetobacter baumanii* by the mechanism of conjugation (Tanner et al. 2017) but remarkably this phenomenon was not observed in their planktonic counterpart (Potron et al. 2011). Conjugative transposons are used for the mechanism of genetic exchange taking place within the biofilm. *Streptococcus spp* has been observed to acquire transposons from *Veillonella dispar* either by the process of conjugation with *V. dispar* or by the process of transformation of purified *V. dispar* DNA (Hannan et al. 2010). The exchange of extracellular DNA among the members of the mixed species is the process of development of antimicrobial resistance. It has been observed that the development of multi-drug resistant *S. aureus* takes place by the interspecies interaction taking place within the biofilm (Weigel et al. 2007). Resistant gene *vanA* present within *Enterococcus faecium* is responsible for the development of resistance within methicillin-resistant *S. aureus*. Moreover, the tetracycline resistance gene (tetU) has been observed in *S. aureus* which is also likely to be transferred from *E. faecium* within the mixed species of biofilm.

7.4 THE PROTECTION OF NEIGHBOURING MICROBES BY Β-LACTAMASE PRODUCING STRAINS

Various bacterial species possess the ability to produce β-lactamase enzyme that acts as a key inactivator of antibiotics. They possess the ability for hydrolysing the β-lactam rings of the antibiotics (Bush 2010). This enzyme is encoded chromosomally or via plasmids and plays a vital role in the development of multi-drug resistance among the Gram-negative bacteria (Bush 2010). β-lactamases are either located within the periplasmic space of the cells producing the enzyme or they are released within the extracellular space. They are also found located within the outer membrane vesicles (OMVs) of various types of the Gram-negative bacteria (Ciofu et al. 2000) as well as within the extracellular vesicles of the Gram-positive bacteria such as *S. aureus* (Lee et al. 2013). It has been observed that the bacterial cells producing the enzyme not only protect themselves from the antibiotic but also various other cells are located in their vicinity (Sorg et al. 2016). Bacteria producing β-lactamase enzyme alter the antimicrobial sensitivity of all the microbial species dwelling within the polymicrobial biofilm. β-lactamases are found to alter the sensitivity of multispecies biofilm infections such as otitis caused by the organisms *Streptococcus pneumonia*, *Moraxella catarrhalis* and *Haemophilus influenza.* β-lactamases have been found within the polymicrobial biofilm of cystic fibrosis (CF) (Giwercman et al. 1992) and otitis media (Brook and Yocum 1995). Studies have also found that the failure of amoxicillin was observed in bacteria exhibiting β-lactamase activity.

7.5 ANTIMICROBIAL RESISTANCE IN A POLYMICROBIAL BIOFILM: THE MECHANISM

Although several studies have been reported in the development of antimicrobial resistance within planktonic cells, recent studies have shown the mechanism of development of antimicrobial resistance with interspecies interaction among the sessile microbial colonies existing within the polymicrobial biofilm.

7.6 INTERSPECIES EXCHANGE OF GENES RESULTING IN ANTIMICROBIAL RESISTANCE

Although the mutation is a cause of development of antimicrobial resistance, the process of horizontal gene transfer (HGT) is another potent mechanism in the development of resistance in the polymicrobial biofilm. With the increase in the rate of conjugation, the rate of HGT increases along with the rate of transformation (Madsen et al. 2012). The presence of ordered structure and high density of cells persisting within the biofilm helps in promoting a better rate of conjugation (Madsen et al. 2012). The process of conjugation is confined to cellular sub population during the time of the initial stages of development of the biofilm (Christensen et al. 1996).

The rate of transformation taking place amongst the cells is found to increase within the biofilm. It has been observed that transformation frequency within *Streptococcus mutans* was found to be 10 to 600 times higher in the biofilm than their planktonic counterpart (Li et al. 2001).

7.7 PRODUCTS SECRETED BY MICROBES ALTERING ANTIBIOTIC TOLERANCE

Quorum sensing is the process by which bacterial cells in the biofilm communicate with each other. An autoinducer (AI) is a chemical signalling molecule that encourages communication between cells within a biofilm. Due to quorum sensing individual bacteria within the colonies can coordinate collective activities by interacting with neighbours such as by bioluminescence, exoenzyme secretion, sporulation, conjugation and biofilm formation. Also, a biofilm's population density can be determined based on the signalling molecules secreted by bacteria within the community, due to QS. Now let's look at an example to understand how QS helps the bacteria or any other microbes in a biofilm. Suppose a disease-causing bacterium enters our body and it tries to attack us but being a single pathogenic bacterium, it cannot do more harm to us. So, the bacterium keeps growing and dividing until there are enough of them and they communicate with each other through QS. Thus, when there are enough of them so that they recognize the assemblage of signalling molecules they release the toxins altogether and make us sick. Examples of this are with *Staphylococcus aureus*, *Bacillus cereus*, and *Pseudomonas aeruginosa*.

7.8 BACTERIAL RESISTANCE AGAINST ANTIBIOTICS

Various mechanisms associated with the actions of antibiotics include interference with the formation of the cell wall synthesis, nucleic acid synthesis, inhibition of synthesis of proteins, inhibition of various types of metabolic pathways and the disruption of the cell membrane (Table 7.1).

TABLE 7.1
List of Antibiotics, Their Action and Development of Resistance

Name of the antibiotic	Structure	Mechanism of development of resistance	Group of antibiotics	Mode of action	Reference
β-Lactum antibiotics	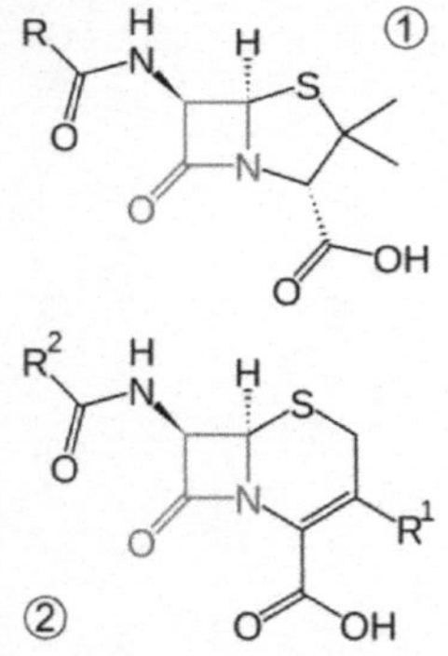	β-lactamases production by the bacterial cells result in the degradation of the β-Lactam group of antibiotics, carbapenem hydrolyzing enzymes, AmpC enzymes and expression of various types of antibiotic resistant genes like *bla shv*, *bla tem*, *bla ctx-m*	Monobactams, Cephalosporins, carbapenems and β-Lactum	They inhibit the bacterial cells by interacting with the mechanism of cell wall formation.	(Shlaes, 2010)
Tetracyclines	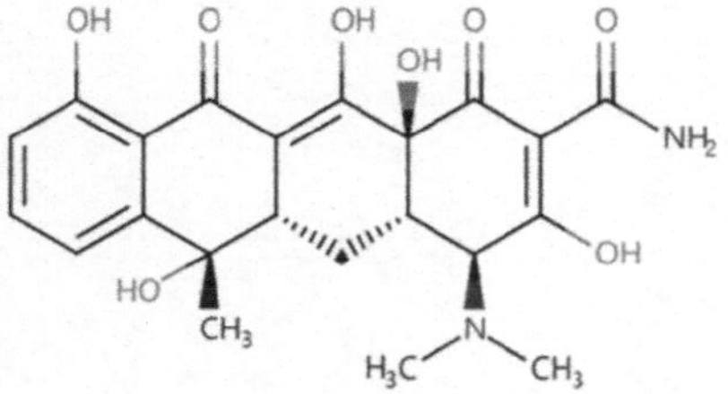	The development of resistance to tetracyclines takes place by ribosomal modifications or due to the presence of ABC-efflux pumps. It can also occur by the presence of tetracycline inactivating enzymes.	Tetracyclines	Causes inactivation of 30S ribosomal subunits	(Schaack *et al.*, 2010)
Carbepenems	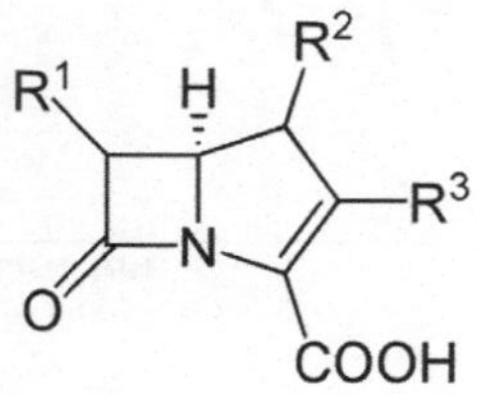	The internal factors associated with certain bacterial species like the presence of metallo-beta-lactamase, development of carbapenem resistance due to the mechanism of horizontal gene transfer, tripartitite efflux pumps and the presence of carbapenem degrading enzymes.	Carbapenem	It acts on the penicillin binding proteins (PBP)	(Meletis, 2016)

(*Continued*)

TABLE 7.1 (Continued)
List of Antibiotics, Their Action and Development of Resistance

Name of the antibiotic	Structure	Mechanism of development of resistance	Group of antibiotics	Mode of action	Reference
Glycopeptide	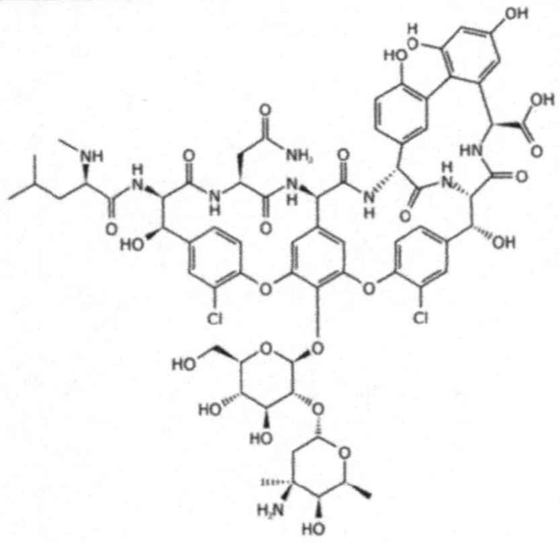	It inhibits the mechanism of cell wall biosynthesis by the mechanism of binding with D-Ala-D-Ala dipeptides of the peptidoglycan thereby preventing it to act as a substrate for the penicillin binding proteins and transglycosylases. Five types of vancomycin resistant genes are responsible for the development of resistance to the antibiotics.	Telcoplanin, vancomycin	Peptidoglycan units	(Lina *et al.*, 1999)
Colistin	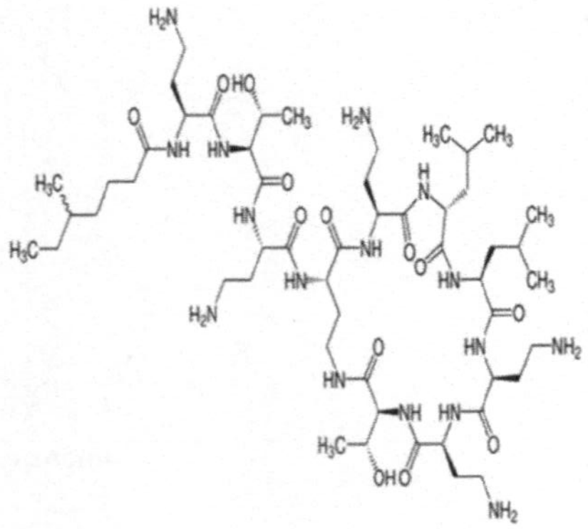	The polymyxins possess the ability to displace the stabilizing magnesium and the calcium ions by the mechanism of electrostatic interaction that is associated with the lipopolysaccharides being present on the outer layer of Gram-negative bacteria. This is responsible for rapid cell death. But the resistance genes prevent the attack of the antibiotics within the organisms.	Polymyxin E and Polymyxin B	Lipopolysaccharides	(Liu *et al.*, 2016)

Quinolone	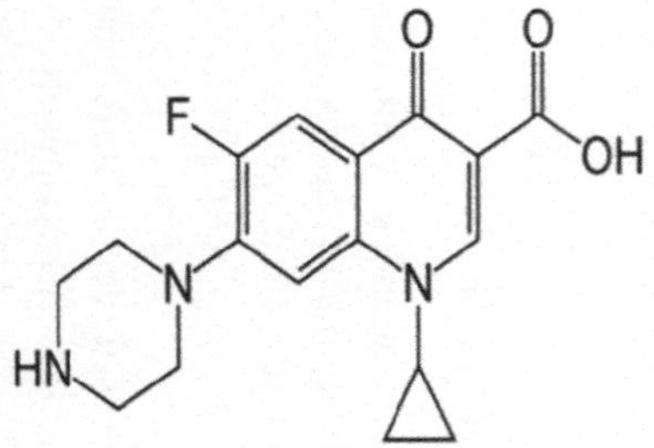	The development of resistance take place by the mechanism of target modifications caused by the mutation of parC plasmid mediated gene *gnrA* and *gyrA*. The development of resistance can also take place by the activation of the fluoroquinolone efflux pumps.	Levoflcxacin, lomefloxacin lomefloxacin, gatif:oxacin, mox:floxacin, norfloxacin, ciprofloxacin, cinoxacin	Topoisomerase I and II	(Aldred *et al.*, 2014)
Sulfonamide	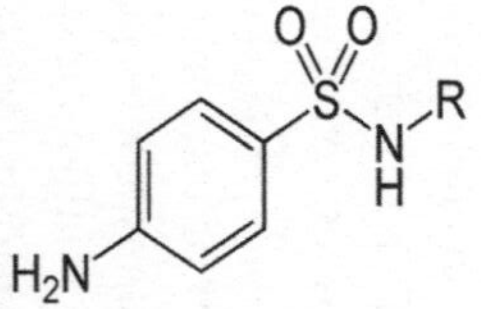	This brings about the inhibition of the enzyme dihydropteroate synthetase that is absent within human. This in turn prevents the replication of DNA thereby act as bacteriostatic drug to Gram-positive and Gram-negative bacteria			(Fair and Tor, 2014)
Aminoglycosides	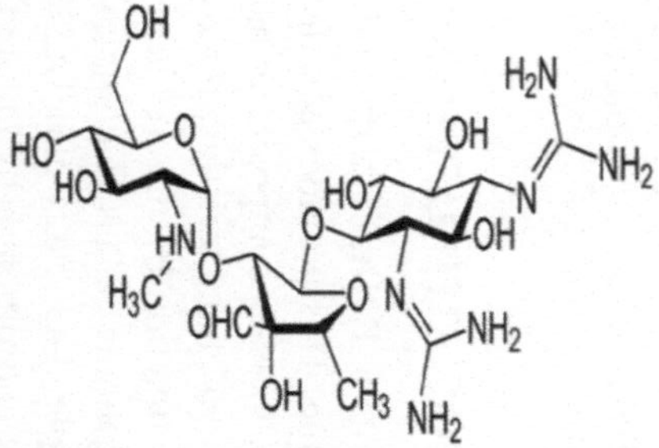	Development of resistance take place by bringing about modifications to aminoacyl transferase, nucleotidyl transferase and phosphotransferase.	Gentamycin, Amikacin, arbekacin, tobramycin, streptomycin	Inhibits the synthesis of cell wall and membrane associated proteins	(Fair and Tor, 2014)

7.9 THE GENETIC BASIS FOR THE DEVELOPMENT OF ANTIBIOTIC RESISTANCE

The origin of bacterial species on the planet dates back to more than a billion years ago and gradually their capacity to resist antimicrobial drugs has increased with the emergence of various antimicrobial drugs and antibiotics. This efficacy is attributed to their genetic makeup and unique ability to transfer the ARG through various ways (Woodford *et al.*, 2011). These transfers of antimicrobial resistant genes is accomplished by a section of mobile genetic elements comprising of transposons, plasmids and integrons. The presence of integrons and transposons has brought about genomic plasticit,y contributing robustness in the bacterial species to act against various types of environmental stresses. The transfer of integrons takes place by the plasmids thereby enhancing resistance within the various types of microbial species (Xu *et al.*, 2011). Integrons, the assembly platforms, were found to be involved with the initial occurrences of multidrug resistance and in the year 1989, various mechanisms for the transfer of integrons were reported in a process of constant scrutinization (Mazel, 2006). It has been further observed that the transfer of the mobile genetic elements plays an important role in evolution and genetic diversity amongst microbial species.

7.10 PLASMID MEDIATED DEVELOPMENT OF RESISTANCE

The global spread of resistance takes place through the plasmids that are responsible for the movement of various types of plasmid borne genes (Carattoli, 2013). The mobilization of the resistance plasmid takes place from one cell to another through the mechanism of conjugation. These conjugative plasmids possess the ability to exhibit both broad and narrow host ranges. The mechanism of homologous recombination is responsible for excision, integration and recombination. The mechanism of the transfer of the conjugative plasmid has been studied within the agar plates and well-mixed broth conditions that do not properly provide the conditions for the transfer of plasmids within the biofilm. The mechanism of horizontal transfer of genes take place between the cells via the mechanism of plasmid transfer through the conjugation process. This mechanism of the transfer of plasmids is more common in sessile cells in comparison to their planktonic counter-parts because the cells are placed in close proximity (Savage *et al.*, 2013). It has been observed that the transfer frequency of the multi-drug resistant plasmids within *Staphylococcus aureus* is 10 000 times faster within the biofilm in comparison with the sessile forms (Savage e*t al.*, 2013). The IncP-1 plasmid found within *E. coli* is responsible for the development of resistance against diverse groups of antibiotics (Król *et al.*, 2011). It has been further observed that the plasmids with *E. faecalis* results in a manyfold increase of antibiotic resistance.

7.11 TRANSPOSON ASSOCIATED DEVELOPMENT OF RESISTANCE

Transposons, the small fragments of DNA, have the ability to alter the DNA sequences by jumping to different loci within a genome. This provides additional flexibility to the genome and provides an altered genetic background. Class I composite transposons comprise of resistance genes with identical structural and functional characteristics, for example, Tn5, Tn9, Tn10, Tn903, Tn1525, and Tn2350 (Table 7.2). They function by the mechanism of transcription and reverse transcription before entering into the genome. Class II complex transposons possess three dissimilar but interrelated families, for example, Tn3, Tn21 and Tn2501. Unlike class I, class II transposons do not involve RNA intermediates and function by cut and paste methods. This involves transposase enzymes that cut the target DNA to produce sticky ends followed by its transposition to the intended site. Transposons have also been reported to jump from chromosomal DNA to plasmid DNA and vice versa for the development of resistance (Sultan *et al.*, 2018).

TABLE 7.2
Transposons That Are Associated with the Development of Antimicrobial Resistance

Class of transposons	Family	Resistant gene	Organisms associated resistance	Mode of resistance	Reference
Class 1 (IntI1)	Tn402	*aadA*–encoding streptomycin and spectinomycin resistance, *dfrA*	Gram-positive bacteria includes *Enterococcus, Streptococcus, Corynebacterium, Aerococcus, Brevibacterium,* and *Staphylococcus* whereas the Gram-nagative organisms include *E.coli, Aeromonas, Pseudomonas, Klebsiella, Stenotrophomonas, Providencia, Enterobacter, Burkholderia, Mycobacterium, Campylobacter, Alcaligenes, Acinetobacter, Citrobacter, Vibrio,* and *Salmonella, Serratia, Shigella*	The retrons that are associated with the genome of the bacterial species are responsible for the synthesis of reverse transcriptase. It has its potent role in the mechanism of enhancing the antimicrobial resistance. These are also responsible in enhancing the resistance of the bacterial cells to phages.	(Toro and Nisa-Martínez, 2014)
Class 2 (IntI3)	Tn1825, Tn1826, and Tn4132	*attI2*–encoding recombination site, *tnsA, tnsB, tnsC, tnsD* and *tnsE*) associated with movement, streptothricin acetyltransferase (*sat1*), adenyltransferase (*aadA1*), dihydrofolate reductase (*dfrA1*)	Gram-positive bacteria includes *E. faecalis* whereas the Gram-nagative organisms include Salmonella, *Enterobacteriaceae, Acinetobacter*, and *Psuedomonas*	The Fanzor proteins that are expressed by this class of transposons are responsible for the development of antimicrobial resistance	(Bao and Jurka, 2013)
Class 3 (IntI3)	Tn7	*attI3*–encoding recombination site,	Gram-negative bacteria includes *Klebsiella pneumoniae Acinetobacter, Alcaligenes, Citrobacter, Escherichia, Pseudomonas, Salmonella,* and *Serratia*	They act as carrier of various types of antibiotic resistant genes and develops resistance to spectinomycin, trimethoprim and streptomycin	(Waddell and Craig, 1988)

(Continued)

TABLE 7.2 (Continued)
Transposons That Are Associated with the Development of Antimicrobial Resistance

Class of transposons	Family	Resistant gene	Organisms associated resistance	Mode of resistance	Reference
	Insertional sequence	*ISAba1*	*Acinetobacter baumannii, P. aeruginosa*	They are responsible for the development of mutation and resulting in exhibiting various types of phenotypes by getting transmitted within the genome of the bacterial cells. The transmission of the insertional sequences among the bacterial cells results in the development of resistance. They are also responsible for the upstream regulation of resistance genes to penicillin, cephamycin, metronidazole and clindamycin. The PrD protein expressed by the Insertional sequences results in the development of resistance to carbapenems in *Pseudomonas aeruginosa.*	(Babakhani and Oloomi, 2018)
	Tn5	ble^R, kan^R, ble^R	Gram-negative bacterial cells include *Agrobacterium, Methylobacterium, Caulobater, Pseudomonas* and *Acinetobacter.*	Tn5 remains associated with the insertional sequences and results in the exhibiting resistance to kanamycin or neomycin (kan^R), streptomycin (str^R) and bleomycin (ble^R)	(Ross *et al.*, 2014)
	Tn10	*tetR, tetA,tetC, tetD*	*Vibrio, Pseudomonas, Proteus, Salmonella* and *Klebsiella*	*tetA* associated with Tn10 is responsible for the tetracycline efflux pump, *tetR* results in the repression of protein for tetracyclines. It also helps in the activation of efflux system within certain group of bacterial species.	(Teo *et al.*, 2002)
	Tn9	*cat* resistance gene		*cat* genes are associated with the development of resistance.	(Partridge, 2011)
	Tn6	kan^R		Results in the development of resistance to kanamycin or neomycin	(Mollet *et al.*, 1985)
	Tn1	*bla*		Results in the development of resistance to β-lactum group of drugs	(Bailey *et al.*, 2011)

TABLE 7.3
Integrons that are associated with the development of antimicrobial resistance

Class of integrons	Description	Reference
Class I	They are especially found within the groups of non-pathogenic bacterial species that include *Azoarcus, Aquabacterium, Hydrogenophaga* and *Thauera.* They are predominantly found within the bacteria found in the environment and provides them with resistance to various types of extrernal stresses. They are also responsible for the activation of the efflux pump which grateful influence the development of resistance within the organism.	(Gillings, 2014)
Class II	This integrin remain associated with Tn7 transposons. They are also responsible for the development of resistance to different types of antibiotics.	(Rodríguez-Minguela *et al.*, 2009)
Class III	They are also found to originate from the environmental bacterial species. They are responsible for the expression of proteins via the cassette of genes whose function is still not known.	(Gillings, 2014)

7.12 INTEGRONS IN THE DEVELOPMENT OF ANTIBIOTIC RESISTANCE

Integrons are the genetic elements that help in the expression and capture of various types of exogenous genes. They help bacteria to grow and survive in stressed environments (Table 7.3). They provide resistance to a wide range of antimicrobials by harbouring antibiotic resistance genes within chromosomes and plasmids (Gillings, 2014). They can be classified based on the amino acid sequences of integrase IntI. IntI1 are referred as class 1, IntI2 as class 2, IntI3 as class 3and so forth. Integrase IntI1, IntI2, and IntI3 are reported as mobile genetic elements, while IntI4 was linked with chromosomal integrons.

7.13 BIOFILM AND THE DEVELOPMENT OF ANTIBIOTIC RESISTANCE

Biofilms are the shield for microbial colonies against various types of chemical and environmental stressors such as antibiotics, due to the presence of EPS (Høiby *et al.*, 2010). It has been observed the microbial species dwelling within the biofilm shows many-fold enhanced resistance to various types of antimicrobial agents in comparison to the planktonic forms (Lahiri, Nag, Ghosh, *et al.*, 2021). The enhancement in the resistance to antibiotics takes place due to the improper penetration of drugs via the polymeric matrix associated with the biofilm, the presence of cells showing the phenotype of antibiotic resistance termed as persisters and the presence of various types of non-growing cells or cells that help in triggering various types of stress responses during unfavourable conditions (Stewart, 2002). The expression of ARGs or various types of extrachromosomal elements along with the presence of these conventional factors results in the development of antibiotic resistance within the organisms (Lahiri, Nag, Ghosh, *et al.*, 2021). The bacterial cells possessing the ability to produce β-lactamase show an increased level of resistance to the β-lactam groups of antibiotics (Anderl *et al.*, 2000). It has been observed that ampC genes that are associated with *Pseudomonas aeruginosa* show an enhanced resistance to antibiotics such as imipenem (Bagge *et al.*, 2004).

Biofilm acts as an additional mode of defense mechanism against various types of antibiotics. Studies have shown that sub-inhibitory concentrations of aminoglycosides result in the induction of biofilm by *Escherichia coli* and *Pseudomonas aeruginosa* as a mechanism of defense response (Hoffman *et al.*, 2005). It has also been observed that *Bacillus cereus* biofilm treated with azithromycin showed a considerable amount of revival (Lahiri *et al.*, 2020). It was further observed that the biofilm associated with *Alcaligenes faecalis* showed negligible degradation in the presence of the antibiotic, amoxicillin (Lahiri, Nag, Dutta, Mukherjee, *et al.*, 2021). Study further showed that the conventional antibiotics showed negligible degradation of the biofilm formed by *P. aeruginosa* in comparison to natural compounds (Lahiri, Nag, Dutta, Dey, *et al.*, 2021). The mechanism associated with the development of antibiotic resistance within the sessile communities existing within the biofilm results in the development of various types of chronic infections.

7.14 POLYSACCHARIDE ASSOCIATED DEVELOPMENT OF RESISTANCE WITHIN THE BIOFILM

The EPS associated with the biofilm is chemically comprised of repeating units of D-glucose, L-rhamnose and L-rhamnose (Byrd *et al.*, 2009). Psl produced by *P. aeruginosa* is an exopolysaccharide that plays a predominant role in the development of resistance against ciprofloxacin, polymyxin B, tobramycin and colistin at various stages of the development of biofilm (Billings *et al.*, 2013). It has been observed that the Psl is responsible for the development of resistance to various types of non-ionic surfactants that are responsible for the hindrance of the development of biofilm (Zegans *et al.*, 2012). It has been observed that Psl is associated with the development of resistance against various types of antibiotics that are cationic in nature comprising of tobramycin and colistin (Lahiri, Nag, Ghosh, *et al.*, 2021). The glycocalyx present within the biofilm formed by *P. aeruginosa* is expressed by the gene located at *pelABCDEFG* which not only provides structural support but also aids in the development of resistance (Franklin *et al.*, 2011).

7.15 INTRINSIC RESISTANCES, ACQUIRED AND ADAPTIVE RESISTANCE

The reduction in the susceptibility of the bacterial cells to any antibiotics is associated with the presence of various types of intrinsic factors. Gram-negative bacterial species develop instrinsic resistance with alteration in the semi-permeability of the outer membrane thereby preventing the uptake of antibiotics. The asymmetric nature of the outer membrane of the bacterial cells is due to the presence of polyanionic polysaccharide along with a layer of phospholipids (Lahiri, Nag, Ghosh, *et al.*, 2021). Reduction in the permeability of the membrane of *P. aeruginosa* acts as a barrier to the penetration of the antibiotics such as quinolones and lactams. The prsA gene is associated with the regulation of transcription, and upregulation of this gene results in the decrease of the susceptibility of the microbial cells to the antibiotics (Lahiri, Nag, Ghosh, *et al.*, 2021). The PhoQ gene is responsible for the development of biofilm and thereby enhances its resistance to antibiotics and various types of antimicrobials.

7.16 MUTATION

The sessile microbial species shows enhanced mutation in comparison to its planktonic counterpart (Driffield *et al.*, 2008) along with the enhancement of the process of horizontal transfer of genes within the biofilm (Molin and Tolker-Nielsen, 2003). These genetic changes are responsible for the development of resistance to various types of drugs. Various newly formed bacterial cells encode different types of enzymes that are responsible for the degradation of antibiotics as well as alterations on the targeted sites. Cystic fibrosis is caused by a mutation taking place within the population of the bacterial species (Høiby *et al.*, 2011). Cystic-fibrosis comprises of hypermutable phenotypes of *P. aeruginosa* that bring about alteration in the repair of DNA that forms a major part in the mismatch repair system. The genes that are responsible for the repair of the DNA comprise of *mutM* and *mutT* (Mandsberg *et al.*, 2009). These mutations are responsible for the development of antibiotic resistance as they result in the enhancement of the multidrug efflux pump (Maciá *et al.*, 2005). The mutatations are also responsible for the enhancement of the reactive oxygen species causing an imbalance between oxidative burden and antioxidant defense (Driffield *et al.*, 2008).

TABLE 7.4
Genes Responsible for the Development of Antibiotic Resistance

The genes associated	Antibiotics	Functionality	Mechanism of protection	References
pqqC, erbR exaA,	Tobramycin	Helps in the mechanism of ethanol oxidation	Unknown	(Kohanski *et al.*, 2010)
ndvB	Ciprofloxacin Tobramycin, Gentamicin,	Glucosyltransferase	Helps in the upregulation of ethanol oxidation and sequestration of antibiotics	(Thomas and Nielsen, 2005)
brlR	Chloramphenicol, norfloxacin, tobramycin, tetracycline, trimethoprim, kanamycin	Transcriptional regulator	Upregulation of multidrug efflux pumps	(Spoering and Lewis, 2001)
pslABCDEFGHIJKLMNO	Tobramycin, Colistin, ciprofloxacin polymyxin B,	Associated with the biosynthesis of Psl	Unknown	(Zheng and Stewart, 2004)
sagS	Norfloxacin Tobramycin,	It is a hybrid of two components	The increase in c-di-GMP results in the activation of BrlR	(Webb *et al.*, 2003)
PA2070	Gentamicin Tobramycin,	Receptor is depependent on TonB	Unknown	(Lechner *et al.*, 2012)
rapA	Gentamicin, norfloxacin, penicillin G, chloramphenicol	They encodes helicase-like proteins	Associated with the synthesis of EPS and upregulation of *YhcQ*	(Van Acker and Coenye, 2016)
yafQ	Cefazolin Tobramycin,	Acts as a toxin	It results in the development of antibiotic resistant organisms	(Kohanski *et al.*, 2010)
dltABCD	Gentamicin	Teichoic acid D-alanylation take place.	Responsible for the reduction in the negative charge upon the cell wall.	(Van Acker and Coenye, 2016)
PA0756-0757	Gentamicin Tobramycin,	Two-component system	Unknown	(Kohanski *et al.*, 2010)
hcp1 *tssC1,*	Ciprofloxacin, gentamicin, tobramycin,	Type VI secretion components	Unknown	(Webb *et al.*, 2003)
PA1875-1877	Tobramycin, gentamicin, ciprofloxacin	Efflux of antibiotics associated with specific biofilm	Antibiotic efflux from the cells	(Zheng and Stewart, 2004)
pelABCDEFG	Tobramycin, gentamicin	Pel biosynthetic enzymes	Unknown	(Webb *et al.*, 2003)
epaI	Daptomycin	Glycolsyltranferase	Unknown	(Whiteley *et al.*, 2001)
fsrC fsrA,	Gentamicin, daptomycin, linezolid	Quorum-sensing players	Unknown	(Thomas and Nielsen, 2005)
PA5033	Tobramycin, gentamicin	Hypothetical proteins	Unknown	(Webb *et al.*, 2003)
epaOX	Gentamicin	Glycolsyltranferase	Maintenance of cell wall integrity	(Whiteley *et al.*, 2001)
gelE	Daptomycin, gentamicin, linezolid	Gelatinase	Unknown	(Spoering and Lewis, 2001)
relA, spoT	Ofloxacin, meropenem, colistin, gentamicin	Players in the stringent response	Upregulate antioxidant defences and downregulate pro-oxidants	(Zheng and Stewart, 2004)

TABLE 7.5
Genes Regulating the Various Mechanisms of Antibiotic Resistance

Type	Gene	Gene(s) included	Mechanism	Genera
A-1	*catA1*	*pp-cat, cat, catI,*	Helps in enzyme inactivation	*Escherichia, Acinetobacter,*
A-2	*catA2*	*catII, cat,*	Helps in enzyme inactivation	*Agrobacterium, Aeromonas, Haemophilus, Escherichia, Salmonella Photobacterium*
A-3	*catA3*	*catIII, cat,*	Helps in enzyme inactivation	*Edwardsiella, Actinobacillus, Shigella, Mannheimia, Klebsiella, Pasteurella*
A-4	*Cat*		Helps in enzyme inactivation	*Proteus*
A-5	*Cat*		Helps in enzyme inactivation	*Streptomyces*
A-6	*cat86*		Helps in enzyme inactivation	*Bacillus*
A-7	*cat(pC221)*	*catC, cat,*	Helps in enzyme inactivation	*Enterococcus,Bacillus, Staphylococcus, Streptococcus, Lactobacillus*
A-8	*cat(pC223)*	*Cat*	enzyme inactivation	*Lactococcus,Enterococcus, Staphylococcus, Listeria, Streptococcus*
A-9	*cat(pC194)*	*cat, cat-TC*	enzyme inactivation	*Enterococcus,Bacillus, Staphylococcus, Lactobacillus, Streptococcus*
A-10	*Cat*		enzyme inactivation	*Bacillus*
A-11	*catP*	*catD*	enzyme inactivation	*Neisseria, Clostridium*
A-12	*catS*		enzyme inactivation	*Streptococcus, Aeromonas,*
A-13	*Cat*		enzyme inactivation	*Campylobacter*
A-14	*Cat*		enzyme inactivation	*Proteus, Photobacterium, Listonella,*
A-15	*catB*		enzyme inactivation	*Clostridium*
A-16	*catQ*		enzyme inactivation	*Streptococcus, Clostridium,*
B-1	*catB1*	*Cat*	enzyme inactivation	*Agrobacterium*
B-2	*catB2*		enzyme inactivation	*Aeromonas, Acinetobacter, Escherichia, Bordetella, Pasteurella, Klebsiella,*
B-3	*catB3*	*catB8, catB6, catB4, catB5,*	enzyme inactivation	*Acinetobacter, Aeromonas, Bordetella, Enterobacter,*
B-4	*catB7*		enzyme inactivation	*Pseudomonas*
B-5	*catB9*		enzyme inactivation	*Vibrio*
B-6	*catB10*		enzyme inactivation	*Pseudomonas*
E-1	*cmlA1*	*cmlA2,cmlA, cmlA4, cmlA5, cmlA6, cmlA7, cmlA8, cmlA10,*	Efflux of antibiotics	*Aeromonas,Acinetobacter, Escherichia, Klebsiella, Pseudomonas, Salmonella, Arcanobacterium,*
E-2	*cml*		Efflux of antibiotics	*Escherichia*
E-3	*floR*	*cmlA-like, flo, pp-flo, cmlA9*	Efflux of antibiotics	*Acinetobacter, Aeromonas, Bordetella, Pasteurella, Salmonella, Stenotrophomonas, Vibrio*
E-4	*fexA*		Efflux of antibiotics	*Bacillus, Staphylococcus, Corynebacterium,*
E-5	*cml*		Efflux of antibiotics	*Pseudomonas*
E-6	*cmlv*		Efflux of antibiotics	*Staphylococcus*
E-7	*cmrA*	*Cmr*	Efflux of antibiotics	*Uncultured*
E-8	*cmr*	*Cmx*	Efflux of antibiotics	*Shigella, Escherichia, Acinetobacter, Salmonella, Klebsiella, Serratia,*

7.17 EXTRACELLULAR DNA DEPENDENT DEVELOPMENT OF RESISTANCE

Extracellular DNA or eDNA is one of the most important components that remains associated with the matrix of the biofilm. The process of quorum sensing is responsible for the endogenous production of eDNA, also being produced by the polymorphonuclear leukocytes at the site of infection (Allesen-Holm *et al.*, 2006). The eDNA has a potent role in the development of the antimicrobial resistance (Lahiri, Nag, Ghosh, *et al.*, 2021). Studies have shown that various types of sub-inhibitory concentrations of antibiotics like vancomycin result in the enhancement of eDNA thereby enhancing antimicrobial resistance (Lahiri, Nag, Ghosh, *et al.*, 2021). Chelation of cations like magnesium can be achieved by the use of eDNA thereby causing a marked decgrease of Mg^{2+} ions within the environment (Lahiri, Nag, Ghosh, *et al.*, 2021). The reduction in the Mg^{2+} results in the activation of environmental signals that further activate PmrAB and PhoPQ systems with *P. aeruginosa* and *S. enterica* (McPhee *et al.*, 2006). The eDNA accumulation within the matrix of *P. aeruginosa* results in the creation of an acidic environment thereby reducing the concentration of Mg^{2+} subsequently causing the activation of PhoPQ and PmrAB signaling pathways (Lahiri, Nag, Ghosh, *et al.*, 2021). The eDNA within the biofilm of *Myxococcus xanthus* interacts with the polysaccharide, resulting in the enhancement of resistance to any sort of biological stress (Hu *et al.*, 2012).

7.18 THE INSERTION SEQUENCE (IS)

ISs are the groups of mobile elements responsible for carrying more than one transposase gene (tnp). They comprise of a strong region of promoters that are responsble for for the development of intrinsic factor responsible for the development of resistance (Lahiri, Nag, Ghosh, *et al.*, 2021).

7.19 GENES RESPONSIBLE FOR THE DEVELOPMENT OF ANTIBIOTIC RESISTANCE

*Genes play an integral role in the mechanism of development of resistance to antibiotics (*Table 7.4*). The genes that are associated with the integrons result in the deactivation of the efflux pumps, inactivation of the enzymes, genetic modification and alteration in the target sites (*Table 7.5*) (*Munita *and Arias, 2016).*

7.20 CONCLUSION

The rampant use of antibiotics in present day conditions has resulted in the development of antibiotic resistance that has brought about a growing and significant threat to the environmental and public health. This problem on global and local scales needs to be properly dealt with by understanding the mechanism of the development of antibiotic resistance. It has been observed that the biofilms, being the syntrophic consortia of microbial species serve as reservoirs of various types of antibiotic resistant genes exerting various resistance mechanisms. Biofilm matrix acts as a potent barrier to the diffusion of antibiotics participating in the development of antibiotic resistance. Several genes constituting the resistome suitable for inhibition of antibiotic entry, antibiotic destruction and antibiotic binding to protein synthetic machinery together make the biofilm totally resistant to antimicrobials. Understanding and specific blocking of these resistomes will definitely make the microbes susceptible and overcome the risk of antibiotic resistance.

REFERENCES

Acker, H. and Coenye, T. (2016) The role of efflux and physiological adaptation in biofilm tolerance and resistance. *J Biol Chem.* **291**:12565–12572.

Aldred, K. J., Kerns, R. J., and Osheroff, N. (2014) Mechanism of quinolone action and resistance. *Biochemistry.* **53**:1565–1574.

Allesen-Holm, M., Barken, K. B., Yang, L., Klausen, M., Webb, J. S., Kjelleberg, S., Molin, S., Givskov, M. (2006) A characterization of DNA release in *pseudomonas aeruginosa* cultures and biofilms. *Mol Microbiol.* **59**:1114–1128.

Anderl, J. N., Franklin, M. J., and Stewart, P. S. (2000) Role of antibiotic penetration limitation in *klebsiella pneumoniae* biofilm resistance to ampicillin and ciprofloxacin. *Antimicrob Agents Chemother.* **44**:1818–1824.

Babakhani, S. and Oloomi, M. (2018) Transposons: The agents of antibiotic resistance in bacteria. *J Basic Microbiol.* **58**:905–917.

Bagge, N., Schuster, M., Hentzer, M., Ciofu, O., Givskov, M., Greenberg, E. P., and Høiby, N. (2004) Pseudomonas aeruginosa biofilms exposed to imipenem exhibit changes in global gene expression and beta-lactamase and alginate production. *Antimicrob Agents Chemother.* **48**:1175–1187.

Bailey, J. K., Pinyon, J. L., Anantham, S., and Hall, R. M. (2011) Distribution of the blaTEM gene and blaTEM-containing transposons in commensal *escherichia coli. J Antimicrob Chemother.* **66**:745–751.

Bao, W. and Jurka, J. (2013) Homologues of bacterial TnpB_IS605 are widespread in diverse eukaryotic transposable elements. *Mob DNA.* **4**:12.

Bartscht, K., Cypionka, H., and Overmann, J. (1999) Evaluation of cell activity and of methods for the cultivation of bacteria from a natural lake community. *FEMS Microbiol Ecol.* **28**:249–259.

Billings, N., Millan, M., Caldara, M., Rusconi, R., Tarasova, Y., Stocker, R., and Ribbeck, K. (2013) The extracellular matrix component psl provides fast-acting antibiotic defense in *pseudomonas aeruginosa* biofilms. *PLoS Pathog.* **9**:e1003526–e1003526.

Boisvert, A. A., Cheng, M. P., Sheppard, D. C., Nguyen, D. (2016). Microbial biofilms in pulmonary and critical care diseases. *Ann. Am Thorac Soc.* **13**:1615–1623.

Brook, I., Yocum, P. (1995). Bacteriology and beta-lactamase activity in ear aspirates of acute otitis media that failed amoxicillin therapy. *Pediatr. Infect Dis J.* **14**:805–807.

Burmølle, M., Ren, D., Bjarnsholt, T., Sørensen, S. J. (2014) Interactions in multispecies biofilms: Do they actually matter?. *Trends Microbiol.* **22**(2):84–91.

Bush, K. (2010). Bench-to-bedside review: The role of β-lactamases in antibiotic-resistant gram-negative infections. *Crit Care.* **14**:224. DOI: 10.1186/cc8892

Byrd, M. S., Sadovskaya, I., Vinogradov, E., Lu, H., Sprinkle, A. B., Richardson, S. H., Ma, L., Ralston, B., Parsek, M. R., Anderson, E. M, Lam, J. S., Wozniak, D. J. (2009) Genetic and biochemical analyses of the *pseudomonas aeruginosa* psl exopolysaccharide reveal overlapping roles for polysaccharide synthesis enzymes in psl and lps production. *Mol Microbiol.* **73**:622–638.

Carattoli, A. (2013) Plasmids and the spread of resistance. *Int J Med Microbiol.* **303**:298–304.

Chekabab, S. M., Silverman, R. J., Lafayette, S. L., Luo, Y., Rousseau, S., Nguyen, D. (2015) *Staphylococcus aureus* inhibits IL-8 responses induced by *pseudomonas aeruginosa* in airway epithelial cells. *PLoS One.* **10**:e0137753. DOI: 10.1371/journal.pone.0137753

Christensen, B. B., Sternberg, C., Molin, S. (1996) Bacterial plasmid conjugation on semi-solid surfaces monitored with the green fluorescent protein (GFP) from *aequorea victoria* as a marker. *Gene.* **173**:59–65. DOI: 10.1016/0378-1119(95)00707-5

Cianfanelli, F. R., Monlezun, L., Coulthurst, S. J. (2016) Aim, load, fire: The type VI secretion system, a bacterial nanoweapon. *Trends Microbiol.* **24**(1):51–62.

Ciofu, O., Beveridge, T. J., Kadurugamuwa, J., Walther-Rasmussen, J., Høiby, N. (2000) Chromosomal beta-lactamase is packaged into membrane vesicles and secreted from *pseudomonas aeruginosa. J Antimicrob Chemother.* **45**:9–13. DOI: 10.1093/jac/45.1.9

Cvitkovitch, D. G., Li, Y-H., Ellen, R. P. (2003) Quorum sensing and biofilm formation in streptococcal infections. *J Clin Invest.* **112**:1626–1632.

Driffield, K., Miller, K., Bostock, J. M., O'Neill, A. J., and Chopra, I. (2008) Increased mutability of *pseudomonas aeruginosa* in biofilms. *J Antimicrob Chemother.* **61**:1053–1056.

Duan, K., Dammel, C., Stein, J., Rabin, H., Surette, M. G. (2003) Modulation of *pseudomonas aeruginosa* gene expression by host microflora through interspecies communication. *Mol Microbiol.* **50**:1477–1491. DOI: 10.1046/j.1365-2958.2003.03803.x

Fair, R. J. and Tor, Y. (2014) Antibiotics and bacterial resistance in the 21st century. *Perspect Medicin Chem.* **6**:25–64.

Franklin, M. J., Nivens, D. E., Weadge, J. T., and Howell, P. L. (2011) Biosynthesis of the *pseudomonas aeruginosa* extracellular polysaccharides, Alginate, Pel, and Psl. *Front Microbiol.* **2**:167.

Gillings, M. R. (2014) Integrons: Past, present, and future. *Microbiol Mol Biol Rev*. **78**:257–277.

Giwercman, B., Meyer, C., Lambert, P. A., Reinert, C., Høiby, N. (1992) High-level beta-lactamase activity in sputum samples from cystic fibrosis patients during antipseudomonal treatment. *Antimicrob Agents Chemother*. **36**:71–76. DOI: 10.1128/aac.36.1.71

Hannan, S., Ready, D., Jasni, A. S., Rogers, M., Pratten, J., Roberts, A. P. (2010) Transfer of antibiotic resistance by transformation with eDNA within oral biofilms. *FEMS Immunol Med Microbiol*. **59**:345–349. DOI: 10.1111/j.1574-695X.2010.00661.x

Hoffman, L. R., D'Argenio, D. A., MacCoss, M. J., Zhang, Z., Jones, R. A., and Miller, S. I. (2005) Aminoglycoside antibiotics induce bacterial biofilm formation. *Nature*. **436**:1171–1175.

Høiby, N., Bjarnsholt, T., Givskov, M., Molin, S., and Ciofu, O. (2010) Antibiotic resistance of bacterial biofilms. *Int J Antimicrob Agents*. **35**:322–332.

Høiby, N., Ciofu, O., Johansen, H. K., Song, Z., Moser, C., Jensen, P. Ø., Molin, S., Givskov, M., Tolker-Nielsen, T., Bjarnsholt, T. (2011) The clinical impact of bacterial biofilms. *Int J Oral Sci*. **3**:55–65.

Hu, W., Li, L., Sharma, S., Wang, J., McHardy, I., Lux, R., Yang, Z., He, X., Gimzewski, J. K., Li, Y., Shi, W. (2012) DNA builds and strengthens the extracellular matrix in *myxococcus xanthus* biofilms by interacting with exopolysaccharides. *PLoS One*. **7**:e51905.

Kohanski, M. A., Dwyer, D. J., and Collins, J. J. (2010) How antibiotics kill bacteria: From targets to networks. *Nat Rev Microbiol*. **8**:423–435.

Król, J. E., Nguyen, H. D., Rogers, L. M., Beyenal, H., Krone, S. M., and Top, E. M. (2011) Increased transfer of a multidrug resistance plasmid in *escherichia coli* biofilms at the air-liquid interface. *Appl Environ Microbiol*. **77**:5079–5088.

Kanger K, Guilford N. G. H, Lee H, Nesbø C. L., Truu J, Edwards E. A. (2020) Antibiotic resistome and microbial community structure during anaerobic co-digestion of food waste, paper and cardboard. *FEMS Microbiol Ecol*. Feb 1; **96**(2):fiaa006.

Kolenbrander, P. E. (2000) Oral microbial communities: Biofilms, interactions, and genetic systems. *Annu Rev Microbiol*. **54**:413–437.

Lahiri, D., Dash, S., Dutta, R., and Nag, M. (2019) Elucidating the effect of anti-biofilm activity of bioactive compounds extracted from plants. *J Biosci*. **44**:52.

Lahiri, D., Nag, M., Dutta, B., Dash, S., Ghosh, S., and Ray, R. R. (2020) Synergistic effect of quercetin with allicin from the ethanolic extract of *allium cepa* as a potent anti-quorum sensing and anti-biofilm agent against oral biofilm. In *Advances in Bioprocess Engineering and Technology*. Singapore: Springer Singapore, pp. 69–81.

Lahiri, D., Nag, M., Dutta, B., Dey, S., Mukherjee, D., Joshi, S. J., and Ray, R. R. (2021) Antibiofilm and anti-quorum sensing activities of eugenol and linalool from *ocimum tenuiflorum* against *pseudomonas aeruginosa* biofilm. *J Appl Microbiol*. **n/a**:

Lahiri, D., Nag, M., Dutta, B., Mukherjee, I., Ghosh, S., Dey, A., Banerjee, R., Rani Ray, R. (2021) Catechin as the most efficient bioactive compound from *azadirachta indica* with antibiofilm and anti-quorum sensing activities against dental biofilm: An in vitro and in silico study. *Appl Biochem Biotechnol*. **193**:1617–1630.

Lahiri, D., Nag, M., Ghosh, A., Das, D., Dey, A., Mukherjee, D., et al. (2021) Biofilm and antimicrobial resistance. In *Biofilm-Mediated Diseases: Causes and Controls*. Ray, R. R., Nag, M., and Lahiri, D. (eds). Singapore: Springer Singapore, pp. 183–208.

Lechner, S., Lewis, K., and Bertram, R. (2012) *Staphylococcus aureus* persisters tolerant to bactericidal antibiotics. *J Mol Microbiol Biotechnol*. **22**:235–244.

Lee, J., Lee E. Y., Kim, S. H., Kim, D. K., Park, K. S., Kim, K. P., Kim, Y. K., Roh, T. Y., Gho, Y. S. (2013) *Staphylococcus aureus* extracellular vesicles carry biologically active β-lactamase. *Antimicrob Agents Chemother*. **57**:2589–2595. DOI: 10.1128/AAC.00522-12

Lee, K., Kim, D. W., Lee, D. H. Kim, Y. S., Bu, J. H., Cha, J. H., Thawng, C. N., Hwang, E. M., Seong, H. J., Sul, W. J., Wellington, E. M. H., Quince, C., Cha, C. J. (2020) Mobile resistome of human gut and pathogen drives anthropogenic bloom of antibiotic resistance. *Microbiome*. **8**(2).

Lewis, K. (2008) Multidrug tolerance of biofilms and persister cells. *Curr Top Microbiol Immunol*. **322**:107–131.

Li, Y. H., Lau, P. C., Lee, J. H., Ellen, R. P., Cvitkovitch, D. G. (2001) Natural genetic transformation of *Streptococcus mutans* growing in biofilms. J Bacteriol **183**:897–908. DOI: 10.1128/JB.183.3.897-908.2001

Li, Y-H., Tian, X. (2012) Quorum sensing and bacterial social interactions in biofilms. *Sensors (Basel)* 12:2519–2538.

Lina, G., Piémont, Y., Godail-Gamot, F., Bes, M., Peter, M. O., Gauduchon, V., Vandenesch, F., Etienne, J. (1999) Involvement of Panton-Valentine leukocidin-producing *Staphylococcus aureus* in primary skin infections and pneumonia. *Clin Infect Dis an Off Publ Infect Dis Soc Am* **29**:1128–1132.

Liu, Y. Y., Wang, Y., Walsh, T. R., Yi, L. X., Zhang, R., Spencer, J., Doi, Y., Tian, G., Dong, B., Huang, X., Yu, L. F, Gu, D., Ren, H., Chen, X., Lv, L., He, D., Zhou, H., Liang, Z., Liu, J. H., Shen, J. (2016) Emergence of plasmid-mediated colistin resistance mechanism MCR-1 in animals and human beings in China: A microbiological and molecular biological study. *Lancet Infect Dis.* **16**:161–168.

Maciá, M. D., Blanquer, D., Togores, B., Sauleda, J., Pérez, J. L., and Oliver, A. (2005) Hypermutation is a key factor in development of multiple-antimicrobial resistance in *pseudomonas aeruginosa* strains causing chronic lung infections. *Antimicrob Agents Chemother.* **49**:3382–3386.

Madsen, J. S., Burmølle, M., Hansen, L. H., Sørensen, S. J. (2012) The interconnection between biofilm formation and horizontal gene transfer. *FEMS Immunol Med Microbiol.* **65**:183–195. DOI: 10.1111/j.1574-695X.2012.00960.x

Mah, T. F. C., O'Toole, G. A. (2001) Mechanisms of biofilm resistance to antimicrobial agents. *Trends Microbiol.* 9:34–39.

Mandsberg, L. F., Ciofu, O., Kirkby, N., Christiansen, L. E., Poulsen, H. E., and Høiby, N. (2009) Antibiotic resistance in *pseudomonas aeruginosa* strains with increased mutation frequency due to inactivation of the DNA oxidative repair system. *Antimicrob Agents Chemother.* **53**:2483–2491.

Mazel, D. (2006) Integrons: Agents of bacterial evolution. *Nat Rev Microbiol.* **4**:608–620.

McPhee, J. B., Bains, M., Winsor, G., Lewenza, S., Kwasnicka, A., Brazas, M. D., Brinkman, F. S. L., Hancock, R. E. W. (2006) Contribution of the PhoP-PhoQ and PmrA-PmrB two-component regulatory systems to Mg2+-induced gene regulation in *pseudomonas aeruginosa. J Bacteriol.* **188**:3995–4006.

Meletis, G. (2016) Carbapenem resistance: Overview of the problem and future perspectives. *Ther Adv Infect Dis.* **3**:15–21.

Mitri, S., Foster, K. R. (2013) The genotypic view of social interactions in microbial communities. *Annu. Rev. Genet.* 47:247–273.

Molin, S. and Tolker-Nielsen, T. (2003) Gene transfer occurs with enhanced efficiency in biofilms and induces enhanced stabilisation of the biofilm structure. *Curr Opin Biotechnol.* **14**:255–261.

Mollet, B., Clerget, M., Meyer, J., and Iida, S. (1985) Organization of the Tn6-related kanamycin resistance transposon Tn2680 carrying two copies of IS26 and an IS903 variant, IS903. *B. J Bacteriol.* **163**:55–60.

Munita, J. M. and Arias, C. A. (2016) Mechanisms of antibiotic resistance. *Microbiol Spectr.* **4**:10.1128/microbiolspec.VMBF-0016–2015.

Oh, M, Pruden, A, Chen, C, Heath, L. S., Xia, K., Zhang, L. (2018) MetaCompare: A computational pipeline for prioritizing environmental resistome risk. *FEMS Microbiol Ecol.* Jul 1; 94(7):fiy079.

Organisation, W. H. (2014) *Antimicrobial Resistance: Global Report on Surveillance*, Geneva PP–Geneva: World Health Organization.

Partridge, S. R. (2011) Analysis of antibiotic resistance regions in gram-negative bacteria. *FEMS Microbiol Rev.* **35**: 820–855.

Percival, S. L., Hill, K. E., Williams, D. W., Hooper S. J., Thomas, D. W., Costerton, J. W. (2012) A review of the scientific evidence for biofilms in wounds. *Wound Repair Regen.* **20**:647–657.

Potron, A., Poirel, L., Nordmann, P. (2011) Plasmid-mediated transfer of the *bla*(NDM-1) gene in gram-negative rods. *FEMS Microbiol Lett.* **324**:111–116. DOI: 10.1111/j.1574-6968.2011.02392.x

Rodríguez-Minguela, C. M., Apajalahti, J. H. A., Chai, B., Cole, J. R., and Tiedje, J. M. (2009) Worldwide prevalence of class 2 integrases outside the clinical setting is associated with human impact. *Appl Environ Microbiol.* **75**:5100–5110.

Ronald, A. (2003) The etiology of urinary tract infection: Traditional and emerging pathogens. *Dis Mon.* **49**:71–82.

Ross, J. A., Trussler, R. S., Black, M. D., McLellan, C. R., and Haniford, D. B. (2014) Tn5 transposition in *escherichia coli* is repressed by Hfq and activated by over-expression of the small non-coding RNA SgrS. *Mob DNA.* **5**:27.

Savage, V. J., Chopra, I., and O'Neill, A. J. (2013) Staphylococcus aureus biofilms promote horizontal transfer of antibiotic resistance. *Antimicrob Agents Chemother.* **57**:1968–1970.

Schaack, S., Gilbert, C., and Feschotte, C. (2010) Promiscuous DNA: Horizontal transfer of transposable elements and why it matters for eukaryotic evolution. *Trends Ecol Evol*. **25**: 537–546.

Shlaes, D. M. (2010) *Antibiotics: The Perfect Storm*, 1st ed. Springer Netherlands.

Sørensen, S. J, Bailey, M., Hansen, L. H., Kroer, N., Wuertz, S. (2005) Studying plasmid horizontal transfer *in situ*: A critical review. *Nat Rev Microbiol*. **3**:700–710

Sorg, R. A., Lin, L., van Doorn, G. S., Sorg, M., Olson, J., Nizet, V., Veening, J. W. (2016) Collective resistance in microbial communities by intracellular antibiotic deactivation. *PLoS Biol*. **14**:e2000631. DOI: 10.1371/journal.pbio.2000631

Spoering, A. L. and Lewis, K. (2001) Biofilms and planktonic cells of *pseudomonas aeruginosa* have similar resistance to killing by antimicrobials. *J Bacteriol*. **183**:6746–6751.

Stewart, P. S. (2002) Mechanisms of antibiotic resistance in bacterial biofilms. *Int J Med Microbiol*. **292**: 107–113.

Sultan, I., Rahman, S., Jan, A. T., Siddiqui, M. T., Mondal, A. H., and Haq, Q. M. R. (2018) Antibiotics, resistome and resistance mechanisms: A bacterial perspective. *Front Microbiol*. **9**:2066.

Tanner, W. D., Atkinson, R. M., Goel, R. K., Toleman, M. A., Benson, L. S., Porucznik, C. A., VanDerslice, J. A. (2017) Horizontal transfer of the *bla*NDM-1 gene to *pseudomonas aeruginosa* and *acinetobacter baumannii* in biofilms. *FEMS Microbiol Lett*. **364**:fnx048. DOI: 10.1093/femsle/fnx048

Teo, J. W. P., Tan, T. M. C., and Poh, C. L. (2002) Genetic determinants of tetracycline resistance in *vibrio harveyi*. *Antimicrob Agents Chemother* **46**:1038–1045.

Teschler, J. K., Zamorano-Sánchez, D., Utada, A. S., Warner, C. J. A., Wong, G. C. L., Linington, R. G., and Yildiz, F. H. (2015) Living in the matrix: Assembly and control of *vibrio cholerae* biofilms. *Nat Rev Microbiol*. **13**:255–268.

Thomas, C. M. and Nielsen, K. M. (2005) Mechanisms of, and barriers to, horizontal gene transfer between bacteria. *Nat Rev Microbiol*. **3**:711–721.

Toro, N. and Nisa-Martínez, R. (2014) Comprehensive phylogenetic analysis of bacterial reverse transcriptases. *PLoS One*. **9**:e114083.

Waddell, C. S. and Craig, N. L. (1988) Tn7 transposition: Two transposition pathways directed by five Tn7-encoded genes. *Genes Dev*. **2**:137–149.

Watkins, R. R. and Bonomo, R. A. (2016) Overview: Global and local impact of antibiotic resistance. *Infect Dis Clin North Am*. **30**:313–322.

Webb, J. S., Thompson, L. S., James, S., Charlton, T., Tolker-Nielsen, T., Koch, B., Givskov, M., Kjelleberg, S. (2003) Cell death in *pseudomonas aeruginosa* biofilm development. *J Bacteriol*. **185**:4585–4592.

Weigel, L. M., Donlan, R. M., Shin, D. H., Jensen, B., Clark, N. C., McDougal, L. K., Zhu, W., Musser, K. A., Thompson, J., Kohlerschmidt, D., Dumas, N., Limberger, R. J., Patel, J. B. (2007) High-level vancomycin-resistant *staphylococcus aureus* isolates associated with a polymicrobial biofilm. *Antimicrob Agents Chemother*. **51**:231–238. DOI: 10.1128/AAC.00576-06

Weinstein, M. P., Murphy, J. R., Reller, L. B., Lichtenstein, K. A. (1983) The clinical significance of positive blood cultures: A comprehensive analysis of 500 episodes of bacteremia and fungemia in adults. II. Clinical observations, with special reference to factors influencing prognosis. *Clin Infect Dis*. **5**:54–70. DOI: 10.1093/clinids/5.1.54

Whiteley, M., Bangera, M. G., Bumgarner, R. E., Parsek, M. R., Teitzel, G. M., Lory, S., and Greenberg, E. P. (2001) Gene expression in *pseudomonas aeruginosa* biofilms. *Nature*. **413**:860–864.

Woodford, N., Turton, J. F., and Livermore, D. M. (2011) Multiresistant Gram-negative bacteria: The role of high-risk clones in the dissemination of antibiotic resistance. *FEMS Microbiol Rev*. **35**:736–755.

Xu, H., Broersma, K., Miao, V., and Davies, J. (2011) Class 1 and class 2 integrons in multidrug-resistant gram-negative bacteria isolated from the salmon river, british Columbia. *Can J Microbiol*. **57**:460–467.

Zegans, M. E., Wozniak, D., Griffin, E., Toutain-Kidd, C. M., Hammond, J. H., Garfoot, A., and Lam, J. S. (2012) Pseudomonas aeruginosa exopolysaccharide Psl promotes resistance to the biofilm inhibitor polysorbate 80. *Antimicrob Agents Chemother*. **56**:4112–4122.

Zheng, Z. and Stewart, P. S. (2004) Growth limitation of staphylococcus epidermidis in biofilms contributes to rifampin tolerance. *Biofilms*. **1**:31–35.

8 Nanotechnology Approaches to Treating Antimicrobial Resistant Infections Caused by Biofilms

Research Needs for Translation from In Vitro to In Vivo Applications

P. Sriyutha Murthy, Arindam Das, Bharathi Ramalingam, Raju Sekar and Sergey Dobretsov

8.1 INTRODUCTION

Biofilms are ubiquitous and are responsible for 80% of chronic infections associated with infected tissues (cystic fibrosis, urinary tract infection (UTI), periodontitis, meningitis and arthritis), wounds, skin surfaces, implants and prostheses, nonsurgical indwelling medical devices such as catheters, in clinical settings. As such, the biofilm mode offers higher resistance to antibiotics/antimicrobials as a result in various modes including those of modified phenotype, namely: increased growth, production of extracellular polymeric substances (EPS), physiological states (low metabolic activity), upregulation of drug efflux pumps, point mutation, expression of resistance genes, persister cells, quorum sensing abilities and horizontal gene transfer. Increased spread and detection of multi-drug resistant [MDR] bacteria over the last decade has established the antimicrobial resistance phenomenon which has been induced naturally due to the increased use of antibiotics and other factors. The presence of MDR bacteria in biofilms makes it even more challenging to treat biofilms serving as reservoirs, responsible for the spread of MDR strains.

Biofilms have been shown to be recalcitrant to antibiotics necessitating alternatives. Several alternate antimicrobial approaches have been investigated amongst which, the use of nanoparticles gained prominence due to its superior antimicrobial properties. Several studies have correlated the size, shape, surface charge, metal ion release properties of metals and metal oxide nanoparticles [MON], (Ag, Au, Fe, ZnO, MgO, Gallium, Se, Cu, CuO, TiO_2, Ni, Tellurium) as well as polymeric (chitosan, solid lipid–micro and nanospheres, PMMA nanospheres, Lignin) and inorganic (SiO_2, carbon graphene oxide, CNT) nanoparticles on their antibacterial properties. The superior antibacterial activity of MON's is chiefly due to mechanical damage to cell wall, release of metal ions which interfere with several enzyme pathways, intracellular free radical generation, DNA damage and alteration of membrane permeability. MON's and MON loaded polymer nanocomposites have found a high degree of success in combating microbial biofilms in environmental settings such

 DOI: 10.1201/9781003281238-8

as desalination membranes, protective exterior paints, marine antifouling coatings, antimicrobial textiles, coatings for public hygiene, and the like, but their application has been limited for *in-situ* clinical settings due to biocompatibility with tissue cells. In comparison, polymeric and inorganic nanoparticles have shown better biocompatibility and have been applied *in-situ* for targeted drug delivery and for combating biofilm infections.

Increased detection of MDR strains from implants, catheters, ventilators with invariably about 25–50% of intravascular devices, found to be colonized upon removal, necessitates a viable antibiofilm strategy for these devices. Numerous studies have demonstrated excellent antibacterial properties of MON incorporated nanocomposites under *in-vitro* conditions but very few have translated for in-situ applications on implants, prosthesis and indwelling devices. Hence the search for a less cytotoxic and a broad-spectrum antimicrobial nanoparticle is still in vogue for *in-situ* application. Alternatively, polymer nanoparticles (polylactic acid, polyglycolic acid, polycaprolactone nanospheres) and MON's have been used as encapsulation material (lipid molecules/lignin/chitosan loaded with antibiotics/antimicrobials) which have been used for targeting biofilm based and intracellular bacterial infections. Encapsulation enables intracellular delivery and accumulation of antimicrobials. Biogenically synthesized MON's have shown higher biocompatibility and antimicrobial activity under in-vitro conditions which needs to be taken forward. Photo thermal and photodynamic therapies using gold nanoparticles have attracted attention to combat AMR strains. Synergistic use of nanocarriers like liposomes, cyclodextrins, solid lipid nanoparticles, natural products and plant-based antimicrobials along with nanoparticles to combat MDR strains has been effective for *in-situ* applications. In addition, surface modification of metallic and polymeric coatings as well as nanogels for implants has yielded better biocompatibility and has found in-situ applications.

In summary, the use of metal oxide nanoparticles as therapy for elimination of biofilms under *in-vivo* conditions still presents a challenge and more research is needed to overcome the biocompatibility issue. In this review an overview of antimicrobial resistance in clinically relevant bacterial biofilms, application of nanoparticles and alternative approaches to combat AMRI and their *in-vitro* and *in–vivo* studies with different nanomaterials as antimicrobials will be dealt. The focus would be on the efficacy of different types of nanoparticles inhibiting biofilm formation and contact killing of MDR cells and on the efficacy and advancements of different types of nano based antimicrobial coatings for implants; prostheses and non-surgical indwelling medical devices.

8.2 BIOFILMS: THE WAY OF LIFE OF MICROBIAL CELLS

Microorganisms attach to any surface whether it is living or non-living, leading to the formation of layered structures called biofilm. Biofilm forming bacteria play a significant role in chronic infections of various tissues and organs in humans and animals. A well-known example is biofilm formation by an opportunistic pathogen, *Pseudomonas aeruginosa*, which results in a lung infection known as cystic fibrosis (CF). Similarly, biofilm forming bacteria cause gum disease, as well as infections in catheters, artificial heart-valves, the urinary tract, the middle-ear and in bones. The infections and diseases caused by the biofilm-forming organisms are hard to cure due to their persistence and tolerance to antimicrobial compounds. It has been reported that the 'stationary phase-like dormancy' cells in the biofilm, causing low nutrient penetration through the biofilm matrix, cause a high level of tolerance to antibiotics (Slonczewski and Foster, 2014). The production of extracellular polymeric substances (EPS) as well as quorum sensing genes and regulators are important for biofilm formation. Breaking the cell-to-cell communication between the bacteria is one of the effective ways of controlling biofilm formation and this is believed to be an important target for new antimicrobial therapies (Slonczewski and Foster, 2014). Biofilms have been responsible for 65% of all bacterial infections (Potera, 1999). The biofilm mode offers the cells protection against host immune systems, phagocytosis, against hydrodynamic forces and antibiotic treatments (Costerton, 1999; Costerton et al., 1999).

Biofilms offer increased resistance to biocides and antibiotics by reduced diffusion of molecules by the EPS matrix. The anionic nature of the biofilm matrix offers an environment for continuous replenishment of nutrients, oxygen, increased cell density in biofilms, the existence of different physiological, metabolic state cells in a population, slow growing cells in the population, altered membrane permeability due to porin synthesis, horizontal gene transfer and the presence of MDR operons, the activation of efflux pumps, quiescence, and the production of enzyme degrading antimicrobials (Drenkard et al., 2003; Pan et al., 2006; Lewis et al., 2008; Shi and Zhu, 2009; Hoiby et al., 2010). The antibiotic treatment of biofilms aids in reducing the cell density in biofilms but does not serve in complete eradication of biofilms (Fernandez-Barat et al., 2017) which serves as a reservoir aiding in the regeneration of cells. Studies have also demonstrated that young biofilms are more susceptible than older biofilms (Stewart et al., 2015).

8.3 ANTIMICROBIAL RESISTANCE MECHANISMS IN BIOFILMS

Antimicrobial resistance (AMR) is a global threat as it can result in the development and spread of drug-resistant microorganisms. In particular, the serious concerns are with the organisms that cause infections and diseases. Antimicrobial compounds such as antibiotics, antivirals, antifungal and antiparasitics are used to treat infections and diseases caused by bacteria, viruses, fungi and parasites, respectively. In fact, the term "antibiotic resistance" is ancient and now it is being increasingly referred to as "Resistome" (Aslam and Baloch, 2018). Once a bacterial cell acquires resistance, they impart this character to other surrounding cells and offspring (Arzanlou et al., 2017). In particular, there are serious concerns with organisms that cause infections and diseases. Biofilms have been demonstrated to be 1000 times more resistant to antimicrobials than their planktonic counterparts (Nickel et al., 1985; Macia et al., 2014). Antimicrobial compounds such as antibiotics, antivirals, antifungals and antiparasitics are used to treat infections and diseases caused by bacteria, viruses, fungi and parasites, respectively. Clear demarcation of terminology is an important criterion to classify a strain as "drug tolerant" or "drug resistant". Drug resistance is defined as the continuation of the replication of a bacterial cell to a particular concentration and hence is a relative component with respect to the concentration of the drug. Hence the terminology of drug resistance is absolute. Alternately this may be referred to as an increase in the MIC values of a drug. In comparison drug tolerance results due to a change in physiology and is considered as phenotypic whereas the former is considered to be genetic. Drug tolerance may also be defined as the efficacy of the antibiotic to inhibit growth of the bacterium but failure to kill it (Tuomanen et al., 1990). In simple terms this can be explained as huge differences between MIC and MBC. The concept has been explained using the antibiotic isoniazid [INH] which acts on the cell wall. In general MIC values of 4 µg/mL are effective and a requirement of higher concentrations and indicate resistance to the drug (Schaaf et al., 2007). Intermittent antibiotic exposures have been demonstrated to exhibit tolerance in strains and that tolerance preceded resistance (Levin-Reisman and Balaban, 2017).

Various mechanisms (which are discussed elsewhere in this chapter) and practices such as the use and overuse of these antimicrobial compounds have led to the development of resistance in these organisms. Even though antimicrobial resistance (AMR) is a threat to humanity and has dire consequences, it has been viewed as a Darwinian process of *"selection of the fittest"* (Rosenblatt-Farrell, 2009). The World Health Organisation (WHO, 2021) reported that antibiotics are increasingly ineffective due to the development and spread of drug-resistant bacteria. Interestingly Canton and Morosini (2011) revealed that in a wild type population of around <10^{-9} cells are not affected by antimicrobial treatment. This subpopulation has been termed as the mutation selection window (MSW) which has mutations which are manifested as AMR in clinical settings. The MIC values correspond to the susceptible cells in the population whereas higher concentrations are required to kill the non-affected cells which represent the mutant prevention concentration (MPC). Both MIC and MPC are required to totally eliminate the entire wild population. The study stressed the importance of combining the pharmokinetic/pharmacodynamic (PK/PD) index along with the MPC concept which would be effective in eliminating both susceptible as well as mutant resistant strains and can also be used as a method to identify resistant mutant cells.

Antibiotics generally act on a bacterium through a variety of mechanisms such as damage to cell walls (penicillin and β-lactam, polymixins), inhibiting cell wall [peptidoglycan] synthesis (β-lactam, cephalosporin, carbapenems, glycopeptides), damaging bacterial DNA, RNA, proteins and metabolites (Rosenblatt-Farrell, 2009), Efflux pump and porin inactivation (β-lactam), A-site decoding the region of 16s RNA (aminoglycosides), inhibition of DNA replication [DNA gyrase enzyme] (quinolone), nucleic acid synthesis (sulfamide), bacterial 30s ribosomal RNA (tetracyline), large ribosomal subunit (Macrolide), LPS molecule (Polymixin), β-subunit of RNA polymerase RpoB (Rifampicin), large 50s ribosomal subunit (chloramphenicol), inhibition of folate synthesis (sulfonamides, trimethoprim), translation inhibitors (aminoglycosides, tetracyclines, oxazolidinones, macrolides), transcription inhibitors (rifampicin) (Diene and Rolain, 2013). Rifampicin has been known to act on DNA gyrase and topoisomerase IV enzymes involved in the transcription process (Koteva et al., 2018; Rodriguez–Martinez et al., 2011). β-lactam antibiotics such penicillins, carbapenems and cephalosporins inhibit bacterial growth by acting on transpeptidase (Rice et al., 2012). Lincosamides, macrolides, tetracycline, aminoglycosides have been demonstrated to block protein synthesis in ribosomes (Tenson et al., 2003). Compared to antibiotics, biocides have multiple target sites on a bacterium. Quaternamonium compounds (QAC) [act on cell membranes], biguanides, phenols, alcohols [act on proteins], per oxygens [act on nucleic acids], halogens [act on cell walls and nucleic acids] have been the most commonly used biocides in hospital and biomedical environments (Gnanadhas and Chakravortty, 2013). Biocides have been shown to confer resistance or cross resistance to antibiotics (Gnanadhas and Chakravortty, 2013). Reduced susceptibility to biocides in a bacterium is often mediated by nonspecific means wherein the biocide is either neutralized by extracellular components or by reducing their intracellular concentrations either through permeability changes or efflux pumps. A classical example of this concept is the study by McBain and Allisson (2003) on the widely used hospital biocide triclosan. Compared to antimicrobial resistance, susceptibility of microbes to biocides has also been found to be decreased and a reduced susceptibility to triclosan, a widely used disinfectant has been a cause of worry in the early 2000s. However, studies by McBain and Allisson (2003) have demonstrated that this reduced susceptibility to these biocides is more related to environmental factors rather than microbes developing resistance to the biocide. Similarly, the occurrence of persister cells in the environment has been found to be more related to the frequency of environmental changes and less on selective pressure of the environment (Kussell and Leibler, 2005).

The drug-resistant bacteria display a variety of resistant mechanisms (Figure 8.1). The resistances are primarily classified into three categories: intrinsic, acquired and adaptive resistance. Intrinsic resistance refers to the ability of the organisms to reduce the efficacy of the antibiotic or other compounds through their actions. As most antimicrobial compounds are naturally produced, the organisms that exist in environments with such compounds develop mechanisms to overcome the attacks by those compounds in order to survive. The intrinsic resistance mechanisms include antibiotic inactivation by the production of antibiotic degrading enzymes, antibiotic target site modifications by antibiotic altering enzymes, as well as decreased

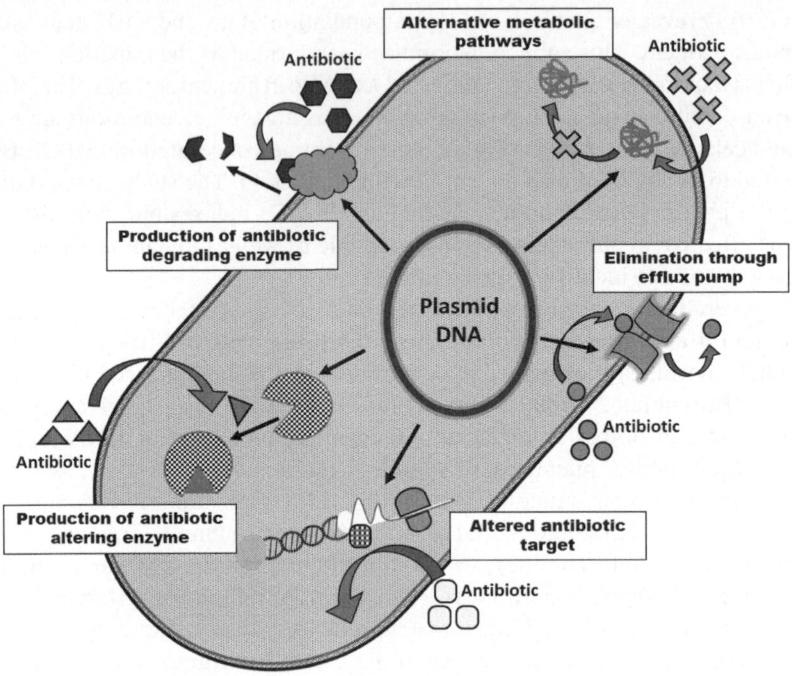

FIGURE 8.1 Antimicrobial resistance mechanisms in bacteria.

permeability and reduced efficiency of the efflux pumps of the antibiotics. (Munita and Arias, 2016; Pang et al., 2019). Efflux pumps have been one of the major contributors to bacterial resistance in which the active molecule structurally resembles the pump's natural substrate (Hernando-Amado et al., 2016). The acquired resistance transfers the antibiotic resistance from resistant organisms to susceptible organisms co-existing in an environment. The genetic bases for this type of resistance are mutations and horizontal gene transfers (HGT) and the latter takes place through transformation, conjugation and transduction (Munita and Arias, 2016; Pang et al., 2019). Through such mutational changes, the organisms are able to develop resistance to antimicrobial compounds through the reduction in the uptake of the antibiotics, target site modifications, increased efflux and antibiotic inactivating enzymes (Munita and Arias, 2016). Through HGT, the antibiotic resistance genes (ARGs) carried on the plasmids, integrons, and transposons are transferred from one bacterium to another (Breidenstein et al., 2011). The third category is adaptive resistance by which the bacteria overcome the action of antibiotic or other stressful conditions. The biofilm mediated resistance and the development of persister cells are examples of adaptive resistance through which the bacteria tolerate high levels of antibiotics but are not genetically resistant (Pang et al., 2019). Current understanding of AMR mechanism points to genetic changes through acquiring resistance genes, or mutations to processes components on which antibiotics act. However, Corona and Martinez (2013) have pointed out to the existence of phenotypic resistance, which may be attributed to the situation of growth arrest. This has been demonstrated by Lee (1944) where resting cells were less susceptible to penicillin which was later named as *"drug indifference"* (McDermott, 1958).

As an example, ciprofloxacin is commonly used as an antibiotic to treat urinary tract infections. The pathogenic *Escherichia coli* and *Klebsiella pneumonia* have been reported to show 8.4–92.9% and 4.1–79.4% resistance against this antibiotic in various countries (WHO, 2021). Similarly, colistin is used as a last-resort antibiotic to treat infections caused by carbapenem-resistant *Enterobacteriaceae* (which include *E. coli, Klebsiella* and other pathogens) and in several parts of the world, colistin-resistant bacteria have been detected. Drug resistance in many bacteria, viruses, fungi and malaria parasites has been widely reported (WHO, 2021) and an increase in drug-resistant microorganisms raises a serious concern for the future of antibiotics and other antimicrobial compounds. It has been estimated that approximately 700 000 people die every year due to bacterial infections, HIV, tuberculosis and malaria caused by the drug-resistant organisms, and the AMR-related death toll is likely to increase up to 10 million per year if the actions are not taken to control the drug-resistant organisms (O'Neil, 2016). Therefore, new antimicrobial compounds and therapeutic methods are urgently needed.

In the early 1990s Brown and Gilbert, have summarized the reduced susceptibility of biofilms essentially due to impermeable exopolysaccharide glycocalyces, mass transfer limitation by detoxifying enzymes, altered physiology and reduced metabolic activity and the growth rate of the bacteria. Toumanen et al., (1986) summarized the reduced effect of B-lactam antibiotics as a function of cell density and specific growth rate of the bacterial population. McDonnell and Russel (1999) have outlined different possible mechanisms in bacteria viz: intrinsic (exclusion, enzyme inactivation, phenotypic modulation, efflux pumps) and extrinsic mechanisms like mutation, inducible efflux, plasmid mediated processes. Conjugation represents a most common form of horizontal gene transfer (HGT) in biofilms which have evolved mechanisms to circumvent host defense mechanisms and compete with co -resident plasmids (Shen and Liu, 2022). Most of the studies on HGT and conjugal plasmids have focused on the F plasmid being involved in conferring the AMR in some of the ESKAPE pathogens. However, the studies in this direction have been carried out with environmental isolates and extensive studies are still required to investigate the role of conjugal plasmids in AMR using clinically relevant bacterial strains. Probiotics on the other hand have shown prophylactic activity both in-vitro and in-vivo against vancomycin, carbapenems and rifampicin resistant MRSA strains of *S. aureus* (Nataraj and Mallappa, 2021), however once again here the concept of horizontal gene transfer of resistance to probiotic cells is also not ruled out. On the other hand, exposure to antibiotics has been found to result in persister cell formation by inducing the TiSB toxin in *E. coli* (Dorr and Lewis, 2010). TiSB is a membrane protein which responds to stepwise exposure to ciprofloxacin which decreases proton motive force and ATP levels leading to persister formation in the bacterium. The DNA damaging antibiotic has been found to induce multi drug resistance in the bacterium. Currently there is no established method to prevent the formation of persister cells in a population. A study by Allison and Collins (2011) demonstrated that aminoglycosides along with metabolites were effective in killing both Gram negative *E. coli* and Gram-positive *S. aureus* by generation of proton motive force (PMF). The study highlights the involvement of metabolic environment in antibiotic treatment. The formation of persister in cells in the environment has been attributed to the increase of toxic substances or, alternatively, the decrease in environmental fitness (Hoon and Wood, 2012). Exposure to hydrogen peroxide treatment increased the level of persisters in the wild type *E. coli* population ultimately resulting in the whole population transforming into a persister state.

8.3.1 Quorum Sensing Inhibitors and Nanoparticles to Control MDR Biofilms

Biofilm formation by bacteria is controlled by various mechanisms including quorum sensing (QS). QS is a cell-mediated response to bacterial load in a given environment and is the ability of bacteria to communicate with each other and coordinate their behaviour via the production of signal molecules. The Gram-negative bacteria produce acyl homoserine lactone (AHL) and Gram-positive bacteria produce oligopeptides as signal molecules (Miller and Bassler, 2001). The breaking the communication between the bacteria by interrupting the QS is considered to be one of the main strategies to control biofilm formation in various systems (Ramussen and Givskov, 2006; Ramalingam, 2012; Papenfort and Bassler, 2016). In particular, QS inhibitors are known to interrupt QS mechanism and thus disrupt biofilm formation (Kalia et al., 2019). The antimicrobial effect of QS autoinducers has been studied previously. Shahbandeh et al. (2022) studied the antimicrobial effect of QS autoinducers

(C12-HSL and C4-HSL) of *P. aeruginosa* against MDR *S. aureus* isolates and the study found that these autoinducers had antimicrobial activity against *S. aureus* whereas they had no activity against *A. baumannii*, which was used as a control.

Combining nanoparticles with suitable QS inhibitors on a biomaterial can be an alternative and effective method for killing bacteria in surgical implants. Titanium and titanium alloys are commonly used as biomedical materials in the treatment of human hard tissues such as dentistry and orthopedics due to their excellent biocompatibility. However, the bacterial biofilms associated with titanium-based implants cause bacterial infections in the human body which leads to post-operative infections (Hu et al., 2020). Recent research synthesized mesoporous polydopamine (MPDA) nanoparticles loaded with QS inhibitors (luteolin) to form MPDA-LUT with calcium phosphate was coated on the surface of MPDA-LUT to form MPDA-LUT@CaP NPs. This nano system was placed on the surface of a Ti biomaterial device and tested in vitro and in vivo with MDR bacterium *S. aureus* and a rat humeral knee joint infectious model, respectively. The results showed that the luteolin was effective in disrupting the QS mechanism in the *S. aureus* bacterium and thus the bacterium failed to form biofilm. Newly modified Ti implants irradiated with near infra-red (NIR) effectively eliminated and killed the planktonic bacterium. The calcium phosphate from the surface of the bioimplant ensured the biocompatibility and thus this new nano delivery system could be used in the manufacturing of orthopaedic implants (Hu et al., 2022).

8.3.2 Biocompatibility Requirements for In-vivo Applications of Nanomaterials

For successful application of nanoparticles in bacterial infection therapy under in-vivo conditions, interaction with tissues, clearance and degradability of nanoparticles are of major concerns. Around 30–99% of administered nanoparticles reach the liver which sequesters the nanoparticles from the bloodstream (Yu and Zheng, 2015). The sequestered nanoparticles are degraded by the hepatocyte cells by endocytosis and breakdown and release along with bile juices (Longmire et al., 2008). Studies by Guo et al., (2013), demonstrated that administered nanoparticles (hollow spheres of CuS) could be could eliminated by hepatobilary excretion within one month of administration. Silver nanoparticles incubated at sub toxic concentrations did not cause changes in osteoclast differentiation and podosomal structures, but the particles were accumulated in the endo-lysosomal compartments. Increased concentration of Ag nanoparticles resulted in oxidative stress and decrease in clathrin dependent endocytosis at the mRNA level (Pauksh et al., 2014).

8.4 NANOTECHNOLOGY BASED ALTERNATE APPROACHES FOR COMBATING AMR

The emergence and existence of super bugs or bad bugs which are multi drug/extremely drug resistant/totally drug resistant strains have necessitated the search for alternative antimicrobial compounds. Among these nanoparticles antibiotic formulations, photodynamic therapy seems to be the future viable/promising alternative (Sharma and Khan, 2019) currently. Nanoparticles/nanomaterials refers to unit particles with dimensions in the nano range between 1 and 100 nm in one direction (Elias et al., 2017). Nanomaterials based therapy/treatment of microbes offers alternate mechanisms to overcome the otherwise existing resistant mechanisms developed by microbes against conventional antibiotics (Makabenta et al., 2021). Over the last two decades a lot of interest has been generated in nanoparticle synthesis through novel routes, which would yield different size, shape, charge, optical properties, reactivity and good stability under aqueous conditions. Based on their shape, nanoparticles have been classified as 0D (holospheres and core-shell quantum dots), 1D (nanowires, nanotubes, nanorods), 2D (thin and multilayer films), 3D (crystalline NP) (Prajitha et al., 2019). For detailed reviews on the application of nanoparticles as antimicrobial and drug delivery agents refer to (Hugh et al., 2011; Rudaraju et al., 2020). For extensive review of approaches to treat MRSA with nanoparticles refer to Vanamala and Iyer (2020). Basically, nanoparticles can be classified into inorganic nanoparticles including quantum dots metal oxide nanoparticles [MONs] such as TiO_2, SiO_2, ZnO NP and metal nanoparticles such as silver and gold (Matteis, 2017). In addition, carbon fibres and carbon nanotubes constitute another class of nanoparticles (De La Fuente, 2012). Graphene based nanomaterials have also found application as antibacterial, antibiofilm and drug delivery agents (Jiang and Liu, 2020).

Most of the studies synthesizing nanomaterials have correlated nanoparticle properties to antimicrobial activity. When it comes to the therapeutic use of nanoparticles, they fall into two broad areas namely, antibacterial/antibiofilm activity and nanocarriers for drugs. Again, most of the research initiatives have been towards the use of polymeric based nanoparticles (natural and synthetic) which represent natural polymers like chitosan, hyaluronic acid, and albumin. Synthetic polymers such as poly (glycolyic acid), acrylic acid, poly (lactic acid) have also been used as drug delivery agents. Most of these have been shaped as nanocapsules, dendrimers, liposomes or as micelles for drug delivery. Parallel research directions have been initiated into developing lipid nanoparticles, namely, liposomes, niosomes, ethosomes and transferosomes for drug delivery and therapeutic uses. Extensive research and developments has been carried out on the antimicrobial properties of metals and metal oxide nanoparticles [MON], namely, silver, gold, iron oxide, zinc oxide, gallium, selenium, copper oxide, magnesium oxide, titanium dioxide, nickel, and so forth. The metal oxide nanoparticles have demonstrated excellent antibacterial activity against a variety of micro-organisms ranging from bacteria to fungi to virus and multi drug resistant strains [MDR] but have faced biocompatibility issues for in-situ applications. Some metal oxide nanoparticles [MON's] such as silver and zinc oxide have found applications for topical use. In addition, another set of research and development work is being carried out on graphene, carbon nano tubes and nanofoils, fullerenes, carbon black (Thambirajoo and Fauzi, 2021).

The mechanism of action of metal oxide nanoparticles on biofilms has been well studied and understood (Figure 8.2). MONs are easily able to penetrate the biofilm matrix due to their size, shape and cause cell lysis by disrupting the cell membrane of most of the microbes. Their attraction towards the cell and binding to the cell membrane is also influenced by surface charges (Lahir, 2020). Nanoparticles to treat bacteria gained importance due to the ineffectiveness of the current range of antibiotics. Bactericidal activity of various nanoparticles against different bacterial species has been found to be influenced by nanoparticle size, charge, which determines the MIC and MBC values (Seil and Webster, 2012). MONs has also been demonstrated to damage DNA structure, inhibit enzyme and cellular functions (Raghunath and Perumal, 2017). Yet some MONs have been shown to target ATP synthesis, with the most common mechanism being that of reactive oxygen species [ROS] generation (Marambio-Jones et al., 2010). Reactive oxygen species also generates holes in the bacterial cell membrane resulting in the discharge of intracellular fluids (Song et al., 2019; Ali et al., 2019). Liposomes have also been demonstrated to possess positive charge and penetrate biofilms at low concentrations and release drugs which enter the cells in the biofilm matrix (Lahir, 2020). Positively charged MONs bind more easily to the negatively charged phospholipid bilayer resulting in an increase in oxidative stress leading to membrane disruption, increased permeability and impairment of cellular function (Gold et al., 2018). Silver ions have been shown to exhibit this electrostatic binding property to the bacterial cell wall. Amongst the MONs, silver nanoparticles have been shown to be effective against both Gram-positive and Gram-negative bacteria exhibiting a broad-spectrum antimicrobial activity (Din et al., 2015) with damage to intracellular components resulting in "Ghost cell effect" (Salleh et al., 2020). Alternatively, many MONs such as silver, zinc, copper, titanium, magnesium, and selenium leach metal ions which pass through cell membranes as these are required in micro quantities for cellular functions. Increased uptake of these metal ions also has an antibacterial effect on microbes. In the case of carbon black nanoparticles, these have been found to bind to the cell wall by electrostatic charges and penetrate the cell membrane and interfere with DNA replication (Lin et al., 2019).

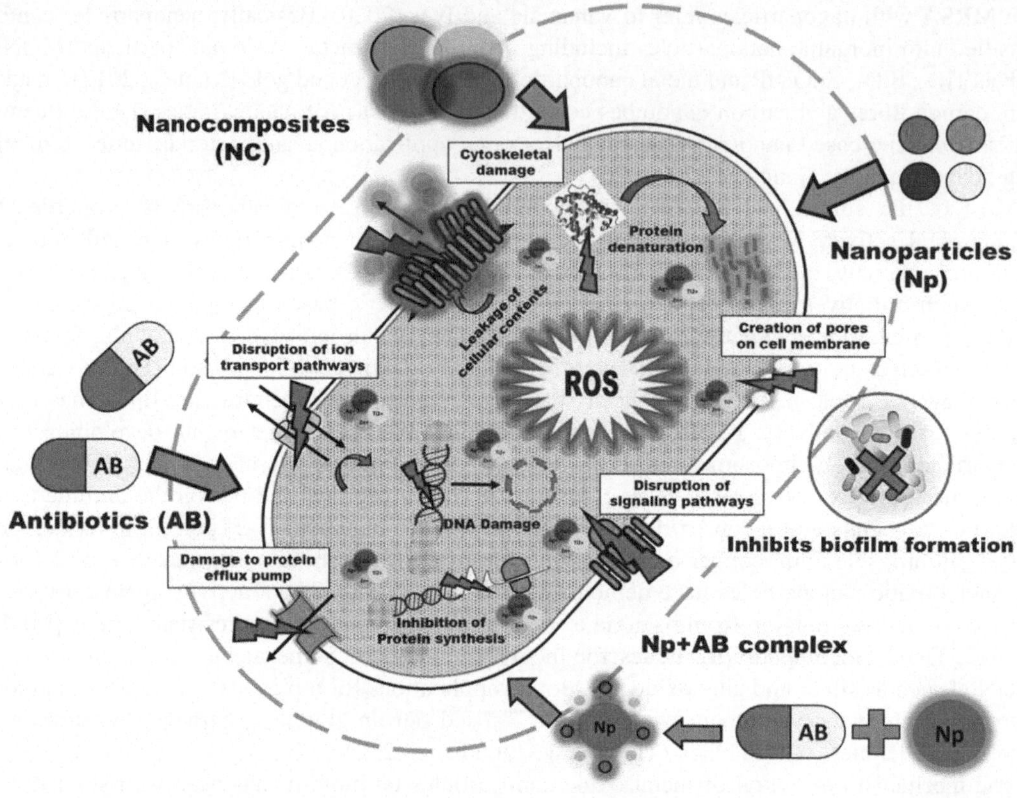

FIGURE 8.2 Mechanism of action of Nanoparticles on bacteria.

Penetration of nanoparticles into biofilms has been reported to be facilitated easily with nanoparticles less than 350 nm and of an uncharged nature. Whereas cationic nanoparticles have been shown to possess homogenous distribution in biofilms (Liu et al., 2009; Peulen et al., 2011; Gupta et al., 2018). Dispersion of cells from biofilms is a natural phenomenon and results in the release of bacterial cells into the blood stream which will be taken care of by the immune system. However sudden spikes in the dispersion of biofilms cannot be handled by the immune system and results in sepsis. Nanoparticles have also been shown to possess intrinsic dispersion properties, for example, MONs, carbon dots, micelles and peptides (Tian et al., 2021). Cationic nanoparticles have been shown to have better penetration power into the negatively charged biofilm matrix. Hence an approach of using dispersants such as DNase1, dispersion B along with chitosan and lipid micelles has been adopted in some cases. Alternately the intrinsic dispersal properties of certain nanoparticles such as citrate capped gold nanoparticles, citrate coated platinum nanoparticles and PEG coated iron oxide nanoparticles have also been used effectively in dispersing biofilms. Once the biofilm has been dispersed it is easy to target the residual biofilm with antimicrobials.

Metal oxide nanoparticles [MONs] once considered an alternate approach to combat MDR strains as effective antimicrobials is currently being demonstrated to exhibit bacterial tolerance/resistance mechanisms. More and more studies are now reporting resistance mechanisms of different MONs against microbes. Examples of these are biofilm adaptation (Wang et al., 2016); efflux pumps (Yang et al., 2012); electrostatic repulsions (Abbaszadegan et al., 2015); mutations, alteration of morphology (Nino-Martinez et al., 2019); down regulation of genes coding for porin proteins (Li et al., 1997); envelope stress response (Jordan et al., 2008); oxidation of nanoparticles by organics (Li et al., 2010); sublethal exposure to NP (Graves et al., 2015); biomolecule corona around NP (Siemer et al., 2019); pigment production (Ellis et al., 2011); flagellin production (Panacek et al., 2018); EPS production (Faghihzadeh et al., 2018); and Plasmids with resistant genes (Wang et al., 2018). The efficacy of nanoparticles against bacterial growth as well as biofilm inhibition depends on minimum inhibitory concentrations which may be MIC_{50} or MIC_{90} and the minimum bactericidal concentration, namely, the MBC. The MIC values are vital in ascertaining whether a strain is susceptible, tolerant and resistant. Similarly, it is vital that the MIC's of synthesized nanoparticles fall in the range of MIC's exhibited by antibiotics. In this context Table 8.1 summarizes some selected studies which show that MICs of nanoparticles range as low as 1–2048 µg/mL. In general MIC's ~1.0 or <1 µg/ml are ideal and can be taken up for pre-clinical animal trials and clinical trial. The varying MIC's shown in Table 8.1 for various nanoparticles is a function of its synthesis route, size shape, charge, and so forth, which affects antibiofilm and antimicrobial activity.

8.4.1 Silver

Silver has been used as an antimicrobial agent for more than a century. The first synthesis of silver nanoparticles was reported as early as 1889 where particles of 7–9 nm had been synthesized (Lea, 1889). In vitro studies used silver-lignin NPs which were approximately 20 nm in diameter and did not agglomerate after one year of storage at 4°C. They were able to inhibit the growth of a panel of MDR clinical isolates, including *Staphylococcus aureus, Staphylococcus epidermidis, Pseudomonas aeruginosa, Klebsiella pneumoniae, and Acinetobacter baumannii* (Slavin and Bach, 2021). These nanoparticles, apart from antimicrobial induced anti-inflammatory response and in *P. aeuriginosa* genes coding for efflux pumps and metal resistance were upregulated. Earlier studies by Mahmoudi et al., (2012) and Rai et al., (2012) have demonstrated the superior activity of silver nanoparticles against multi drug resistant bacteria. Plasmid mediated resistance has been observed for *Pseudomonas stutzeri*, *Enterobacteriaceae*, and *Citrobacter sp.* (Russell et al., 1997). There was a tenfold reduction in preformed biofilms of *Candida albicans* biofilms with a concentration of 100 µg/ml of silver nanoparticles. Their efficacy was higher (100 - fold) reduction with the fungus *C. globrata* (Silva and Azerado, 2013). Silver nanoparticles have been shown to be effective against the MDR strain of *K. pseumoniae* inducing a triclosan type bactericidal mechanism by inhibiting fatty acid biosynthesis (Pareek and Fanning, 2021).

Nano silver has been shown to possess potential antibacterial activity against a broad spectrum of Gram-positive and Gram-negative bacteria including non-clinical strains (Siddique et al., 2020). Most antibacterial studies have used nano silver compared to the ionic forms due to the controlled release of silver ions and sustained antibacterial activity (Liu et al., 2010). Nano silver dissolved in an aqueous phase has been shown to undergo oxidative dissolution and release of silver ions (Gunawan et al., 2009; Gunawan et al., 2020). However bacterial resistance to silver has been reported as early as 1998 (Gupta, 1998; Gupta et al., 1998). Nanosilver has been reported to have multiple modes of action on a bacterial cell, namely, targeting peptidoglycan and phospholipid molecules (Gunawan et al., 2020); disruption of respiratory enzymes including iron-sulfur clusters in many enzymes (Lemire et al., 2013); and generation of ROS through disruption of respiratory enzymes using Fenton active Fe^{2+} ions (Gunawan et al., 2020). However, Gunawan et al., (2013) reported the growth of environmental *Bacillus sp.* to nano silver at otherwise toxic concentrations following prolonged exposure to the nanoparticle. Such resistance to nanosilver has also been reported by Graves et al., (2015) and Valentin et al., (2020) for the uropathogens *E. coli* and S. *aureus*. The increased tolerance to nanosilver by *E. coli* has been attributed to the production of the flagella protein, flagellin, which passivated the nanoparticle (Panacek et al., 2018). Exposure to nano sliver induces gene mutation in *S. aureus* to develop stable resistance even after discontinuation to nano silver exposure (Valentin et al., 2020). Mijnendonckx et al., (2013) have attributed the resistance/tolerance to nanosilver in the presence of silver ion efflux system like the chromosomally encoded Omp/EnvZ porin and Cus efflux mechanisms as well as the plasmid coded sil mechanism. A peptide dephosphorylation

TABLE 8.1
Summary of Minimum Inhibitory Concentrations against Different Bacterial Strains Using Metal Oxide and Polymeric Nanoparticles

Nanoparticles	Micro-organism	MIC values and % reduction	Reference
Ag	*E. coli, V. Cholerae, S. typhi, P. aeuriginosa*	100% at 75 µg/mL	Morones et al.,2005
Ag	*E. coli*	99% at 0.1 µg/mL	Pal et al., 2007
Ag	*E. coli, V. cholera, P. aeuriginosa*	75 µg/mL	Morones et al., 2005
Ag	*E. coli*	70% at 10 µg/mL	Sondi et al.,2004
Ag	*S. aureus, E. coli*	3.56–0.356 µg/L	Kim et al.,2007
Ag	*P. intermedia, P. gingivalis, F. nucleatum and A. actinomycetemcomitans*	100–250 µg/mL	Vargas-Reus et al., 2012
Commercial Ag Silver nitrate Ag sulfadiazine Chitosan nano Ag PVA nano Ag Citrate nano Ag	oxacillin–resistant *Staphylococcus aureus*, vancomycin-resistant *Enterococcus spp.*, carbapenem–and polymyxin B-resistant *A. baumannii*, carbapenem-resistant *P. aeruginosa and* carbapenem-resistant *Enterobacteriaceae*	≥10 µg/mL 1.6–13.5 µg/mL 6.7–≥ 27 µg/mL 1.6–13.5 µg/mL 6.7–≥ 54 µg/mL 3.3–13.5 µg/mL	Cavassin et al., 2015
Ag	*CoNS S. epidermidis S. haemolyticus*	Biofilm inhibitions 90.9–91.9% at 20 µg/mL	Thomas et al., 2015
Ag	*C. albicans, C. glabrata*	0.4–3.3 µg/mL	Monteiro et al., 2011
Ag	*S. aureus* (MRSA) *S. epidermidis* (MRSE)	11.25–45 µg/mL 100% biofilm inhibition at 50 µg/mL	Ansari et al., 2015
Ag-ZnO	*P. intermedia, P. gingivalis, F. nucleatum and A. actinomycetemcomitans*	362.5 µg/mL	Vargas-Reus et al., 2012
ZnO	*S. aureus Vancomycin*	VRSA–625 µg/mL 73.95% biofilm reduction at 10 mg/mL	Jasim and Rheima 2020
ZnO	*S. pyogenes MTCC442* *B. cereus MTCC122* *E. coli MTCC1687* *P. aeruginosa MTCC4673*	2 µg/mL 20 µg/mL 50 µg/mL 1000 µg/mL	Kaur et al., 2020
ZnO	*H. pylori*	100% inhibition 4 mg/mL	Medina-Ramirez et al., 2020
ZnO	*S. aureus*	128–2048 µg/mL Biofilm inhibition 76.47% at sub-MIC of 68–1000 µg/mL	Abdelraheem et al., 2021
ZnO	*S. aureus*	95% at 80 µg/mL	Reddy et al.,2007
ZnO	*P. intermedia, P. gingivalis, F. nucleatum and A. actinomycetemcomitans*	437.5 µg/mL	Vargas-Reus et al., 2012
ZnO	*S. aureus*	50% at 400 µg/mL	Jones et al., 2008

TABLE 8.1 (Continued)
Summary of Minimum Inhibitory Concentrations against Different Bacterial Strains Using Metal Oxide and Polymeric Nanoparticles

Nanoparticles	Micro-organism	MIC values and % reduction	Reference
ZnO	*Metallo-βlactamase producing P. aeuriginosa*	100–500 µg/mL	Vincent and Murugan 2014
ZnO	*S. aureus, E. coli*	99% at 400 µg/mL	Nair et al.,2009
ZnO	*E. coli*	90% at 400 µg/mL	Padmavathy et al.,2008
ZnO	*E. coli, K. pneumoniae, P. aeruginosa, and S. aureus*	0.5–1.0 mg/mL MIC 1.0–1.5 mg/mL	Ahmed and Musarrat 2018
Zn ions	*S. aureus, P. aeuriginosa, C. albicans*	100% at 1917, 9 and 39 µg/mL respectively	McCarthy et al.,1992
ZnO-PK NP	*S. aureus* (MRSA)	ZnO–PK NP 4.5 log reduction at 75 µg/mL Vancomycin 0.5–2 µg/mL Vancomycin + ZnO–PK NP ¼ MIC 100% growth inhibition	Bannerjee et al., 2020
Biogenic ZnO	*A. baumannii*	0.04–0.19 µg/mL	Hou and Zhang 2020
Chitosan Zinc Oxide micelles	MDR *Enterococcus faecium*	50% biofilm reduction at 200 µg/mL Chitosan alone - 15.6 at 200 µg/mL ZnO alone - 13.9% at 200 µg/mL	Mehta et al., 2019
Sulfonated hafnium oxide and CuO nanocomposite	MDR *Enterobacter asburiae, E. coli, P. aeruginosa S. aureus ATCC6538*	Cu/SHfNP 1 mM decreased cell growth E. asburiae SC4 (74%) and E. coli SE4 (85%) and 2 mM decreased cell growth of ATCC6538 by 86%	Manoharan and Ahn 2021
CuO and Fe_2O_3	MDR *E. coli* MRSA	CuO MIC 350–45 µg/mL Fe_2O_3–30 (*E. coli*) and 40 (MRSA) µg/mL	Agarwala et al., 2014
CuO	*P. intermedia, P. gingivalis, F. nucleatum and A. actinomycetemcomitans*	250–500 µg/mL	Vargas-Reus et al., 2012
Cu_2O	*P. intermedia, P. gingivalis, F. nucleatum and A. actinomycetemcomitans*	100–1000 µg/mL	Vargas-Reus et al., 2012
Cu	*E. coli, Bacillus subtilis*	28.2–33.4 µg/mL	Yoon et al., 2007
Chitosan	*E. coli, S. aureus*	100% at 4 and 8 µg/mL	Qi et al.,2004
WO_3	*P. intermedia, P. gingivalis, F. nucleatum and A. actinomycetemcomitans*	2500 µg/mL	Vargas-Reus et al., 2012
SeNP	*P. aeuriginosa, P. aeuriginosa, P. aeuriginosa, P. aeuriginosa, S. maltophila, S. maltophila, S. aureus, S. aureus, S. haemolyticus, S. epidermidis*	8, 32, 64, 32, 64, 128, 4, 16 and 4 µg/mL respectively for the individual strains	Cremonini et al., 2018

(Continued)

TABLE 8.1 (Continued)
Summary of Minimum Inhibitory Concentrations against Different Bacterial Strains Using Metal Oxide and Polymeric Nanoparticles

Nanoparticles	Micro-organism	MIC values and % reduction	Reference
SeO_2 NP	MRSA, MDR, *K. pneumonia*	8.6 and 26.2 µg/mL	Huang et al., 2020
Selenium pellets	MDR, MRSA, *E. coli*	2.35 and 14.25 ppm	Geoffrion et al., 2020
Biogenic SeNP	*S. aureus,* MRSA, *E. coli, P. aeuriginosa*	75–150 µg/mL	Cruz et al., 2018
TiO_2	*P. intermedia, P. gingivalis, F. nucleatum and A. actinomycetemcomitans*	250–2500 µg/mL 250–2500 µg/mL	Vargas-Reus et al., 2012
TiO_2	*Metallo-βlactamase producing P. aeuriginosa*	100–500 µg/mL	Vincent and Murugan 2014
TiO_2 powder TiO_2 nanoparticle	MDR *P. aeuriginosa*	With TiO_2 powder 11 isolates 1–1024 µg/mL with 2 strain 2.0 µg/mL With TiO_2 Np 8–64 µg/mL and 2 strains <1.0 µg/mL	Ahmed and Waly 2021
TiO_2	MDR *Mycobacterium tuberculosis*	10–100 µg/mL	Ramalingam and Rajaram 2019
TiO_2	*Gram positive MDR oral cocci Georgenia sp., S. saprophyticus, and Rothia mucilaginosa*	250 -2000 µg/mL	Fatima et al., 2021
Synergistic use of TiO_2, Ag, Fe_2O_3	*S. mutans and S. sanguinis*	Synergistic use 0.019 µg/mL TiO_2 - 0.078; Ag - 0.156; Fe_2O_3 - 0.312 µg/mL	Lavee and Modaresi 2016
TiO_2	*E. coli*	35% at 100 µg/mL and 80% at 500 µg/mL	Simon-Deckers et al.,2009
SiO_2	*E. coli, Bacillus subtilis*	70% reduction at 20 µg/mL	Jiang et al.,2009
Fe_3O_4	*S. epidermidis*	65% at 2 mg/mL	Taylor et al.,2009
Fe_3O_4	*S. aureus*	3 mg/mL	Tran et al.,2010
Spions	MR *S. epidermidis RP62a*	Methicillin 2 µg/mL for planktonic and 500 µg/mL for biofilms 100% biofilm eradication with Spions 40 µg/mL and 20 µg/mL of methicillin	Geilich et al., 2017
Al_2O_3	*E. coli*	35, 70 and 68% at 10, 100, 500 µg/mL	Simon-Deckers et al., 2009
Al_2O_3	*E. coli, Bacillus subtilis*	36 and 57% at 20 µg/mL	Jiang et al.,2009
Al_2O_3	MDR *A. baumannii*	MIC and MBC 125–1000 µg/mL 11–70% inhibition of biofilms Non-toxic up to 120 µg/mL to HeLa cell lines	Muzammil et al., 2020
Nisin Au NP	*E. faecalis, S. aureus*	156–312 µg/mL for EF and 39.0–78.1 µg/mL	Pradeepa et al., 2017
Baicalein Au NP	*P. aeruginosa*	58–76% reduction in biofilm at 100 g/mL	Rajkumari et al., 2017
Au NP	MDR *E. coli, MRSA*	>128 µg/mL	Wang and Jiang 2020

mechanism in response to nanosilver has been observed in *E. coli* and *Salmonella typhi* strains which was not observed in treatments with ionic silver (Kedziora et al., 2018).

Mann and Gunawan (2021) have attributed this tolerance to being an adaptive feature of the biofilm mode of this organism, similar to those reported in its planktonic counterparts. The study has indicated the formation of persister cells upon prolonged exposure to ionic silver and also cross resistance to silver in a gentamycin resistant *P. aeuriginosa* biofilm.

8.4.2 Gallium

Gallium ions are mistaken by microbes as chemical constituents essential for the bacterial processes and their increased influx affects bacterial biochemical metabolic pathways resulting in mortality (Godoy-Gallardo et al., 2021).

8.4.3 Selenium

Selenium is a trace element and forms a constituent of selenoproteins, cell signal transduction, metabolic process and immune regulation (macrophage and T effector cell activation) in humans (Labunskyy et al., 2014). Selenium nanoparticles exhibit low systemic toxicity, ease of degradation, superior anticancer, antimicrobial and antiviral properties (Hosnedlova et al., 2018). Selenium nanoparticles find increased use in tumor treatment and drug delivery due to increased selectivity to cancer cells, drug solubility and targeting effects (Lin et al., 2021). Selenium NPs have also demonstrated antimicrobial activity against MDR strains (Lin et al., 2019) as well as fungicidal effects against *C. albicans* (Lara et al., 2018). However, there is a need to look into the mechanistic aspects of antimicrobial activity of selenium NP. Even though selenium is required in trace quantities, long term exposure and toxicity induced by selenium nanoparticles to normal cells has still not been clearly investigated. Truong et al., (2021) have reviewed the antimicrobial potential of different synthesized nanoparticles such as chemical and biogenic, and the study outlines the superior antimicrobial activity of biogenic synthesized Se nanoparticles.

8.4.4 Copper oxide

Copper oxide NPs have been shown to be very effective in inhibiting fungal biofilms by interfering with the quorum sensing mechanism by breaking the signalling pathways and have been shown to be more effective than conventional drugs such as fluconazole (Rasool et al., 2019).

8.4.5 Nickel

Nickel oxide nanoparticles have also been shown to act on the bacterial cell walls by altering the structural components and replication of protein and DNA to halt the cellular processes (Maruthupandy et al., 2020). The study also demonstrated very low cytotoxicity to the brine shrimp *Artemia franciscana*. Very low concentrations (0.001 mg/mL) of nickel nanoparticles have been shown to be very effective against clinical isolates of *S. epidermidis* (Vahedi et al., 2017).

8.4.6 Bimetallic and Hybrid Nanoparticles

The use of bimetallic nanoparticles such as ZnO and MgO into hydroxypropyl methylcellulose have been shown to be very effective against the biofilms of *Proteus mirabilis,* a device associated urinary pathogen at very low concentrations of 0.0011% ZnO: MgO (Iribarnegaray et al., 2019). The use of polyacrylic iron oxide nanoparticles activated by an AC current supply penetrated the biofilm matrix and adhered to bacterial cells leading to ROS production which can be utilized as an alternate method to combat antibiotic resistant biofilms (Nie et al., 2020).

8.4.7 Chitosan

Chitosan nanoparticles [CS NP] are the most commonly preferred drug delivery agents due to their ease of alteration and blending with other polymers like polyvinyl alcohol and PMMA to enhance its mechanical and barrier characteristics (Garg et al., 2019). It has been found to be suitable in environments ranging from neutral to higher pH with ability to solubilize in acidic conditions (Mohammed et al., 2017). CS NPs combined with oxacillin and DNase have been demonstrated to be effective against biofilms through the action of the antibiotic and the enzyme neutralizing the eDNA of the bacteria (Tan et al., 2018). Solid lipid nanospheres [SLNs] with anacardic acid, chitosan and DNase I as matrix material was effective in combating biofilms of *S. aureus* wherein chitosan aids in fusing SLNs to the negatively charged biofilm matrix (Anjum et al., 2020). Downregulation of quorum sensing mechanisms using chitosan nanomaterials offers a promising approach to tackle resistant biofilms as chitosan offers an excellent carrier vehicle (Nag and Ray, 2021).

8.4.8 Polylactic Acid

Polylactic acid [PLA] forms a good biocompatible material with less immune response and is often used as a drug carrier by conjugating with enzymes, peptides, chelating agents' metals and antibiotics (Scaffaro et al., 2019). PLA NPs have been approved by the Food and Drug Administration [FDA] for administration as a parenteral route carrier for drugs (Cruz et al., 2017; Byda et al., 2018). A combination of a natural and synthetic polymer, namely, octenyl succinic anhydride modified HA (OSA-HA) and poly (lactic-co-glycolic acid) (PLGA) eradicated the biofilms of *P. aeuriginosa* formed on mucosal lining of lungs under in-vitro conditions (Kłodzinska et al., 2019). Another in-vitro study demonstrated a superior antibiofilm activity of PLA NPs against *Dermatophytes* and *Candida sp.* by encapsulation of the ketoconazole drug up to 45% loading into a 188.5 nm PLA NP (Endo et al., 2020). A PLGA NP loaded with peptides has been demonstrated to show potent in-vitro antimicrobial activity against MRSA, *E. coli* and *P. aeuriginosa* with excellent biocompatible and biodegradable activity (Cruz et al., 2017).

8.4.9 Polycaprolactone

Polymer micelles have been fabricated essentially to enhance the solubility of hydrophobic drugs (Lin et al., 2020). Micelles offer good absorbency and small sized particles to retain drugs for targeted delivery. Polyethylene glycol with polycaprolactone encapsulated nanoparticles (PEG-cu-PU) offers excellent delivery systems with the PEG acting as a storage vesicle and the positive charge of the PU aiding in penetration into the biofilm matrix with polycaprolactone degrades to release the payload at the targeted site (Su et al., 2018).

8.4.10 Nanoparticles as Nanocarriers for Drug Delivery to Treat *In-Situ* Bacterial Infections

Natural and synthetic polymeric nanoparticles have been used as drug carriers for targeted delivery. These nanoparticles are manufactured for controlled release of drugs, biocompatibility to host cells, quench toxicity effects (Nascimento et al., 2016). The nanoparticles for drug delivery are usually made as nanospheres with a centrally oily matrix with polymeric shells for leaching of drugs into the external environment (Zielinska et al., 2020). These nanoparticles are manufactured to release drugs by either of the following principles: diffusion, solvent, chemical interaction and stimulated release. The stimulus often would be pH, temperature, magnetic attraction and ultrasound (Son et al., 2017; Ma et al., 2018). Mostly degradable polymers are preferred to non-degradable polymers and complete dissolution of the nanoparticle occurs after eradication of the biofilm. Bovine serum albumin [BSA] NPs have been used as drug carriers for various biofilm infections (Karami et al.,

2019). LL-37, an antibiofilm agent against *S. aureus* and *P. aeruginosa* when administered is easily neutralized by certain proteolytic enzymes (trypsin, pepsin, elastase) in the human system as well as enzymes V8 secreted by *S. aureus* itself (Galdiero et al., 2019). BSA NPs have been shown to stabilize the LL-37 peptide from degradation by the enzymes and its release in low concentrations (Yang et al., 2020). BSA NPs have been shown to have dual activity as they have been observed to block the proliferation of bacterial cells which reduce the production of inflammatory cytokines by macrophages whereas the peptide kills the biofilm cells. This property of BSA NPs makes it an excellent nanocarrier and a promising future material. Bacterial cell surface lectins such as fucose binding lectin LecB (PA-IIL) and galactose-binding lectin LecA (PA-IL) of *P. aeruginosa* have been shown to be inactivated by using polymeric nanoparticles coupled with D-galactose releasing the drug (Flockton et al., 2019). Photothermal activation of polymer vesicles of porphyrin alternating copolymer P(TPP-a-BDE) have been effective in treating wound biofilms (Chen et al., 2020). Kappa-Carrageenan capped ZnO NPs effectively inhibited both MRSA and *P. aeuriginosa* biofilms at concentrations of 100 µg/mL (Vijaykumar et al., 2019).

8.4.11 Magnetic Nanoparticulates for Antibiofilm Therapy

Magnetically targetable nanoparticles with or without antimicrobial carriers have emerged as an alternate method of biofilm control (Quan and Peterson, 2021). This approach finds application in chemotherapeutics associated with tumor sites; delivery of drugs; and localized heat generation to kill biofilms. An obstacle for direct in-vivo applications involves reticuloendothelial uptake and physical obstruction in organs which is being addressed by surface modification approaches to reduce toxicity. This approach originally designed for targeted delivery of drugs and treatment of tumor cells in the centimeter range is now being researched actively to tackle biofilm infections which are in the micron range (Quan and Busscher, 2019). Metals such as iron, cobalt, nickel and iron oxides which are easily magnetized (100 to 218 emu g^{-1}) have been actively used (Chao et al., 2019). A magnetic, biocompatible core-shell nanocomposite material of ZnO@Fe_3O_4 has shown excellent antibiofilm activity against both Gram-negative (*H. pylori*) and Gram-positive (*S. aureus*) (Medina-Ramirez et al., 2020) by ion lixiviation.

8.4.12 Nano Based Photothermal (PTT) and Photodynamic Therapy (PDT) against MDR biofilms

Photothermal therapy [PTT] was devised by the thermal stress caused by irradiation of light of specific wavelengths for cancer therapy (Cheng et al., 2014). Alternately PTT can be called as "NIR-triggered" antimicrobial strategy which is dependent on light without inducing resistance. Significant progress has been achieved in the use of PTT for clinical bacteriostatic applications (Yin and Zheng, 2016). Several different approaches to target bacteria by using photothermal therapy [PTT] have been designed which have been extensively reviewed by (Xu and Hu, 2021). The approaches involve use of: plain and antibiotic loaded nanoparticles, pH-sensitive surface charge change, enzyme response bacterial targeting, metal-organic frameworks and hydrogels. Some areas of concern regarding photothermal therapy are preventing dispersion of photosensitizers, phototoxicity of NIR wavelength usage on cytotoxicity, increasing deep penetration of NIR and sensitivity, for which NIR II is currently in vogue. The use of blue light has been shown to have a broad-spectrum antimicrobial activity against six ESKAPE pathogens (Halstead et al., 2016). Wavelengths of 415 ± 10 nm have been administered for the treatment of wound infections with minimal damage to tissue cells (Katayama et al., 2018). Graphene NP has been used a vehicle for drug delivery (Li et al., 2015). Graphene based nanomaterials are used in PTT as they possess excellent optical absorption and photothermal conversion with NIR light (Xu et al., 2019). Combinations of graphene using PTT and PDT along with photosensitizers have also been used (Wu et al., 2016). For an extensive review on the antibacterial and antibiofilm properties of graphene and graphene oxide refer to (Yousefi et al., 2017).

There has been much advancement in photothermal therapy [PTT] as a result of limited antibacterial efficacy observed. Combinatorial/synergistic approaches have been tried out along with PTT like combining with photodynamic therapy. PTT-PDT combined treatments were effective due to the heat generated by the PTT and the ROS generated by the PDT (Zhao and Zhang, 2029; Tan et al., 2020). Photothermal therapy [PTT] coatings have been developed on the surface of bone implants to eliminate bacterial adhesion through electrostatic and hydrophobic interactions between the surface and the bacteria (Zhao et al., 2020). In addition, chemo dynamic antibacterial therapy [CDT] has also been used along with PTT which again uses the ROS produced by CDT agents to generate hydroxyl radicals. Several combinations using metal ions, nitric oxide release, CDT, polymeric polydopamine nanoparticles have been used in combination with PTT for antimicrobial activity. Amongst the nanoparticle-based photo thermal therapy, the absorbance spectra of gold nanostructures could be easily tuned by changing the nano-morphology and hence AuNPs are more suitable for PTT (Zheng et al., 2017).

New approaches to combat MDR pathogens are the use of a combinatorial approach such as antimicrobial photodynamic therapy [aPDT] along with antibiotics and antifungals (Perez-Laguna et al., 2021), which ensures prevention of regrowth of treated bacteria as well as not contributing to the selection of antimicrobial resistance. In comparison to PTT, PDT is an FDA approved therapeutic method which uses singlet oxygen and ROS generated by photosensitizer molecules to generate cell apoptosis or necrosis (Agostini et al., 2011; Yang et al., 2013). The design of different photosensitizer molecules for antibacterial activity using PDT has been reviewed by (Pucelik and Dąbrowski, 2022). Antimicrobial photodynamic therapy uses a photosensitizer molecule [PS] which, in the presence of oxygen and activation by light, generates reactive oxygen species [ROS] which have been shown to be very effective in treating antimicrobial resistant *Staphylococcus aureus* in skin and soft tissue infections [SSTI] (Kharkwal et al., 2011; Perez-Laguna et al., 2018). Some of the common photosensitizer molecules (5-ALA: 5-aminolevulinic acid; TAPP: meso-tetra (4-aminophenyl) porphine; TMP: tetra-substituted N-methyl-pyridyl-porphine; MB: methylene blue; RB: Rose Bengal; ATAZTMPo; 2-aminothiazolo[4,5-c]-2,7,12,17-tetrakis (methoxyethyl) porphycene; have been reviewed extensively by Perez-Laguna et al., (2021). The effectiveness of PDT against methicillin resistant *S. aureus* in response to various photosentizers has been reviewed by (Dharmaratne et al., 2020).

8.5 BIOFILM FORMING MULTI-DRUG RESISTANT BACTERIA AND THEIR MECHANISMS OF RESISTANCE

In 2017, the WHO published a list of priority bacteria for which more research and antibiotics were required. The list includes 11 bacteria in three categories (critical, high and medium) priority. The drug-resistant and pathogenic *Acinetobacter baumannii*, *Pseudomonas aeruginosa*, *Enterobacteriaceae* are on the top of the list (WHO, 2017). The first two bacteria and the member of *Enterobacteriaceae* are part of the ESKAPE pathogens (*Enterococcus faecium*, *Staphylococcus aureus*, *Klebsiella pneumoniae*, *A. baumannii*, *P. aeruginosa*, and *Enterobacter sp.*) which are multi-drug resistant, highly virulent and nosocomial infection causing pathogens (Mulani et al., 2019). As these pathogens exhibit resistance to multiple antibiotics, alternative therapies are needed, and the application of nanoparticles is one of the alternative methods to combat these pathogens.

Many of the ESKAPE pathogens are known to produce biofilms and cause serious infections and diseases. The highly pathogenic and multi-drug resistant Gram-negative bacterium, *A. baumannii* is known to produce biofilm and cause ventilator-associated pneumonia, catheter-related and bloodstream infections, as well as urinary and respiratory tract infections (Gedefie et al., 2021). The ability of this bacterium to produce biofilms, multiple virulence factors and the antibiotic resistance mechanisms makes antibiotic-related therapy ineffective. The main resistance mechanisms involved in *A. baumannii* include target sites modifications, multidrug efflux pumps, degradation of antibiotics, permeability deficiencies and the presence of large genome islands with multiple antibiotic resistance genes (Gorden et al., 2010; Kim et al., 2012).

8.5.1 Enterococcus faecium

Enterococci are the third leading pathogens causing nosocomial infection, with 17% surgical related infection, 11.5% bacteraemia and 14.3% urinary tract infections (Richards et al., 2000). *Enterococci* constitute natural microorganisms present in the oral cavity, but has also been reported to occur in the bloodstream, the urinary tract, the intestinal tract and the genital tracts. Among the *Enterococci* genus, *Enterococcus faecalis* constitutes about 80–90% of infections in humans (Jones et al., 2004). *E. faecalis* constitutes around 7–10% of bloodstream infections in hospitals (Sandoe et al., 2003). The bacterium infects the root canals resulting in periodontitis (Pei et al., 2002) and colonized central venous catheters (Mohamed et al., 2007). The bacterium is one of the most antibiotic tolerant and heat resistant pathogens (Fischer et al., 2009). Factors responsible for adhesion and biofilm formation in *Enterococcus faecalis* have been characterized involving adhesins-*Esp* (Toledo-Arana et al., 2001); QS signal transduction systems (Hancock and Peredo, 2004); *GelE*–the secreted metalloprotease (Qin et al., 2001); *BoPD*–the sugar transcriptional regulator (Hufnagel et al., 2004); *ebp*–encoded pili (Nallapareddy et al., 2006); and bee locus (Tendolkar et al., 2006).

8.5.2 Staphylococcus aureus

Staphylococcus aureus is a major pathogen associated with skin and soft tissue infections [SSTI]. Rajaduraipandi et al., (2006) have reported the prevalence of MRSA strains from clinical and carrier samples which exhibited 99.6% resistance to Penicillin, 93.6% (ampicillin), 63.3% (gentamycin, co-trimoxazole, cephalexine, erythromycin, cefotaxamine). The susceptibility of MRSA strains to clindamycin, vancomycin and fluoroquinolones has also found to be diminishing (Siberry et al., 2003). *Staphylococcus aureus* is a Gram-positive, commensal and opportunistic pathogen normally associated with skin and mucus membranes. This bacterium causes infections ranging from minor skin infections to severe bacteraemia, and sepsis (Tong et al., 2015). Particularly, the infections caused by methicillin resistant or multi-drug resistant *S. aureus* are more serious due to their high prevalence in hospitals and other environments, their ability to form biofilms on medical devices and implants and antibiotic tolerance in the persister cells of *S. aureus* strains (Karimaei et al., 2021). The antibiotic resistant and virulent methicillin resistant *S. aureus* (MRSA) has been found to be associated with ocular infections and ocular damage (Kathirvel et al., 2021). The main antibiotic resistance mechanisms involved in *S. aureus* include antibiotic inactivation by enzymes, alteration of target sites due to reduced affinity to the antibiotic, efflux pumps and antibiotic trapping (Pantosti et al., 2007).

8.5.3 Klebsiella pneumoniae

Klebsiella pneumoniae is a gram-negative fastidious bacillus that occurs commonly in the environment. *Klebsiella pneumoniae* is also Gram-negative opportunistic pathogen which has been found to be associated with blood stream infections, sepsis, urinary tract infections and hospital/ventilator associated pneumonia (Bengoechea and Sa Pessoa, 2019). It is a significant hospital acquired infection resulting in respiratory tract, bile ducts, urinary tract and surgical wound infections (DeRosa et al., 2015). It has been found to persist on epithelial cells and on the surfaces of medical devices with the biofilm mode being resistant to many antimicrobials. Multi drug resistant *Klebsiella pneumoniae* has been the main source of blood stream infections [BSIs] caused by extended spectrum β-lactamase [ESBL] production (Viale and Tumbarello, 2013) and has been reported to be fatal in approximately 50% of cases. The use of efflux pump inhibitors [EPIs] has increased the antibiofilm activity of antibiotics in most of the ESKAPE pathogens by increasing the internalization of the drugs. One such example is alkaloids such as reserpine and berberine at low concentrations along with antibiotics have been found to be effective against MDR varieties of this strain (Mahesh et al., 2013; Zhou et al., 2016). Alkoxyquionolines have also increased the susceptibility of antibiotics to resistant *K. pneumoniae* biofilms (Chevalier et al., 2004). This bacterium is a known biofilm-forming organism and shows resistance to multiple antibiotics (Nirwati et al., 2019). A strong correlation between the antibiotic resistance and the expression of efflux pumps and biofilm genes in *K. pneumonia* was observed in previous studies (Tang et al., 2020; Mirzaie and Ranjbar, 2021).

8.5.4 Acinetobacter baumannii

Acinetobacter baumannii ranks high among the nosocomial pathogens due to the existence of MDR and pan drug resistant strains and has been found to colonize the upper respiratory tract, skin epidermis, urinary tract, to invade the bloodstream and open wounds and to become a major organism for hospital acquired infection (HAI) causing blood stream infections, secondary meningitis and ventilator associated pneumonia (Dijkshoorn and Seifert, 2007). The bacterium has been demonstrated to exhibit remarkable adaptability to different environments and selective pressure which makes it a dangerous microbe of concern in the hospital environment. The bacterium forms biofilms essentially through pilus with a hydrophobic sugar in the O side chain moiety of the lipopolysaccharide (Haseley et al., 1997). The bacterium has become problematic in ICU environments and has developed broad spectrum antimicrobial resistance (Meric et al., 2008). The outer membrane protein of *A. baumannii* has been involved in cytotoxicity to the epithelial cells (Choi et al., 2007). In-spite of infections (meningitis) caused by a MDR strain of *A. baumannii* which was resistant to imipenem and β-lactam drugs the patients responded to treatments with ampicillin and sulbactam at a concentration of 2g/1g for period of 9–12 days (Sun and Li, 2017). Antibiotic sensitivity to clinically isolated *A. baumannii* from Turkey medical (Altun et al., 2014) revealed MIC_{50} values of netilmicin (4 µg/mL), tigecycline (3 µg/mL), sulbactam (8 µg/mL), amikacin (128 µg/mL), meropenem (64 µg/mL) and colistin (1 µg/mL), whereas the MIC_{90} values varied netilmicin (512 µg/mL), tigecycline (8 µg/mL), sulbactam (12 µg/mL), amikacin (1024 µg/mL), meropenem (128 µg/mL) and colistin (1 µg/mL) respectively.

8.5.5 Pseudomonas aeuriginosa

Pseudomonas aeuriginosa (PAO1) is an opportunistic nosocomial pathogen manifesting in hospital acquired pneumonia in immunocompromised patients (Degi et al., 2021). *Pseudomonas aeruginosa* is a Gram-negative opportunistic pathogen and a well-known biofilm-forming bacterium. *P. aeuriginosa* exhibits resistance to multiple drugs due to the presence of multi drug efflux pumps and the presence of chromosomal AmpC β-lactamase (Henwood et al., 2001). It has been found to be sensitive to fourth generation cephalosporins, monobactams, carbapenems and fluoroquinolones. Paβ (MC207-110) a broad-spectrum efflux pump inhibitor has been shown to facilitate the action of fluoroquinolones, levofloxacin (Mahmood et al., 2016). It is more commonly associated with nosocomial infections, and ventilator-associated pneumonia and also causes cystic fibrosis (CF) lung infections, leading to high morbidity and mortality in CF and immunocompromised patients (Sadikot et al., 2005). This bacterium shows resistance to multiple antibiotics including β-lactam, quinolones and aminoglycocides, and the resistant mechanisms involved include decreased outer-membrane permeability, reduced efficiency of efflux systems, inactivation of antibiotics by the production of enzymes, and the presence of multiple ARGs and adaptive resistance through strong biofilm formation and development of persister cells (Pang et al., 2019). Due to the complexity associated with this pathogen and its various antibiotic resistance mechanisms, alternative therapies are needed.

8.5.6 Enterobacter aerogenes

Enterobacter aerogenes is a leading organism causing nosocomial and respiratory tract infections (Dupont et al., 2004). It exhibits a resistance to β-lactam antibiotics by production of hydrolytic enzymes and membrane permeability resulting in efflux of intracellular concentration of drugs. Carbapenem is another antibiotic which has become resistant to many strains of *Enterobacteriaceae*. Several Enterobacter species have developed intrinsic resistance to ampicillin and cephalosporins

8.5.7 Comparison of Efficacy of Antibiotic and Nanoparticle on Biofilm Formation of ESKAPE Pathogens

ESKAPE pathogens (*Enterococcus faecium, Staphylococcus aureus, Klebsiella pneumoniae, Acinetobacter baumannii, Pseudomonas aeruginosa*, and *Enterobacter species*) have developed resistance to antibiotics and have been increasingly isolated from community acquired infections in hospitals globally (Pendelton et al., 2013). When using the terminology of tolerance/resistance for any molecule/compound/nanoparticle the MIC_{50}, MIC_{90} and MBC values are of importance. These values form the basis for classifying resistant/tolerant organisms under investigation. Literature reports for different nanoparticles have exhibited a range of MIC_{50}, MIC_{90} and MBC values starting from 1 µg/mL to 2048 µg/mL. This wide range of MIC has been found to be influenced by the route of synthesis of nanoparticles which influence the physicochemical properties such as shape, size, charge, ion leaching ability which has a major say on antimicrobial and antibiofilm efficacy. Same nanomaterials (for example CuO/ZnO/Ag) synthesized through different precursors and reducing agents exhibit different MIC and MBC values depending on their shape, size and charge. In general, there has been some correlation of antimicrobial activity with size of the NP, with lesser sized NP (1–20 nm) showing low MIC values. Similarly, literature scans reveal the minimum bactericidal concentrations [MBC] of antibiotic to range from 0.3 to 50 µg/mL, with systemic (therapeutic) administration dose ranging from 100–500 mg with time periods ranging for 6–24 hours. The corresponding peak level serum concentrations of antibiotics range from 0.3 to 40 mg/L (MacGregor and Graziani, 1997). The minimum inhibitory concentration [MIC] of antibiotics is a major pharmacodynamic parameter used for the design and optimization of antibiotic treatment protocols using the pharmacokinetic PK/PD index (Udekwu and Levin, 2009). The classical MIC of antibiotics, biocides and disinfectants is described by CLSI (2005) protocol based on optical density measurements at 600vnm on an exponentially growing microbial population in an enriched media. Apart from this, several other factors are also known to influence the antibiotic/biocide susceptibility and the efficacy of the antimicrobials is based on these factors, namely, the physiological and metabolic status of the bacteria growing in the presence of antibiotics (Hobby and Dawson, 1944), morbid non growing bacteria resembling persister cells (Bigger, 1944), biofilm formation (Brooun and Lewis, 2000), and the pharmacodynamic function based on the growth/death of bacteria in the presence of an antibiotic concentration (Regoes, 2004). Based on these earlier investigations the bactericidal and bacteriostatic MIC concentrations have come into vogue (CLSI, 2002; Wayne-CLSI, 2011).

The objective of the present investigation was to demonstrate the efficacy of some of the known potent antibiotics against biofilm formation, metabolic activity of the ESKAPE pathogens and to compare the minimum biofilm inhibition concentrations [MBIC] of antibiotics to that of the MBICs of nanoparticles. In the present study a commercial ZnO nanoparticle (Figure 8.3) has been chosen for investigation which is a broad-spectrum nanomaterial used widely in biomedical applications.

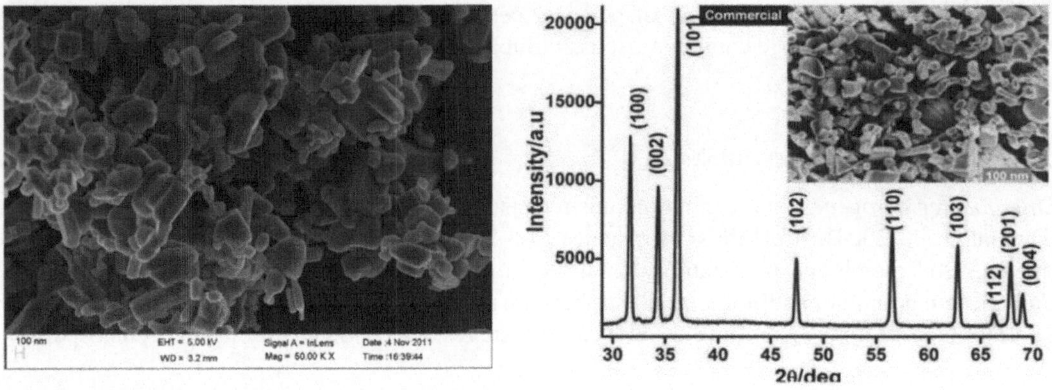

FIGURE 8.3 FESEM of Zinc oxide Nanoparticles and XRD of commercial ZnO nanoparticles.

Similarly general antibiotics towards which ESKAPE pathogens had exhibited susceptibility has been chosen for the study.

8.5.7.1 Strains, Culture Conditions and Evaluation of MBIC

The strains were maintained in the respective medium from which they were isolated, namely, (*Enterococcus faecium* - TSA agar, *Staphylococcus aureus* - TSA agar, *Klebsiella pneumonia* - LB agar, *Acinetobacter baumannii* - LB agar, *Pseudomonas aeruginosa* - LB agar and *Enterobacter aerogenes* - TSA agar). Overnight grown culture of the strains in the respective broths were used for the MIC and MBC assays. MIC and MBC assays were carried out in 96 well microtitre plates using the broth dilution assay method. The efficacy of a procured ZnO nanoparticle was compared with different antibiotics, namely, ampicillin, rifampicin, tetracycline, gentamycin and kanamycin. The test concentrations were prepared based on literature values for these antibiotics ranging from 25–1.6 µg/mL. Four replicates of each concentration were assayed. Parallel controls without antibiotics and nanoparticles along with plain media were also assayed. Similarly, the test concentration of the nanoparticles was prepared ranging from 1024–64 µg/mL. The microtitre plates were incubated under static conditions in an incubator at 37°C. After 24 hours the planktonic OD_{600} of the control and treatments were measured using a multimode reader (BIOTEK-SYNERGY, USA), [Data not shown]. The planktonic cells were also checked for viability using the live dead staining method using the BacLight® staining procedure. The planktonic cells were aspirated and the microtitre wells were washed twice with sterile 1X PBS (pH = 7.4) to remove any loosely bound cells and stained with 200 µL of 0.1% crystal violet (CV) for 10 minutes (Djordjevic et al., 2002). The crystal violet solution was then discarded, and the plates were washed twice with PBS to remove any unbound CV and the plates were air dried. Subsequently 200 µL of 33% glacial acetic acid was added to cv each of the wells to solubilise the cell bound CV for 10 min and the optical density of solubilised CV was measured at OD_{570} nm using a hybrid multimode reader (BIOTEK-SYNERGY, USA).

8.5.7.2 Metabolic Activity of Biofilm Cells - XTT Reduction Assay

The metabolic status of biofilm cells exposed to different concentrations of antibiotics and nanoparticles were assayed at the concentrations mentioned above. The metabolic activity of biofilm cells was evaluated using the XTT [2,3-bis(2-methoxy-4-nitro-5-sulfophenyl)–2H-tetrazolium-5-carboxanilide sodium salt] reduction assay (Ramage et al., 2001; Nett et al., 2011). An XTT reduction kit (Hi media Laboratories) was used to measure the metabolic activity of planktonic cells. XTT and Menadione solutions were mixed in a ratio of 20:1 and added to the biofilm cell suspension and incubated at 37°C for 3 hours in the dark. The coloured supernatant obtained was transferred to new microtitre plates and the fluorescence intensity was measured at 560 nm (excitation) and 590 nm (emission) wavelengths respectively.

A 50% reduction in biofilm (MBIC) was observed at a concentration of 3.1vμg/mL for the antibiotic ampicillin against *E. faecium* with the highest tested concentrations of 25 μg/mL exhibiting 63% reduction in biofilm formation compared to untreated controls (Figure 8.4). In comparison with the ZnO nanoparticles, a concentration of 356 μg/mL was required for 50% reduction in the biofilm (MBIC) of this strain with a corresponding 69% reduction observed at a concentration of 512 μg/mL. The highest tested concentration of 1024 μg/mL elicited only 74% reduction in biofilm formation. Similarly, a 35% reduction in metabolic activity was observed at the MBIC value of 3.1 μg/mL for the antibiotic with the highest concentration of 25 μg/mL eliciting a 43% reduction in metabolic activity. A 40% reduction in metabolic activity was observed at a concentration of 356 μg/mL with the highest concentration eliciting 67% reduction in metabolic activity.

A 50% reduction in biofilm (MBIC) was observed at a concentration of 6.3 μg/mL for the antibiotic rifampicin against *S. aureus* with the highest tested concentrations of 25 μg/mL exhibiting 72% reduction in biofilm formation compared to untreated controls (Figure 8.5). In comparison with the ZnO nanoparticles the highest concentration of 1024 μg/mL tested elicited only 40% reduction

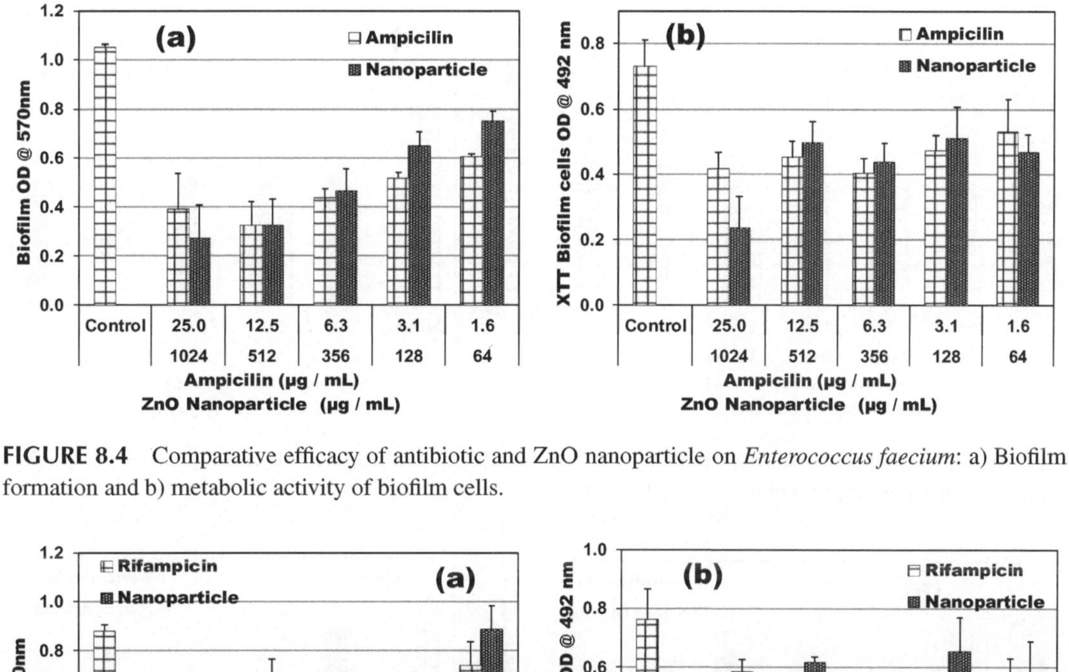

FIGURE 8.4 Comparative efficacy of antibiotic and ZnO nanoparticle on *Enterococcus faecium*: a) Biofilm formation and b) metabolic activity of biofilm cells.

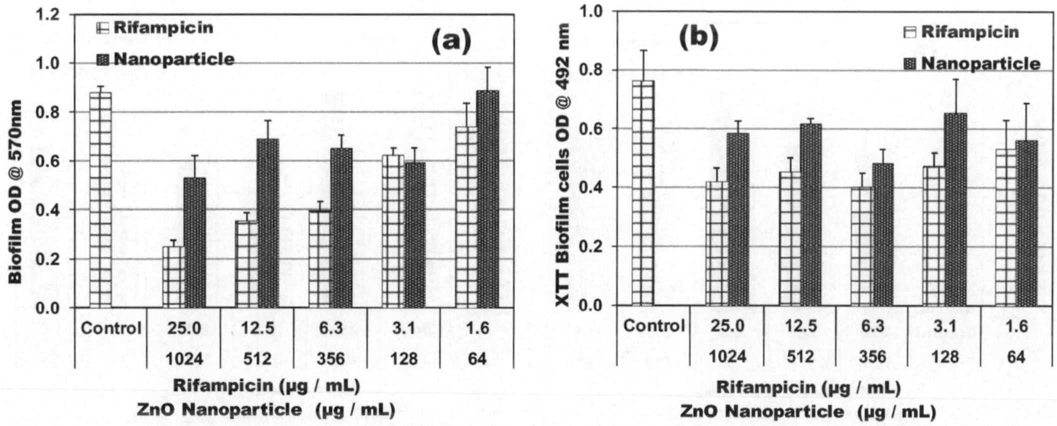

FIGURE 8.5 Comparative efficacy of antibiotic and ZnO nanoparticle on *Staphylococcus aureus*: a) Biofilm formation and b) metabolic activity of biofilm cells.

in the biofilm (MBIC) formation of this strain. Similarly, a 47% reduction in metabolic activity was observed at an MBIC value of 6.3 µg/mL for the antibiotic with the highest concentration of 25 µg/mL eliciting a 45% reduction in metabolic activity. A 23% reduction in metabolic activity was observed at a concentration of 1024 µg/mL for the nanoparticle.

A 50% reduction in biofilm (MBIC) was observed at a concentration of 12.5 µg/mL for the antibiotic tetracycline against *K. pneumoniae* with the highest tested concentrations of 25 µg/mL exhibiting 51% reduction in biofilm formation compared to untreated controls (Figure 8.6). In comparison with the ZnO nanoparticles a concentration of 512 µg/mL was required for 50% reduction in the biofilm (MBIC) for this strain. The highest tested concentration of 1024 µg/mL elicited only 51.5% reduction in biofilm formation. Similarly, a 37% reduction in metabolic activity was observed at the highest concentration of 25 µg/mL tested for the antibiotic. In comparison only 18.6% reduction in metabolic activity was observed at a concentration of 1024 µg/mL for the ZnO nanoparticle.

A 50% reduction in the biofilm (MBIC) was observed at a concentration of 12.5 µg/mL for the antibiotic gentamycin against *A. baumannii* with the highest tested concentrations of 25 µg/mL exhibiting 52% reduction in biofilm formation compared to untreated controls (Figure 8.7). In comparison with the ZnO nanoparticles a reduction of 27% biofilm formation was observed at the highest

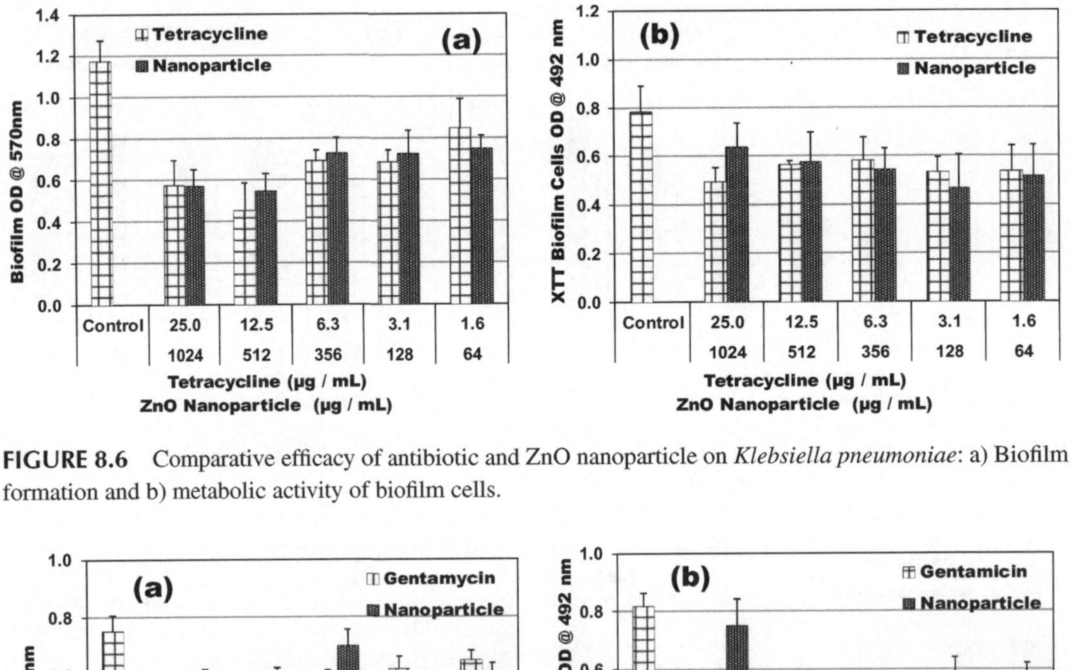

FIGURE 8.6 Comparative efficacy of antibiotic and ZnO nanoparticle on *Klebsiella pneumoniae*: a) Biofilm formation and b) metabolic activity of biofilm cells.

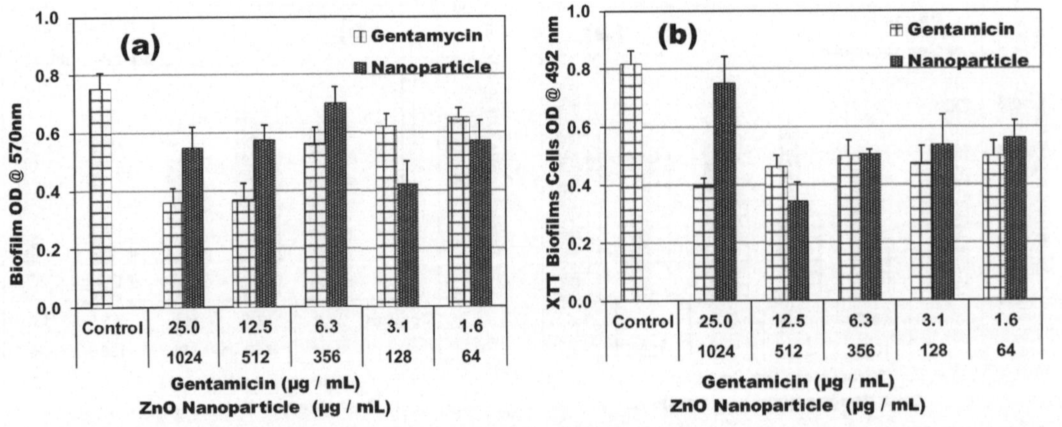

FIGURE 8.7 Comparative efficacy of antibiotic and ZnO nanoparticle on *Acinetobacter baumannii*: a) Biofilm formation and b) metabolic activity of biofilm cells.

tested concentration of 1024 μg/mL. Similarly, a 50% reduction in metabolic activity was observed at the highest concentration of the antibiotic tested, namely, 25 μg/mL. In comparison a 58% reduction in metabolic activity was observed at a concentration of 512 μg/mL for the nanoparticle.

The highest concentration of the antibiotic tetracyline tested against the bacterium *P. aeuriginosa* elicited only 9.4% reduction in biofilm formation at the highest test concentration of 25 μg/mL compared to untreated controls (Figure 8.8). In comparison with the ZnO nanoparticles the highest concentration of 1024 μg/mL tested elicited a was required for 59% reduction in biofilm formation in this strain. In comparison to low levels of biofilm inhibition observed with the antibiotic tetracyline for this strain, the antibiotic was able to cause 47% reduction in metabolic activity. Interestingly the decrease in metabolic activity did not result in decreased biofilm formation for the strain. The highest concentration of the ZnO nanoparticles tested elicited 4.1% reduction in metabolic activity.

A 50% reduction in the biofilm (MBIC) was observed at a concentration of 3.1 μg/mL for the antibiotic kanamycin against *Enterobacter aerogenes* with the highest tested concentrations of 25 μg/mL exhibiting 68.3% reduction in biofilm formation compared to untreated controls (Figure 8.9). In comparison with the ZnO nanoparticles a concentration of 128 μg/mL was required for a 50% reduction in the biofilm (MBIC) of this strain with a corresponding 7.49% reduction observed at a

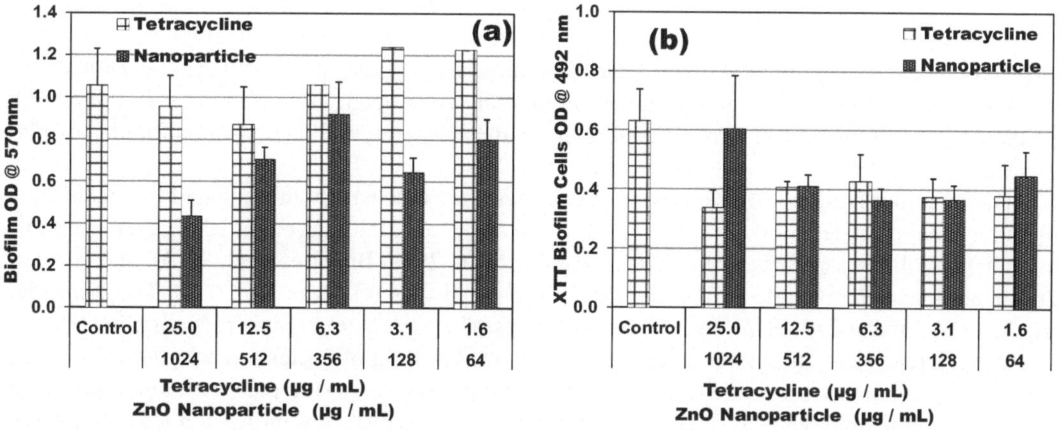

FIGURE 8.8 Comparative efficacy of antibiotic and ZnO nanoparticle on *Pseudomonas aeruginosa*: a) Biofilm formation and b) metabolic activity of biofilm cells.

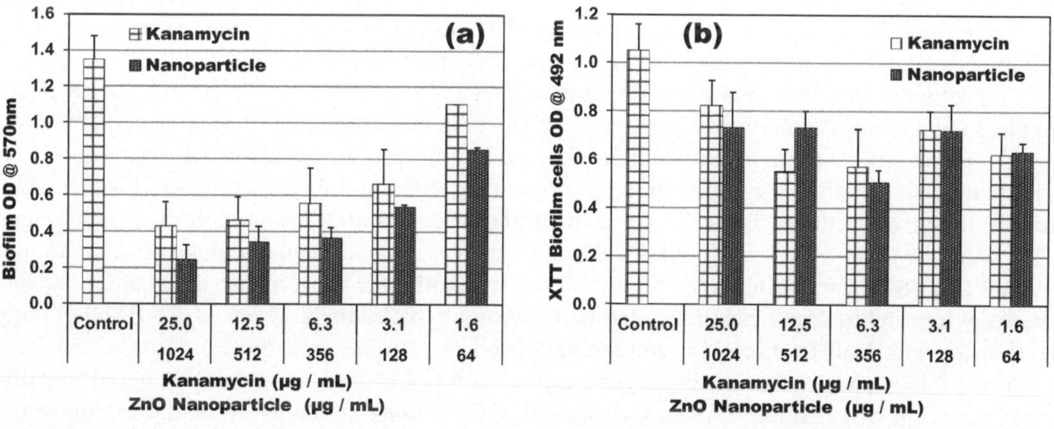

FIGURE 8.9 Comparative efficacy of antibiotic and ZnO nanoparticle on *Enterobacter aerogenes*: a) Biofilm formation and b) metabolic activity of biofilm cells.

concentration of 512 μg/mL. The highest tested concentration of 1024 μg/mL elicited only 81.9% reduction in biofilm formation. Similarly, a 31.5% reduction in metabolic activity was observed at the MBIC value of 3.1 μg/mL for the antibiotic with the highest concentration of 25 μg/mL eliciting a 21.7% reduction in metabolic activity. A 30.5% reduction in metabolic activity was observed at a concentration of 1024 μg/mL.

8.6 IMPACT OF MDR MICROBIAL BIOFILMS ON IMPLANTS AND ADVANCES IN PROTECTIVE ANTIMICROBIAL COATINGS

Nosocomial infections contribute a burden of USD 94 billion per year in the USA associated with biofilm growth including that of the pathogen *P. aeuriginosa* (Romling et al., 2014). Biofilm resistant infections pose a significant problem in treating and managing infected patients clinically. The widespread presence of multi-drug resistant (MDR) and pathogenic bacteria is posing a severe threat to healthcare industries. The surgical tools, implants and medical device-associated infections are of serious concern. The major problem linked with MDR bacteria is the formation of biofilms and its resistance to antibiotics which makes it difficult to treat infections or diseases caused by these bacteria.

The application of nanotechnology in biomaterials and medical devices to control MDR biofilms is explored in this chapter. Different kinds of nanoparticles (NPs) such as silver (Ag), copper (Cu), zirconium (Zr), zinc oxide (ZnO), and magnetic nanoparticles are commonly used as antibacterial and antimicrobial properties against bacteria, fungi and enveloped viruses. In order to improve the efficiency of antibiofilm and antibacterial activity, the NPs are combined with antibiotics or plant extracts.

In the previous studies, different types of coatings were extensively investigated to treat biofilms including nanoparticles (Lewisoscar et al., 2021, Huang et al., 2020, Du et al., 2022), and in combination with antibiotics (Shaaban et al., 2017, Pandey et al., 2019, Banerjee et al., 2020, Scutera et al., 2021), antimicrobial peptides (Ye et al., 2022) and metal ions (Quan et al., 2021). The three main mechanisms involved in the antimicrobial properties of these coatings are a reduction in adhesion, a controlled release of antimicrobial agents and a contact killing of bacteria. Though these coatings were successful in in vitro studies on antimicrobial properties, in vivo studies were not as promising due to coating technologies as chemical and mechanical stability, scalability and toxicity must be taken into consideration (Ye et al., 2022; Zhao et al., 2009).

The mechanism of NPs involved in antibacterial properties is the reduction in the adhesion of bacteria and an increase in the internal ion concentration, thereby allowing the antibiotics to penetrate into the biofilm and kill the bacteria. The biofilm forming and pathogenic bacteria *S. aureus, E. coli, P. aeruginosa, K. pneumonia* are known to be resistant to multiple drugs. These bacteria are commonly found associated with surgical implants and other medical devices.

Contact lenses are widely used as implants and their long-term usage leads to bacterial accumulation and biofilm formation on the lens. The most common organisms found on contact lenses are S. *aureus and P. aeruginosa,* which are known biofilm-producing and MDR bacteria. A previous study showed that the application of copper (Cu) NPs on polyvinyl alcohol (PVA) contact lenses effectively reduced bacterial attachment and also inhibited the growth of *S. aureus* and *P. aeruginosa* (Kharagami et al., 2019). Surgical devices such as scalpels, bone plates, vascular clamps and other surgical tools and devices used in health care must be sterile when in use and these devices and tools are usually coated with zirconium based metallic beads (Zr-BMGs). Long term usage can lead to microbial contamination on the surface and these microbes can form biofilms. Silver (Ag) NPs with hydrogen peroxide (H_2O_2) coated on zirconium based metallic glass beads showed antibacterial activity against MDR *S. aureus* due to a reduction in the adhesion on the surface of biomaterial and this approach was found to be suitable for Zr-BMGs coated surgical devices (Du et al., 2022).

Various approaches were used in NPs coating to effectively eradicate the MDR bacteria associated with medical devices. The combination of NPs with antibiotics is one such strategy to kill the MDR bacteria. In a previous study, imipenem loaded poly ε-caprolactone (PCL) and polylactide-co-glycolide (PLGA) nanocapsules were formulated and tested against *K. pneumonia* and *P. aeruginosa* for anti-adhesion and antibacterial properties (Shabaan et al., 2017) and they found that the imipenem loaded nanocapsules effectively eliminated the bacterial attachment and biofilm formation on the implants. Similarly, Eudragit RL 100 encapsulated gentamicin (E-G-S) on the catheter effectively killed the *E. coli* and it was found that the slow release of antibiotics in low concentrations through NPs was effective in entering the biofilm and killing the bacteria whilst also reducing the cytotoxicity (Pandey et al., 2019). This finding suggested that the Eudragit RL100 mediated drug delivery system provides an effective and suitable way to control biofilm formation on catheters.

In general, all the bacteria live in a planktonic state followed by attachment to surface and formation of biofilm. D-cysteine, an amino acid widely used in the food, pharmaceutical and personal care industries is known to inhibit biofilm formation/aggregation thus keeping the bacteria in the planktonic state. Silver NPs are effective in killing bacteria in the free-living state. D-cysteine embedded into Ag NPs coated on polydopamine stainless steel acted as a strong antimicrobial agent against *E. coli, P. aeruginosa* and *S. aureus* and thus was found to be suitable for coating surgical instruments and implants (Huang et al., 2020).

Biomaterial-associated infections can occur following surgical implantation, thereby limiting their success rates. Though it is normally difficult to eliminate bacteria located within biofilm on biomaterial surfaces, the use of magnetic NPs provides an alternative solution. Magnetic NPs coated on biomaterial surfaces can be easily removed by applying an external magnetic field. The magnetic field pulls the magnet NPs, forming a water channel through the biofilm, making way for antibiotics to enter and kill the bacteria within the biofilm. This strategy would be a very useful technique for application in surgical implants (Quan et al., 2022)

Previous work on antimicrobial peptides (AMP) in combination with Ag NP nanocomposites had strong antimicrobial property due to its potential for permeabilization of the bacterial membrane by AMPs which facilitates the Ag NPs to act upon the colonized bacteria (*S. aureus, Streptococcus gordinii* and *P. aeruginosa*) within the biofilm (Ye et al., 2022).

Nanoparticles and in combination with antibiotics are widely used to treat MDR biofilms and to manufacture biomaterials for implants. Table 8.2 provides a summary of NPs and antibiotics based new methods developed and applied in the health care industry.

8.7 CONCLUSIONS

The current pace at which new antibiotics are being discovered is insufficient to combat the alarming pace of development of antibiotic resistance, necessitating the need for alternate antimicrobial approaches. Currently some of the alternate approaches such as antimicrobial peptides, small molecules and photodynamic therapy are currently being used synergistically along with conventional antibiotics to overcome/treat persistent chronic MDR biofilm infections. In the search for alternate antimicrobials, molecules which can target the central dogma as well as attacking the stress related pathways exhibited by MDR bacterial strains needs to be extensively investigated. Compared to MONs polymeric and lipid-based nanoparticles have found increasing in-vivo applications in targeted drug delivery and combating biofilms in-vivo. Biocompatibility issues, systemic safety and the long-term effects of nanoparticles have been a hindrance for in-vivo clinical applications. Studies on the pharmacokinetic properties of nanoparticles are required to understand the effects in the human body. Currently investigations of nanomaterials undergoing preclinical and clinical trials are very limited and voluminous in-vitro studies exist. Compared to nanoparticles synthesized through chemical routes, biological synthesized nanoparticles have shown better efficacies as antimicrobials

TABLE 8.2
Nanoparticles and Its Combination with Antibiotics Used against MDR Bacterial Biofilms

S. no	Bacterial pathogens	NPs (and with antibiotics)	Effect on the bacteria and biofilm formation	Applications/ Implants	Advantages	References
1	*Staphylococcus aureus*	MPDA-LUT@ CaP with Ti	Disruption of cell membrane of *S. aureus* by the NPS and NIR treatment	Orthopedic implant	Controlled release of QS inhibitor (LUT), release of Ca^{2+} and $(PO_4)^{3-}$from the surface of implant promotes osteogenic differentiation in later stages.	Hu et al., (2021)
2	*Pseudomonas aeruginosa*	AgNPs with antibiotics	Antimicrobial and anti-biofilm properties, inhibition of biofilm	Dental implant	Inhibitory action on biofilm formation and plaque formation	El Telbany and El-Sharaki (2022)
3	*Staphylococcus aureus and Pseudomonas aeruginosa*	AgNPs with antibiotics	Disruption and detachment established biofilms.		Eradication of biofilms	Feizi et al., (2022)
4	*Serratia marcescens*	ZnO NPs	Inhibition of biofilm formation, disruption of QS mechanism	Bactericidal solution against pathogenic bacteria	Inhibition of biofilm formation, antibacterial, ant virulence activity	Pugazhendi et al.,(2022)
5	*Pseudomonas aeruginosa*	AgNPs	Inhibit or kill the cells through various pathways like inhibition of DNA replication, expression of ribosomal/ other proteins, and interruption of the electron transport chain	Urinary catheter	Inhibition of biofilm formation.	Lewisoscar et al., (2021)
6	*Pseudomonas aeruginosa and Staphylococcus aureus*	Silver and copper PVA nanoparticles	Inhibition of microbial growth	Contact lens	Inhibition of adherence of bacteria to contact lens and less cytotoxic	Kharaghani et al.,(2019)

TABLE 8.2 (Continued)
Nanoparticles and Its Combination with Antibiotics Used against MDR Bacterial Biofilms

S. no	Bacterial pathogens	NPs (and with antibiotics)	Effect on the bacteria and biofilm formation	Applications/ Implants	Advantages	References
7	*Pseudomonas aeruginosa*	Selenium NPs with antibiotics	Oxidative stress and inhibition of QS system	Dietary supplement and chemoprevention therapy	Selenium nutritional supplement, adjunct treatment for PAO1	Kong et al., (2021)
8	*Staphylococcus aureus*	AgNPs	decrease surface adhesion	Zirconium based metallic beads in surgical tools and orthopedic implants	antibacterial activity	Du et al., (2022)
9	*Pseudomonas aeruginosa Staphylococcus aureus*	MNPs with antibiotics	Inhibition of biofilm formation	biomaterial coating	anti-biofilm property	Quan et al., (2021)
10	*Pseudomonas aeruginosa, Staphylococcus aureus and E.coli*	AgNPs	Inhibit formation of mature biofilm	Surgical devices and medical implants	disperse and then reduce the bacterial activity	Huang et al., (2020)
11	*Staphylococcus aureus and E.coli*	AgNPs with Ti/Tix multilayer coatings	Inhibition biofilm growth and reduction of adhesion	Surgical device coating material/ spinal implants	No cytotoxicity	Kasperkiewicz et al., (2021)
12	*Streptococcus gordinii S. aureus P. aeruginosa*	Ag NPS with AMP NP		Antimicrobial coating on orthopedic and dental implants	Antimicrobial activity	Ye et al., (2022)
13	*Staphylococcus aureus and P. aeruginosa*	Magnetic NPs	Control biofilm adherence to the surface	Biomaterial implants	Reduction of biofilm formation	Quan et al., (2021)
14	*Staphylococcus aureus*	Hydroxyapatite NPs on polyether ether ketone (PEEK)	Inhibition of biofilms	Medical implants with biocoating	Stable release of drugs	Abouaitah et al.,(2021)
15	*P. aeruginosa, K. pneumonia*	Imipenem loaded poly e-caprolactone (PCL) NPS	Prevent bacterial attachment and biofilm formation	Medical implants	Enhanced antimicrobial activity of Imipenem	Shaaban et al., (2017)

(*Continued*)

TABLE 8.2 (Continued)
Nanoparticles and Its Combination with Antibiotics Used against MDR Bacterial Biofilms

S. no	Bacterial pathogens	NPs (and with antibiotics)	Effect on the bacteria and biofilm formation	Applications/ Implants	Advantages	References
16	*E. coli*	Eudragit RL100 encapsulated gentamicin sulfate (E-G-S) NPs	Eradication of biofilms	Catheter biomaterials	10–20 times more effective in biofilm reduction	Pandey et al., (2019)
17	*MRSA*	ZnO NPs-PK with vancomycin	Increased sensitivity to vancomycin, targets cell membranes	Treatment for swine dermis application	Nontoxic and selective antibacterial action	Banerjee et al., (2020)
18	*E.coli, B. subtilis and Candida albicans*	Zn/Mg NPs with ciprofloxacin and penicillin	Biofilm formation *by E. coli and B. subtilis* was inhibited.	Coating on dental implants	Reduction of bacterial growth	Rosenberg et al., (2020)
19	*P, aeruginosa*	Ag@MON with Caftrazidime	Inhibition and degradation of biofilms	Alginate reduction and eradication of biofilm in CF patients	Synergistic effect in biofilm reduction and antibacterial therapy	Wan et al., (2020)
20	*A. baumanii K. pneumonia*	Col/haNPs with Colistin	Inhibition of biofilm formation	Colistin nanocarrier with high antimicrobial activity on MDR GNB	Decline of bacterial growth overtime	Scutera et al., (2021)

with targeted delivery and better biocompatibility with less toxicity. According to literature reports biogenic nanoparticles offer a more promising viable alternative due to less toxicity which needs to be further studied. Alternatively, carbon nanoparticles have shown high potential in in-vitro studies but also have biocompatibility issues which have limited possible applications. Literature reports a higher efficacy of MONs against biofilms of different species and shows promising results with each study unique and each study drawing different conclusions based on the shape, size, charge, and so forth. Voluminous data exists on the antimicrobial activity of nanoparticles under in-vitro conditions which needs investigation for clinical translation. Collaborative interdisciplinary fundamental and translational research is required to enable the clinical use of nanoparticles. Results obtained through in-vitro studies have to be investigated using in-vivo models with very few studies investigating using animal models and a handful of studies investigating human trials. Similarly for successful clinical translations there is a requirement for standardized guidelines for testing antimicrobial and toxicological activity.

Currently, monotherapy approach using antibiotics has limitations as major known antibiotics have been shown to be resistant to different bacterial forms over the last century of their use. However, antibiotics can still not be done away with or ignored as they still are our first line of treatment strategy. There is a paradigm shift towards a combinatorial approach, like use of polymeric nanoparticles to deliver drugs which have shown better efficacies in treating biofilm infections in-vivo. Even with polymeric nanoparticles several studies have investigated different nano formulations under in-vitro models which all have intrinsic issues such as premature drug release before the target site, drug retention and payload, systemic toxicity, biocompatibility, carrying higher concentration of drugs and its controlled/sustained release so that the surrounding tissues are not affected. However, even with these disadvantages some models have been successful under in-vivo conditions warranting extensive studies in this direction in improving these formulations.

REFERENCES

Ali, K. Ahmed, B. Ansari, S.M. Saquib, Q. Al-Khedhairy, A.A. Dwivedi, S. Alshaeri, M. Khan, M.S. Musarrat, J. 2019. Comparative in situ ROS mediated killing of bacteria with bulk analogue, *Eucalyptus* leaf extract (ELE)-capped and bare surface copper oxide nanoparticles. *Mater Sci Eng C* 100:747–758

AbouAitah, K. Bil, M. Pietrzykowska, E. Szałaj, U. Fudala, D. Woźniak, B. Nasiłowska, J. Swiderska-Sroda, A. Lojkowski, M. Sokołowska, B. Swieszkowski, W. Lojkowski. W. 2021 Drug-releasing antibacterial coating made from Nano-Hydroxyapatite using the sonocoating method. *Nanomaterials* 11(7):1690.

Abbaszadegan, A. Ghabramani, Y. Gholami, A. Hemmateenejad, B. Dorostkar, S. Nabavizadeh, M. Sharghi, H. 2015. The effect of change at the surface of silver nanoparticles on antimicrobial activity against gram-positive and gram-negative bacteria: a preliminary study. *J Nanomater* 720654:8.

Agarwala, M. Choudhury, B. Yadav, R. N. S. 2014. Comparative study of antibiofilm activity of copper oxide and iron oxide nanoparticles against multidrug resistant biofilm forming uropathogens. *Indian J Microbiol* 54(3):365–368.

Ahmed, B. Solanki, B. Zaidi, A. Khan, M.S. and Musarrat, J. 2018. Bacterial toxicity of biomimetic green zinc oxide Nano antibiotic: Insights into ZnO NP uptake and Nanocolloid–bacteria interface. *Toxicol Res* 8(2): 246–261.

Ahmed, F.Y. Aly, U.F. Abd El-Baky, R.M. Waly, N.G.F.M. 2021. Effect of titanium dioxide nanoparticles on the expression of efflux pump and quorum-sensing genes in MDR *Pseudomonas aeruginosa* isolates. *Antibiotics* 10(6):625.

Allison, K.R. Brynildsen, M.P. and Collins, J.J. 2011. Metabolite-enabled eradication of bacterial persisters by Aminoglycosides. *Nature* 12, 473(7346): 216–220.

Altun, H.U. Yagci, S. Bulut, C. Sahin, H Kinikli, S Adiloglu, S.K. Demiroz A.P. 2014. Antimicrobial susceptibilities of clinical *Acinetobacter baumannii* isolates with different genotypes *Jundishapur J Microbiol* 7(12):e13347.

Ansari, M.A. Khan, H.M. Khan, A.A. Cameotra, S.S. Alzohairy, M.A. 2015. Anti-biofilm efficacy of silver nanoparticles against MRSA and MRSE isolated from wounds in a tertiary care hospital. *Indian Journal of Medical Microbiology*, 33(1):101–109.

Anjum, M. Patel, K.K. Dehari, D. Pandey, N. Tilak, R. Agrawal, A.K. Singh, S. 2020. Anacardic acid encapsulated solid lipid nanoparticles for *Staphylococcus aureus* biofilm therapy: Chitosan and DNase coating improves antimicrobial activity. *Drug Deliv Transl Res* 11(5):305–317.

Arzanlou M. Chai, W.C. Venter, H. 2017. Intrinsic, adaptive and acquired antimicrobial resistance in Gram-negative bacteria. *Essays Biochem* 61(1):49–59.

Aslam, B. Wang, W. Arshad, M.I. Khurshid, M. Muzammil, S. Rasool, M.H. Nisar, M.A. Alvi, R.A. Aslam, M.A. Qamar, M.U. Salamat, M.K.F. Baloch, Z. 2018 Antibiotic resistance: a rundown of a global crisis. *Infection and Drug Resistance* 11:1645–1658.

Banerjee, S. Vishakha, K. Das, S. Dutta, M. Mukherjee, D Mondal, J. Mondal, S. Ganguli, A. 2020. Antibacterial, anti-biofilm activity and mechanism of action of pancreatin doped zinc oxide nanoparticles against methicillin resistant *Staphylococcus aureus*. *Colloids and Surfaces B: Biointerfaces* 190:(110921).

Bayda, S. Hadla, M. Palazzolo, S. Riello, P. Corona, G. Toffoli, G. Rizzolio, F. 2018. Inorganic nanoparticles for cancer therapy: A transition from lab to clinic. *Curr Med Chem* 25(34):4269–4303.

Bengoechea, J.A. Sa Pessoa, J. 2019. *Klebsiella pneumoniae* infection biology: Living to counteract host defences. *FEMS Microbiology Reviews* 43(2):123–44.

Bigger, J. 1944. Treatment of *Staphylococcal* infections with penicillin. *Lancet* 244(6320):497–500.

Brown, M.R.W. Collier, P.J. Gilbert P. 1990. Influence of growth rate on susceptibility to antimicrobial agents: modification of the cell envelope and batch and continuous culture studies. *Antimicrobial Agent and Chemotherapy* 34: 1623–1628.

Breidenstein, E.B. de la Fuente-Nunez, C. Hancock, R.E. 2011. *Pseudomonas aeruginosa*: All roads lead to resistance. *Trends Microbiology* 19: 419–426.

Brooun, A. Liu, S. Lewis, K. 2000. A dose–response study of antibiotic resistance in *Pseudomonas aeruginosa* biofilms. *Antimicrob Agents Chemotherapy* 44:640–646.

Canton, R. and Morosini M.I. 2011. Emergence and spread of antibiotic resistance following exposure to antibiotics. *FEMS Microbiol Rev* 35:977–991.

Cavassin, E.D. de Figueiredo, L.F.P. Otoch, J.P. Seckler, M.M. Oliveira, R.A.D. Franco, F.F. Marangoni, V.S. Zucolotto, V Levin, A.S.S. and Costa, S.F. 2015. Comparison of methods to detect the in vitro activity of silver nanoparticles (AgNP) against multidrug resistant bacteria. *J Nanobiotechnol* 13:64.

Chao, Y. Chen, G. Liang, C. Xu, J. Dong, Z. Han, X. Wang, C. Liu, Z. 2019. Iron nanoparticles for low-power local magnetic Hyperthermia in combination with immune checkpoint blockade for systemic antitumor therapy. *Nano Letters* 9:4287–4296.

Chen, C. Chu, G. Qi, M. Liu, Y. Huang, P. Pan, H. Wang, Y. Chen, Y. Zhou, Y. 2020. Porphyrin alternating copolymer vesicles for photothermal drug-resistant bacterial ablation and wound disinfection. *ACS Appl Bio Mater* 3(12):9117–9125.

Cheng, L. Wang, C. Feng, L. Yang, K. Liu, Z. 2014. Functional nanomaterials for phototherapies of cancer. *Chem Rev* 114(21):10869–10939

Chevalier, J. Mahamoud, A. Malle, M. Barbe, J. Page, J. 2004. Inhibitors of antibiotic efflux in resistant *Enterobacter aerogenes* and *Klebsiella pneumoniae* strains. *Antimicrob Agents Chemotherapy* 48(3):1043–1046.

Choi, C.H. Hyun, S.H., Lee, J.Y., Lee, J.S. Lee, Y.S. Kim S.A. Chae, J.P. Yoo, S.M. Lee, J.C. 2008. *Acinetobacter baumannii* outer membrane protein a targets the nucleus and induces cytotoxicity. *Cell Microbiol* 10(2):309–19.

Clinical and Laboratory Standards Institute. 2005. *Performance Standards for Antimicrobial Susceptibility Testing*. Fifteenth Informational Supplement. CLSI, Wayne, PA.

Clinical and Laboratory Standards Institute [CLSI]. 2002. Reference method for broth dilution antifungal susceptibility testing of yeasts; approved standard, CLSI document M27-A2. Wayne: Clinical and Laboratory Standards Institute.

Corona, F. and Martinez, J.L. 2013. Phenotypic resistance to antibiotics. *Antibiotics* 2:237–255.

Costerton, J.W. 1999a. Introduction to biofilm. *Int J Antimicrob Agents* 1999; 11:217–221.

Costerton, J.W. Stewart, P.S. Greenberg, E.P. 1999b. Bacterial biofilms: A common cause of persistent infections. *Science* 284:1318–1322.

Cremonini, E. Boaretti, M. Vandecandelaere, I. Zonaro, E. Coenye, T. Lleo, M.M. Lampis, S. and Vallini, G. 2018. Biogenic selenium nanoparticles synthesized by *Stenotrophomonas maltophilia* SeITE02 loose antibacterial and antibiofilm efficacy as a result of the progressive alteration of their organic coating layer. *Microbial Biotechnology* 11(6):1037–1047.

Cruz, J. Flórez, J. Torres, R.T.R. Urquiza, M. Gutiérrez, J.A. Guzmán, F. Ortiz, C.C. 2017. Antimicrobial activity of a new synthetic peptide loaded in polylactic acid or poly(lactic-co-glycolic) acid nanoparticles against *Pseudomonas aeruginosa*, *Escherichia coli* O157:H7 and methicillin resistant *Staphylococcus aureus* (MRSA). *Nanotechnology* 28(13):135102.

Cruz, D.M. Mi, G. Webster, T.J. 2018. Synthesis and characterization of biogenic selenium nanoparticles with antimicrobial properties made by Staphylococcus aureus, methicillin-resistant Staphylococcus aureus (MRSA), *Escherichia Coli*, and *Pseudomonas Aeruginosa*. *J Biomed Mater Res Part A* 106(5):1400–1412.

Daw, K. Baghdayan, A.S. Awasthi, S. and Shankar, N. 2012. Biofilm and planktonic *enterococcus faecalis* elicit different responses from host phagocytes in vitro *FEMS. Immunol Med Microbiol* 65(2):270–282.

Dégi, J. Motco, O.A. Dégi, D.M. Suici, T. Mares, M. Imre, K. Cristina, R.T. 2021. Antibiotic susceptibility profile of *Pseudomonas aeruginosa* canine isolates from a multicentric study in Romania. *Antibiotics* 10:846.

De Rosa, F.G. Corcione, S. Cavallo, R. Di Perri, G. Bassetti, M. 2015. Critical issues for Klebsiella pneumoniae KPC-carbapenemase producing *K. pneumoniae* infections: A critical agenda. *Future Microbiol* 10(2):283–294.

De la Fuente, J.M.; Grazu, V. Nanobiotechnology: Inorganic nanoparticles vs. *Organic Nanoparticles; Frontiers of Nanoscience*, Elsevier: Oxford, UK, 2012; pp. 1–520, ISBN 978-012-415769-9.

De Matteis, V. 2017. Exposure to inorganic nanoparticles: Routes of entry, immune response, biodistribution and in vitro/in vivo toxicity evaluation. *Toxics* 5(4):29.

Diene, S.M. and Rolain, J.M. 2013. Investigation of antibiotic resistance in the genomic era of multidrug-resistant gram-negative bacilli, especially *Enterobacteriaceae, Pseudomonas* and *Acinetobacter. Expert Rev. Anti-Infect Ther* 11(3):277–296.

Dijkshoorn, L. Nemec, A. and Seifert, H. 2007. An increasing threat in hospitals: Multidrug-resistant *Acinetobacter baumannii. Nature Reviews Microbiology* 5:939–951.

Din, L.B. Mie, R. Samsudin, M.W. Ahmad, A. Ibrahim, N. 2015. Biomimetic synthesis of silver nanoparticles using the lichen *Ramalina dumeticola* and the antibacterial activity. *Malaysian J Anal Sci* 19:369–376.

Dharmaratne, P. Sapugahawatte, D.N. Wang, B. Chan, C.L. Lau, K.M. Lau, C.B.S. Fung, K.P. Ng, D.K.P. Margaret, I.P. 2020. Contemporary approaches and future perspectives of antibacterial photodynamic therapy (aPDT) against methicillin-resistant *Staphylococcus aureus* (MRSA): A systematic review. *European Journal of Medicinal Chemistry* 200:11234.1

Djordjevic, D. Wiedmann, M. McLandsborough, L.A. 2002. Microtiter plate assay for assessment of *Listeria monocytogenes* biofilm formation. *Appl Environ Microbiol* 68(6):2950–2958. doi: 10.1128/AEM.68.6.2950-2958.2002. PMID: 12039754; PMCID: PMC123944.

Drenkard E. 2003. *Antimicrobial Resistance of Pseudomonas Aeruginosa Biofilms Microbes and Infection* 5:1213–1219.

Dorr, T. Vulic, M. Lewis K. 2010. Ciprofloxacin Causes Persister Formation by Inducing the TisB toxin in *Escherichia coli. PLoS Biol* 23:8(2):e1000317.

Dupont, M. De, E. Chollet, R. Chevalier, J. Jean-Marie Pages, J.M. 2004. *Enterobacter Aerogenes OmpX, A cation-Selective Channel Mar–and Osmo-Regulated Pages FEBS Letters* 569(1–3):27–30.

Du, C Yang, Y. Zheng, L. Zhang, T. Zhao, X. and Wang, C. 2022. Structure-element Surface modification strategy enhances the antibacterial performance of Zr-BMGs. *ACS Applied Materials and Interfaces* 14(7):8793–8803.

Ealias, A.M. Saravanakumar, M.P. 2017. A review on the classification, characterisation, synthesis of nanoparticles and their application. *IOP Conf Ser Mater Sci Eng* 263(3):032019.

El-Telbany, M. El-Sharaki, A. 2022. Antibacterial and anti-biofilm activity of silver nanoparticles on multi-drug resistance *Pseudomonas aeruginosa* isolated from dental-implant. *Journal of Oral Biology and Craniofacial Research* 12(1):(199–203).

Ellis, D.H. Maurer-Gardner, E.I. Sulentic, C.E. Hussain, S.M. 2018. Silver nanoparticle antibacterial efficacy and resistance development in key bacterial species. *Biomed Phys Eng Express* 5:015013.

Endo, E.H. Makimori, R.Y. Companhoni, M.V.P. Ueda-Nakamura, T. Nakamura, C.V. Filho, B.P.D. 2020. Ketoconazole-loaded poly-(lactic acid) nanoparticles: Characterization and improvement of antifungal efficacy in vitro against *Candida* and dermatophytes. *J Mycol Méd* 30(3):101003.

Fairoz Ali Al-Wrafy, Adel Ali Al-Gheethi, Senthil Kumar Ponnusamy, Efaq Ali Noman, Shaima Abdul Fattah, 2022. Nanoparticles approach to eradicate bacterial biofilm-related infections: A critical review. *Chemosphere* 288(2):(132603).

Faghihzadeh, F. Anaya, N. Astudillo-Castro, C. Oyanedel-Craver, V. 2018. Kinetic, metabolic and macromolecular response of bacteria to chronic nanoparticle exposure in continuous culture. *Environ Sci Nano* 5:1386–1396.

Fatima, S.; Ali, K.; Ahmed, B.; Al Kheraif, A.A.; Syed, A.; Elgorban, A.M.; Musarrat, J.; Lee, J. 2021 Titanium dioxide nanoparticles induce inhibitory effects against planktonic cells and biofilms of human oral cavity isolates of Rothia mucilaginosa, Georgenia sp. and Staphylococcus saprophyticus. *Pharmaceutics*, 13, 1564.

Feizi, S. Cooksley, C.M. Nepal, R. Psaltis, A.J. Wormald, P.J. Vreugde, S. 2021. Silver nanoparticles as a bio adjuvant of antibiotics against biofilm-mediated infections with methicillin-resistant *Staphylococcus aureus* and *Pseudomonas aeruginosa* in chronic rhinosinusitis patients. *Pathology* S0031–3025(21)00493-1

Fernandez-Barat, L. Ciofu, O. Kragh, K. N. Pressler, T. Johansen, U. Motos, A. et al. 2017. Phenotypic shift in *Pseudomonas aeruginosa* populations from cystic fibrosis lungs after 2-week antipseudomonal treatment. *J Cyst Fibros* 16:222–229.

Fisher, K., Phillips, C. 2009. The ecology, epidemiology and virulence of *Enterococcus*. *Microbiology-SGM* 155:1749e57.

Flockton, T.R. Schnorbus, L. Araujo, A. Adams, J. Hammel, M. Perez, L.J. 2019. Inhibition of *Pseudomonas aeruginosa* biofilm formation with Surface modified polymeric nanoparticles. *Pathogens* 8(2):55.

Galdiero, E. Lombardi, L. Falanga, A. Libralato, G. Guida, M. Carotenuto, R. 2019.Biofilms: Novel strategies based on antimicrobial peptides. *Pharmaceutics* 11(7):322.

Garg, U. Chauhan, S. Nagaich, U. Jain, N. 2019. Current advances in Chitosan nanoparticles based drug delivery and targeting. *Adv Pharmeceutical Bull* 9(2):195–204.

Gedefie, A. Demsis, W. Ashagrie, M. Kassa, Y. Tesfaye, M, Tilahun, M. Bisetegn, H, and Sahle, Z. 2021. *Acinetobacter baumannii* biofilm formation and its role in disease pathogenesis: A review. *Infection and Drug Resistance* 14: 3711–3719.

Geilich, B.M. Gelfat, I. Sridhar, S. van de Ven, A.L. Webster, T.J. 2017. Superparamagnetic iron oxide-encapsulating polymersome nanocarriers for biofilm eradication, *Biomaterials*

Geoffrion, L.D. Hesabizadeh, T. Medina-Cruz, D. Kusper, M. Taylor, P. Vernet-Crua, A. Chen, J. Ajo, A. Webster, T.J. Guisbiers, G. 2020. Naked selenium nanoparticles for antibacterial and anticancer treatments. *ACS Omega* 5(6):2660–2669.

Gnanadhas, D.P. and Marathe, Sandhya and Dipshikha, C 2012. Biocides—resistance, cross-resistance mechanisms and assessment. *Expert Opinion on Investigational Drugs* 22.

Gordon, N.C. Wareham, D.W. 2010. Multidrug-resistant *acinetobacter baumannii*: Mechanisms of virulence and resistance. *Int J Antimicrob Agents* 35(3):219–226.

Godoy-Gallardo, M. Eckhard, U. Delgado, L.M. Puente, Y.J.D.R. Hoyos-Nogués, M. Gil, F.J. Perez, R.A. 2021. Antibacterial approaches in tissue engineering using metal ions and nanoparticles: From mechanisms to applications. *Bioactive Materials* 6(12):4470–4490.

Gold, K. Slay, B. Knackstedt, M. Gaharwar, A.K. 2018. Antimicrobial activity of metal and metal-oxide based nanoparticles. *Advanced Therapeutics* 1(3):1700033.

Graves, J.L. Jr, Tajkarimi, M. Cunningham, Q. Campbell, A. Nonga, H. Harrison, S.H. Barrick, J.E. 2015. Rapid evolution of silver nanoparticle resistance in *Escherichia coli*. *Front Genet* 6:42.

Gupta, A, Silver, S. 1998a. Silver as a biocide: Will resistance become a problem? *Nat Biotechnol* 16(10):888.

Gupta, A. Maynes, M. Silver, S. 1998b. Effects of halides on plasmid-mediated silver resistance in *Escherichia coli*. *Appl Environ Microbiol* 64(12):5042–5.

Gupta, A. Landis, R.F. Li, C.H. Schnurr, M. Das, R. Lee, Y.W. Yazdani, M. Liu, Y Kozlova, A and Rotello, V.M. 2018. Engineered polymer nanoparticles with unprecedented antimicrobial efficacy and therapeutic indices against multidrug–resistant bacteria and biofilms. *J. Am. Chem. Soc.* 140,12137–12143.

Gunawan C, Faiz MB, Mann R, Ting SRS, Sotiriou GA, Marquis CP, Rose, A. 2020. Nanosilver targets the bacterial cell envelope: The link with generation of reactive oxygen radicals. *ACS Appl Mater Interfaces* 12(5):5557–5568.

Gunawan, C. Teoh, W.Y. Marquis, C.P. Lifia, J. Amal, R. 2009. Reversible antimicrobial photoswitching in nanosilver. *Small* 5(3):41–44.

Gunawan, C. Teoh W.Y. Marquis, C.P. Amal, R. 2013. Induced adaptation of *Bacillus* sp. to antimicrobial nanosilver. *Small* 9(21):3554–60.

Guo, L. Panderi, I. Yan, D.D. Szulak, K. Li, Y. Chen, Y.Y. Ma, H. Niesen, D.B. Seeram, N. Ahmed, A. Yan, B. Pantazatos, D. and Lu, W. 2013. A Comparative study of hollow copper sulfide nanoparticles and hollow gold nanospheres on degradability and toxicity. *ACS Nano* 7(10):8780–8793.

Halstead, F.D. Thwaite, J.E. Burt, R. Laws, T.R. Raguse, M. Moeller, R. 2016 Antibacterial activity of blue light against nosocomial wound pathogens growing planktonically and as mature biofilms. *Appl Environ Microbiol* 82(13): 4006–4016.

Hancock, L.E, and Perego, M. 2004. The *Enterococcus faecalis* fsr two-component system controls biofilm development through production of gelatinase. *J Bacteriol* 186(17):5629–5639.

Haseley, S.R. Pantophlet, R. Brade, L. Holst, O. and Brade, H. 1997. Structural and serological characterisation of the O-antigenic polysaccharide of the lipopolysaccharide from *Acinetobacter junii* strain 65. *Eur. J. Biochem.* 245:477–481.

Henwood, C.J. Livermore, D.M. James, D. Warner, M. 2001. Antimicrobial susceptibility of *Pseudomonas aeruginosa*: Results of a UK survey and evaluation of the british society for antimicrobial chemotherapy disc susceptibility test. *Journal of Antimicrobial Chemotherapy* 47(6):789–799

Hosnedlova, B. Kepinska, M. Skalickova, S. Fernandez, C. Ruttkay-Nedecky, B. Peng, Q. Baron, M. Melcova, M. Opatrilova, R. Zidkova, J. Bjorklund, G. Sochor, J Kizek, R. 2018. Nano-Selenium and its nanomedicine applications: A critical review. *Int J Nanomedicine* 13:2107–2128.

Huang, T. Holden, J.A. Reynolds, E.C. Heath, D.E. O'Brien-Simpson, N.M. O'Connor, A.J. 2020. Multifunctional antimicrobial polypeptide-selenium nanoparticles combat drug-resistant bacteria. *ACS Appl Mater Interfaces* 16(12):55696–55709.

Hobby, G. Dawson, M. 1944. Effect of growth rate of bacteria on action of penicillin. *Proc Soc Exp Biol* 56:181–4.

Hoiby, N. Bjarnsholt, T. Givskov, M. Molin, S. and Ciofu, O. 2010. Antibiotic resistance of bacterial biofilms. *Int J Antimicrob Ag* 35(4): 322–332.

Hong, S.H. Wang, X. O'Connor, H.F. Benedik M.J. and Wood, T.K. 2012. Bacterial persistence increases as environmental fitness decreases. *Micor Biotechnology* 5(4):509–22.

Hou, Y. Hou, Y. Ren, Y. Shi, Y. Jin, X Dong, Y and Zhang, H. 2020. *C. aromaticus* leaf extract mediated synthesis of zinc oxide nanoparticles and their antimicrobial activity towards clinically multidrug-resistant bacteria isolated from pneumonia patients in nursing care. *Mater Res* Express 7 095015

Hu, C. Ashok, D Nisbet, D.R. Gautam, V. 2019. Bioinspired surface modification of orthopedic implants for bone tissue engineering. *Biomaterials* 219:(119366)

Hu, J. Ding, Y. Tao, B. Yuan, Z. Yang, Y. Xu, K. Li, Y. liu, P. Cai. K. 2022. Surface modification of titanium substrate via combining photothermal therapy and quorum-sensing-inhibition strategy for improving osseointegration and treating biofilm-associated bacterial infection. *Bioactive Materials* 18:(228–241).

Huh, A.J. Kwon, Y.J. 2011. "Nanoantibiotics": A new paradigm for treating infectious diseases using nanomaterials in the antibiotic's resistant era. *J Control Rel* 156(2):128–145.

Huang, L. Lou, Y. Zhang, D. Ma, L. Qian, H. Hu, Y Ju, P. Xu, D. Li. L. 2020. D-Cysteine functionalised silver nanoparticles surface with a "disperse-then-kill" antibacterial synergy. *Chemical Engineering Journal* 381(1):122662

Iribarnegaray, V. Navarro, N. Robino, L. Zunino, P. Morales, J. Scavone, P. 2019. Magnesium-doped zinc oxide nanoparticles alter biofilm formation of *Proteus mirabilis*. *Nanomedicine* 4(12):1551–1564.

Jasim, N.A. Al-Gasha, F.A. Al-Marjani, M.F. Al-Rahal, A.H. Abid, H.A. Al-Kadhmi, N.A. Jakaria, Md. Rheima, A.M. 2020. ZnO nanoparticles inhibit growth and biofilm formation of vancomycin-resistant *S. aureus* (VRSA) *Biocatalysis and Agricultural Biotechnology* 29:101745.

Jiang, W. Mashayekhi, H. Xing, B. 2009. Bacterial toxicity comparison between nano–and micro-scaled oxide particles. *Environ Pollut* 157(5):1619–1625.

Jiang, C. Zhao, H. Xiao, H. Wang, Y. Liu, L. Chen, H. Shen, C. Zhu, H. and Liu, Q. 2020. Recent advances in graphene-family nanomaterials for effective drug delivery and phototherapy. *Expert Opinion on Drug Delivery*.

Jones, M.E. Draghi, D.C. Thornsberry. C. Karlowsky, J.A. Sahm, D.F. Wenzel, R.P. 2004. Emerging resistance among bacterial pathogens in the intensive care unit e a European and North American surveillance study (2000e2002). *Ann Clin Microbiol Antimicrob* 3:14.

Jones, N. Ray, B. Ranjit, K.T. Manna, A.C. 2008. Antibacterial activity of ZnO nanoparticle suspensions on a broad spectrum of microorganisms. *FEMS Microbiol Lett* 279(1):71–6.

Jordan, S. Hutchings, M.I. Mascher, T. 2008. Cell envelope stress response in gram-positive bacteria. *FEMS Microbiol Rev* 32:107–146.

Kalia, V.P. Patel, S.K.S. Chan Kang, Y Lee, J.K. 2019. Quorum sensing inhibitors as antipathogens: Biotechnological applications. *Biotechnology Advances* 37(1):(68–90).

Karami, E. Behdani, M. Kazemi-Lomedasht, F. 2019. Albumin nanoparticles as nanocarriers for drug delivery: Focusing on antibody and nanobody delivery and albumin-based drugs. *J Drug Deliv Sci Technol* 55:101471.

Karimaei, S. Aghamir, S.M. Aghamir, S.M.K. Foroushani, A.R. Pourmand, M.R. 2021. Antibiotic tolerance in biofilm persister cells of *Staphylococcus aureus* and expression of toxin-antitoxin system genes. *Microbial Pathogenesis* 159: 105126.

Katayama, B. Ozawa, T. Morimoto, K. Awazu, K. Ito, N. Honda, N. Oiso, N. Tsuruta, D. 2018. Enhanced sterilization and healing of cutaneous pseudomonas infection using 5-aminolevulinic acid as a photosensitizer with 410-nm LED light. *J Dermatol Sci* 90(3):323–331.

Kaur, T. Putatunda, C. Vyas, A. and Kumar, G. 2020. Zinc oxide nanoparticles inhibit bacterial biofilm formation via altering cell membrane permeability, *Preparative Biochemistry and Biotechnology* 51(4):309–319.

Kasperkiewicz K, Major R, Sypien A, Kot M, Dyner M, Major Ł, Byrski A, Kopernik M, Lackner JM. 2021. Antibacterial optimization of highly deformed titanium alloys for spinal Implants. *Molecules* 26(11):3145.

Kathirvel, K. Rudhra, O. Rajapandian, S.G.K. Prajna, N.V. Lalitha, P. Devarajan B. 2021. Characterization of antibiotic resistance and virulence genes of ocular methicillin-resistant *Staphylococcus aureus* strains through complete genome analysis. *Experimental Eye Research* 212: 108764.

Kędziora, A. Speruda, M. Krzyżewska, E. Rybka, J. Łukowiak, A, Bugla-Płoskońska, G. 2018. Similarities and differences between silver ions and silver in nanoforms as antibacterial agents. *Int J Mol Sci* 19(2):444.

Kharkwal, G.B. Sharma, S.K. Huang, Y.Y. Dai, T. Hamblin, M.R. 2011. Photodynamic therapy for infections: Clinical applications. *Lasers Surg Med* 43(7):755–767.

Kharaghani, D. Dutta, D. Gitigard, P. Tamada, Y. Katagiri, A. Phan, D.M. Willcox, M.D.P. Kim, I.S. 2019. Development of antibacterial contact lenses containing metallic nanoparticles. *Polymer Testing* 79:106034.

Kim, Y.J., Kim, S.I. Kim, Y.R. Hong, K.W. Wie, S.H. Park, Y.J. Jeong, H. Kang M.W. 2012. Carbapenem-resistant *acinetobacter baumannii*: Diversity of resistant mechanisms and risk factors for infection. *Epidemiol Infect* 40(1):137–145.

Kim, J.S. Kuk, E. Yu K.N. Kim, J.H. Park, S.J. Lee, H.J. Kim, S.H. Park, Y.K. Park, Y.H. Hwang, C.Y. Kim, Y.K. Lee, Y.S. Jeong, D.H. Cho, M.H. 2007. Antimicrobial effects of silver nanoparticles. *Nanomedicine* 3(1):95–101.

Kłodzinska, S.N. Wan, F. Jumaa, H. Sternberg, C. Rades, T. Nielsen, H.M. 2019. Utilizing nanoparticles for improving anti-biofilm effects of azithromycin: A head-to-head comparison of modified hyaluronic acid nanogels and coated poly (lactic-co-glycolic acid) nanoparticles. *J Colloid Interface Sci* 1(555):595–606.

Kong, W. Tian, Q. Yang, Q. Liu, Y. Wang, G. Cao, Y. Wang, L. Xia, S. Sun, Y. Zhao, C. Wang, S. 2021. Sodium selenite enhances antibiotics sensitivity of *Pseudomonas aeruginosa* and deceases its pathogenicity by inducing oxidative stress and inhibiting quorum sensing system. *Antioxidants* 10(12):1873.

Koteva, K. Cox, G. Kelso, J.K. Surette, M.D. Zubyk, H.L. Ejim, L. Stogios, P. Savchenko, A. Sørensen, D. and Wright, G.D. 2018. Rox, a Rifamycin resistance enzyme with an unprecedented mechanism of action. *Cell Chemical Biology* 25(4):403–412.e405.

Kussell, E. Kishony, R. Balaban, N.Q. and Leibler S. 2005. Bacterial persistence: A model of survival in changing environments. *Genetics* 169(4):1807–1814.

Labunskyy, V.M. Hatfield, D.L. and Gladyshev, V.N. 2014. Selenoproteins: Molecular pathways and physiological roles. *Physiol* Rev 94(3):739–777.

Lara, H.H. Guisbiers, G. Mendoza, J. Mimun, L.C. Vincent, B. Lopez-Ribot, J.L. Nash, K.L. 2018. Synergistic antifungal effect of chitosan-stabilized selenium nanoparticles synthesized by pulsed laser ablation in liquids against *Candida Albicans* biofilms. *Int J Nanomedicine* 13:2697–2708.

Lahir, Y.K. 2020. Interactions at interface between nanomaterial's and biofilm: A general survey. *Adv Clin Toxicol* 5.

Lavaee, F. Faez, K. Faez, K. Hadi, N. Modaresi, F. 2016. Antimicrobial activity of silver, titanium dioxide and iron nanoparticles. *Americal Journal of Dentistry* 29(6):316–320. PMID: 29178718.

Lea, M.C. 1889. On allotropic forms of silver. *Am J Sci* 37:476.

Lee, S.W. Foley, E.J. Epstein, J.A. 1944. Mode of Action of penicillin: I. Bacterial growth and penicillin activity-*Staphylococcus aureus* FDA. *J Bacteriol* 48:393–399.

Lemire, J.A. Harrison, J.J. Turner, R.J. 2013. Antimicrobial activity of metals: Mechanisms, molecular targets and applications. *Nat Rev Microbiol* 11(6):371–84.

Levin-Reisman, I. Ronin, I. Gefen, O. Braniss, I. Shoresh, N. Balaban N.Q. 2017. Antibiotic tolerance facilitates the evolution of resistance. *Science* 24,355(6327):826–830.

Lewis, K. 2008. Multidrug tolerance of biofilms and persister cells. *Current Top Microbiol. Immunol* 322:107–131.

Li, X.Z. Nikaido, H. Williams, K.E. 1997. Silver-resistant mutants of *Escherichia coli* display active efflux of Ag^+ and are deficient in porins. *J Bacteriol* 179(19):6127–6132.

Li, Z. Greden, K. Alvarez, P.J. Gregory, K.B. Lowry, G.V. 2010. Adsorbed polymer and NOM limits adhesion and toxicity of nano scale zerovalent iron to *E. coli*. *Environ Sci Technol* 44(9):3462–3467.

Li, Y, Dong, H. Li, Y. Shi, D. 2015. Graphene-based Nano vehicles for photodynamic medical therapy. *Int J Nanomedicine* 10:2451–2459.

Liu, L. Xu, K. Wang, H. Tan, P.K.J. Fan, W. Venkatraman, S.S. Li L. and Yang Y.Y. 2009. Self–assembled cationic peptide nanoparticles as an efficient antimicrobial agent. *Nat. Nanotechnology* 4:457–463.

Liu, J. Sonshine, D.A. Shervani, S. Hurt, R.H. 2010. Controlled release of biologically active silver from nanosilver surfaces. *ACS Nano* 4(11):6903–6913.

Lin, A. Liu, Y. Zhu, X. Chen, X. Liu, J. Zhou, Y. Qin, X. Liu J. 2019. Bacteria–responsive biomimetic selenium nanosystem for multidrug-resistant bacterial infection detection and inhibition. *ACS Nano* 13(12):13965–13984.

Lin, F. Bao, Y.W. Wu, F.G. 2019. Carbon dots for sensing and killing microorganisms. *C Journal of Carbon Research* 5(2), 33.

Lin, M. Dai, Y. Xia, F. Zhang, X. 2020. Advances in non-covalent crosslinked polymer micelles for biomedical applications. *Mater Sci Eng C* 119:111626.

Lin, W. Zhang, J. Xu, J.F. and Pi, J. 2021. The advancing of selenium nanoparticles against infectious diseases. *Front Pharmacol* 12:682284.

Longmire, M. Choyke, P.L. and Kobayashi, H. 2008. Clearance properties of nano-sized particles and molecules as imaging agents: Considerations and caveats. *Nanomedicine* 3(5):703–717.

Ma, X. Williams, R.O. 2018. Polymeric nanomedicines for poorly soluble drugs in oral delivery systems: An update. *Journal of Pharmaceutical Investigation* 48(1):61–75.

Macia, M.D. Rojo-Molinero, E. and Oliver, A. 2014. Antimicrobial susceptibility testing in biofilm-growing bacteria. *Clin Microbiol Infect* 20:981–990.

Magesh, H. Kumar, A. Alam, A. Sekar, U. 2013. Identification of natural compounds which inhibit biofilm formation in clinical isolates of *Klebsiella pneumoniae*. *Indian J Exp Biol* 51(9):764–772.

Makabenta, J.M.V. Ahmed Nabawy, A. Li, C.H. Schmidt–Malan, S. Patel, R. and Rotello, W.M. 2021. Nanomaterial-based therapeutics for antibiotic-resistant bacterial infections. *Nature Reviews Microbiology* (1):23–36.

Mahmoudi, M. Serpooshan V. 2012. Silver-coated engineered magnetic nanoparticles are promising for the success in the fight against antibacterial resistance threat. *ACS Nano* 6(3):2656–2664.

Mann, R. Holmes, A. McNeilly, O. Cavaliere, R. Sotiriou, G.A. Rice, S.A. and Gunawan, C. 2021. Evolution of biofilm-forming pathogenic bacteria in the presence of nanoparticles and antibiotic: Adaptation phenomena and cross-resistance. *Nanobiotechnology* 9(1):291.

Marambio-Jones, C. Hoek, E.M.V. 2010. A review of the antibacterial effects of silver nanomaterials and potential implications for human health and the environment. *J Nanoparticle Research* 12:1531–1551.

Maruthupandy, M. Rajivgandhi, G.N. Quero, F. Li, W.J. 2020. Anti-quorum sensing and anti-biofilm activity of nickel oxide nanoparticles against *Pseudomonas aeruginosa*. *J Environ Chem Eng* 8:104533.

McBain, A.J. Gilbert P. and Allison D.G. 2003. Biofilms and biocides: Are there implications for antibiotic resistance? school of pharmacy and pharmaceutical sciences, university of manchester, oxford road, manchester M13 3PL, *UK Reviews in Environmental Science and Bio/Technology* 2: 141–146, 2003. 2004 Kluwer Academic Publishers. Printed in the Netherlands.

Mc Dermott, W. 1958. Microbial persistence. *Yale J. Biol. Med* 30:257–291.

McCarthy, T.J. Zeelie, J.J. Krause, D.J. 1992. The antimicrobial action of zinc ion/antioxidant combinations. *J Clin Pharm Ther* 17(1):51–54.

McDonnell, G. and Russell, D. 1999. Antiseptics and disinfectants: Activity, *Action, and Resistance Clinical Microbiology Reviews* 12 (1):147–179.

Mijnendonckx, K. Leys, N. Mahillon, J. Silver, S. Van Houdt, R. 2013. Antimicrobial silver: Uses, toxicity and potential for resistance. *Biometals* 26(4):609–21.

Mirzaie, A. and Ranjbar, R. 2021.Antibiotic resistance, virulence-associated genes analysis and molecular typing of *Klebsiella pseumoniae* strains recovered from clinical samples. *AMB Express* 11:122.

Mohamed, J.A. Huang, D.B. 2007. Biofilm formation by *enterococci*. *J Med Microbiol* 56(12):1581e8.

Morones, J.R. Elechiguerra, J.L. Camacho, A. Katherine, H. Kouri, J.B. Ramírez, J.T. Yacaman, M.J. 2005. The bactericidal effect of silver nanoparticles. *Nanotechnology* 16(10):2346–2353.

Mahmood, H.Y. Jamshidi, S. Sutton, J.M. Rahman, K.M. 2016. Current advances in developing inhibitors of bacterial multidrug efflux pumps. *Curr Med Chem* 23:1062–1081.

Medina-Ramírez, I.E. Díaz de León-Macias, C.E. Pedroza-Herrera, G. Gonzáles-Segovia, R. Zapien, J.A. Rodríguez-López, J.L. 2020. Evaluation of the biocompatibility and growth inhibition of bacterial biofilms by ZnO, Fe_3O_4 and ZnO@Fe_3O_4 photocatalytic magnetic materials. *Ceramics International* 46:8979–8994.

Mehta, M. Allen-Gipson, D. Mohapatra, S. Kindy, M. Limayem, A. 2019. Study on the therapeutic index and synergistic effect of chitosan-zinc oxide nanomicellar composites for drug-resistant bacterial biofilm inhibition, *International Journal of Pharmaceutics* 30(565):472–480.

Meric, M. Kasap, M. Gacar, G. Budak, F. Dundar, D. Kolayli F, Eroglu C. and Vahaboglu, H. 2008. Emergence and spread of carbapenem-resistant *Acinetobacter baumannii* in a tertiary care hospital in Turkey. *FEMS Microbiol Lett* 282:214–218.

Miller, M.B. and Bassler, B.L. 2001. Quorum sensing in bacteria. *Annual Review of Microbiology* 55:165–199.

Mohammed, M.A. Syeda, J.T.M. Wasan, K.M. Wasan, E.K. 2017. An overview of chitosan nanoparticles and its application in non-parenteral drug delivery. *Pharmaceutics* 9(4):53.

Monteiro, D.R. Gorup, L.F. Silva, S. Negri, M. de Camargo, E.R. Oliveira, R. Barbosa D.B. and Henriques, M. 2011. Silver colloidal nanoparticles: Antifungal effect against adhered cells and biofilms of *Candida albicans* and *Candida glabrata*. *Biofouling* 27(7):711–719.

Mulani, M.S. Kamble, E.E. Kumkar, S.N. Tawre, M.S. Pardesi, K.R. Tawre, M.S. and Pardesi, K.R. 2019. Emerging strategies to combat ESKAPE pathogens in the era of antimicrobial resistance: A review. *Front Microbiol* 10(539):1–24.

Munita, J.M. Arias, C.A. 2016. Mechaisms of antibiotic resistance. *Microbiol Spectr* 4(2).

Muzammil, S. Khurshid, M. Nawaz, I. Siddique, M.H. Zubair, M. Nisar, M.A. Imran, M. and Hayat, S. 2020. Aluminium oxide nanoparticles inhibit EPS production, adhesion and biofilm formation by multidrug resistant *Acinetobacter baumannii*. *Biofouling* 36(4):492–504.

Nair, S. Sasidharan, A. Divya Rani, V.V. Menon, D. Nair, S. Manzoor, K. Raina, S. 2009. Role of size scale of ZnO nanoparticles and microparticles on toxicity toward bacteria and osteoblast cancer cells. *J Mater Sci: Mater Med* 20:235.

Nag, M. Lahiri, D. Mukherjee, D. Banerjee, R. Garai, S. Sarkar, T. Ghosh, S. Dey, A. Ghosh, S. Patnaik, S. Edinur, H.A. Kari, Z.A. Pati, S.A. and Ray, R.R. 2021. Functionalized chitosan nanomaterials: A jammer for quorum sensing. *Polymers* 13:2533.

Nallapareddy, S.R. Singh, K.V. Sillanpaa, J. Garsin, D.A. Hook, M. Erlandsen, S.L. and Murray, B.E. 2006. Endocarditis and biofilm-associated pili of *Enterococcus faecalis*. *J Clin Invest* 116(10):2799–2807.

Nascimento, T.G.D. Da Silva, P.F. Azevedo, L.F. Da Rocha, L.G. Porto, I.C.C.D.M. Moura, T. Basílio-Júnior, I.D. Grillo, L. Dornelas, C.B. Fonseca, E.J.D.S. Oliveira, E.D.J. Zhang, A.T. Watson, D.G. 2016. Polymeric nanoparticles of brazilian red propolis extract: Preparation, characterization, antioxidant and leishmanicidal activity. *Nanoscale Res. Letters* 11(1):301.

Nataraj, B.H. Mallappa, R.H. 2021. Antibiotic resistance crisis: An update on antagonistic interactions between probiotics and methicillin-resistant *Staphylococcus aureus* (MRSA). *Current Microbiology* 78(6):2194–2211.

Nett, J.E, Cain, M.T, Crawford, K. and Andes, D.R. 2011. Optimizing a candida biofilm microtiter plate model for measurement of antifungal susceptibility by tetrazolium salt assay. *J Clin Microbiol* 49:1426–1433.

Nickel, J.C. Ruseska, I. Wright, J.B. Costerton, J.W. 1985. Tobramycin resistance of *Pseudomonas aeruginosa* cells growing as a biofilm on urinary catheter material. *Antimicrob. Agents Chemother* 27(4): 619–624.

Nie, L. Chang, P. Ji, C. Zhang, F. Zhou, Q. Sun, M. Sun, Y. Politis, C. Shavandi, A. 2020. Poly(acrylic acid) capped iron oxide nanoparticles via ligand exchange with antibacterial properties for biofilm applications. *Colloids Surf B Biointerfaces* 197:111385.

Nino-Martinez, N. Orozco, M.F.S. Martinez-Castanon, G.A. Mendez, F.T. Ruiz, F. 2019. Molecular mechanisms of bacterial resistance to metal and metal oxide nanoparticles. *Int J Mol Sci* 20:2808.

O'Neill, J. 2016. *The Review on Antimicrobial Resistance*. Tackling drug-resistant infections globally. UK Government report. P84.

Oscar, F.L. Nithya, C. Vismaya, S. Arunkumar, M. Pugazhendhi, A. Nguyen-Tri, P. Alharbi, S.A. Alharbi, N.S. Thajuddin, N. 2021. In vitro analysis of green fabricated silver nanoparticles (AgNPs) against *Pseudomonas aeruginosa* PA14 biofilm formation, their application on urinary catheter, *Progress in Organic Coatings* 151:106058.

Padmavathy, N. Vijayaraghavan, R. 2008. Enhanced bioactivity of ZnO nanoparticles–an antimicrobial study. *Sci Technol Adv Mat*. 9(3):035004.

Pal, S. Tak, Y.K. Song, J.M. 2007. Does the antibacterial activity of silver nanoparticles depend on the shape of the nanoparticle? A study of the Gram-negative bacterium *Escherichia coli*. *Appl Environ Microbiol* 73(6):1712–1720.

Pan, Y. Breidt, F. and Kathariou, S. 2006. Resistance of *listeria monocytogenes* biofilms to sanitizing agents in a simulated food processing environment. *Appl. Environ. Microbiol* 72(12): 7711–7717.

Pang, Z. Raudonis, R. Glick, B.R. Lina, T.J. Cheng, Z. 2019. Antibiotic resistance in *Pseudomonas aeruginosa*: Mechanisms and alternative therapeutic strategies. *Biotechnology Advances* 37:177–192.

Pantosti, A. Sanchini, A. and Monaco, M. 2007. Mechanisms of antibiotic resistance in *Staphylococcus aureus*. *Future Microbiology* 2(3):323–34.

Panacek, A. Kvitek, L. Smekalova, M. Vecerova, R. Kolar, M. Roderova, M. Dycka, F. Sebela, M. Prucek, R. Tomanec, O. Zboril, R. 2018. Bacterial resistance to silver nanoparticles and how to overcome it. *Nature Nanotechnology* 13:65–71.

Pandey, V.K. Srivastava, K.R. Ajmal, G. Thakur, V.K. Gupta, V.K. Upadhyay, S.N. Mishra, P.K. 2019. Differential susceptibility of catheter biomaterials to biofilm-associated infections and their remedy by drug-encapsulated Eudragit RL100 nanoparticles. *International Journal of Molecular Sciences* 20(20):5110.

Papenfort, K. Bassler, B. 2016. Quorum sensing signal–response systems in Gram-negative bacteria. *Nat Rev Microbiol* 14:576–588.

Parastan, R. Kargar, M. Solhjoo, K. Kafilzadeh, F. 2020. *Staphylococcus aureus* biofilms: Structures, antibiotic resistance, inhibition, and vaccines. *Gene Reports* 20:100739.

Pareek, V. Devineau, S. Sivasankaran, S.K. Bhargava, A. Panwar, J. Srikumar, S. and Fanning, S. 2021 Silver nanoparticles induce a Triclosan-Like antibacterial action mechanism in multi-drug resistant *Klebsiella pneumoniae*. *Front Microbiol* 12:638640.

Paukscha, L. Rohnke, M. Schnettler, R. Lipsa, K.S. 2014. Silver nanoparticles do not alter human osteoclast genesis but induce cellular uptake. *Toxicology Reports* 18:1:900–908.

Pei, X. Lu, X. Liu, J. Liu, D. Yang, Y. Ostrikov, K. et al. 2012. Inactivation of a 25.5mm *Enterococcus faecalis* biofilm by a room-temperature, battery-operated, handheld air plasma jet. *J Phys D Appl Phys* 45:165205.

Peulen, T.O. and Wilkinson, K.J. 2011. Diffusion of nanoparticles in a biofilm. *Environ. Sci. Technol* 45:3367–3373.

Pendleton, N.J. Gorman, S.P. Gilmore, B.F. 2013. Clinical relevance of the ESKAPE pathogens. *Expert Rev Anti Infect Ther* 11(3):297–308.

Pérez-Laguna, V. García-Malinis, A.J. Aspiroz, C. Rezusta, A. Gilaberte, Y. 2018. Antimicrobial effects of photodynamic therapy: An overview. *G Ital Dermatol Venereol* 153(6):833–846.

Perez-Laguna, V. García-Luque, I. Ballesta, S. Rezusta, A. Gilaberte, Y. 2021. Photodynamic therapy combined with antibiotics or antifungals against microorganisms that cause skin and soft tissue infections: A planktonic and biofilm approach to overcome resistances. *Pharmaceuticals* 14:603.

Potera, C. 1999. Forging a link between biofilms and disease. *Science* 283:1837–1839.

Pradeepa, K. Bhat, U. Vidya, S.M. 2017. Nisin gold nanoparticles assemble as potent antimicrobial agent against *Enterococcus faecalis* and *Staphylococcus aureus* clinical isolates *Journal of Drug Delivery Science and Technology* 37:2 0e27

Prajitha, N. Athira, S. Mohanan, P. 2019. Bio-interactions and risks of engineered nanoparticles. *Environmental Research* 172:98–108.

Pucelik B and Dąbrowski, J.M. 2022 Photodynamic inactivation (PDI) as a promising alternative to current pharmaceuticals for the treatment of resistant microorganisms. *Advances in Inorganic Chemistry* 12:003. ISSN 0898-8838

Pugazhendhi, A. Mathimani, T. Alharbi, S.A. Chinnathambi, A. Karuppusamy, I. Brindhadevi, K. Kheawfu, K. Pikulkaew, S. 2022. In vitro efficacy of green synthesized ZnO nanoparticles against biofilm and virulence of *Serratia marcescens*. *Progress in Organic Coatings* 166:106781.

Qi, L. Xu, Z. Jiang, X. Hu, C. Zou, X. 2004. Preparation and antibacterial activity of chitosan nanoparticles. *Carbohydr Res* 339(16):2693–2700.

Qin, X. Singh, K.V. Weinstock, G.M. and Murray, B.E 2001. Characterization of fsr, a regulator controlling expression of gelatinase and serine protease in *Enterococcus faecalis* OG1RF. *J Bacteriol* 183(11):3372–3382.

Quan, K. Zhang, Z. Chen, H. Ren, X. Ren, Y. Peterson, B.W. van der Mei, H.C. Busscher, H.J. 2019. Artificial channels in an infectious biofilm created by magnetic nanoparticles enhanced bacterial killing by antibiotics. *Small* 15(39):1902313.

Quan, K. Zhang, Z. Ren, Y. Busscher, H. van der Mei, H.C. and Peterson, B.W. 2021. Possibilities and impossibilities of magnetic nanoparticle use in the control of infectious biofilms. *Journal of Materials Science and Technology* 69(10):69–78.

Quan, K. Zhang, Z. Ren, Y. Busscher, H.J. van der Mei, H.C. Peterson, B.W. 2021. On-demand pulling-off of magnetic nanoparticles from biomaterial surfaces through implant-associated infectious biofilms for enhanced antibiotic efficacy. *Materials Science and Engineering: C* 131:(112526).

Rai, M.K Deshmukh, S.D. Ingle, AP, Gade, A.K. 2012. Silver nanoparticles: The powerful nano-weapon against multidrug resistant bacteria. *J Appl Microbiol* 112(5):841–852.

Raghunath, A.; Perumal, E. 2017. Metal oxide nanoparticles as antimicrobial agents: A promise for the future. *Int J Antimicrob Agents* 49(2):137–152.

Rajaduraipandi, K. Mani, K.R. Panneerselvam, K. Mani, M. Bhaskar, M. Manikandan, P. 2006. Prevalence and antimicrobial susceptibility pattern of methicillin resistant *Staphylococcus aureus*: A multicentre study. *J IAM* 24:34–8.

Rajkumari, J. Busi, S. Vasu, A.C. Reddy P. 2017. Facile green synthesis of baicalein fabricated gold nanoparticles and their antibiofilm activity against *Pseudomonas aeruginosa* PAO1 *Microbial Pathogenesis* 107:261e269

Ramage, G. Vande Walle, K. Wickes, B.L. and Lopez-Ribot, J.L. 2001. Standardized method for in vitro antifungal susceptibility testing of *Candida albicans* biofilms. *Antimicrob Agents Chemotherapy* 45(9): 2475–2479.

Ramalingam, B. (2012). *The Role of Cell-to-Cell Interactions and Quorum Sensing in Formation of Biofilms by Drinking Water Bacteria*. PhD Thesis, University of Sheffield, UK

Ramalingam, V. Sundaramahalingam, S. and Rajaram, R.2019. Size-dependent antimycobacterial activity of titanium oxide nanoparticles against *Mycobacterium tuberculosis*. *Mater Chem B* 7:4338–4346.

Manoharan, R.K. .Gangadaran, P. Ayyaru, S. Ahn, B.C and Ahn, Y.H. 2021. Self-healing functionalization of sulfonated hafnium oxide and copper oxide nanocomposite for effective biocidal control of multidrug-resistant bacteria† *New J. Chem RSC of chemistry* 45:9506

Rasmussen TB, Givskov M. 2006. Quorum-sensing inhibitors as anti-pathogenic drugs. *International Journal of Medical Microbiology* 296(2–3):149–161.

Rasool, U. Sah, S.K. Hemalatha, S. 2019. Effect of biosynthesized copper nanoparticles (CUNPS) on the growth and biofilm formation of fluconazole-resistant *Candida albicans*. *J Microbiol Biotechnol Food Sci* 9:21–24.

Regoes, R.R. Wiuff, C. Zappala R.M. Garner, K.N. Baquero, F. Levin B.R. 2004. Pharmacodynamic functions: A multiparameter approach to the design of antibiotic treatment regimens. *Antimicrob Agents Chemotherapy* 48:3670–6.

Reddy, K.M. Feris, K. Bell, J. Wingett, D.G. Hanley, C. Punnoose, A. 2007. Selective toxicity of zinc oxide nanoparticles to prokaryotic and eukaryotic systems. *Appl Phys Lett* 90(213902):2139021–2139023.

Rice, L.B. 2012. Mechanisms of resistance and clinical relevance of resistance to β-lactams, glycopeptides, and fluoroquinolones. *Clinic Proceedings* 87(2):198–208.

Richards, M.J. Edwards, J.R. Culver, D.H. and Gaynes, R.P, 2000. Nosocomial infections in combined medical-surgical intensive care units in the United States. *Infect Control Hosp Epidemiol* 21(8):510–515.

Rodrıguez-Martınez, J.M. Cano, M.E. Velasco, C. Martınez-Martınez, L. Pascual, A.J. 2011. Plasmid-mediated quinolone resistance: an update. *Journal of Infection and Chemotherapy* 17(2):149–182.

Romling, U. Kjelleberg, S. Normar, S. Nyman, L. Uhlin, B.E. Åkerlund, B. 2014. Microbial biofilm formation: A need to act. *J Intern Med* 276:98–110.

Rosenblatt-Farrell, N. 2009. The landscape of antibiotic resistance. *Environ. Health Persp* 117(6): A245–A250

Rosenberg, M. Visnapuu, M. Vija, H. Kisand, V. Kasemets, K. Kahru, A. and Ivask, A. 2020. Selective antibiofilm properties and biocompatibility of nano-ZnO and nano-ZnO/Ag coated surfaces. *Scientific Reports* 10:13478.

Russell, A.D. 1997. Plasmids and bacterial resistance to biocides. *J Appl Microbiol* 83(2):155–65.

Ruddaraju, L.K. Pammi, S.V.N. Guntuku, G.S. Padavala, V.S. Kolapalli, V.R.M. 2020. A review on antibacterials to combat resistance: From ancient era of plants and metals to present and future perspectives of green nano technological combinations. *Asian Journal of Pharmaceutical Sciences* 15:42–59

Sadikot, R.T. Blackwell, T.S. Christman, J.W. Prince, A.S. 2005. Pathogen-host interactions in *Pseudomonas aeruginosa* pneumonia. *Am J Respir Crit Care Med* 171:1209–1223.

Salleh, A. Naomi, R. Utami, N.D. Mohammad, A.W. Mahmoudi, E. Mustafa, N. Fauzi, M.B. 2020. The potential of silver nanoparticles for antiviral and antibacterial applications: A mechanism of action. *Nanomaterials* 10(8):1566.

SaraHernando-Amado, Blanco, P. Alcalde-Rico, M. Corona, F. Reales-Calder6n, J.A. Sánchez, M.B. Martínez, J.L. 2016. Multidrug efflux pumps as main players in intrinsic and acquired resistance to antimicrobials. *Drug Resistance* 28:13–27.

Scaffaro, R. Lopresti, F. Marino, A. Nostro, A. 2018. Antimicrobial additives for poly (lactic acid) materials and their applications: Current state and perspectives. *Appl Microbiol Biotechnol* 102(18): 7739–7756.

Schaaf HS, Victor TC, Engelke E, Brittle, W. Marais, B.J. Hesseling, A.C. van Helden, P.D. Donald P.R. 2007. Minimal inhibitory concentration of isoniazid in isoniazid-resistant *Mycobacterium tuberculosis* isolates from children. *Eur J Clin Microbiol Infect Dis* 26:203–5.

Scutera S, Argenziano M, Sparti R, Bessone F, Bianco G, Bastiancich C, Castagnoli C, Stella M, Musso T, Cavalli R. 2021. Enhanced antimicrobial and antibiofilm effect of new colistin-loaded human albumin nanoparticles. *Antibiotics* 10(1):57.

Sharma, D. Misba, L. and Khan, A.U. 2019. Antibiotics versus biofilm: An emerging battleground in microbial communities. *Antimicrobial Resistance and Infection Control* 8:76.

Shaaban, M.I. Shaker, M.A. and Mady, F.M. 2017. Imipenem/cilastatin encapsulated polymeric nanoparticles for destroying carbapenem-resistant bacterial isolates. *J Nanobiotechnol* 15:29.

Shahbandeh M. Moghaddam, M.M. Golmohammadi, R. and Mirnejad, R. 2022. The antimicrobial effect of quorum sensing autoinducers of Pseudomonas aeruginosa, C12-HSL and C4-HSL, against MDR *Staphylococcus aureus* isolates. *Comparative Immunology, Microbiology and Infectious Diseases* 81:101747.

Siberry, G.K. Tekle, T. Carroll, K. Dick, J. 2003. Failure of clindamycin treatment of methicillin-resistant *Staphylococcus aureus* expressing inducible clindamycin resistance in vitro. *Clin Infect Dis* 37(9):1257–60.

Siel, T. Webster, T.J. 2012. Antimicrobial applications of nanotechnology: Methods and literature. *International Journal of Nanomedicine* 7:2767–2781

Sandoe, J.A.T. Witherden, I.R. Cove, J.H. Heritage, J. Wilcox, M.H. 2003. Correlation between enterococcal biofilm formation in vitro and medical-device-related infection potential in vivo. *J Med Microbiol* 52(7):547e50.

Siddique MH, Aslam B, Imran M, Ashraf A, Nadeem H, Hayat S, Khurshid, M. Afzal, M. Malik, I.R. Shahzad, M Qureshi, U. Khan, Z.H.U. and Muzammil, S. 2020. Effect of silver nanoparticles on biofilm formation and EPS production of multidrug-resistant *Klebsiella pneumoniae*. *Biomed Res International*, 9. Article ID 6398165

Siemer, S. Westmeier, D. Barz, M. Eckrich, J. Wünsch, D. Seckert, C. Thyssen, C. Schilling, O. Hasenberg, M. Pang, C 2019. Biomolecule-corona formation confers resistance of bacteria to nanoparticle–induced killing: Implications for the design of improved nano antibiotics. *Biomaterials* 92:551–559.

Silva, S. Pires, P. Monteiro, D.R. Negri, M. Gorup, L.F. Camargo, E.R. Barbosa, D.B. Oliveira, R. Williams, D.W. Henriques, M. Azeredo J. 2013. The effect of silver nanoparticles and nystatin on mixed biofilms of *Candida glabrata* and *Candida albicans* on acrylic. *Med Mycol* 51(2):178–84.

Simon-Deckers, A. Loo, S. Mayne-Lhermite, M. Herlin-Boime, N. Menguy, N. Reynaud, C. Gouget, B. Carrière, M. 2009. Size-, composition–and shape-dependent toxicological impact of metal oxide nanoparticles and carbon nanotubes toward bacteria. *Environ Sci Technol* 43(21):8423–8429.

Slonczewski, J.L. and Foster, J.W. 2014. *Microbiology: An Evolving Science*, 3rd Edition. W.W.Norton and Company Inc, London. p.1184.

Song, Z. Wu, Y. Wang, H. Han, H. 2019. Synergistic antibacterial effects of curcumin modified silver nanoparticles through ROS mediated pathways. *Mater. Sci. Eng. C* 99:255–263.

Song, J. Liu, H. Lei, M. Tan, H. Chen, Z. Antoshin, A. Payne, G.F. Qu X. and Liu, C. 2020. Redox-Channelling Polydopamine-Ferrocene (PDA-Fc) coating to confer context-dependent and photothermal antimicrobial activities. *ACS Appl Mater Interfaces* 12(7):7855–7868.

Son, G.H. Lee, B.J. Cho, C.W. 2017. Mechanisms of drug release from advanced drug formulations such as polymeric-based drug-delivery systems and lipid nanoparticles. *Journal of Pharmaceutical Investigations* 47:287–296.

Sondi, I. Salopek-Sondi, B. 2004. Silver nanoparticles as antimicrobial agent: A case study on *E. coli* as a model for Gram-negative bacteria. *J Colloid Interf Sci* 275(1):177–182.

Stewart, P.S. 2015. Antimicrobial tolerance in biofilms. *Microbiol Spectr* 3:1–30.

Su, Y. Zhao, L. Meng, F. Qiao, Z. Yao, Y. Luo, J. 2018. Triclosan loaded polyurethane micelles with pH and lipase sensitive properties for antibacterial applications and treatment of biofilms. *Mater Sci Eng C* 93:921–930.

Sun, L. Wang, X. and Li 2017 2018 Z. Successful treatment of multidrug-resistant *Acinetobacter baumannii meningitis* with ampicillin sulbactam in primary hospital. *British Journal of Neurosurgery* 32:6.

Tan, S. Wu, X. Xing, Y. Lilak, S. Wu M. and Zhao, J.X. 2020. Enhanced synergetic antibacterial activity by a reduce graphene oxide/Ag nanocomposite through the photothermal effect. *Colloids Surf B* 185(1):110616.

Tan, Y. Ma, S. Leonhard, M. Moser, D. Haselmann, G.M. Wang, J. Eder, D. Schneider-Stickler, B. 2018. Enhancing antibiofilm activity with functional chitosan nanoparticles targeting biofilm cells and biofilm matrix. *Carbohydrate Polymers* 200:35–42.

Tang, M. Wei, X. Wan, X. Ding, Z. Ding, Y. Liu, J. 2020. The role and relationship with efflux pump of biofilm formation in *Klebsiella pneumonia*. *Microbial Pathogenesis* 147: 104244.

Taylor, E.N. Webster, T.J. 2009. The use of superparamagnetic nanoparticles for prosthetic biofilm prevention. *Int J Nanomedicine* 4:145–152.

Tendolkar, P.M. Baghdayan, A.S. and Shankar, N. 2005. The N-terminal domain of enterococcal surface protein, esp, is sufficient for esp-mediated biofilm enhancement in *Enterococcus faecalis J Bacteriol* 187(17):6213–6222.

Tenson, T. Lovmar, M. Ehrenberg, M.J. 2003. The mechanism of action of Macrolides, Lincosamides and Streptogramin B reveals the nascent peptide exit path in the ribosome. *Mol Biol* 330(5):1005–14.

Thomas, R. Soumya, K.R. Mathew, J. Radhakrishnan, E.K. 2015. Inhibitory effect of silver nanoparticle fabricated urinary catheter on colonisation efficiency of coagulase negative *Staphylococci*. *Journal of Photochemistry and Photobiology B: Biology* 149:68–77.

Thambirajoo, M. Maarof, M. Lokanathan, Y. Katas, H. Ghazalli, N.F. Tabata, Y. Fauzi, M.B. 2021. Potential of nanoparticles integrated with antibacterial properties in preventing biofilm and antibiotic resistance. *Antibiotics* 10(11):1338.

Tian, S. van der Mei, H.C. Ren, Y. Busscher, H.J. and Shi, L. 2021. Recent advances and future challenges in the use of nanoparticles for the dispersal of infectious biofilms. *Journal of Materials Science and Technology* 84:208–218.

Tong S.Y., Davis J.S., Eichenberger E. Holland, T.L. Fowler V.G. Jr. 2015. *Staphylococcus aureus* infections: epidemiology, pathophysiology, clinical manifestations, and management, *Clinical Microbiology Review* 28(3):603–61.

Toledo-Arana, A. Valle, J. Solano, C. Arrizubieta, M.J. Cucarella, C. Lamata. M. Amorena, B. Leiva, J. Penades, J.R. and Lasa, I. 2001. The enterococcal surface protein, esp, is involved in *Enterococcus faecalis* biofilm formation. *Appl Environ Microbiol* 67(10):4538–4545.

Tran, N. Mir, A. Mallik, D. Sinha, A. Nayar, A. Webster, T.J. 2010. Bactericidal effect of iron oxide nanoparticles on *Staphylococcus aureus*. *Int J Nanomedicine* 15(5):277–283.

Truong, L.B. Medina-Cruz, D. Mostafavi, E. Rabiee, N. 2021. Selenium nanomaterials to combat antimicrobial resistance. *Molecules* 26(12):3611.

Tuomanen, E, Tomasz, A. 1990. Mechanism of phenotypic tolerance of nongrowing *pneumococci* to beta-lactam antibiotics. *Scand J Infect Dis Suppl* 74:102–112.

Udekwu, K.I. Parrish, N. Ankomah, P, Baquero, F. and Levin, B.R. 2009. Functional relationship between bacterial cell density and the efficacy of antibiotics. *Journal of Antimicrobial Chemotherapy* 63:745–757

Vahedi, M. Hosseini-Jazani, N. Yousefi, S. Ghahremani, M. 2017. Evaluation of anti-bacterial effects of nickel nanoparticles on biofilm production by *Staphylococcus epidermidis*. *Iran J Microbiol* 9(3): 160–168.

Valentin E, Bottomley AL, Chilambi GS, Harry EJ, Amal R, Sotiriou GA, Rice, S.A. and Gunawan, C. 2020. Heritable nanosilver resistance in priority pathogen: A unique genetic adaptation and comparison with ionic silver and antibiotics. *Nanoscale* 12:2384–92.

Vanamala, K. Tatiparti, K. Bhise, K. Sau, S. Scheetz, M.H. Rybak, M.J. David Andes and Arun K. Iyer 2020 Novel approaches for the treatment of methicillin-resistant *Staphylococcus aureus*: Using nanoparticles to overcome multidrug resistance. *Drug Discovery Today* (1):31–43.

Vargas-Reus, M.A. Memarzadeh, K. Huang, J. Ren, G.G. Allaker, R.P. 2012. Antimicrobial activity of nanoparticulate metal oxides against peri-implantitis pathogens. *Int J Antimicrob Agents* 40(2):135–9.

Viale, P. Giannella, M. Lewis, R. Trecarichi, E.M. Petrosillo, N and Tumbarello, M. 2013. Predictors of mortality in multidrug-resistant *Klebsiella pneumoniae* bloodstream infections. *Expert Rev Anti Infect Ther* 11(10):1053–1063.

Vijayakumar, S. Saravanakumar, K. Malaikozhundan, B. Divya, M. Vaseeharan, B. Durán-Lara, E.F. Wang, M.H. 2019. Biopolymer K-carrageenan wrapped ZnO nanoparticles as drug delivery vehicles for anti MRSA therapy. *International Journal of Biological Macromolecules* 144:9–18.

Wang, Q. Kang, F. Gao, Y. Mao, X. Hu, X. 2016. Sequestration of nanoparticles by an EPS matrix reduces the particle-specific bactericidal activity. *Sci Rep* 6:21379.

Wang, X. Yang, F. Zhao, J. Xu, Y. Mao, D. Zhu, X. Luo, Y. Alvarez, P. 2018. Bacterial exposure to ZnO nanoparticles facilitates horizontal transfer of antibiotic resistance genes. *NanoImpact* 10:61–67.

Wang, L. Natan, M. Zheng, W. Zheng, W. Liu, S. Jacobi, G. Perelshtein, I. Gedanken, A. Banin, B and Jiang X. 2020. Small molecule-decorated gold nanoparticles for preparing antibiofilm fabrics *Nanoscale Adv* 2:2293–2302

Wan, B, Zhu, Y. Tao, J. Zhu, F. Chen, J. Li, L. Zhao, J. Wang, L. Sun, S. Yang, Y. Zhang, X. and Zhang, Y. 2020. Alginate Lyase guided silver nanocomposites for eradicating *Pseudomonas aeruginosa* from lungs. *ACS Applied Materials and Interfaces* 12(8):9050–9061.

Wedad, M.A, Khairy, R.M.M. Zaki, A.I. and Zaki, S.H. 2021. Effect of ZnO nanoparticles on methicillin, vancomycin, linezolid resistance and biofilm formation in *Staphylococcus aureus* isolates. *Ann Clin Microbiol Antimicrob* 20:54.

Wayne, A. Clinical and Laboratory Standards Institute, CLSI. 2011. *Performance standards for antimicrobial susceptibility testing. 20th Informational Supplement.* CLSI document M100-S21.

World Health Organization (WHO) 2021. *Antibiotic Resistance.* www.who.int/news-room/ fact-sheets/detail/ antimicrobial-resistance.

World Health Organization (WHO) 2017. *WHO Published List of Bacteria for Which New Antibiotics are Urgently Needed.* www.who.int/news/item/27-02-2017-who-publishes-list-of-bacteria-for-which-new-antibiotics-are-urgently-needed.

Wu, C. Zhu, A. Li, D. Wang, L. Yang, H. Zeng, H. Liu, Y. 2016. Photosensitizer-assembled PEGylated graphene-copper sulfide nanohybrids as a synergistic near-infrared phototherapeutic agent. *Expert Opin Drug Deliv* 13(1):155–165.

Xu, J.W. Yao, K. Xu, Z.K. 2019. Nanomaterials with a photothermal effect for antibacterial activities: An overview. *Nanoscale* 11(18):8680–8691.

Xu, M Li, L. and Hu, Q 2021. The recent progress in photothermal-triggered bacterial eradication *Biomater Science* 1995–2008

Yael, N. Ivanova, S.K. Hoyo, J. Perelshtein, I. Owen, G. Haegert, A Lin, Y.Y. LeBihan, S. Gedanken, A. Häfeli, U.O. Tzanov, T. and Bach H. 2021. Novel Lignin-Capped Silver Nanoparticles against Multidrug-Resistant Bacteria This. *ACS Appl Mater Interfaces* 13:22098–22109.

Yang, Y. Mathieu, J.M. Chattopadhyay, S. Miller, J.T. Wu, T, Shibata, T. Guo, W, Alvarez P.J. 2012. Defense mechanisms of *Pseudomonas aeruginosa* PAO1 against quantum dots and their released heavy metals. *ACS Nano* 6(7):6091–6098.

Yang, L. Liu, Y. Wang, N. Wang, H. Wang, K. Luo, X.L. Dai, R.X. Tao, R.J. Wang, H.J. Yang, J.W. 2020. Albumin-Based LL37 peptide nanoparticles as a sustained release system against *Pseudomonas aeruginosa* lung infection. *ACS Biomater Sci Eng* 7(5):1817–1826.

Ye, Z. Sang, T. Li, K. Fischer, N.G. Mutreja, I. Echeverría, C. Kumar, D. Tang, Z. Aparicio, C. 2022. Hybrid nanocoatings of self-assembled organic-inorganic amphiphiles for prevention of implant infections, *Acta Biomaterialia* 140:338–349.

Yin, W. Yu, J. Lv, F. Yan, L. Zheng, L.R. Gu Z. and Zhao, Y. 2016. Functionalized Nano-MoS_2 with peroxidase catalytic and near-infrared photothermal activities for safe and synergetic wound antibacterial applications. *ACS Nano* 10:11000–11011.

Yoon, K.Y. Byeon, J.H. Park, J.H. Hwang, J. 2007. Susceptibility constants of *Escherichia coli* and *Bacillus subtilis* to silver and copper nanoparticles. *Sci Total Environ* 373(2–3):572–575.

Yousefi, M. Dadashpour, M. Hejazi, M. Hasanzadehd, M. Behnam, B. Guardia, M.D.L. Shadjou, N. Mokhtarzadeh, A. 2017. Anti-bacterial activity of graphene oxide as a new weapon nanomaterial to combat multidrug-resistance bacteria. *Materials Science and Engineering C* 74 568–581.

Yu, M. and Zheng, J. 2015. Clearance pathways and tumor targeting of imaging nanoparticles. *ACS Nano* 9: 6655–6674.

Zhao, L. Chu, P.K. Zhang, Y. Wu, Z. 2009.Antibacterial coatings on titanium implants. *Journal of Biomedical Materials Research Part B Applied Biomaterials* 9(1):470–80.

Zhao, Y. Yu, C. Yu, Y. Wei, X. Duan, X. Dai, X. and Zhang, X. 2019. Bioinspired heteromultivalent Ligand-Decorated nanotherapeutic for enhanced photothermal and photodynamic therapy of antibiotic-resistant bacterial *Pneumonia. ACS Appl Mater Interfaces* 11(43):39648–39661.

Zhou, X.Y. Ye, X.G. He, L.T. Zhang, S.R. Wang, R.L. Zhou, J. He, Z.S. 2016. In vitro characterization and inhibition of the interaction between ciprofloxacin and berberine against multidrug-resistant *Klebsiella pneumoniae. J Antibiotics* 69(10):741–746.

Zheng, K. Setyawati, M.I. Leong, D.T. and Xie, J. 2017. Antimicrobial Gold Nanoclusters. *ACS Nano* 11:6904–6910.

Zielinska, A. Carreiró, F. Oliveira, A.M. Neves, A. Pires, B. Venkatesh, D.N. Durazzo, A. Lucarini, M. Eder, P. Silva, A.M. Santini, A, Souto, E.B. 2020. Polymeric nanoparticles: Production, characterization, toxicology and ecotoxicology. *Molecules* 25(16):3731.

Ziąbka, M. Dziadek, M. Królicka, A. 2019. Biological and physicochemical assessment of middle ear prosthesis. *Polymers* 11(1):79.

9 Antimicrobial Peptides for Treating Biofilm Associated Infections

Dipro Mukherjee, Bandita Dutta, Archisman Bhunia, Moupriya Nag, Debarati Basu, Rina Rani Ray, Dibyajit Lahiri and Debasmita Bhattacharya

9.1 INTRODUCTION

"Bacterial resistance" has become the hot-topic in the focus of the modernizing world that has been the cause of the perishing of millions of lives, especially in the past decades. The hospitals have been the epicenter of the scenario of death occurrence as a result of development of lethal bacterial resistance amongst non-eliminated petrogenic microorganisms during the treatment of patients [1, 2]. Hence, the microorganisms survive the stressed microenvironment under antibiotic conditions and later begin to proliferate. Such a scenario results in the development of resistant variants amongst several bacterial species against different drugs, allowing a massive degree of proliferation of them, inculcating in them the ability to form biofilms and making them more resistant and difficult to kill. The development of resistance within the bacterial strains may be innate and/or acquired depending upon the stimuli of their surrounding microenvironment. The presence of such resistance genes within, not only benefits them in survival, but also allows horizontal transfer, promoting communal survival and expansion. Naturally, the presence of such biofilm forming bacteria is in every nook and crevice [3]. Their mechanism of resistance initiates with biofilm formation by adherence to the abiotic and/or biotic surfaces. This consortium of mono/poly-species of microbes comprising the biofilm is surrounded and protected by a matrix, called an extracellular polymeric substance (EPS). The EPS layer is reported to provide the ultimate shape to the biofilm and is debated to be a conglomerate of eDNA, polysaccharides, proteins, and cell debris. The disease caused by bacteria upon biofilm formation occurs slowly allowing the microbes in adaptation of the environment and proliferate subsequently, resulting in several local inflammations. To counter the scenario of biofilm formation, researchers came up with a development of antimicrobial peptides (AMPs) as compounds acting against the biofilm forming bacterial species that are potent in multidrug resistant (MDR) and/or extensively drug resistant (XDR) in nature [4–8].

9.2 ANTIMICROBIAL RESISTANCE

Although effective, antimicrobial resistance (AMR) can be a setback for modern medicine because of its challenges in the control of series of pathogenic infections, with limitations in the options for treatment against these resistant variants. This has caused AMRs to be in the spotlight

DOI: 10.1201/9781003281238-9

of challenges in the health sector of this era [9–11]. The United Nations Interagency Coordinating Group has held AMRs to be the prime cause for the loss of about 700 000 lives worldwide every year, and is debated to rise to about ten million by 2050. This persistence in bacterial proliferation is also due to the inefficient measures of standard treatment, associating the downside of the antibacterial treatment and development of resistant variants with low spectrum of action and/or high toxicity. The pattern of living of an individual influences the development of systemic bacterial infections, increased rates of recurrence, development of opportunistic infections and chronicity by antibiotic-resistant organisms such as pan-drug-resistant (PDR) strains, XDR, and MDR [12–16]. Hence, AMR is a serious issue impacting the pharmaceutical industry in terms of socioeconomic factors. This setback is in terms of time limitation and the involvement of expenses for research and for development of antimicrobials and commercializing them. This development of resistance property within the bacterial strains is also considered to be a consequence of low investment and the long times needed for research and development of the antibiotics, accompanied with long-term-usage of the same drug compounds. Such circumstances have brought about deterioration in the scenario worldwide and urge the need for the development of new drug compounds, and often of higher doses. Antimicrobial administration to an individual may result in bacteriolytic, bacteriostatic, or bactericidal effects. Bacteriostatic effects are a result of the use of drugs that cause the cellular activity of the bacteria to become dormant without killing them [17–22]. The bacteriolytic effect is the result of complete dissolution and destruction of a bacterial strain by antimicrobial compounds. Being among those to survive for billions of years in the earth's changing environment, bacterial species have developed strong adaptive abilities for their genomic flexibility that has helped them to sustain once toxic chemicals in the environment [23–30]. Alongside adaptivity, the potency for dormancy and to transfer and/or exchange genes to surrounding species also inculcated in them the potency for antimicrobial resistance. The development of antimicrobial resistance usually occurs in the presence of a fragile balance between the bacteria and drug interactions, initiating a selective pressure upon the bacterial strain prone to developing a resistant gene within them. This ends up surpassing the relationship between the bacterial-resistance/drug interactions. The collective events of selective pressure and interaction within the bacterial ecology, involving the genetic material grouping within the introns, can develop an entire ecology with resistant and tolerant traits [31–36].

9.3 AMR MECHANISM

Antibiotics are potent towards targeting different sites depending upon their structure and the presence of complementary sites within the bacteria in order to check the bacterial proliferation. The target can be of a diverse range from cell wall to cell membranes, inhibition of essential protein structures and nucleic acid synthesis or other metabolic pathways. Antimicrobial resistance can be simply justified as running away from the action of earlier incomplete treatment with antibiotics, dose interval errors, missed doses, and improper composition of the antibiotics. The present world is facing a great public health challenge termed "multi-resistant forms", which has been developed within several bacterial strains and shows resistance against different types of antibiotics [37–40]. Intrinsic resistance is a notable AMR mechanism within the bacterial species. This mechanism results from the natural resistance of some bacteria against antibiotics to which they have never been exposed, but tolerate because of the inner cellular defense mechanism against them. Mutation is another primary cause of the development of AMR. Though a natural phenomenon in proliferating cells, an increase in the number of bacterial cells with resistant parameters as a result of mutation may only occur due to the imposition of selective pressure from the environment [41–46]. This can also develop due to precise use of antibiotics over the long term. Within the bacterial system, the enzymatic activity may also be a reason behind the suppression of antibiotic efficacy and its mobilization to the target site. Antibiotics having a β-lactam ring are cleaved by β-lactamase enzyme within the biotic system, hence preventing them from this mode of action. Such types of resistance occurs against cephalosporin, carbapenem, monobactams and penicillin [47–50]. Polymyxin antibiotics are known to form pores within the cell membranes, resulting in leakage and death of the cells. This class of antibiotic is used as a last line of defense against MDR strains for their development of AMR [51–54]. The mechanism against polymyxins may occur by the loss of a negative charged group on the bacterial membrane, reducing its affinity to the antibiotic, which in turn may be related to rigidity of the cell membrane as a response to the stressed environment. Researchers also debate on the horizontal transfer of the gene among the bacterial species to be a cause of AMR development. Transduction, transformation, and conjugation are the mechanisms involved in such horizontal gene transfer. Last but not the least, the other mechanisms involved in the development of AMR are those that rise in the degree of efflux pump on the bacterial membrane, and the biofilm formation.

9.4 BIOFILMS

Biofilms are simply defined as a consortium of microorganisms appearing from the same/diverse niches to form a three-dimensional structure performing unidirectional functions resembling social cooperation [55–56]. These phenomena are a result of the evolutionary capacity of the microorganisms to create a resistance against the environmental threats that a plankton cannot tolerate [57–62]. The biofilm comprises of the microorganisms themselves, and are embedded in a matrix known as extracellular polymeric substance (EPS). The matrix is made up of substances that act as nutrient providers and which also provide basic structure to the colony. The structure is formed from cell-cell and cell-substance adherence and bears a varying architecture and chemical composition. The EPS can hold substances produced by the microorganisms within the biofilm, in general, organic compounds, salts and minerals, and substances from the debris of infected host like saliva derivatives, serum and excreted fluids [63–69].

Biofilm as a whole possesses a distinct characteristic of gene expression, synthesis of protein and degree of proliferation. Besides, the resistance mechanism within the biofilm is also significantly different from that of the plankton, such as in their tolerance to the surrounding stress. The composition of the EPS may differ and is decided upon by the sustaining microenvironment and/or the nature of the species. This EPS contributes to proliferation aside from maintaining a protection system and supply of nutrients. It also contributes to signal interaction, namely, quorum sensing, horizontal gene transfer and the exchange of molecules between non-descendant cells [70–73]. Certainly, the treatment to counter this phenomenon requires drugs at higher doses to kill the biofilm structure and reach the target bacteria.

9.5 CONTROL MECHANISM

As the biofilm uplifts the degree of bacterial pathogenicity and its ability to resist antibiotics, there is an urge to formulate mechanisms potent against bacterial growth and biofilm formation. Such a pathogenic system within the bacteria is often found to develop resistance against invasion of the host immune system by altering the activation of the complement system and phagocytes, bypassing the immune response [74–76].

Among many strategies put into practice, use of antibiotics is a promising measure, although it has cytotoxic effects upon patients, and it is challenging to attain the minimum inhibitory concentration (MIC) of antibiotic in-vitro. Accompanying that, biofilm formation also increases the AMR to a higher level, making the chances of suppressing an infection associated with biofilm more difficult [77–80]. A report has claimed that the rise in biofilm associated infectious diseases is from the use of foreign bodies, and it can be cured by treating the infected object or by replacement of the object, or by administration of antibiotics. Previous research came up with antibiotic tolerance development against the compounds due to selective sensitivity of the microorganisms and had questioned the potency of the drugs in penetrating the biofilm matrix. Some strategies that came forth are in the use of hydrophobic polymer coatings upon the surfaces to prevent biofouling of the microbes that occurs upon adhesion. These polymer coatings are generally comprised of a mixture of disinfectants or antibiotics. Photoactive dyes are also used in irradiation in the presence of oxygen [81–83]. The composition is also reported to mix with antibiofilm and/or biofilm-dissolving molecules, and shows bactericidal effects. This method of treatment is termed photodynamic therapy. Research has found most antibiofilm molecules to be polyphenols, peptides, antibiotics, and enzymes, since their mechanism of action is interrupting the signal transduction pathway, namely, quorum quenching. The bacteriophage cocktail is another strategy of biofilm prevention that targets the host receptors and delays the formation of phage-resistance bacteria. Phages with varying lytic activity are used to expand the host range and number-target pathogens. Upon combining with antibiotics, the phages improve their potency against biofilm inhibition.

A wide spectrum of anti-biofilm substances have been characterized from natural sources and also synthetically and also comprise of antibiotics and chelating agents [84–86]. Hence, there exists a great diversity in the mechanism of anti-biofilm action. Some of the mechanisms are as follows: the inhibition of quorum sensing, the use of quorum quenchers to check colony-wide communication and motility, and population control by inhibiting the LuxR transcriptional activator protein. The protein is responsible for regulation of the synthesis of LuxI-like synthase protein, that synthesizes the signaling molecules [87–89]. Peptidoglycan cleavage, usually using tannic acid, a polyphenolic compound, is used in this method of inhibiting biofilm formation in *Staphylococcus aureus*, but does

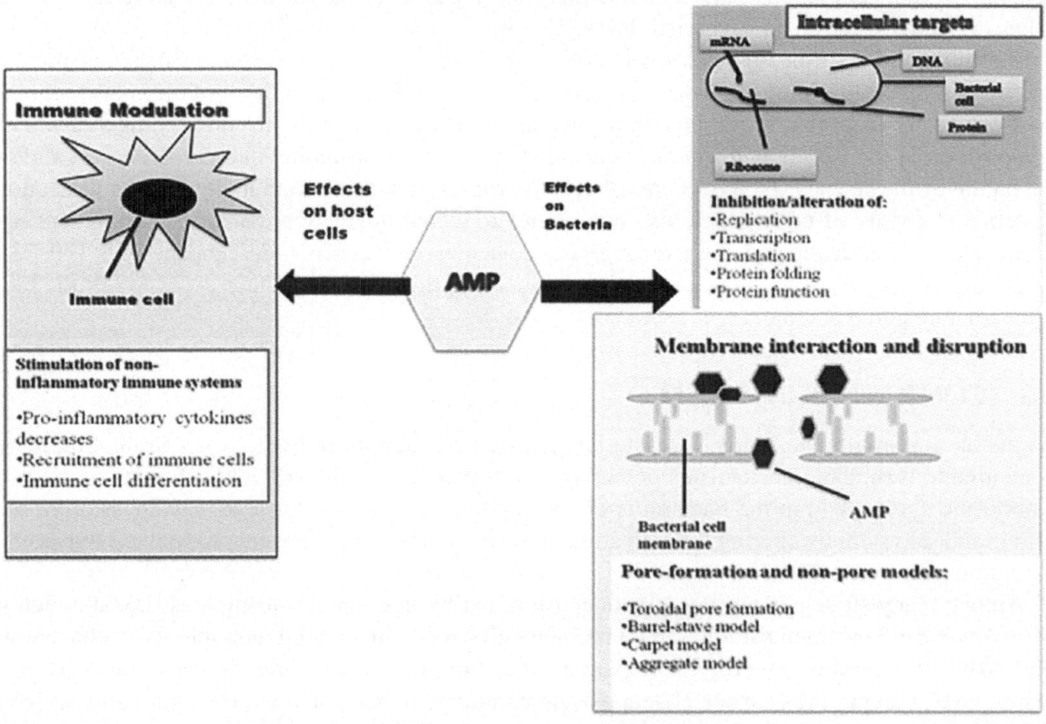

FIGURE 9.1 Antibacterial activities of AMPs against planktonic bacteria.

not alter bacterial growth. It alters the mechanism in various ways such as inhibiting the release of signalling molecules that trigger the genes for biofilm formation, and it alters the composition of teichoic acid and other components responsible for cell wall formation [90–92] (Figure 9.1).

9.6 ANTIMICROBIAL PEPTIDES (AMPS) AND THEIR CLASSIFICATIONS

Antimicrobial peptides or AMPs are macromolecules consisting of 12–50 amino acid sequences and vary in their lengths [93–98]. They are well known for their antibacterial, antifungal, and antiviral properties and are potent to eliminate bacterial load bypassing the resistance by associating rapidly with membranes. Though dominantly hydrophobic, AMPs comprise of both hydrophilic and hydrophobic regions. This allows the establishment of an interaction with the membrane lipid tails of the membrane layer via van der Waals interaction.

Depending upon the amino acid residues and their structural conformation AMPs are classified into four major classes which are: α-helices, β-sheet, extended peptides and loop. These first two classes are the most common and are best studied. Magainins, magainin 2 and PGLa are the α-helical peptides that are isolated from the skin of *Xenopus laevis,* an African frog. Cathelicidins and defensins are the host defense peptides present in humans. Leukocyte mediated Cathelicidins are the AMPs that have cathelin, namely, the amino-terminal cathepsin L inhibitor domain [99]. Defensins are high complexity open ended cystine rich peptides, namely, widely distributed in both vertebrates

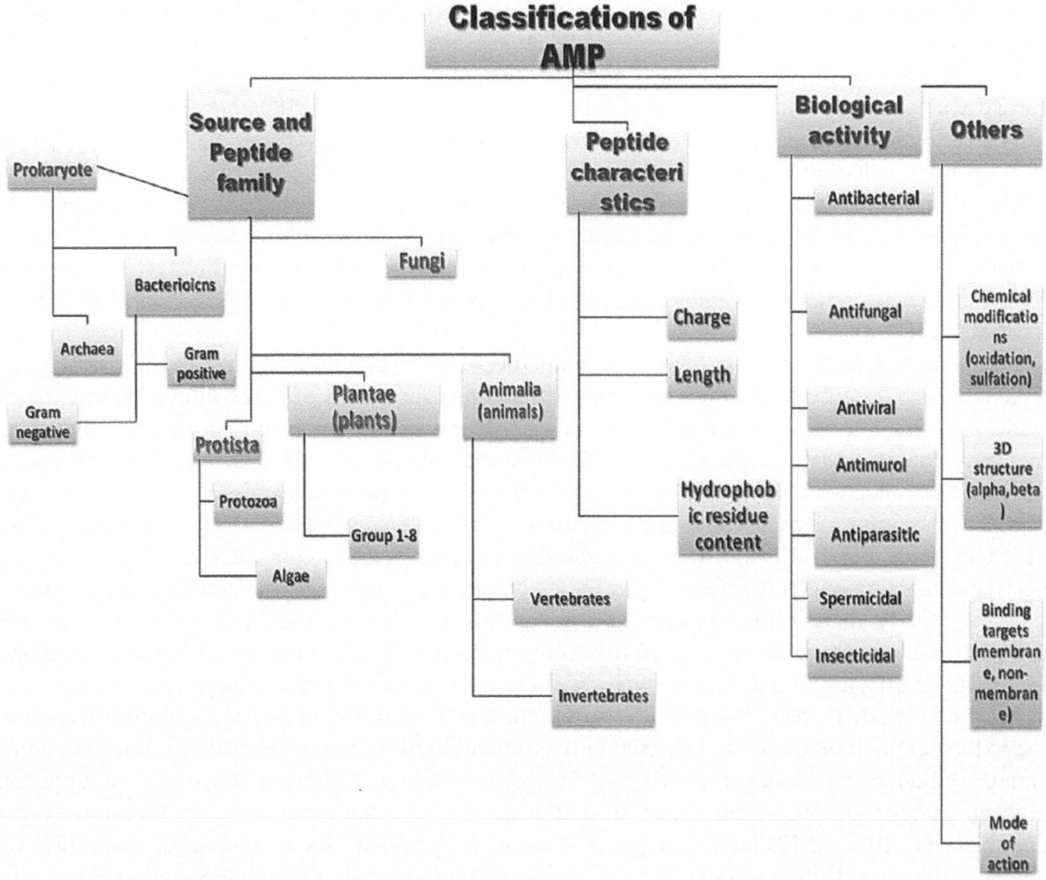

FIGURE 9.2 Classification of Aantimicrobial peptides.

and invertebrates. Mammalian defensins are classified into α, β and θ defensins depending upon the size and pattern of disulfide bonding (Figure 9.2).

9.7 AMPS AS ANTIBIOFILM AGENTS

AMPs originated from Bacteria are post-transcriptionally modified class I bacteriocins, namely, lantibiotics and nisin, and unmodified class II bacteriocins, namely, mersacin which manifests antimicrobial activity against both the Gram-positive and Gram-negative bacteria via targeting cell membranes [83]. Bacteriocins and their derivatives have therapeutic potential against the MDR infections. Polymyxins possess antibiofilm activities, and are considered to be "last resort antibiotics" from the 1940s. Polymyxin is a circular polypeptide which is produced by *Paenibacillus polymyxa*, amphipathic in nature with a cationic heptapeptide ring connected with a hydrophobic fatty acid chain through additional tripeptide. Polymyxins have antimicrobial activity mainly against Gram-negative bacteria via the integration of negatively charged bacterial membranes leading to membrane destabilization and neutralization of lipopolysaccharide (LPS) for electrostatic interactions. Extensive use of polymyxins lead to neuro and nephrotoxicity at the early discovery of polymyxins in 1947. Polymyxins were found to be "last hope antibiotics" when the increase of multidrug resistant Gram-negative bacteria mediated infections namely *Pseudomonas aeruginosa* transmitted chronic lung disease and multi drug resistant *Acinetobacter baumanii* transmitted wound infections. Current studies have reported antibiofilm activities along with the bactericidal functions of polymyxins. The killing activity of polymyxin E (colisitn) with 80% reduction of viable cells against *A. baumanii*, *P. aeruginosa* and *S. maltophilia* biofilms has been determined through various biofilm assays *in vitro*.

80% survival of infected rats that imitate *P. aeruginosa* mediated chronic lung disease cystic fibrosis was observed with the intratracheal application of a 64 fold MIC concentration of colistin compared to the 92% demise of untreated infected rats. Tobramycin escalates the bactericidal effects of colistin by reducing biofilm forming cells within seven days from treatment. It was observed that continual administration of colistin through a urinary device for seven days was apt for complete reduction of MDR *A. baumanii* mediated urinary infections without any detrimental effects to the patients. Although all of the clinical reports have suggested that colistin is effective against Gram negative bacteria it is not potent against Gram positive bacteria and MRSA and VRE biofilm associated infections.

AMPs have been found to be more advantageous over novel antibiotics due to their wide range of activity and specific mechanism of actions including emergence of pores in cytoplasmic membrane with low frequency of bacterial resistance. AMPs are mostly stable on a large scale of pH and temperature. AMPs have low toxic effects on the eukaryotic cells thus presenting an opportunity for therapeutics. AMPs with inhibitory as well as disruptive properties are able to eliminate entrenched biofilm even in a low concentration. AMPs are used to show synergic effects along with antibiotics. For example, AMPs can neutralize endotoxin and become active in animal models. Resistance against AMPs is infrequent as they get attracted to the bacterial cell membrane, specifically negatively charged lipid bilayer. AMPs are effective against both the slow growing as well as non-growing bacteria due to their ability for penetration and pore formation in the cytoplasmic membrane. AMPs have rapid killing dynamics compared to conventional antibiotics.

Most of the AMPs have the potential to function against different stages of biofilm formation due to their explicit modes of action, including elimination of biofilm at the primary stage via inhibition of cell adhesion and down regulations of quorum sensing. AMPs are active against multidrug resistant bacteria although their ability to inhibit quorum sensing is not properly investigated yet. The majority of the AMPs have emerged for oppressive applications against acne, cystic fibrosis, catheter infections, wound associated biofilms and many more only following pre-clinical and clinical trials. In phase III clinical trials both the AMPs, named omiganan and pexiganan, have shown antibiofilm efficacy. Although neither of the AMPs are yet approved for clinical use they are able to develop topical gels against catheter linked infections and a wide range of novel antibiotics against foot ulcer infections respectively.

Selectively Targeted Antimicrobial Peptides (STAMPs) are a novel group of antimicrobial peptides with enhanced potency for killing mechanisms. A semi synthetic lipoglycopeptide oritavancin is undergoing clinical trials against Gram positive bacterial infections such as methicillin-susceptible (MSSA), methicillin resistant (MRSA), and vancomycin-resistant *S. aureus* (VRSA).

9.8 MODE OF ACTIONS OF AMPS

AMPs have specific mechanisms of actions which usually target the cytoplasmic membrane for penetration leading to cell lysis. Structural properties of the AMPs such as size, amino acid sequences, cationic nature and amphipathicity are responsible for their mode of actions. AMPs target the transmembrane potential through the perfusion of microbial membranes resulting in cell death. Whereas conventional antibiotics are intended to act towards the inhibition of cell walls as well as DNA, RNA and protein synthesis, AMPs usually penetrate the cell membrane through the formation of membrane spanning pores which as exemplified in the barrel-stave model whereas the carpet model is mostly recognized for the mediated membrane destabilization of AMPs. Along with membrane permeabilization, AMPs are responsible for lipopolysaccharide (LPS) neutralization or disaggregation. Alamethicin is an AMP, which is apt for transmembrane pore formation similar to the barrel-stave model. It was observed that alamethicin acquired alpha helical configuration in order to be inserted into the lipid bilayer with a moistened surface. In the case of the carpet model, the AMPs usually

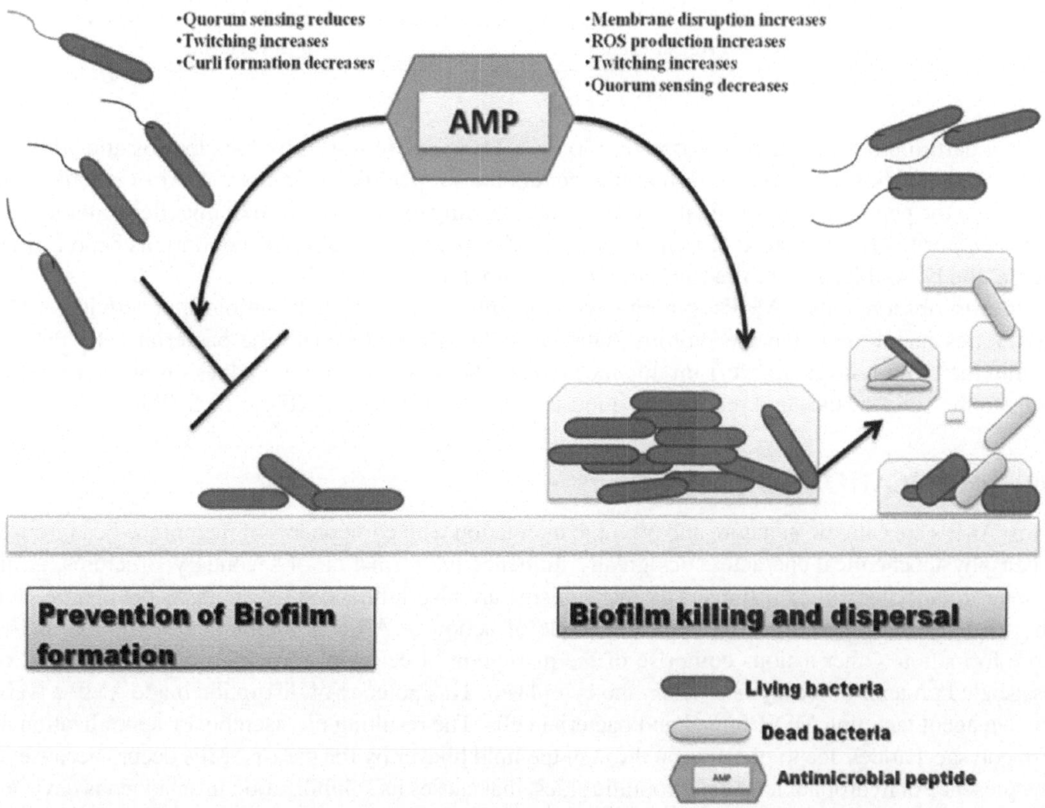

FIGURE 9.3 Antibiofilm effect of AMPs and involved mechanisms.

TABLE 9.1
Advantages of AMPs over Antibiotics

Properties	AMP	Antibiotics
Activity Spectrum	Generally broad spectrum	Generally narrow spectrum
Anti-persister activity	Demonstrated for many AMPs	Poor or none
Immuno-modulatory capacity	Demonstrated for many AMPs	Poor or none
Wound-healing activity	Demonstrated for many AMPs	Poor or none
Manipulations prone	Easily manipulated for improving antimicrobial activity/reduce toxicity	Difficulty in manipulation
Action against beneficial flora	Possibly able to target AMPs against specific pathogens, leaving undisturbed the normal flora	Active against beneficial flora Induction of resistance
Induction resistance	Generally low; in some cases induction of resistance after several passages in vitro. In the case of polymicrobial infections, possibility of insurgence of community-based AMP-resistance mechanisms	Resistance is easily induced. In the case of polymicrobial infections, interspecies interactions may affect the antibiotic susceptibility of individual organisms.
Active concentrations	The need to use increased concentrations as compared to mono-species biofilms has been reported with a consequent risk of cytotoxicity	Therapeutic concentrations against susceptible strains highly optimized
Approval by drug agencies	Only a few AMPs are approved for medical use. It is very difficult.	Approval Of new antibiotics are slower than required.

form a partition between acidic and zwitterionic membranes followed by the electrostatic addition of the anionic phospholipid head group that covers the surface like a membrane. After reaching the threshold the peptides form toroidal transient holes leading to the formation of micelles. Interactions between positively charged AMPs and negatively charged lipids enable the continuous bend formation at the bottom leading to the formation of a toroidal hole.

It was observed that AMPs synergistic with antibiotics manifest antibiofilm activities. The antibiotics mainly derange the biofilm matrix and the AMPs dissipate the bacterial cells present within the biofilm layer. Apart from this mechanism the AMPs inhibit the adhesion of the bacterial cells to the solid surface and restrict the quorum sensing mechanisms (Figure 9.3, Table 9.1).

9.9 APPLICATIONS OF AMPS

Most AMPs are cationic in nature and promote interaction with components of membrane headgroups. Their physicochemical characteristics greatly influence the formation of secondary structures, influencing the mechanism of actions. The mechanisms are also influenced by aphinity, net charge, and the sequence of amino acid. The primary mode of action of AMP is membrane disruption and/or pore formation. Other actions comprise of the disruption of cell wall formation by the inhibition of associated proteins and enzymes within the cytoplasm. This potency of disruption made AMP a well-known agent targeting MDR fungal and bacterial cells. The resulting disassembly or neutralization of lipopolysaccharides due to penetration through the lipid bilayer by the use of AMPs occurs because of the presence of hydrophobic and hydrophilic sides, that causes its solubilization in an aqueous environment. AMPs electrostatically bind to the lipopolysaccharides (LPS) and infiltrate the biofilm, causing its death. The interaction involves heads of groups of two cationic amino acids, arginine and lysine, leading to the formation of a complex. This complex performs the role of pore formation within the LPS, destabilizing and impairing the structural integrity of the cell membrane. This interruption occurs as the complex is primarily stabilized by the hydrophobic interactions between the fatty acyl chain of the LPS and hydrophobic amino acids of the peptide sequence.

In contrast to the conventional antibiotics, AMPs have proved to exhibit promising dominance against resistant bacterial strains. This efficacy is due to the ability of a non-specific targeting mechanism that bypasses the likelihood of resistance development. Several AMPs are also noted to have anti-multi-resistant biofilm activity that disrupts the adhesion property at the initial stage of biofilm formation and/or deceases mature biofilms formed (Table 9.2).

TABLE 9.2
Synergistic Antibiofilm Activities of AMPs and Other Compounds

Tested compounds		Microorganisms	
Compound A (AMP)	**Compound B (AMP or other)**	***In vitro* antibiofilm activity**	***In vivo* antibiofilm activity**
Bacitracin	Anprocide	*S. aureus, S. epidermidis*	
Citropin 1.1	• Rifampin • Monocycline	*S. aureus*	*S. aureus*
DD_{13}-RIP • **DD_{13} [K_4-S4(1-13)$_a$** • **RIP**	Rifampin		*S. aureus, S. epidermidis*
G10KHc	Tobramycin	*P. aeruginosa*	
IB-367	Linezolid	*S. aureus, E. faecalis*	*S. aureus, E. faecalis*
Indolicidin	• Daptomycin • Linezolid • Teichoplanin • Ciprofloxacin • Azithromycin • Rifampicin • Ceftazidim	*S. aureus*	
Lactoferrin	• Amikacim • Ciprofloxacin • Tobramycin	*B. cenocepacia, B. multivorans, B. dolosa*	
	• Xylitol	*P. aeruginosa*	
	• Xylitol + Silver	*P. aeruginosa, S. aureus*	
	• Xylitol + Farnesol	*S. epidermidis, E. faecalis*	
Cecropin (1-7)–Melittin A (2–amide	• Daptomycin • Linezolid • Teichoplanin • Ciprofloxacin • Azithromycin,	*S. aureus*	
HBD-3	• DNAseI	*Nontypeable H. influenzae*	
	• Ultrasound-targeted Microbubble Destruction	*S. aureus, S. epidermidis*	
BMAP-28	• Vancomycin		*S. aureus*
	• Q/D (Quinupristin/ Dalfopristin)		*E. faecalis*
	• Linezolid		

(*Continued*)

TABLE 9.2 (Continued)
Synergistic Antibiofilm Activities of AMPs and Other Compounds

Tested compounds		Microorganisms	
Compound A (AMP)	**Compound B (AMP or other)**	***In vitro* antibiofilm activity**	***In vivo* antibiofilm activity**
Colistin	• Tobramycin • Fosfomycin • Levofloxacin • Clarithromycin • Ciprofloxacin • Tetracycline	*P. aeruginosa* *S. maltophilia* *A. baumannii*	*P. aeruginosa* *E. coli*

9.10 APPLICATIONS OF AMPS AGAINST BIOFILM MEDIATED INFECTIONS

9.10.1 Oral Infections

Biofilm associated oral disease is a periodontal infection that empowers the conversion from reversible mild gingivitis to chronic periodontitis. Periodontal disease is an example of a mixed bacterial biofilm disease. This involves a large number of clinical indications that range from gentle, reversible gum disease to extreme, constant periodontitis, which might cause damage of bone and connective tissue in the periodontal region, with subsequent tooth loss. It is generally acknowledged that particular groups of oral microorganisms under the "red complex" (for example, *P. gingivalis, T. forsythia*), leads to the formation of periodontitis by attacking periodontal tissues and secreting various virulence factors. Similarly, a significant pathogenesis enables an uncontrolled host post-inflammatory reaction to bacterial invasion, which incorporates the upregulation of proinflammatory cytokines, matrix metalloproteinases, and responsive oxygen species. All of them are responsible for tissue damage and loss of teeth, namely, related to the periodontitis.

Microbial equilibrium and oral health is maintained by the various naturally obtained AMPs in the oral cavity which contribute to the innate immune system.

Numerous AMPs have shown great action against oral microorganisms and their utilization to forestall/treat oral diseases appears to be encouraging, despite the fact that their antimicrobial power in the oral depression may be tested by the presence of spit or on the other hand crevicular liquid, because of high salt focus, the presence of proteases of host/bacterial beginning, or sequestration by the macromolecules present in such liquids. Wang and collaborators detailed the capacity of an engineered cationic AMP, Nal-P-113, to apply a critical bactericidal movement against oral microorganisms, namely, *Streptococcus gordonii, Fusobacterium nucleatum* and, furthermore, *P. gingivalis*, in both planktonic and polymicrobial biofilm states . The peptide is the ameliorated subordinate of another peptide, P-113 (AKRHHGYKRKFH-NH2), into which histidine deposits were supplanted with the amino acid–naphthylalanine, enabling high salt defiance. Nal-P-113 showed over 85% integrity after 8 h when incubated with phosphate buffer saline (PBS), there is slight harm to typical oral cells (1.28 mg/mL), Nal-P-113 had the option to kill triple strain biofilms of *S. gordoni, F. nucleatum* and *P. gingivalis*, while the base biofilm destruction centralizations of penicillin and metronidazole were 2 mg/mL and 80 mg/mL, separately. It is vital that numerous AMPs (for example, beta-defensins, human neutrophil defensins, the human cathelicidin LL-37) have shown lipopolysaccharide (LPS)–killing exercises against periodontopathogens, causing the hindrance of the IL-1, IL-8, and intercellular grip atom 1 (ICAM-1) articulation instigated by LPS from *P. intermedia* and *T. forsythia* in THP-1 cells and human gingival fibroblasts. Thus, these outcomes suggest that AMPs might be considered as preventive and restorative specialists against blended bacterial diseases, for example, periodontitis, by killing the microorganisms as well as by diminishing the action of LPS and sickness related irritation.

9.10.2 Respiratory Tract Infections

It is observed that patients suffering from ventilator related pneumonia, or chronic pulmonary diseases, these types of lung infections are polymicrobial in nature. The key airway secretions as seen in the host AMPs within the innate defense of the lungs include defensins and cathelicidins.

They can add to host protection in the lung by killing the microbes as well as by regulating the host inflammatory reaction. These beneficial properties have led to research on their exogenous organization to forestall/treat infections. A few AMPs have been tried in lung polymicrobial diseases. One of these peptides is tachyplesin III, which is a sheet peptide obtained from the hemocytes of the horseshoe crab (*Tachypleus tridentatus*), and which has been tried in bacterial co-disease pneumonia. As an analyzed tomono-bacterial disease, the intranasal co-contamination of mice with MDR *Pseudomonas aeruginosa* and *A. baumannii* caused a more serious infection, with expanded favorable to fiery cytokines (IL-1, IL-6, TNF) and chemokines (MCP-1/MIP-2) and increases death. The pretreatment of mice with a solitary portion of tachyplesin III (10 mg/kg, i.v.) could draw out mice that survive and essentially decrease the complete bacterial inclusion in the bronchoalveolar lavage liquid, when contrasted with the untreated or meropenem-treated control groups of mice.

Moreover, the peptide was likewise found to diminish the serum level of the pro-inflammatory cytokines IL-1, IL-6,and TNF-, and to diminish inflammatory cell penetration, vascular leakage, and alveolar disturbance in the tachyplesin III-pretreated group when contrasted with the co-infected bunch or the meropenem-treated bunch. The peptide showed the capability to increase the phagocytic function of alveolar macrophages when they were in vitro tested. It further suggested that prophylactic adequacy might occur due to multimodal mechanisms of activity.

9.10.3 Sepsis

Sepsis is a dangerous condition whose symptoms involve excess systemic inflammation and also an infection of the blood stream. Sepsis causes the greatest number of deaths during stays at hospitals world-wide. It the most common public health infection globally. It is observed that almost 60% of the hospital related BSI of all types is caused due to devices which are vascular operated such as bloodstream infections related to catheters (CRBSI). These types of devices are commonly used mainly in severely ill patients primarily requiring fluid administration, antibiotics or other nutritional solutions, and chemotherapy. The development of modern medical practice has led to the use of intravascular catheters. These catheters are susceptible to growth of skin microbes leading to biofilm formation on both the internal and external surface of the body. The maturation of the biofilm is followed by the detachment of the biofilm particles or single microbes and finally accessing the bloodstream causing polymicrobial BSI or CRSBI, both of which are becoming epidemiologically important. It is significant as it is observed that between 5–38% of them develop into dangerous sepsis at an increasing rate compared to monomicrobial bacteremia.

The potential of Epinecidin-1 has been evaluated in a mice model where polymicrobial sepsis and endotoxemia has been incorporated through cecal ligation and puncture (CLP).

After 30 mins of surgery, Epinecidin-1 has been introduced to the mice intraperitoneally in the amount of 50 mg/kg. Compared to the control group, namely, CLP along with mice injected with saline Epinecidin-1 injected mice showed exorbitant viability. The novel three synthetic AMPs usually bind to the lipid. One of those peptides, 19–2.5 (Pep 2.5) is able to significantly raise substantial activity of mice compared to the control, namely, 24 h CIP sepsis induced mice. Suffusion of Pep 2.5 in septic animals enable a notable increase of IL-6, IL-10 and monocyte chemoattractant protein serum levels with an articulation of CD14 mRNA in heart, lung and spleen. Some of the peptides are involved in the prevention of cardiac sarcoplasmic reticulum Ca2+-ATP-ase (SERCA2) downregulation emerging from the attenuation of cardiac dysfunction.

9.10.4 Wound Infections

With the increase of world civilization, the number of diseased patients has also increased. Diseases like cancer, cardiovascular diseases, diabetes or immunosuppressants lead to chronic wound infections appended diabetic foot ulcers (DFU), pressure ulcers or venous leg ulcers over 1–2% of the whole population. Biofilm formation through bacterial colonization at the side of severe wound infections affects mitigation. Chronic wound infections are mainly associated with *S. aureus* and *P. aeruginosa* . Skin AMPs provide the first line of congenital defence mechanisms against any skin wound infections. Cathelicidins (human hCAp18/LL37), chemerin, defensins, RNase 7 and secretory leukocyte protease inhibitors (SLPI) are some skin AMPs. It was observed that hCAP18/LL37 restrains the upregulation of epidermis that leads to the skin infections. Deficiency of the murine equivalent in hCAP18/LL37 (CRAMP) enables intrathecal streptococcal infections in mice. AMPs have specific sanative roles against wound infections for their comprehensive activities as antibiofilm, immunomodulator and angiogenic agents. Synergism between several discrete AMPs or other antimicrobial agents has increased the antimicrobial as well as antibiofilm efficacy due to the reduction of uprising resistance and cytotoxicity. Amalgamation of nisin and pexiganan is competent in controlling polymicrobial diabetic foot infections. Compared to the Gram negative *P. aeruginosa* the binary effect of AMPs is greater against Gram positive *S. aureus*. Pexiganan, along with nisin, has the combined effect of the inhibition of cell wall biosynthesis via binding to lipid II on cytoplasmic membrane. In the case of Gram negative bacteria this AMP combination has a reduced effect due to the presence of an outer membrane. AMPs immobilized in a bio gel matrix has intensified antibacterial activity against *S. aureus* and null activity against *P. aeruginosa*.

9.11 CONCLUSION

Concluding the entire scenario, there is an urgent need for research into potent drugs for the well-being of public health to act against antibiotic resistance and multi-drug resistant bacterial strains. The review further focuses upon the bacteria causing the biofilm formation. To date, researchers have characterized a wide range of mechanisms involved in microbial resistance, but biofilm formation is the most concerning one. Additionally, the mechanism of biofilm formation involves diverse particularity in its structural and molecular level of mechanism. Hence, it too requires a broad spectrum of agents potent in inhibiting it. Aside from this, the antimicrobial properties that have been characterized in these AMPs, they are also promising inhibitors of biofilm formation. It has also been found that AMPs show synergistic effects upon combination with obsolete conventional antibiotics in eradicating the bacteria as well as their biofilm formation. Nanotechnology is yet another promising strategic tool to eradicate XRD/MRD bacteria, biofilm. This happens by high specificity, decreased degree of toxicity, controlled release of peptides and drugs, and increased bioavailability. Bioconjugation has also shown significant outcomes countering XRD and MRD biofilm-forming bacteria. Hence, it is concluded that the use of AMPs in drug-discovery can be used as alternatives against MDR-bacterial infections. More intense research is needed around the biofilm-forming bacteria to prevent their adversity.

REFERENCES

1. Gristina, A.G., Oga, M., Webb, L.X., and Hobgood, C.D. Adherent bacterial colonization in the pathogenesis of osteomyelitis. *Science*. 1985; 228: 990–3.
2. Hyde, J.A., Darouiche, R.O., and Costerton, J.W. Strategies for prophylaxis against prosthetic valve endocarditis:A review article. *J Heart Valve Dis*. 1998; 7: 316–26.

3. Bjarnsholt, T., Kirketerp-Moller, K., Jensen, P.O., *et al.* Why chronic wounds will not heal: a novel hypothesis. *Wound Repair Regen.* 2008; 16: 2–10.
4. Costerton, W., Veeh, R., Shirtliff, M., Pasmore, M., Post, C., and Ehrlich, G. The application of biofilm science to the study and control of chronic bacterial infections. *J Clin Invest.* 2003; 112: 1466–77.
5. Costerton, J.W., Stewart, P.S., and Greenberg, E.P. Bacterial biofilms: a common cause of persistent infections. *Science.* 1999; 284: 131822.
6. Siddiq, D.M., and Darouiche, R.O. New strategies to prevent catheter associated urinary tract infections. *Nat Rev Urol.* 2012; 9: 305–14.
7. Rajan, S., and Saiman, L. Pulmonary infections in patients with cystic fibrosis. *Semin Respir Infect.* 2002; 17: 47–56.
8. Davey, M.E., and O'Toole, G.A. Microbial biofilms: From ecology to molecular genetics. *Microbiol Mol Biol Rev.* 2000; 64: 847–67.
9. Drenkard, E., and Ausubel, F.M. *Pseudomonas* biofilm formation and antibiotic resistance are linked to phenotypic variation. *Nature.* 2002; 416: 740–3.
10. Stewart, P.S., and Costerton, J.W. Antibiotic resistance of bacteria in biofilms. *Lancet.* 2001; 358: 135–8.
11. Donlan, R.M., and Costerton, J.W. Biofilms: Survival mechanisms of clinically relevant microorganisms. *Clin Microbiol Rev.* 2002; 15: 167–93.
12. Hoiby, N., Bjarnsholt, T., Givskov, M., Molin, S., and Ciofu, O. Antibiotic resistance of bacterial biofilms. *Int J Antimicrob Agents.* 2010; 35: 322–32.
13. Allen, H.K., Donato, J., Wang, H.H., Cloud-Hansen, K.A., Davies, J., and Handelsman, J. Call of the wild: Antibiotic resistance genes in natural environments. *Nat Rev Microbiol.* 2010; 8: 251–9.
14. Hoiby, N., Ciofu, O., Johansen, H.K., *et al.* The clinical impact of bacterial biofilms. *Int J Oral Sci.* 2011; 3: 55–65.
15. Fux, C.A., Costerton, J.W., Stewart, P.S., and Stoodley, P. Survival strategies of infectious biofilms. *Trends Microbiol.* 2005; 13: 34–40.
16. de la Fuente-Nunez, C., Korolik, V., Bains, M., *et al.* Inhibition of bacterial biofilm formation and swarming motility by a small synthetic cationic peptide. *Antimicrob Agents Chemother.* 2012; 56: 2696–704.
17. Siddiqui, A.R., and Bernstein, J.M. Chronic wound infection: Facts and controversies. *Clin Dermatol.* 2010; 28: 519–26.
18. Whitehouse, J.D., Friedman, N.D., Kirkland, K.B., Richardson, and W.J., Sexton, D.J. The impact of surgical-site infections following orthopedic surgery at a community hospital and a university hospital: Adverse quality of life, excess length of stay, and extra cost. *Infect Control Hosp Epidemiol.* 2002; 23: 183–9.
19. Jenssen, H., Hamill, P., and Hancock, R.E. Peptide antimicrobial agents. *Clin Microbiol Rev.* 2006; 19: 491–511.
20. Brogden, K.A. Antimicrobial peptides: pore formers or metabolic inhibitors in bacteria?. *Nat Rev Microbiol.* 2005; 3: 238–50.
21. Hancock, R.E. Peptide antibiotics. *Lancet.* 1997; 349: 418–22.
22. Boman, H.G. Peptide antibiotics and their role in innate immunity. *Annu Rev Immunol.* 1995; 13: 61–92.
23. Hancock, R.E., Nijnik, A., and Philpott, D.J. Modulating immunity as a therapy for bacterial infections. *Nat Rev Microbiol.* 2012; 10: 24354.
24. Afacan, N.J., Yeung, A.T., Pena, O.M., and Hancock, R.E. Therapeutic potential of host defense peptides in antibiotic-resistant infections. *Curr Pharm Des.* 2012; 18: 807–19.
25. Jacobsen, A.S., and Jenssen, H. Human cathelicidin LL-37 prevents bacterial biofilm formation. *Future Med Chem.* 2012; 4: 1587–99.
26. Elssner, A., Duncan, M., Gavrilin, M., and Wewers, M.D. A novel P2X7 receptor activator, the human cathelicidin-derived peptide LL37, induces IL-1 beta processing and release. *J Immunol.* 2004; 172: 4987–94.
27. Mookherjee, N., Hamill, P., Gardy, J., *et al.* Systems biology evaluation of immune responses induced by human host defence peptide LL-37 in mononuclear cells. Mol Biosyst. 2009; 5: 483–96.
28. Davidson, D.J., Currie, A.J., Reid, G.S., *et al.* The cationic antimicrobial peptide LL-37 modulates dendritic cell differentiation and dendritic cell-induced T cell polarization. *J Immunol.* 2004; 172: 1146–56.
29. Hancock, R.E., and Sahl, H.G. Antimicrobial and host-defense peptides as new anti-infective therapeutic strategies. *Nat Biotechnol.* 2006; 24: 1551–7.

30. Nicolas P. Multifunctional host defense peptides: Intracellulartargeting antimicrobial peptides. *FEBS J*. 2009; 276: 6483–96.
31. Mok, W.W., and Li, Y. *Therapeutic Peptides: New Arsenal Against Drug Resistant Pathogens*. Curr Pharm Des 2013.
32. Hurdle, J.G., O'Neill, A.J., Chopra, I., and Lee, R.E. Targeting bacterial membrane function: An underexploited mechanism for treating persistent infections. *Nat Rev Microbiol*. 2011; 9: 62–75.
33. Lohner, K. New strategies for novel antibiotics: Peptides targeting bacterial cell membranes. *Gen Physiol Biophys*. 2009; 28: 105–16.
34. Laverty, G., Gorman, S.P., and Gilmore, B.F. The potential of antimicrobial peptides as biocides. *Int J Mol Sci*. 2011; 12: 6566–96.
35. Fjell, C.D., Jenssen, H., Hilpert, K., *et al*. Identification of novel antibacterial peptides by chemoinformatics and machine learning. *J Med Chem*. 2009; 52: 2006–15.
36. Takahashi, D., Shukla, S.K., Prakash, O., and Zhang, G. Structural determinants of host defense peptides for antimicrobial activity and target cell selectivity. *Biochimie*. 2010; 92: 1236–41.
37. Yeamanm, M.R., and Yount, N.Y. Mechanisms of antimicrobial peptide action and resistance. *Pharmacol Rev*. 2003; 55: 27–55.
38. Bessalle, R., Haas, H., Goria, A., Shalit, I., and Fridkin, M. Augmentation of the antibacterial activity of magainin by positive-charge chain extension. *Antimicrob Agents Chemother*. 1992; 36: 313–7.
39. Matsuzaki, K., Harada, M., Funakoshi, S., Fujii, N., and Miyajima, K. Physicochemical determinants for the interactions of magainins 1 and 2 with acidic lipid bilayers. *Biochim Biophys Acta*. 1991; 1063: 162–70.
40. Dathe, M., Nikolenko, H., Meyer, J., Beyermann, M., and Bienert, M. Optimization of the antimicrobial activity of magainin peptides by modification of charge. *FEBS Lett*. 2001; 501: 146–50.
41. Strom, M.B., Haug, B.E., Skar, M.L., Stensen, W., Stiberg, T., and Svendsen, J.S. The pharmacophore of short cationic antibacterial peptides. *J Med Chem*. 2003; 46: 1567–70.
42. Wieprecht, T., Dathe, M., Beyermann, M., *et al*. Peptide hydrophobicity controls the activity and selectivity of magainin 2 amide in interaction with membranes. *Biochemistry*. 1997; 36: 6124–32.
43. Dathe, M., Wieprecht, T., Nikolenko, H., *et al*. Hydrophobicity, hydrophobic moment and angle subtended by charged residues modulate antibacterial and haemolytic activity of amphipathic helical peptides. *FEBS Lett*. 1997; 403: 208–12.
44. Chen, Y., Guarnieri, M.T., Vasil, A.I., Vasil, M.L., Mant, C.T., and Hodges, R.S. Role of peptide hydrophobicity in the mechanism of action of alpha-helical antimicrobial peptides. *Antimicrob Agents Chemother*. 2007; 51: 1398–406.
45. Malina, A., and Shai, Y. Conjugation of fatty acids with different lengths modulates the antibacterial and antifungal activity of a cationic biologically inactive peptide. *Biochem J*. 2005; 390: 695–702.
46. Laverty, G., McLaughlin, M., Shaw, C., Gorman, S.P., and Gilmore, B.F. Antimicrobial activity of short, synthetic cationic lipopeptides. *Chem Biol Drug Des*. 2010; 75: 563–9.
47. Kondejewski, L.H., Jelokhani-Niaraki, M., Farmer, S.W., *et al*. Dissociation of antimicrobial and hemolytic activities in cyclic peptide diastereomers by systematic alterations in amphipathicity. *J Biol Chem*. 1999; 274: 13181–92.
48. da Silva, B.R., de Freitas, V.A., Nascimento-Neto, L.G., *et al*. Antimicrobial peptide control of pathogenic microorganisms of the oral cavity: A review of the literature. *Peptides*. 2012; 36: 315–21.
49. Gorr, S.U., and Abdolhosseini, M. Antimicrobial peptides and periodontal disease. *J Clin Periodontol*. 2011; 38 (11): 126–41.
50. Gilmore, K., Chen, P., and Leung, K.P. Anti-microbial peptides for plaque control and beyond. *J Calif Dent Assoc*. 2009; 37: 779–88.
51. Harris, F., Dennison, S.R., and Phoenix, D.A. Anionic antimicrobial peptides from eukaryotic organisms. *Curr Protein Pept Sci*. 2009; 10: 585–606.
52. Straus, S.K., and Hancock, R.E. Mode of action of the new antibiotic for gram-positive pathogens daptomycin: Comparison with cationic antimicrobial peptides and lipopeptides. *Biochim Biophys Acta*. 2006; 1758: 1215–23.
53. De Smet, K., and Contreras, R. Human antimicrobial peptides: Defensins, cathelicidins and histatins. *Biotechnol Lett*. 2005; 27: 1337–47.

54. Kosciuczuk, E.M., Lisowskim, P., Jarczak, J., *et al.* Cathelicidins: Family of antimicrobial peptides. A review. *Mol Biol Rep.* 2012; 39: 10957–70.
55. Vandamme, D., Landuyt, B., Luyten, W., and Schoofs, L. A comprehensive summary of LL-37, the factotum human cathelicidin peptide. *Cell Immunol.* 2012; 280: 22–35.
56. Duplantier, A.J., and van Hoek, M.L. The human cathelicidin antimicrobial peptide LL-37 as a potential treatment for polymicrobial infected wounds. *Front Immunol.* 2013; 4: 143.
57. Overhage, J., Campisano, A., Bains, M., Torfs, E.C., Rehm, B.H., and Hancock, R.E. Human host defense peptide LL-37 prevents bacterial biofilm formation. *Infect Immun.* 2008; 76: 4176–82.
58. Nagant, C., Pitts, B., Nazmi, K., *et al.* Identification of peptides derived from the human antimicrobial peptide LL-37 active against biofilms formed by *Pseudomonas aeruginosa* using a library of truncated fragments. *Antimicrob Agents Chemother.* 2012; 56: 5698–708.
59. Dean, S.N., Bishop, B.M., and van Hoek, M.L. Susceptibility of *Pseudomonas aeruginosa* biofilm to Alpha-Helical Peptides: Denantiomer of LL-37. *Front Microbiol.* 2011; 2: 128.
60. Kai-Larsen, Y., Luthje, P., Chromek, M., *et al.* Uropathogenic *Escherichia coli* modulates immune responses and its curli fimbriae interact with the antimicrobial peptide LL-37. *PLoS Pathog.* 2010; 6: e1001010.
61. Amer, L.S., Bishop, B.M., and van Hoek, M.L. Antimicrobial and antibiofilm activity of cathelicidins and short, synthetic peptides against *Francisella. Biochem Biophys Res Commun.* 2010; 396: 246–51.
62. Hell, E., Giske, C.G., Nelson, A., Romling, U., and Marchini, G. Human cathelicidin peptide LL37 inhibits both attachment capability and biofilm formation of *Staphylococcus epidermidis. Lett Appl Microbiol.* 2010; 50: 211–5.
63. Dean, S.N., Bishop, B.M., and van Hoek, M.L. Natural and synthetic cathelicidin peptides with antimicrobial and anti-biofilm activity against *Staphylococcus aureus. BMC Microbiol.* 2011; 11: 114.
64. Bergsson, G., Reeves, E.P., McNally, P., et al. LL-37 complexation with glycosaminoglycans in cystic fibrosis lungs inhibits antimicrobial activity, which can be restored by hypertonic saline. *J Immunol.* 2009; 183: 543–51.
65. Pompilio, A., Scocchi, M., Pomponio, S., *et al.* Antibacterial and antibiofilm effects of cathelicidin peptides against pathogens isolated from cystic fibrosis patients. *Peptides.* 2011; 32: 1807–14.
66. Pompilio, A., Crocetta, V., Scocchi, M., *et al.* Potential novel therapeutic strategies in cystic fibrosis: Antimicrobial and antibiofilm activity of natural and designed alpha-helical peptides against *Staphylococcus aureus, Pseudomonas aeruginosa*, and *Stenotrophomonas maltophilia. BMC Microbiol.* 2012; 12: 145.
67. Cirioni, O., Giacometti, A., Ghiselli, R., *et al.* Pre-treatment of central venous catheters with the cathelicidin BMAP-28 enhances the efficacy of antistaphylococcal agents in the treatment of experimental catheter-related infection. *Peptides.* 2006; 27: 210410.
68. Mataraci, E., and Dosler, S. *In vitro* activities of antibiotics and antimicrobial cationic peptides alone and in combination against methicillin-resistant *Staphylococcus aureus* biofilms. *Antimicrob Agents Chemother.* 2012; 56: 6366–71.
69. Steinstraesser, L., Tack, B.F., Waring, A.J., *et al* Activity of novispirin G10 against *Pseudomonas aeruginosa in vitro* and in infected burns. *Antimicrob Agents Chemother.* 2002; 46: 1837–44.
70. Jacobsen, F., Mohammadi-Tabrisi, A., Hirsch, T., *et al.* Antimicrobial activity of the recombinant designer host defense peptide Pnovispirin G10 in infected full-thickness wounds of porcine skin. *J Antimicrob Chemother.* 2007; 59: 493–8.
71. Song, Z., Wu, H., Mygind, P., *et al.* Effects of intratracheal administration of novispirin G10 on a rat model of mucoid *Pseudomonas aeruginosa* lung infection. *Antimicrob Agents Chemother.* 2005; 49: 3868–74.
72. Ammons, M.C., and Copie, V. Mini-review: Lactoferrin: A bioinspired, anti-biofilm therapeutic. *Biofouling.* 2013; 29: 443–55.
73. Hwang, I.S., Hwang, J.S., Hwang, J.H., *et al.* Synergistic effect and antibiofilm activity between the antimicrobial peptide coprisin and conventional antibiotics against opportunistic bacteria. *Curr Microbiol.* 2013; 66: 56–60.
74. Minardi, D., Ghiselli, R., Cirioni, O., *et al.* The antimicrobial peptide tachyplesin III coated alone and in combination with intraperitoneal piperacillin-tazobactam prevents ureteral stent *Pseudomonas infection* in a rat subcutaneous pouch model. *Peptides.* 2007; 28: 2293–8.

75. Schillaci, D., Arizza, V., Parrinello, N., *et al.* Antimicrobial and antistaphylococcal biofilm activity from the sea urchin *Paracentrotus lividus*. *J Appl Microbiol.* 2010; 108: 17–24.
76. Cirioni, O., Giacometti, A., Ghiselli, R., *et al.* Citropin 1.1-treated central venous catheters improve the efficacy of hydrophobic antibiotics in the treatment of experimental staphylococcal catheter-related infection. *Peptides.* 2006; 27: 1210–6.
77. Zhang, R., Zhou, M., Wang, L., *et al.* Phylloseptin-1 (PSN-1) from *Phyllomedusa sauvagei* skin secretion: A novel broad-spectrum antimicrobial peptide with antibiofilm activity. *Mol Immunol.* 2010; 47: 2030–7.
78. Dennison, S.R., Morton, L.H., Shorrocks, A.J., Harris,F., and Phoenix, D.A. A study on the interactions of Aurein 2.5 with bacterial membranes. *Colloids Surf B Biointerfaces.* 2009; 68: 225–30.
79. Choi, H., and Lee, D.G. Antimicrobial peptide pleurocidin synergizes with antibiotics through hydroxyl radical formation and membrane damage, and exerts antibiofilm activity. *Biochim Biophys Acta.* 2012; 1820: 1831–8.
80. Schneider, J.J., Unholzer, A., Schaller, M., Schafer-Korting, M., and Korting, H.C. Human defensins. *J Mol Med (Berl).* 2005; 83: 587–95.
81. Zhu, C., Tan, H., Cheng, T., *et al.* Human beta-defensin 3 inhibits antibiotic-resistant *Staphylococcus* biofilm formation. *J Surg Res.* 2013; 183: 204–13.
82. Huang, Q., Yu, H.J., Liu, G.D., *et al.* Comparison of the effects of human beta-defensin 3, vancomycin, and clindamycin on *Staphylococcus aureus* biofilm formation. *Orthopedics.* 2012; 35: e53–60.
83. Cotter, P.D., Ross, R.P., and Hill, C. Bacteriocins–a viable alternative to antibiotics?. *Nat Rev Microbiol.* 2013; 11: 95–105.
84. Dawson, M.J., Scott, R.W. New horizons for host defense peptides and lantibiotics. *Curr Opin Pharmacol.* 2012; 12: 545–50.
85. Michalopoulos, A., and Falagas, M.E. Colistin and polymyxin B in critical care. *Crit Care Clin.* 2008; 24: 377–91, x.
86. Azzopardi, E.A., Boyce, D.E., Thomas, D.W., and Dickson, W.A. Colistin in burn intensive care: Back to the future? *Burns.* 2013; 39: 7–15.
87. Antoniu, S.A., and Cojocaru, I. Inhaled colistin for lower respiratory tract infections. *Expert Opin Drug Deliv.* 2012; 9: 333–42.
88. Herrmann, G., Yang, L., Wu, H., *et al.* Colistin-tobramycin combinations are superior to monotherapy concerning the killing of biofilm *Pseudomonas aeruginosa. J Infect Dis.* 2010; 202: 1585–92.
89. Marumo, K., Komukai, D., Hirose, M., *et al.* Evaluation *in vitro* of the efficacy of colistin methanesulfonate against biofilm-forming multidrug-resistant *Pseudomonas aeruginosa* (MDRP). *J Infect Chemother.* 2013; 19: 348–51.
90. Wu, K., Yau, Y.C., Matukas, L., and Waters, V. Biofilm compared to conventional antimicrobial susceptibility of *Stenotrophomonas maltophilia* Isolates from cystic fibrosis patients. *Antimicrob Agents Chemother.* 2013; 57: 1546–8.
91. Ozbek, B., and Mataraci, E. *In vitro* effectiveness of colistin, tigecycline and levofloxacin alone and combined with clarithromycin and/or heparin as lock solutions against embedded *Acinetobacter baumannii* strains. *J Antimicrob Chemother.* 2013; 68: 827–30.
92. Volkow-Fernandez, P., Rodriguez, C.F., and Cornejo-Juarez, P. Intravesical colistin irrigation to treat multidrug-resistant *Acinetobacter baumannii* urinary tract infection: A case report. *J Med Case Rep.* 2012; 6: 426.
93. Falagas, M.E., and Kasiakou, S.K. Colistin: The revival of polymyxins for the management of multidrug-resistant gram-negative bacterial infections. *Clin Infect Dis.* 2005; 40: 1333–41.
94. Hinsa, S.M., Espinosa-Urgel, M., Ramos, J.L., and O'Toole, G.A. Transition from reversible to irreversible attachment during biofilm formation by *Pseudomonas fluorescens* WCS365 requires an ABC transporter and a large, secreted protein. *Mol Microbiol.* 2003; 49: 905–18.
95. Zhu, X., Liu, W., Lametsch, R., *et al.* Phenotypic, proteomic, and genomic characterization of a putative ABC-transporter permease involved in *Listeria monocytogenes* biofilm formation. *Foodborne Pathog Dis.* 2011; 8: 495–501.
96. Hou, S., Zhou, C., Liu, Z., *et al.* Antimicrobial dendrimer active against *Escherichia coli* biofilms. *Bioorg Med Chem Lett.* 2009; 19: 5478–81.

97. Hou, S., Liu, Z., Young, A.W., Mark, S.L., Kallenbach, N.R., and Ren, D. Effects of Trp–and Arg-containing antimicrobial-peptide structure on inhibition of *Escherichia coli* planktonic growth and biofilm formation. *Appl Environ Microbiol*. 2010; 76: 1967–74.
98. Brandenburg, L., Merres, J., Albrecht, L., Varoga, D., and Pufe, T. Antimicrobial peptides: Multifunctional drugs for different applications. *Polymers*. 2012, 4, 539e60.
99. Cotter, P.D., Ross, R.P., and Hill, C. Bacteriocins–a viable alternative to antibiotics?. *Nat Rev Microbiol*. 2013, 11, 95–105.

10 Enzymatic Dispersion of Biofilms

Applications, Challenges and Quantification

Anushree R. Chaphalkar

10.1 INTRODUCTION

Bacterial and fungal species display the ability to exist in a complex matrix of self-secreted polymers, called a 'biofilm'. Cells within biofilms demonstrate a distinct and altered phenotype as a consequence of a difference in growth rates and gene transcription (Mikkelsen et al., 2007). Such a multicellular mode of survival confers greater resistance to the constituent cells against various antimicrobial agents, as compared to free-living or planktonic bacteria (Rabin et al., 2015; Mah and O'Toole, 2001). This is particularly because a majority of the mass of the biofilm is formed by extracellular polymeric substance (EPS) which not only provides nutrients to the sessile cells, but also acts as a coat and scaffold, mediating the attachment of the cells to the substrate (Costerton, 1999), amongst its other functions. The anionic and cationic molecules present in the EPS bind to antibiotics and other microbial agents, hindering their contact with the cells within. The EPS also facilitates the absorption of extracellular DNA (eDNA) and promotes horizontal gene transfer and DNA communication (Madsen et al., 2012), aiding the development of antimicrobial resistance.

Biofilm dispersal is a natural and essential event in the lifecycle of a biofilm, regulated by several internal and environmental cues such as the increase or decrease in nutrition availability, oxygen depletion, c-di-GMP, as well as quorum sensing (Karatan and Watnick, 2009). It aids the transmission of the microbes within the film to the environment and in case of pathogens, to potential hosts and cross-hosts. Inside the host, the dispersion of a biofilm formed by pathogenic microbes allows the rapid conquest of new substrates within the body, aiding the spread of infection. Accelerated research on biofilm dispersion has opened new avenues for the development of therapeutic strategies against biofilms. In particular, the use of dispersal-causing agents for the inhibition of biofilm formation and the detachment of pre-formed biofilms is proving to be a promising area, especially in industries and healthcare settings.

Unwanted biofilms in industries can foul cooling towers (Di Pippo et al., 2018), pipelines and food-processing plants (Jass and Walker, 2000). Depending on the system, several physical and chemical strategies can be applied for the removal of unwanted biofilms in industries, including mechanical removal, the use of detergents and antimicrobial agents (Stewart et al., 2000), starvation by sequestering essential nutrients, prevention of biofilm formation and promoting the detachment of cells from existing biofilms. Among these, mechanical cleaning and the use of antimicrobial agents remain popular. Mechanical cleaning has limitations; it is possible only in areas within physical reach in the equipment and is especially unsuitable for medical devices.

Evidence suggests that several chronic bacterial infections are associated with the formation of biofilms within the host and those with the ability to resist the host's immune response (Patel et al., 2014). These films expand their presence to medical devices and implants, rendering them

DOI: 10.1201/9781003281238-10

ineffective. The effectiveness of traditional therapeutic options such as antibiotics is limited by the development of resistance, much to the growing concern of healthcare practitioners across the globe. Further, the use of antibiotics may result in a surface covered by an inactive biofilm. Depriving a biofilm of essential nutrients is a useful strategy (Thormann et al., 2005) but may not always be feasible.

Inducing biofilm-removal by leveraging enzymes and small molecules that are involved in the natural dispersal of biofilms is an attractive option (Xavier et al., 2005). Such an approach reduces the reliance on antimicrobial agents, thereby promoting sustainable chemistry. These agents are also milder alternatives in acting against biofilms formed on medical devices and implants inside the body.

The mechanisms of biofilm dispersion can be classified as active or passive. While active mechanisms are initiated by bacterial cells themselves, passive dispersion of a biofilm often results from external physical forces such as shear, abrasion and forces of biotic origin, such as predator grazing, inter-species interactions within a mixed biofilm, and human intervention. When dispersal happens such that single cells are continuously released through the formation of the biofilm, the process is called 'erosion'. If larger portions of the film come off towards the end of its lifecycle, the process is called 'sloughing'. Both erosion and sloughing may be active or passive. However, there's yet another mode of dispersal, reported to only be active in nature, called 'seeding dispersal' or central hollowing (Kaplan, 2010).

Active dispersal is carried out largely by enzymes, peptides and other small molecules released by indwelling cells in response to internal and external cues. In this chapter, we dwell on the role of enzymes in the degradation of biofilms and their applications in healthcare and industry.

10.2 BIOFILM-DEGRADING ENZYMES

The EPS is comprised of various biopolymers such as polysaccharides, proteins, glycoproteins, glycolipids and eDNA (Flemming et al., 2007). Given the benefits it provides to the indwelling cells, the components of the EPS are great potential targets for biofilm dispersion and anti-biofilm therapies. In the following sections, we discuss the various enzymes that target specific biopolymers within the EPS matrix.

10.2.1 Proteases

Proteases are enzymes of prokaryotic or eukaryotic origin that degrade proteins to form amino acids. Apart from contributing significantly to the EPS dry mass, proteins mediate DNA-binding and bacterial adhesion and are important to maintain the stability of biofilm architecture, making them an attractive target for enzymatic dispersion (Lister and Horswill, 2014). Several biofilm-dispersing proteases have been identified to date. Esp, a serine protease produced by *Staphylococcus*

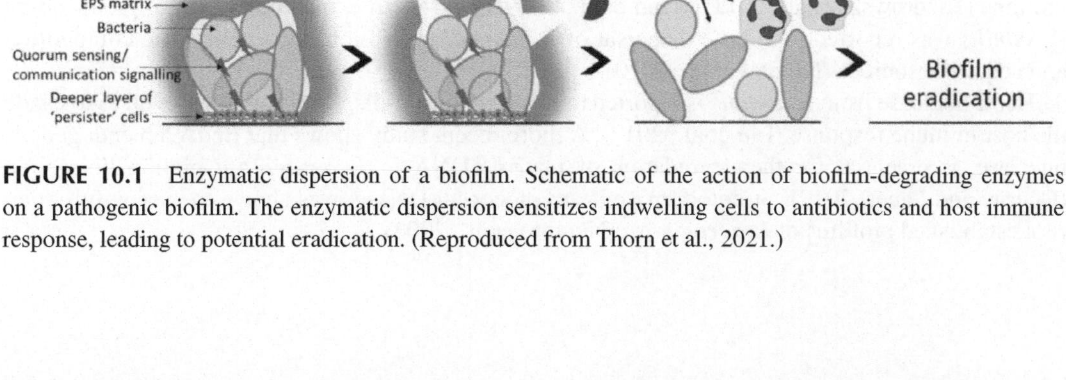

FIGURE 10.1 Enzymatic dispersion of a biofilm. Schematic of the action of biofilm-degrading enzymes on a pathogenic biofilm. The enzymatic dispersion sensitizes indwelling cells to antibiotics and host immune response, leading to potential eradication. (Reproduced from Thorn et al., 2021.)

epidermidis, a human commensal, which is shown to inhibit the formation of *Staphylococcus aureus* biofilms as well as degrade existing biofilms (Iwase et al., 2010). LasB elastase is a protease from *Pseudomonas aeruginosa* that induces the expression of endogenous protease genes in *Staphylococcus aureus* cells in a biofilm, accelerating its dispersal (Park et al., 2012). Yet another enzyme derived from the fungus *Tritirachium album* called Proteinase-K shows anti-biofilm activity in vitro and in vivo (Lauderdale et al., 2010; Nguyen and Burrows, 2014). It mediates dispersal by degrading biofilm-associated protein (Bap) in *bap*-positive *S. aureus* species, an effect that was enhanced when the enzyme was used in combination with antibiotics (Kumar Shukla and Rao, 2013). Similarly, bap-positive *S. aureus* biofilms were also inhibited by degradation of Bap by Aur and SspA, two extracellular proteases from *S. aureus* itself (Martí et al., 2010). By modifying a biofilm matrix component and bacterial adhesion protein LapA, a cysteine protease LapG is shown to disperse *Pseudomonas putida* biofilms (Gjermansen et al., 2010). *S. aureus*-origin extracellular cysteine proteases ScpA and SspB (called Staphopains) were found to inhibit its own biofilm formation and disperse a pre-existing biofilm (Mootz et al., 2013).

An enzymatic cocktail of pancreatic protease type I (PtI) and *B. subtilis*-derived alpha amylase exhibited a strong inhibition effect and moderate biofilm prevention effect on biofilms of *E. coli*, *S. aureus* and methicillin-resistant *S. aureus* (MRSA) in vitro (Jee et al., 2020). Papain, an endolytic cysteine protease from *Carica papaya* is also shown to exhibit anti-biofilm properties in soluble as well as immobilised forms (Baidamshina et al., 2021; Trizna et al., 2016).

A metalloprotease peptidase M16 secreted by *Microbacterium sp.* could efficiently degrade biofilms and sensitize them to antibiotics, at concentrations lower than trypsin, papain and alpha-amylase (Saggu et al., 2019). Two proteases, subtilisin (from *B. subtilis*) and alpha-chymotrypsin when immobilized on polypropylene surfaces, showed an increased efficacy in reducing the adhesion and increasing the dispersion of biofilm-forming *Candida albicans* (a yeast) to these surfaces (Andreani et al., 2017).

Lysostaphin is an interesting endopeptidase that not only degrades *S. aureus* biofilms efficiently, but also displays antimicrobial activity as it cleaves the pentaglycine crossbridges found in the peptidoglycan of *S. aureus* (Wu et al., 2003).

10.2.2 Nucleases

As eDNA forms the structural and binding component of biofilms, nucleases can be used to inhibit biofilm formation, degrade the EPS matrix, as well as to improve the permeability of recalcitrant antibiotic resistant biofilms (Sharma et al., 2019). Numerous studies have demonstrated the utility of DNase I in the dispersion of biofilms and sensitizing them to antimicrobial treatment in *S. aureus* and *P. aeruginosa*, *E. coli*, *Acinetobacter baumannii*, *Haemophilus influenzae* and *K. pneumoniae* (Okshevsky et al., 2015; Tetz et al., 2009; Fleming and Rumbaugh, 2017). However, DNase I is found to be less effective on mixed-species biofilms. The efficiency of DNase in reducing and degrading *P. aeruginosa* biofilms was shown to increase significantly with the concomitant use of magnesium ions (Sharma and Pagedar Singh, 2018). Nuc2, an extracellular nuclease, was shown to inhibit the formation of *Staphylococcus aureus* biofilms in vitro and was also found to partially disperse young biofilms (Kiedrowski et al., 2011; Mann et al., 2009). Nuc B, an extracellular DNase produced by *B. subtilis* was reported to aid the dispersal of *B. licheniformis* biofilms, to eliminate competition and acquire resources (Nijland et al., 2010).

DeoC nuclease from *S. mutans* is reported to possess biofilm-dispersal activity as well as to evade the host immune response (Liu et al., 2017). A more recent study shows that EndA, an endogenous nuclease, is essential for the degradation of matrix eDNA in *P. aeruginosa* biofilm dispersion (Cherny and Sauer, 2019). Streptodornase, a streptococcal DNase, was shown to successfully disrupt established biofilms of *P. aeruginosa* (Nemoto et al., 2003).

Although the enzymatic degradation of eDNA is an effective anti-biofilm strategy in vitro as well as in vivo, large-scale application of these enzymes requires cost-effective methods of production. One way of achieving this could be the replacement of mammalian nucleases with those expressed in bacteria (Okshevsky et al., 2015). The production of NucB was shown to increase fivefold by the addition of manganese ions that trigger sporulation. Additionally, the Plackett and Burman design was applied for optimisation, leading to a 10-fold increase. Phosphate was found to inhibit NucB production (Rajarajan et al., 2013).

10.2.3 Glycoside Hydrolases

The establishment and persistence of biofilms is highly dependent on the presence of exopolysaccharides secreted by indwelling cells. These polymers help the bacterial cells by scaffolding, aiding the evasion of host immune response, adsorbing antimicrobial compounds, offering protection against desiccation and serving as nutrients during starvation (Flemming and Wingender, 2010). Glycoside hydrolases are therefore effective targets for the disruption of biofilms and prevention of biofilm-formation (Fleming and Rumbaugh, 2017).

Dispersin B is the most well-studied glycoside hydrolase that cleaves β(1,6) linkages to degrade poly(1,6)-*N*-acetyl-d-glucosamine (PNAG) which is a common component in the biofilms formed by *S. aureus*, *A. actinomycetemcomitans*, *S. epidermidis*, *A. baumannii*, *K. pneumoniae*, *E. coli*, *Burkholderia* spp., *A. pleuropneumoniae*, *Yersinia pestis* and *Pseudomonas fluorescens* (Itoh et al., 2005).

Alginate lyase targets alginate, a key EPS component, to bring about the degradation and antibiotic-sensitization of mucoid *P. aeruginosa* biofilms (Lamppa and Griswold, 2013). The thermostability and anti-biofilm activity of this enzyme was improved on immobilization using low molecular weight chitosan nanoparticles (Li et al., 2019). Alpha-amylase is an α(1,4) glycosidase of bacterial and mammalian origin that helps disperse mature biofilms of several pathogens including *S. aureus*, *V. cholerae* and *P. aeruginosa* (Kalpana et al., 2012; Craigen et al., 2011), including those in liquid suspension. Amylase is particularly interesting as it displays the anti-biofilm characteristics in addition to those of an inducer (Lahiri et al., 2021). *P. aeruginosa* biofilms have also been successfully degraded by glycosyl hydrolases such as $PelA_h$ and $PslG_h$ (Baker et al., 2016) and alpha-mannosidase and beta-mannosidase (Banar et al., 2016) and nanomolar concentrations of PsIG (Yu et al., 2015). Hyaluronidase has been utilized in vivo against hyaluronic acid-containing biofilms of *S. aureus* and *Streptococcus intermedius* (Ibberson et al., 2016; Pecharki et al., 2008). Cellulase, a β(1,4) glycoside hydrolase breaks down *S. aureus* and *P. aeruginosa* biofilms (Fleming et al., 2017). Biofilms of multidrug-resistant clinical isolates of *P. aeruginosa* were treated with Ahl-1 lactonase originating from *Bacillus weihenstephanensis* and expressed in *E. coli*. The enzyme displayed quorum-quenching abilities and also inhibited the swarming motility and biofilm formation (Sakr et al., 2018). In *Candida albicans* biofilms, β-1,3-glucanase, which can degrade poly-β(1→3)-glucose of these biofilms proved to be an effective anti-biofilm agent that also enhanced antifungal susceptibility (Tan et al., 2018). It was also found to be effective against mixed *Candida species* biofilms (Tan et al., 2017). Polyurethane-immobilized acylase from *Aspergillus melleus* was shown to catalyze the hydrolysis of the quorum-sensing molecules *N*-butyryl-L-homoserine lactone (C4-LHL), *N*-hexanoyl-L-homoserine lactone (C6-LHL), and *N*-(3-oxododecanoyl)-L-homoserine lactone (3-oxo-C12-LHL). This resulted in about 60% reduction in biofilm formation by *Pseudomonas aeruginosa* (Grover et al., 2016). Lysozyme, also called muramidase or N-acetyl-muramic hydrolase, is a component of human body fluids and is known to degrade the peptidoglycan in the cell walls of Gram-positive bacteria, thereby possessing antibacterial properties. In an in vitro study, the effect of lysozyme on biofilms formed by *Staphylococcus aureus* (methicillin sensitive and resistant), *Streptococcus pyogenes*, *Pseudomonas aeruginosa*, and *Gardnerella vaginalis* was assessed. The enzyme could affect the biofilm forming capacity of all the four species in a concentration-specific

TABLE 10.1
Common Bacteriophage-origin Enzymes with Anti-biofilm Activity Are Described

Lysin	Biofilm	Description	Reference
Phi 11	*S. aureus*	Destabilised the biofilm and caused cell lysis	(Sass and Bierbaum, 2007)
LysH5	*S. aureus and S. epidermidis*	Reduced the bacterial population and persister cells in the film; completely eliminated some strains	(Gutiérrez et al., 2014)
$CHAP_K$	*S. aureus*	Peptidase; complete removal of biofilm; prevention of formation	(Fenton et al., 2013)
SAL-1	*S. aureus*	Cell wall degrading enzyme; rapid anti-biofilm activity	(Jun et al., 2013)
SAL-2	*S. aureus*	Cell wall degrading enzyme; broad spectrum anti-biofilm activity	(Son et al., 2010)
PlyC	*S. pyogenes*	In vitro activity against static and dynamic biofilms	(Shen et al., 2013)
Cpl-711	*S. pneumoniae*	Reduced biofilm formation	(Díez-Martínez et al., 2015)
LytA	*S. pneumoniae*	Highly efficient *N*-acetylmuramoyl-L-alanine amidase	(Domenech et al., 2011)

manner (Hukić et al., 2018). Glucose oxidase was tested in combination with lactoperoxidase against model biofilms of *Staphylococcus aureus*, *Staphylococcus epidermidis*, *Pseudomonas fluorescens*, and *Pseudomonas aeruginosa* (Johansen et al., 1997). While a bactericidal activity was noted against indwelling bacteria, the enzymes together could not remove the biofilm from the substrata, corroborated by another study (Cooper, 2013). When oxidoreductases were combined with glycoside hydrolases, complete removal of the biofilm as well as bactericidal activity was observed.

10.2.4 Phage-origin Endolysins

In the last few decades, bacteriophage endolysins (phage lysins), namely, peptidoglycan hydrolases have gained prominence as anti-biofilm agents owing to their (i) antibacterial activity against Gram-positive bacteria, (ii) rapid and efficient lytic activity, (iii) low probability of inducing bacterial resistance and (iv) specificity of action that safeguards commensal bacteria (Fischetti, 2005; Loessner, 2005). These enzymes are naturally produced by phages towards the end of their life cycle to enable their release from within the host bacterial cell. The use of these enzymes is a safer alternative in comparison to phage therapy for the destruction of biofilms in vivo. Some of these lysins are given in Table 10.1.

10.3 THERAPEUTIC USE OF BIOFILM-DISRUPTING ENZYMES

About 80% of chronic and recurrent bacterial infections in humans, are caused and exacerbated by biofilms (Bjarnsholt, 2013). These include otitis media, acne, dental plaque, endocarditis, cystic fibrosis and so on. It is also established that biofilms contribute extensively to infections in medical devices such contact lenses, catheters, cardiac valves, endotracheal tubes, as well as orthopaedic and breast implants. With limited access to infected regions within the body, conventional methods of biofilm removal cannot be applied in therapy. Antimicrobial therapy continues to be the first choice of treatment, but the alarming rise of antimicrobial resistance and the added complication of resistance conferred by the complex biofilm architecture to the indwelling cells, calls for new therapies that can individually or synergistically help eliminate biofilms in vivo. **Table 10.2** lists some of the enzymes applied in clinical settings for treatment of biofilm infections.

TABLE 10.2
Anti-biofilm Enzymes in Healthcare Applications

Organism	Biofilm infection consequences	Major anti-biofilm enzyme
Enterococcus faecalis	Urinary tract infection, periodontitis	DNase I
Streptococcus pyogenes	Impetigo	DNase I, PlyC
Streptococcus mutans	Dental plaque and caries	DNase I, SPRE, Lysozyme, Proteinase K
Staphylococcus aureus	Multidrug-resistant nosocomial infections	DNase I, Proteases, Alpha-amylase
Bacillus subtilis	Endocarditis, septicemia, pneumonia	NucB, Alpha-amylase
Acinetobacter baumannii	Ventilator-associated pneumonia and catheter-related infection	DNase I, Papain
Escherichia coli	Recurrent urinary tract infections	DNase I, Alpha-amylase, Dispersin B
Klebsiella pneumoniae	Gastrointestinal, respiratory and urinary tract infections	DNase I, Dispersin B
Pseudomonas aeruginosa	Cystic fibrosis, chromin obstructive pulmonary disorder, chronic wounds, burn wounds, implant infection	DNase I, Alginate lyase
Candida albicans	Vaginosis	DNase I, Glucanase

Although remarkable research is being conducted on the use of enzymes and other small molecule inhibitors against pathogenic biofilms, there remains a dearth of in vivo studies and clinical trials proving the practical applicability of these enzymes in clinical settings. Most of these studies are carried out in vitro and on monospecies biofilms, the findings from which are hard to extrapolate to mixed, multi-species biofilms that tend to exist in a biotic environment. Potential toxicity and cross-interactions of enzymes like proteases within the host may also be factors restricting their therapeutic use (Fleming and Rumbaugh, 2017). In addition, although certain enzymes are shown to effectively degrade and sensitize bacterial biofilms to antimicrobial therapies and the host's immune system, dispersal may not be the ideal strategy for fungal biofilms. The hyphal nature of fungal biofilms necessitates the containment of these infections instead of their dispersal for better clinical outcomes (Uppuluri and Lopez-Ribot, 2016). In some species, evidence also suggests increased virulence in dispersed bacterial cells as compared to normal planktonic cells (Chua et al., 2014).

The commercial viability of enzymes in clinical practice is yet another challenge as the current modes of anti-biofilm enzyme production are far from cost-effective. Limitations to the clinical use of enzymes are also imposed by their immunogenic potential, limited shelf life and stability, sensitivity to structures changes, pH and temperature, lower bioavailability via oral consumption and susceptibility to proteases in the digestive tract. For anti-biofilm enzymes without antimicrobial activity, the drug delivery system enzymes need to be delivered concomitantly with synergistic antimicrobials to eliminate biofilms in vivo. As biofilm-associated infections are often site-specific and localized in early stages, systemic administration of enzymes is suboptimal. This calls for a need for the development of smart formulations for drug delivery to ensure targeted administration to the site of infection, minimal immunogenicity and unaltered activity of the enzyme (Thorn et al., 2021).

10.4 INDUSTRIAL APPLICATIONS OF BIOFILM-DEGRADING ENZYMES

Studies investigating the structure, physiology and methods of control of biofilms are important especially in the food industry to ensure food safety and minimize operational breaks in food processing. Biofilm-degrading enzymes are a powerful tool for biofilm-control in the food industry given their low toxicity and biodegradability. Pectin methylesterase is an enzyme capable of reducing biofilm formation in bioreactors (Torres et al., 2011) that could be used as a pre-treatment for the various machines and pipes across industry. *E. coli* biofilms from stainless steel surfaces in a dairy plant were successfully eliminated with the use of a mixed cocktail of proteases (Oulahal-Lagsir et al., 2003). An effective alternative to harsh chemical cleaning agents could be such mixtures of enzymes targeting multiple biomolecules in the biofilm matrix, along with detergents, dispersing and chelating agents. Phage endolysins and bacterial origin peptidases are promising alternatives to the use of conventional agents on food-contact surfaces (Nahar et al., 2021; Cha et al., 2019).

Immobilizing enzymes on compatible polymers anchors them and prevents their diffusion into food. Papain immobilized on linear low density polyethylene is shown to significantly reduce biofilms formed by *Acinetobacter spp*., a frequent contaminant of frozen foods and pasteurized milk (Manohar et al., 2015).

10.5 QUANTITATIVE ANALYSIS OF BIOFILMS

The complexity and dynamics of biofilm structure make the measurements of critical biofilm properties challenging. To assess the effect of various antimicrobial agents and macromolecules on the growth and dispersal of biofilms, it is important to rapidly and objectively quantify biofilms.

Several direct and indirect methods exist for biofilm quantitation, that often necessitate the homogenization of the film prior to analysis (Wilson et al., 2017). There are pros and cons to each method, and this necessitates the standardization of protocols and the training of research personnel to enable them to choose and implement appropriate methods for the characterization and analysis of biofilms.

Indirect methods use a proxy marker such as dry mass, total organic carbon, protein, nucleic acids, carbohydrates and metabolites, that helps draw an inference about biofilm quantity. This is assuming that the quantity of the marker is proportional to the quantity of the biofilm and is consistent across indwelling cells. On the other hand, direct methods allow the quantitation of cells that can be cultured. Examples of direct methods are cell counts via colony forming unit or a CFU-based plate count, Coulter count, flow cytometry, photography and microscopy.

10.5.1 Analysing Biofilm Dynamics by Microscopy and Image Processing

Direct imaging via microscopy helps in the detection of microbes and the characterization of the structure, surface attachment, volume, area as well as homogeneity of the biofilm. Light microscopy, scanning electron microscopy, confocal laser scanning microscopy, bioluminescence and autofluorescence imaging have been employed to date to visualize biofilm structure and dynamics. Microscopy is especially useful when combined with automated image analysis for the objective quantitation of biofilm parameters. Automation of image analysis helps not only to reduce manual intervention but also provides unbiased, reproducible measurements of the morphology, quantity and behaviour of the cells in terms of motion, concentration and interaction with the surface or other cells.

Light microscopes make use of light in the visible range (wavelength 400–700 nm) and a combination of lenses to magnify micron-sized objects. In very young biofilms, cell counting can be possible by suspending homogenized biofilms in chamber counting slides, with or without staining of cells, and 2D imaging via light microscopes (Figure 10.2A, B). Traditional label-free microscopy techniques such as differential interference contrast (DIC) (Figure 10.3, *top left*) and phase-contrast allow the observation of cells and large subcellular structures in their native states with minimal exposure to light, thereby preventing phototoxicity. Automated 2D and 3D particle and cell counting and tracking algorithms developed for the analysis of label-free images (Chaphalkar et al., 2021) could be used for biofilm analysis. CMEIAS, short for 'Center for Microbial Ecology Image Analysis System' was developed as a computer program to analyse the morphological diversity in growing microbial communities using images from phase-contrast microscopy (Liu et al., 2001).

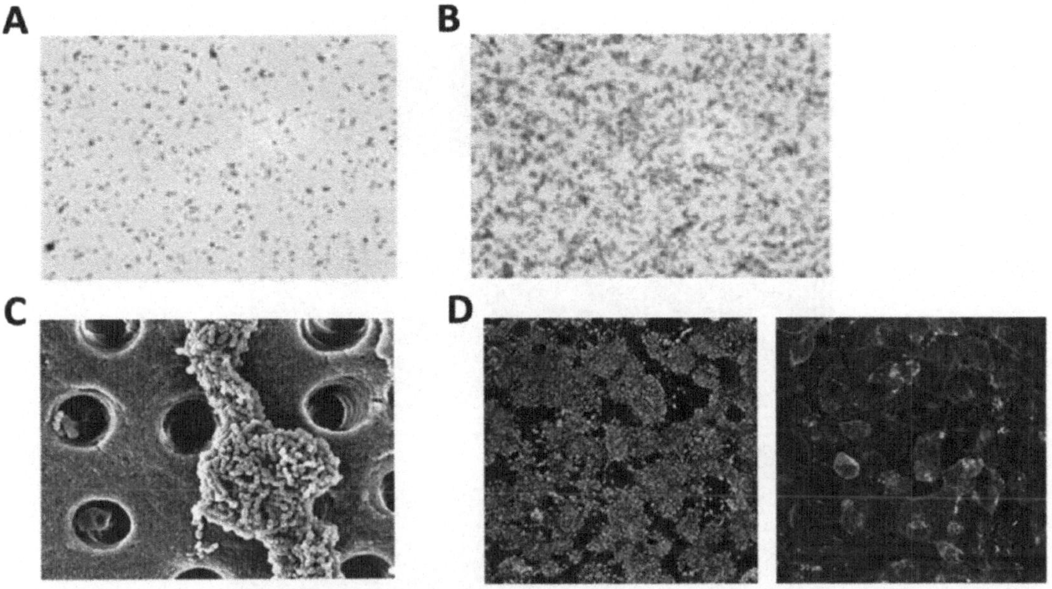

FIGURE 10.2 Microscopy of biofilm growth and dispersal. Crystal violet-stained cells of light P. aeruginosa cells imaged by a light microscope (magnification 5000X) at **(A)** 1 hour and **(B)** 3 hours. Adapted from Wilson et al., 2017. **(C)** SEM image of a section of dentin treated with NaOCl and a cleanser, showing the reduced presence of a bacterial biofilm. Magnification 5000X. Adapted from Clegg et al., 2006. **(D)** *Left*: A 48-hour old biofilm of *A. pleuropneumoniae* grown on SJPL cells in vitro. *Right:* The biofilm was treated with anti-biofilm enzyme dispersin B. (Adapted from Tremblay et al., 2013.)

Fluorescence microscopy makes use of fluorophores or dyes that absorb light of a particular wavelength, reach their excited state and subsequently emit light of a longer wavelength. These fluorophores are used to 'label' the molecules of interest either by expressing them within cells or by physically introducing them into the system. Over the past two decades, the emergence of a plethora of compatible fluorophores, high resolution microscopy and robust computer programs for quantitative analysis of data has popularized the use of fluorescence microscopy for *in vitro* as well as *in vivo* imaging. This technique was utilized to study the localization of a matrix component of *Pseudomonas aeruginosa* biofilm, PsI exopolysaccharide (Ma et al., 2009) (Figure 10.3, *top panel*), with and without cellulase treatment.

A statistical-modelling based segmentation and analysis tool named 'BiofilmQuant' was developed to analyse dental biofilms using quantitative light-induced fluorescence (QLF) images (Mansoor et al., 2014). The use of fluorescent dyes in monitoring dental biofilms may be of aesthetic concern. To address this, autofluorescence imaging of biofilms is also reported to be used in the quantitation of dental biofilms of *Porphyromonas gingivalis* on dental appliances (Garcez et al., 2020).

Scanning electron microscopy (SEM) involves the use of a focused beam of electrons that raster scans the fixed sample to generate an image that provides information about the topography of the surface of the object with resolution ~1 nm. Clegg et al. imaged dentin surfaces with biofilms pre- and post-treatment with anti-biofilm agents (Figure 10.2C) using SEM. A machine-learning based tool for segmentation (detection) and analysis of biofilms from SEM images was developed to investigate dental biofilm disruption (Vyas et al., 2016).

Confocal microscopes provide greater axial and lateral resolution as compared to typical epifluorescence microscopes as they illuminate the samples with focussed laser beams of light and

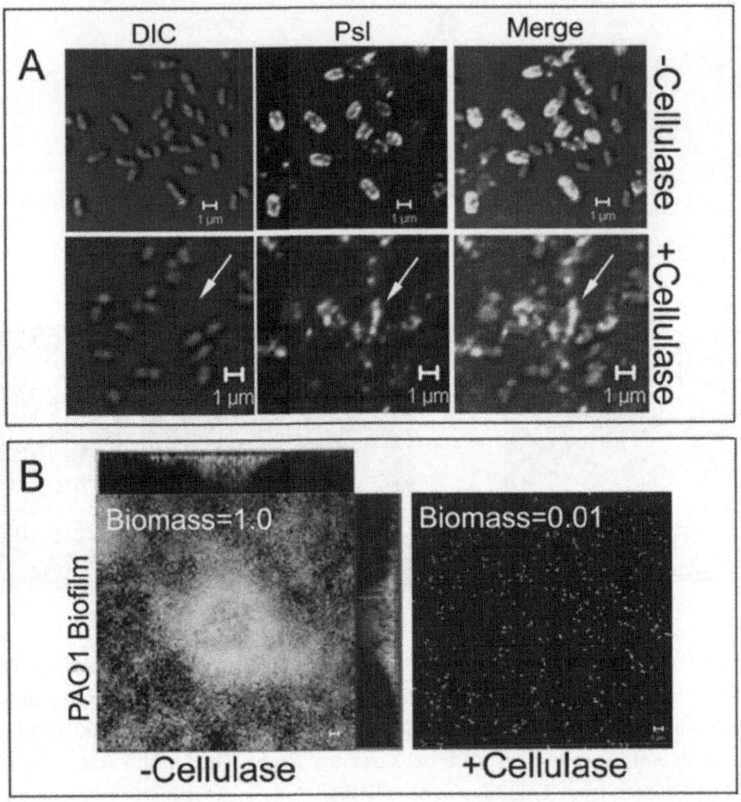

FIGURE 10.3 The effect of cellulase treatment on Psl localization and *P. aeruginosa* biofilm formation. **Top panel:** HHA-FITC (green) staining of surface-attached WFPA801 cells with or without cellulase treatment. **Bottom panel:** The biofilms of GFP-tagged PAO1 grown in flow cells in the presence or absence of cellulase. The total biomass was quantified by image analysis. The arrows reveal Psl that has dissociated from the bacterial surface and adhered with the substratum. Scale bars, 1 µm for panel A and 5 µm for panel B. (Reproduced from Ma et al., 2009.)

further eliminate out-of-focus light by incorporating a pinhole aperture at the detector (Pawley, 1995). The 3D structure of biofilms, cell counts and matrix distribution can be analysed by confocal laser scanning microscopy (CLSM) (Reichhardt and Parsek, 2019). Tremblay et al. imaged via confocal laser scanning microscopy, a 48-hour old biofilm of *A. pleuropneumoniae* grown on SJPL cells (Figure 10.3 D, *left*) and treated it with dispersin B, a matrix-degrading enzyme (Figure 10.2 D, *right*). CLSM is advantageous as it allows the time-lapse monitoring of a live biofilm in a non-destructive manner.

Multiple image analysis tools implemented in MATLAB and ImageJ have been developed to estimate the biomass, surface coverage, surface area, thickness distribution, roughness coefficient, diffusion distance, fractal dimension or porosity of biofilms from CLSM images (Mueller et al., 2006; Daims et al., 2006; Beyenal et al., 2004). COMSTAT, a popular ImageJ based computer program (Heydorn et al., 2000) was developed to analyse image stacks of 3D confocal microscopy to investigate the mono-species biofilms of four *Pseudomonas* species. The structural and spatial analysis based on intensity of the fluorophore could help measure mean thickness, roughness, substratum coverage and surface to volume ratio of each biofilm. The same software was used to measure the total biomass of *P. aeruginosa* biofilms in flow cells using confocal microscopy images (Ma et al., 2009) (Figure 10.3).

More recently, BIAM, short for 'biofilm intensity and architecture measurement', was developed as an ImageJ based open software to supplement CLSM in investigating the spatial dynamics and structure of *P. aeruginosa* biofilms during growth, differentiation and dispersal (Baudin et al., 2017). The latest addition to this set of tools for the analysis of microbial communities is BiofilmQ, an image cytometry software for the automated and high-throughput quantification, analysis and visualization of numerous internal and whole properties of biofilms in 3D space and time (Hartmann et al., 2021).

10.6 CONCLUSION

Increasing antimicrobial resistance emphasizes the need for identifying sustainable, mild and compatible molecules and enzymes that can aid conventional antimicrobial compounds to eliminate biofilms. Direct, rapid and objective quantitation of growth and dispersal of biofilms using microscopy image analysis and other computational methods (An et al., 2021) could help accelerate the identification and assessment of novel therapeutic candidates that can be employed in clinical and industry settings.

REFERENCES

An, A.Y., Choi, K.Y.G., Baghela, A.S., and Hancock, R.E.W. 2021. An overview of biological and computational methods for designing mechanism-informed anti-biofilm Agents. *Front. Microbiol.* 12.

Andreani, E.S., Villa, F., Cappitelli, F., Krasowska, A., Biniarz, P., Łukaszewicz, M., and Secundo, F. 2017. Coating polypropylene surfaces with protease weakens the adhesion and increases the dispersion of Candida albicans cells. *Biotechnol. Lett.* 39:423–428.

Baidamshina, D.R., Koroleva, V.A., Olshannikova, S.S., Trizna, E.Y., Bogachev, M.I., Artyukhov, V.G., Holyavka, M.G., and Kayumov, A.R. 2021. Biochemical properties and anti-biofilm activity of chitosan-immobilized papain. *Mar. Drugs.* 19.

Baker, P., Hill, P.J., Snarr, B.D., Alnabelseya, N., Pestrak, M.J., Lee, M.J., Jennings, L.K., Tam, J., Melnyk, R.A., Parsek, M.R., Sheppard, D.C., Wozniak, D.J., and Howell, P.L. 2016. Exopolysaccharide biosynthetic glycoside hydrolases can be utilized to disrupt and prevent Pseudomonas aeruginosa biofilms. *Sci. Adv.* 2:e1501632.

Banar, M., Emaneini, M., Satarzadeh, M., Abdellahi, N., Beigverdi, R., van Leeuwen, W.B., and Jabalameli, F. 2016. Evaluation of mannosidase and trypsin enzymes effects on biofilm production of Pseudomonas aeruginosa isolated from burn wound infections. *PLoS One.* 11:e0164622.

Baudin, M., Cinquin, B., Sclavi, B., Pareau, D., and Lopes, F. 2017. Understanding the fundamental mechanisms of biofilms development and dispersal: BIAM (Biofilm Intensity and Architecture Measurement), a new tool for studying biofilms as a function of their architecture and fluorescence intensity. *J. Microbiol. Methods.* 140:47–57.

Beyenal, H., Donovan, C., Lewandowski, Z., and Harkin, G. 2004. Three-dimensional biofilm structure quantification. *J. Microbiol. Methods.* 59:395–413.

Bjarnsholt, T. 2013. The role of bacterial biofilms in chronic infections. *APMIS. Suppl.* 1–51.

Cha, Y., Son, B., and Ryu, S. 2019. Effective removal of staphylococcal biofilms on various food contact surfaces by Staphylococcus aureus phage endolysin LysCSA13. *Food Microbiol.* 84:103245.

Chaphalkar, A.R., Jawale, Y.K., Khatri, D., and Athale, C.A. 2021. Quantifying intracellular particle flows by DIC object tracking. *Biophys. J.* 120:393–401.

Cherny, K.E., and Sauer, K. 2019. Pseudomonas aeruginosa requires the DNA-specific endonuclease EndA to degrade extracellular genomic DNA to disperse from the biofilm. *J. Bacteriol.* 201.

Chua, S.L., Liu, Y., Yam, J.K.H., Chen, Y., Vejborg, R.M., Tan, B.G.C., Kjelleberg, S., Tolker-Nielsen, T., Givskov, M., and Yang, L. 2014. Dispersed cells represent a distinct stage in the transition from bacterial biofilm to planktonic lifestyles. *Nat. Commun.* 5:1–12.

Clegg, M.S., Vertucci, F.J., Walker, C., Belanger, M., and Britto, L.R. 2006. The effect of exposure to irrigant solutions on apical dentin biofilms in vitro. *J. Endod.* 32:434–437.

Cooper, R.A. 2013. Inhibition of biofilms by glucose oxidase, lactoperoxidase and guaiacol: The active antibacterial component in an enzyme alginogel. *Int. Wound J.* 10:630–637.

Costerton, J.W. 1999. Introduction to biofilm. *Int. J. Antimicrob. Agents*. 11:217–221.

Craigen, B., Dashiff, A., and Kadouri, D.E. 2011. The use of commercially available alpha-amylase compounds to inhibit and remove Staphylococcus aureus biofilms. *Open Microbiol. J.* 5:21–31.

Daims, H., Lücker, S., and Wagner, M. 2006. Daime, a novel image analysis program for microbial ecology and biofilm research. *Environ. Microbiol.* 8:200–213.

Di Pippo, F., Di Gregorio, L., Congestri, R., Tandoi, V., and S. Rossetti. 2018. Biofilm growth and control in cooling water industrial systems. *FEMS Microbiol. Ecol.* 94:fiy044.

Díez-Martínez, R., De Paz, H.D., García-Fernández, E., Bustamante, N., Euler, C.W., Fischetti, V.A., Menendez, M., and García, P. 2015. A novel chimeric phage lysin with high in vitro and in vivo bactericidal activity against Streptococcus pneumoniae. *J. Antimicrob. Chemother.* 70:1763–1773.

Domenech, M., García, E., and Moscoso, M. 2011. In vitro destruction of streptococcus pneumoniae biofilms with bacterial and phage peptidoglycan hydrolases. *Antimicrob. Agents Chemother.* 55:4144–4148.

Fenton, M., Keary, R., McAuliffe, O., Ross, R.P., O'Mahony, J., and Coffey, A. 2013. Bacteriophage-derived peptidase eliminates and prevents staphylococcal biofilms. *Int. J. Microbiol.*

Fischetti, V.A. 2005. Bacteriophage lytic enzymes: Novel anti-infectives. *Trends Microbiol.* 13:491–496.

Fleming, D., Chahin, L., and Rumbaugh, K. 2017. Glycoside hydrolases degrade polymicrobial bacterial biofilms in wounds. *Antimicrob. Agents Chemother.* 61.

Fleming, D., and Rumbaugh, K.P. 2017. Approaches to dispersing medical biofilms. *Microorganisms*. 5.

Flemming, H.C., Neu, T.R., and Wozniak, D.J. 2007. The EPS matrix: The "house of biofilm cells." *J. Bacteriol.* 189:7945–7947.

Flemming, H.C., and Wingender, J. 2010. The biofilm matrix. *Nat. Rev. Microbiol.* 8:623–633.

Garcez, A.S., Barros, L.C., Fernandes, M.R.U., Fujii, D.N., Suzuki, S.S., and Nepomuceno, R. 2020. Fluorescence image and microbiological analysis of biofilm retained around healthy and inflamed orthodontic miniscrews. *Photodiagnosis Photodyn. Ther.* 30:101707.

Gjermansen, M., Nilsson, M., Yang, L., and Tolker-Nielsen, T. 2010. Characterization of starvation-induced dispersion in pseudomonas putida biofilms: Genetic elements and molecular mechanisms. *Mol. Microbiol.* 75:815–826.

Grover, N., Plaks, J.G., Summers, S.R., Chado, G.R., Schurr, M.J., and Kaar, J.L. 2016. Acylase-containing polyurethane coatings with anti-biofilm activity. *Biotechnol. Bioeng.* 113:2535–2543.

Gutiérrez, D., Ruas-Madiedo, P., Martínez, B., Rodríguez, A., and García, P. 2014. Effective removal of staphylococcal biofilms by the endolysin LysH5. *PLoS One*. 9:e107307.

Hartmann, R., Jeckel, H., Jelli, E., Singh, P.K., Vaidya, S., Bayer, M., Rode, D.K.H., Vidakovic, L., Díaz-Pascual, F., Fong, J.C.N., Dragoš, A., Lamprecht, O., Thöming, J.G., Netter, N., Häussler, S., Nadell, C.D., Sourjik, V., Kovács, Á.T., Yildiz, F.H., and Drescher, K. 2021. Quantitative image analysis of microbial communities with BiofilmQ. *Nat. Microbiol.* 6:151–156.

Heydorn, A., Nielsen, A.T., Hentzer M., Sternberg, C., Givskov, M., Ersbøll, B.K., and Molin, S. 2000. Quantification of biofilm structures by the novel computer program COMSTAT. *Microbiology*. 146 (10):2395–2407.

Hukić, M., Seljmo, D., Ramovic, A., Ibrišimović, M.A., Dogan, S., Hukic, J., and Bojic, E.F. 2018. The effect of lysozyme on reducing biofilms by Staphylococcus aureus, Pseudomonas aeruginosa, and Gardnerella vaginalis: An in vitro examination. *Microb. Drug Resist.* 24:353–358.

Ibberson, C.B., Parlet, C.P., Kwiecinski, J., Crosby, H.A., Meyerholz, D.K., and Horswill, A.R. 2016. Hyaluronan modulation impacts Staphylococcus aureus biofilm infection. *Infect. Immun.* 84:1917–1929.

Itoh, Y., Wang, X., Hinnebusch, B.J., 3rd Preston, J.F., and Romeo, T. 2005. Depolymerization of beta-1,6-N-acetyl-D-glucosamine disrupts the integrity of diverse bacterial biofilms. *J. Bacteriol.* 187:382–387.

Iwase, T., Uehara, Y., Shinji, H., Tajima, A., Seo, H., Takada, K., Agata, T., and Mizunoe, Y. 2010. Staphylococcus epidermidis esp inhibits Staphylococcus aureus biofilm formation and nasal colonization. *Nature*. 465:346–349.

Jass, J., and Walker, J.T. 2000. Biofilms and biofouling. *Ind. Biofouling Detect. Prev. Control*. 1–12.

Jee, S.C., Kim, M., Sung, J.S., and Kadam, A.A. 2020. Efficient biofilms eradication by enzymatic-cocktail of pancreatic protease type-I and bacterial α-amylase. *Polymers (Basel)*. 12.

Johansen, C., Falholt, P., and Gram, L. 1997. Enzymatic removal and disinfection of bacterial biofilms. *Appl. Environ. Microbiol.* 63:3724–3728.

Jun, S.Y., Jung, G.M., Yoon, S.J., Oh, M.D., Choi, Y.J., Lee, W.J., Kong, J.C., Seol, J.G., and Kang, S.H. 2013. Antibacterial properties of a pre-formulated recombinant phage endolysin, SAL-1. *Int. J. Antimicrob. Agents*. 41:156–161.

Kalpana, B.J., Aarthy, S., and Pandian, S.K. 2012. Anti-biofilm activity of α-amylase from Bacillus subtilis S8-18 against biofilm forming human bacterial pathogens. *Appl. Biochem. Biotechnol.* 167:1778–1794.

Kaplan, J.B. 2010. Biofilm dispersal: Mechanisms, clinical implications, and potential therapeutic uses. *J. Dent. Res.* 89:205–218.

Karatan, E., and Watnick, P. 2009. Signals, regulatory networks, and materials that build and break bacterial biofilms. *Microbiol. Mol. Biol. Rev.* 73:310–347.

Kiedrowski, M.R., Kavanaugh, J.S., Malone, C.L., Mootz, J.M., Voyich, J.M., Smeltzer, M.S., Bayles, K.W., and Horswill, A.R. 2011. Nuclease modulates biofilm formation in community-associated methicillin-resistant Staphylococcus aureus. *PLoS One*. 6:e26714.

Kumar Shukla, S., and Rao, T.S. 2013. Dispersal of bap-mediated Staphylococcus aureus biofilm by proteinase K. *J. Antibiot. (Tokyo)*. 66:55.

Lahiri, D., Nag, M., Banerjee, R., Mukherjee, D., Garai, S., Sarkar, T., Dey, A., Sheikh, H.I., Pathak, S.K., Edinur, H.A., Pati, S., and Ray, R.R. 2021. Amylases: Biofilm inducer or biofilm inhibitor?. *Front. Cell. Infect. Microbiol.* 11:660048.

Lamppa, J.W., and Griswold, K.E. 2013. Alginate lyase exhibits catalysis-independent biofilm dispersion and antibiotic synergy. *Antimicrob. Agents Chemother.* 57:137–145.

Lauderdale, K.J., Malone, C.L., Boles, B.R., Morcuende, J., and Horswill, A.R. 2010. Biofilm dispersal of community-associated methicillin-resistant Staphylococcus aureus on orthopedic implant material. *J. Orthop. Res. Off. Publ. Orthop. Res. Soc.* 28:55–61.

Li, S., Wang, Y., Li, X., Lee, B.S., Jung, S., and Lee, M.S. 2019. Enhancing the thermo-stability and anti-biofilm activity of alginate lyase by immobilization on low molecular weight chitosan nanoparticles. *Int. J. Mol. Sci.* 20.

Lister, J.L., and Horswill, A.R. 2014. Staphylococcus aureus biofilms: Recent developments in biofilm dispersal. *Front. Cell. Infect. Microbiol.* 4:178.

Liu, J., Dazzo, F.B., Glagoleva, O., Yu, B., and Jain, A.K. 2001. CMEIAS: A computer-aided system for the image analysis of bacterial morphotypes in microbial communities. *Microb. Ecol.* 173–194.

Liu, J., Sun, L., Liu, W., Guo, L., Liu, Z., Wei, X., and Ling, J. 2017. A Nuclease from Streptococcus mutans facilitates biofilm dispersal and escape from killing by neutrophil extracellular traps. *Front. Cell. Infect. Microbiol.* 7:97.

Loessner, M.J. 2005. Bacteriophage endolysins—Current state of research and applications. *Curr. Opin. Microbiol.* 8:480–487.

Ma, L., Conover, M., Lu, H., Parsek, M.R., Bayles, K., and Wozniak, D.J. 2009. Assembly and development of the Pseudomonas aeruginosa biofilm matrix. *PLoS Pathog.* 5:e1000354.

Madsen, J.S., Burmølle, M., Hansen, L.H., and Sørensen, S.J. 2012. The interconnection between biofilm formation and horizontal gene transfer. *FEMS Immunol. Med. Microbiol.* 65:183–195.

Mah, T.F.C., and O'Toole, G.A. 2001. Mechanisms of biofilm resistance to antimicrobial agents. *Trends Microbiol.* 9:34–39.

Mann, E.E., Rice, K.C., Boles, B.R., Endres, J.L., Ranjit, D., Chandramohan, L., Tsang, L.H., Smeltzer, M.S., Horswill, A.R., and Bayles, K.W. 2009. Modulation of eDNA release and degradation affects Staphylococcus aureus biofilm maturation. *PLoS One*. 4:e5822.

Manohar, C.M., Prabhawathi, V., Sivakumar, P.M., and Doble, M. 2015. Design of a papain immobilized antimicrobial food package with curcumin as a crosslinker. *PLoS One*. 10:e0121665.

Mansoor, A., Patsekin, V., Scherl, D., Robinson, J.P., and Rajwa, B. 2014. BiofilmQuant: A computer-assisted tool for dental biofilm quantification. *Annu. Int. Conf. IEEE Eng. Med. Biol. Soc. IEEE Eng. Med. Biol. Soc. Annu. Int. Conf.* 2014:4244–4247.

Martí, M., Trotonda, M.P., Tormo-Más, M.A., Vergara-Irigaray, M., Cheung, A.L., Lasa, I., and Penadés, J.R. 2010. Extracellular proteases inhibit protein-dependent biofilm formation in Staphylococcus aureus. *Microbes Infect.* 12:55–64.

Mikkelsen, H., Duck, Z., Lilley, K.S., and Welch, M. 2007. Interrelationships between colonies, biofilms, and planktonic cells of Pseudomonas aeruginosa. *J. Bacteriol.* 189:2411–2416.

Mootz, J.M., Malone, C.L., Shaw, L.N., and Horswill, A.R. 2013. Staphopains modulate Staphylococcus aureus biofilm integrity. *Infect. Immun.* 81:3227–3238.

Mueller, L.N., de Brouwer, J.F.C., Almeida, J.S., Stal, L.J., and Xavier, J.B. 2006. Analysis of a marine phototrophic biofilm by confocal laser scanning microscopy using the new image quantification software PHLIP. *BMC Ecol.* 6:1.

Nahar, S., Ha, A.J.W., Byun, K.H., Hossain, M.I., Mizan, M.F.R., and Ha, S.D. 2021. Efficacy of flavourzyme against Salmonella typhimurium, Escherichia coli, and Pseudomonas aeruginosa biofilms on food-contact surfaces. *Int. J. Food Microbiol.* 336:108897.

Nemoto, K., Hirota, K., Murakami, K., Taniguti, K., Murata, H., Viducic, D., and Miyake, Y. 2003. Effect of varidase (streptodornase) on biofilm formed by Pseudomonas aeruginosa. *Chemotherapy*. 49:121–125.

Nguyen, U.T., and Burrows, L.L. 2014. DNase I and proteinase K impair Listeria monocytogenes biofilm formation and induce dispersal of pre-existing biofilms. *Int. J. Food Microbiol.* 187:26–32.

Nijland, R., Hall, M.J., and Burgess, J.G. 2010. Dispersal of biofilms by secreted, matrix degrading, bacterial DNase. *PLoS One*. 5:e15668.

Okshevsky, M., Regina, V.R., and Meyer, R.L. 2015. Extracellular DNA as a target for biofilm control. *Curr. Opin. Biotechnol.* 33:73–80.

Oulahal-Lagsir, N., Martial-Gros, A., Bonneau, M., and Blum, L.J. 2003. "Escherichia coli-milk" biofilm removal from stainless steel surfaces: Synergism between ultrasonic waves and enzymes. *Biofouling*. 19:159–168.

Park, J.-H., Lee, J.H., Cho, M.H., Herzberg, M., and Lee, J. 2012. Acceleration of protease effect on Staphylococcus aureus biofilm dispersal. *FEMS Microbiol. Lett.* 335:31–38.

Patel, I., Patel, V., Thakkar, A., and Kothari, V. 2014. Microbial biofilms: Microbes in social mode. *Int. J. Agric. Food Res.* 3.

Pawley, J.B. (Ed.). 1995. *Handbook of Biological Confocal Microscopy*. 2nd ed.

Pecharki, D., Petersen, F.C., and Scheie, A.A. 2008. Role of hyaluronidase in Streptococcus intermedius biofilm. *Microbiology*. 154:932–938.

Rabin, N., Zheng, Y., Opoku-Temeng, C., Du, Y., Bonsu, E., and Sintim, H.O. 2015. Biofilm formation mechanisms and targets for developing anti-biofilm agents. *Future Med. Chem.* 7:493–512.

Rajarajan, N., Ward, A.C., Burgess, J.G., and Glassey, J. 2013. Use of physiological information and process optimisation enhances production of extracellular nuclease by a marine strain of Bacillus licheniformis. *Bioresour. Technol.* 130:552–558.

Reichhardt, C., and Parsek, M.R. 2019. Confocal laser scanning microscopy for analysis of Pseudomonas aeruginosa biofilm architecture and matrix localization. *Front. Microbiol.*10.

Saggu, S.K., Jha, G., and Mishra, P.C. 2019. Enzymatic degradation of biofilm by metalloprotease from Microbacterium sp. SKS10. *Front. Bioeng. Biotechnol.* 7:192.

Sakr, M.M., Aboshanab, K.M., Elkhatib, W.F., Yassien, M.A., and Hassouna, N.A. 2018. Overexpressed recombinant quorum quenching lactonase reduces the virulence, motility and biofilm formation of multidrug-resistant Pseudomonas aeruginosa clinical isolates. *Appl. Microbiol. Biotechnol.* 102:10613–10622.

Sass, P., and Bierbaum, G. 2007. Lytic activity of recombinant bacteriophage phi11 and phi12 endolysins on whole cells and biofilms of Staphylococcus aureus. *Appl. Environ. Microbiol.* 73:347–352.

Sharma, K., and Pagedar Singh, A. 2018. Anti-biofilm effect of DNase against single and mixed species biofilm. *Foods (Basel, Switzerland)*. 7:42.

Sharma, P., Garg, N., Sharma, A., Capalash, N., and Singh, R. 2019. Nucleases of bacterial pathogens as virulence factors, therapeutic targets and diagnostic markers. *Int. J. Med. Microbiol.* 309:151354.

Shen, Y., Köller, T., Kreikemeyer, B., and Nelson, D.C. 2013. Rapid degradation of Streptococcus pyogenes biofilms by PlyC, a bacteriophage-encoded endolysin. *J. Antimicrob. Chemother.* 68:1818–1824.

Son, J.S., Lee, S.J., Jun, S.Y., Yoon, S.J., Kang, S.H., Paik, H.R., Kang, J.O., and Choi, Y.J. 2010. Antibacterial and biofilm removal activity of a podoviridae Staphylococcus aureus bacteriophage SAP-2 and a derived recombinant cell-wall-degrading enzyme. *Appl. Microbiol. Biotechnol.* 86:1439–1449.

Stewart, P.S., McFeters, G.A., and Huang, C.T. 2000. Biofilm control by antimicrobial agents. *In Biofilms II: Process Analysis and Applications*. 373–405.

Tan, Y., Leonhard, M., Ma, S., Moser, D., and Schneider-Stickler, B. 2017. Dispersal of single and mixed non-albicans Candida species biofilms by β-1,3-glucanase in vitro. *Microb. Pathog.* 113:342–347.

Tan, Y., Ma, S., Leonhard, M., Moser, D., and Schneider-Stickler, B. 2018. β-1,3-glucanase disrupts biofilm formation and increases antifungal susceptibility of Candida albicans DAY185. *Int. J. Biol. Macromol.* 108:942–946.

Tetz, G.V, Artemenko, N.K., and Tetz, V.V. 2009. Effect of DNase and antibiotics on biofilm characteristics. *Antimicrob. Agents Chemother.* 53:1204–1209.

Thormann, K.M., Saville, R.M., Shukla, S., and Spormann, A.M.. 2005. Induction of rapid detachment in Shewanella oneidensis MR-1 biofilms. *J. Bacteriol.* 187:1014–1021.

Thorn, C.R., Howell, P.L., Wozniak, D.J., Prestidge, C.A., and Thomas, N. 2021. Enhancing the therapeutic use of biofilm-dispersing enzymes with smart drug delivery systems. *Adv. Drug Deliv. Rev.* 179:113916.

Torres, C.E., Lenon, G., Craperi, D., Wilting, R., and Blanco, A. 2011. Enzymatic treatment for preventing biofilm formation in the paper industry. *Appl. Microbiol. Biotechnol.* 92:95–103.

Tremblay, Y.D.N., Lévesque, C., Segers, R.P.A.M., and Jacques, M. 2013. Method to grow actinobacillus pleuropneumoniae biofilm on a biotic surface. *BMC Vet. Res.* 9:213.

Trizna, E., Diana, B., Kholyavka, M., Sharafutdinov, I., Hairutdinova, A., Khafizova, F., Zakirova, E., Hafizov, R.G., and Kayumov, A. 2016. Soluble and immobilized papain and trypsin as destroyers of bacterial biofilms. *Genes Cells.* 10:106–112.

Uppuluri, P., and Lopez-Ribot, J.L. 2016. Go forth and colonize: Dispersal from clinically important microbial biofilms. *PLoS Pathog.* 12:e1005397.

Vyas, N., Sammons, R.L., Addison, O., Dhghani, H., and Walmsley, A.D. 2016. A quantitative method to measure biofilm removal efficiency from complex biomaterial surfaces using SEM and image analysis. *Sci. Rep.* 6:32694.

Wilson, C., Lukowicz, R., Merchant, S., Valquier-Flynn, H., Caballero, J., Sandoval, J., Okuom, M., Huber, C., Brooks, T.D., Wilson, E., Clement, B., Wentworth, C.D., and Holmes, A.E. 2017. Quantitative and qualitative assessment methods for biofilm growth: A mini-review. *Res. Rev. J. Eng. Technol.* 6: www.rroij.com/open-access/quantitative-and-qualitative-assessment-methods-for-biofilm-growth-a-minireview-.pdf

Wu, J.A., Kusuma, C., Mond, J.J., and Kokai-Kun, J.F. 2003. Lysostaphin disrupts Staphylococcus aureus and Staphylococcus epidermidis biofilms on artificial surfaces. *Antimicrob. Agents Chemother.* 47:3407–3414.

Xavier, J.B., Picioreanu, C., Rani, S.A., van Loosdrecht, M.C.M., and Stewart, P.S. 2005. Biofilm-control strategies based on enzymic disruption of the extracellular polymeric substance matrix—a modelling study. *Microbiology.* 151:3817–3832.

Yu, S., Su, T., Wu, H., Liu, S., Wang, D., Zhao, T., Jin, Z., Du, W., Zhu, M.J., Chua, S.L., Yang, L., Zhu, D., Gu, L., and Ma, L.Z. 2015. PslG, a self-produced glycosyl hydrolase, triggers biofilm disassembly by disrupting exopolysaccharide matrix. *Cell Res.* 25:1352–1367.

11 CRISPR/Cas Therapeutics in the Inhibition of Biofilm

Bandita Dutta, Debarati Basu, Rina Rani Ray, Tanmay Sarkar, Dibyajit Lahiri, Moupriya Nag and Debasmita Bhattacharya

11.1 INTRODUCTION

Biofilm is a syntropic aggregation of the microorganisms which are pinned in the autogenic matrix. Bacteria which are present inside the biofilm layer are of sessile type in nature and able to bring resistance against antimicrobial treatments. *E. coli* is one of the gram negative anaerobes responsible for intestinal (InPEC) and extra intestinal (ExPEC) infections by forming a combative dense biofilm. In humans some infections such as Crohns disease, enteric syndrome, mastitis, neonatal sepsis and urinary tract infections have been reported, which are caused by *E. coli* (Vogeleer et al., 2014). Various nosocomial infections such as intravascular infections, infections associated with the urinary catheter and the medical implants such as grafts, prosthetic joints, shunts are usually caused by *E.coli*. Biofilm formation is an amalgamation of five distinct steps (initial reversibile attachment, irreversible attachment, microcolony formation, maturation and dispersion) where sets of genes such as bcsA operon, csgD, fimB, hha and pgaC administer the emergence of biofilm and the expression of its virulence factors (Keseler et al., 2013). The production of AidA and TibA auto-transporters act as chaperones in the maturation steps of biofilms and three dimensional mesh-like structures of the biofilms are manifested by the extracellular polymeric substances (EPS) and the molecular interactions via quorum sensing. Quorum sensing is the chemical signaling method where mutual behavior of the bacteria is being regulated in order to form biofilms and mediate pathogenesis. Several genes such as luxS, lsrK, lsrR, csrA, mqsR, motA, fliA, flhD, pfs, qseB, qseC are involved in the quorum sensing mechanism of the bacterial biofilm. The production of autoinducer-2 (AI-2) is usually mediated by the luxS associated active methyl cycle (Vendeville et al., 2005).

A bacterial immune system derived CRISPR acts as a molecular scissors where the Cas9 endonuclease and target identifying CRISPR-RNA duplex (crRNA and tracrRNA) have been engineered in a chimeric form. This chimeric form is called, single guide RNA (sgRNA), which usually binds to the target DNA through base pairing to the complementary sequences adjacent to the short DNA motif, namely, the protospacer adjacent motif (PAM). The PAM is the only impediment to the sgRNA which is essential for Cas9.

11.2 STRUCTURE OF CRISPR-CAS

Viruses are apprised as the key factors for the evolution of life and provide selective pressure to the bacteria present in the different environments in order to enhance the genetic exchange. Around 20–40% of bacteria are killed everyday due to bacteriophage infections. However attainment of

DOI: 10.1201/9781003281238-11

the exogenous DNA through conjugation, horizontal gene transfer (HGT) and transduction is followed by the selection process considered as an array of defence mechanisms in bacteria that distinguish between the foreign DNA and self DNA (Horvath & Barrangou, 2010). Congregated defence genes offer specific selective advantages on the "genomic junkyards" which are specially acquired by the accumulation of HGT, as insertion of these foreign region is detrimental (Koonin, 2018). Association of the Cas protein to the CRISPR arrays forms a CRISPR-Cas immune system (Rath et al., 2015). Information which is stored within the CRISPR array is able to administer sequence specific neutralization of conquered genetic elements comprehended with viruses and plasmids (Jackson et al., 2017). The CRISPR system was first described in 1987 and it has been identified in 85% of archea and 45% of bacteria since the discovery (McGinn & Marraffini, 2019). The localization of the CRISPR system is different as CRISPR loci are found in chromosomes as well as in plasmids (Horvath & Barrangou, 2010). The CRISPR-Cas locus comprises of an operon of Cas genes along with the CRISPR array which is conserved 25–35 bp long sequence specific repeats with uniform unique inserts from extrachromosomal elements called spacers (namely, 26–72 bp long) (Wright et al., 2016). A few of the CRISPR spacer sequences are homologous to the extrachromosomal DNA. These are usually detected in the viral genomes and plasmids called protospacers (Horvath et al., 2008). Spacers are fundamental rudiments of adaptive immunity during previous infections. They are essential for the recognition and neutralization of the subsequent attacks of assailants (Barrangou et al., 2007). In most of the cases, partial palindromic sequences are present in the repeats which confers stability of the highly conserved secondary structures. Usually, palindromic repeats enable the generation of the RNA with stable hairpin loops. One CRISPR locus encompasses conserved repeat sequences though sequences and length of the repeats are polymorphic with respect to the different CRISPR loci (Kunin et al., 2007). At the initial of the CRISPR array the repeat sequence of the DNA called the leader sequence is rich in adenine and thymine (Grissa et al., 2007). Invasive MGEs derived new spacers are integrated into the CRISPR array via the Cas protein activity. The spacer sequences present in the bacterial strain are responsible for the past CRISPR-Cas mediated interactions. Multiple CRISPR loci (18 CRISPR loci in *Methanocaldococcus sp.* FS406-22) with hundreds of repeat-spacer units present in bacteria.

11.3 STRUCTURE OF THE CAS PROTEIN

The CRISPR system comprises of CRISPR arrays, namely, spacers and repeat sequences with adjacent CRISPR associated Cas genes. These Cas genes encode proteins which are involved in immune response and the repair of DNA (Brouns et al., 2008; Makarova et al., 2002).

Various Cas genes were located in various organisms, leading to difficulty in their classification due to the increase of the annotated genomes. Initially four Cas genes were recognized in the CRISPR-Cas system genomes (Jansen et al., 2002). Accumulation of genome sequences based on sequence similarity of the encoded proteins assists to identify 93 Cas genes that are grouped in ~45 different gene families (Haft et al., 2005; Makarova et al., 2015).

Among these 93 Cas genes six (Cas1–Cas6) are widely distributed. The presence of Cas1 and Cas2 are the indication of genomes containing CRISPR loci (Haft et al., 2005). Exploration of Cas1 sequences indicates the presence of various sharp CRISPR-Cas systems (Haft et al., 2005). All these six Cas genes(Cas1–Cas6*)* have distinctive constitutions that also preserve the gene arrangements of *cas*.

Distinct sets of CRISPR loci are being interacted by the Cas proteins (Makarova et al., 2006). Cas proteins and their subtypes were named after delegate organisms containing a singular CRISPR-Cas locus (for instance, Cas proteins, namely, obtained from Escherichia coli were assigned as

Cse1 [CRISPR arrangement of *E. coli* gene1]; other subtypes included *aeropyrum* [Csa], *mycobacterium* [Csm], *neisseria* [Csn], *desulfovibrio* [Csd], *Haloarcula* [Csh], *Thermotoga* [Cst], and *Yersinia* [Csy]) (Haft et al., 2005). The relationship between the homologous Cas proteins, the alteration of Cas operons of the organisms that contain various CRISPR loci are not easily handled by these underlying classifications. The evolutionary relationships between conserved proteins and Cas operons was compelled to a new classification system that professed the association of the CRISPR-Cas framework in two modules: first the adaptation module requires the proteins Cas1 and Cas2, where spacer procurement are required, and second the effector module, that is expected for the handling of primary CRISPR transcripts (crRNA) in order to resist and degrade foreign nucleic acids. The cascade proteins are fabricated by the CRISPR subtypes that are associated with the processing of crRNA. Some of the CRISPR systems require different proteins whereas others require a solitary multifunctional protein. Various protein complexes were distinguished generally in the class 1 CRISPR-Cas system, where the effector complexes of type I and type III are phylogenetically related ribonucleoproteins (Brouns et al., 2008; Makarova et al., 2015).

On the contrary, the class 2 system contains a single interference protein that exists in types II, V and VI (Shmakov et al., 2017). It was found that one RNA recognition motif with unknown function (RAMPs) is associated with the handling of pre-crRNA (Makarova et al., 2011) (Table 11.1).

TABLE 11.1
Name of Some Major Cas Proteins

Protein	Distribution type/subtype	Process	Function	References
Cas 1	Universal	Adaptation	DNase, not sequence specific, can bind RNA and are present in all types; structure available for several cas 1 protein	(Bhaya et al., 2011; Makarova et al., 2011)
Cas 2	Universal	Adaptation	Small RNase specific to U-rich regions; present in all types; structures available from *Thermus thermophiles* and *Sulfolobus solfataricus* and others	(Bhaya et al., 2011) (Makarova et al., 2011)
Cas 3	Type I signature (A, B, C, D, E, F)	Adaptation	DNA helicase and DNase activity; most proteins have a fusion with HD Nuclease Single-stranded DNA nuclease (HD domain) and ATP-dependent helicase	(Bhaya et al., 2011)(Huo et al., 2014)
Cas 4	I (A, B, C, D), II (B), V	Adaptation	RecB-like nuclease with exonuclease activity homologous to RecB	(Bhaya et al., 2011)
Cas 5	I (A, B, C, E), III, IV	Expression	RAMP protein endoribonuclease involved in crRNA biogenesis; part of Cascade	(Makarova et al., 2015)
Cas 6	I (A, B, D, E, F), III (A, B)	Expression	RAMP protein, an endoribonuclease involved in crRNA biogenesis; part of Cascade; structure available from *P. furiosus*	(Sorek et al., 2013)
Cas 7	I (A, B, C, D, E, F)	Adaptation, Expression	RAMP protein, endoribonuclease involved in crRNA biogenesis; part of Cascade	(Makarova & Koonin, 2015)
Cas 8	I (A, B, C)	Adaptation	Subunit of the interference module. Important in targeting of invading DNA by recognizing the protospacer adjacent motif sequence	(Makarova & Koonin, 2015)

TABLE 11.1 (Continued)
Name of Some Major Cas Proteins

Protein	Distribution type/subtype	Process	Function	References
Cas 8	I (A, B, C), IV	Expression	Large protein with McrA-HNH-nuclease domain and RuvC-like nuclease; part of Cascade	(Bhaya et al., 2011)
Cas 9	Type II signature (A, B)	Adaptation, Interference	Large multidomain protein with McrA-HNH nuclease domain and RuvC-like nuclease domain	(Sampson & Weiss, 2013) (Heler et al., 2015)
Cas 10	I (D), III (A, B)	Adaptation, Interference	HD nuclease domain, palm domain, Zn ribbon; some homologies with Cascade Elements	(Richter et al., 2012; Sampson & Weiss, 2013; Sorek et al., 2013)
Cas13a Cas13b Cas13c Cas13d	VI	Interference	RNA-guided RNase	(Cox et al., 2017)

11.4 CRISPR-CAS ADAPTIVE IMMUNITY MECHANISM

The CRISPR-Cas system-mediated mitigation process can be divided into three stages: adaptation, expression, and interference. The initial encounter of the foreign DNA with the cell is considered as the first step. This first step mediates the occurrence of the second and third steps. Protospacer-Adjacent Motif (PAM) is the sequences of protospacers, which are invading DNA relating to the spacers, that are perceived by the system because of a short conserved area inside a couple of bases (2–5 bp) close to them (Gleditzsch et al., 2019; Mojica et al., 2009; Shah et al., 2013).

Following the protospacer is the generation and integration of a new sequence of nucleic acids which is indistinguishable from the protospacer into the CRISPR array as a new spacer. Next to this is the CRISPR array repared by the cellular repair proteins (Hille & Charpentier, 2016; J. Wang et al., 2015), and the proximal repeat duplication (Liu et al., 2017). The spacer is obtained from RNA which are records of a DNA genome of an MGE through reverse transcription. It is mainly by fusing Cas1 protein mainly by reverse transcriptase (Silas et al., 2016, 2017).

Type III CRISPR has variable length spacers whereas most of the CRISPR-Cas systems have profoundly steady system-specific spacer lengths. To avoid alignment at the wrong end of the crRNA target restricting site the PAM sequence inside the pre-spacer substrate permits the right joining of new spacers into the CRISPR array. This process is interrupted by the Cas1–Cas2 complex. The cleavage of the CRISPR array at the leader repeat junction and the approaching spacer in the middle of the repeat strands have the involvement of the Cas1 protein. The nicking and ligation happen slowly, compared to a classical integrase response (Arslan et al., 2014). The anticipated integrase action of the Cas1 protein is a consistent mechanism (Makarova et al., 2006). It is believed by some researchers that the new spacer integration takes place at the insertion site and it includes two cleavages. The primary cut, interceded by Cas1 or Cas2, happens at the CRISPR/leader boundary, while the second nick happens at the leader distal end of the CRISPR, the distance from the preceding cleavage controls it. A complex is formed by the interaction of both Cas1 and Cas2 proteins that appear as a spacer integrase. Two separate DNA-binding sites are present within this biologically active hexameric complex between Cas1 and Cas2 proteins, the approaching protospacer binds to one of them and the CRISPR array binds with the other. The Cas1–Cas2 complex once stacked with the approaching spacer catalyzes two cleavage-ligation responses, first at the leader end of the first repeat, and accordingly at the spacer end of the repeat (McGinn & Marraffini, 2019). The non-Cas proteins associated with DNA repair by the cooperation of Cas1 with key parts of the repair system like RecB, RecC, and RuvB are needed for the acquisition process (Babu et al., 2011). Díez-Villaseñor affirmed this mechanism (César Díez-Villaseñor et al., 2013). Jackson reviewed the cascade complex function details (Jackson et al., 2017).

One more significant characteristic is that spacers are not obtained from host DNA nut are acquired by the foreign DNA, it is done in order to avoid the self attacking mechanism. This self attacking mechanism resembles the autoimmunity of eukaryotes adaptive immune system (Stern et al., 2010). Native CRISPR adaptation refers to those spacers that are obtained from MGEs and not from the host CRISPs. They require pre-spacer substrates created from forign material (Fineran & Charpentier, 2012). The mechanism behind how the different CRISPR-Cas systems balance the native CRISPR adaptation production of pre-spacers from MGEs against the risk of securing spacers from host DNA is still not clear and will require more studies. Spacer acquisition towards MGEs are biased for type II systems. Moreover host and foreign DNA are not distinguished when Cas9 are nuclease-inadequate (Wei et al., 2015).

In the second stage, crRNA is matured after the spacer is transcribed into a precursor of CRISPR RNA (pre-crRNA). The expressed spacer sequence given by the crRNA is remembered to perceive and direct the complex to bind the particular protospacer target. The precrRNA is processed at the expression-processing stage for the generation of mature crRNAs by distinct complex of Cas proteins, a committed processing nuclease (Cas6), a single large Cas protein or an external RNase (Barrangou & Horvath, 2017; Brouns et al., 2008; Mohanraju et al., 2016; Rath et al., 2015; Sorek et al., 2013).

The corresponding protospacer is located and the degradation of the target by specific Cas nucleases are triggered by the complex of crRNA-Cas proteins (Rath et al., 2015).

Interference and crRNA and Cas proteins recognize and destroy the foreign nucleic acids in the third stage. The neutralization of the DNA sequence by the spacer will not occur if the CRISPR spacer and the foreign DNA are not matched. Phages can multiply within the cell which will lead to bacterial lysis and death (Han et al., 2013). Various parts of CRISPR-Cas science were recently reviewed in various articles (Bourgogne et al., 2008; Charpentier et al., 2015; Jackson et al., 2017; Komor et al., 2017; Mohanraju et al., 2016; Nishimasu & Nureki, 2017; Sorek et al., 2013; Wright et al., 2016).

11.5 ROLE PLAYED BY CRISPR-CAS SYSTEM IN THE ANTIBIOTIC-RESISTANT BACTERIA

Some studies show that the CRISPR-Cas system and antibiotic resistance have a reverse relationship between them as observed in some *Enterococci species* (Burley & Sedgley, 2012; Palmer & Gilmore, 2010), at the same time, other studies have shown no such relation, for example, in *Escherichia. coli* (Touchon et al., 2012). Three CRISPR loci are present in *Enterococci* these three loci includes CRISPR-Cas, orphan CRISPR2, lack of Cas genes, and CRISPR-Cas (Palmer & Gilmore, 2010). Results are suggestive that CRISPR1-Cas and CRISPR3-Cas are not present in all strains of *E. faecalis* like that of orphan CRISPR2 which is present amongst all strains. According to Palmer and Gilmore (Palmer & Gilmore, 2010) both CRISPR1-Cas and CRISPR2 are functionally connected to one another. According to certain analysis on CRISPR spacers it is seen that pheromone-response plasmids are considered as a significant factor in the genome plasticity of *Enterococci*, chromosome mobility, and resistance towards antibiotics. CRISPR-Cas displacement and the ability to transfer themselves are also performed by it (Manson et al., 2010; Palmer & Gilmore, 2010).

As experienced by pheromone responsive plasmids, spacers are distributed in such a manner that some of their components have the tendency to include spacers of CRISPR loci, but as the *Enterococci* are antibiotic resistant so Tn916 does not have CRISPR spacers (Wolfgang, 2003). *Escherichia faecalis* contains CRISPR loci (Palmer & Gilmore, 2010) and it is observed that tetM genes are present in it. This tetM gene is disseminated by Tn916 and conjugative transposons (Roberts & Mullany, 2009; Wolfgang, 2003) and the defense is invaded by these conjugative transposons. Vancomycin-resistant genes are disseminated by the Inc 18 plasmid family that does not have spacers and are obtained from *Enterococci* to MRSA (Flannagan et al., 2003; Zhu et al., 2008).

These discoveries might mirror the general shortcomings of element transfer, which does not have the mechanisms for effective pair development or the overall uncommonness of interspecies exchange. CRISPR, holding onto strains of *Streptococcus thermophiles*, exhibit these roles which gained new spacers from the virus and became phage resistant (Barrangou et al., 2007). Bronus reported a sensitivity decrease of the lambda phage. These lambda phage genes that are being targeted by artificial CRISPR-Cas systems have spacers and they are present with the *Escherichia coli* K12 strains (Brouns et al., 2008). As demonstrated by Maraffini and Sontheimer (Marraffini & Sontheimer, 2008), the CRISPR-Cas system limits the plasmid conjugation that is present in *S.epidermidis*.it is suggestive of the significant role played by CRISPR-Cas system in preventing horizontal gene transfer (HGT). *Escherichia coli* contains mainly four CRISPR loci each of which contains different kinds of Cas gene. These CRISPR loci are: CRISPR 1, CRISPR 2, CRISPR 3, and CRISPR 4 (C Díez-Villaseñor et al., 2010; Touchon & Rocha, 2010). *Escherichia species* does not exceed three CRISPR and it is observed that every *Escherichia* genome containing CRISPR 1, CRISPR 2 is missing (Touchon & Rocha, 2010). It is demonstrated by Touchon that there is no association between the presence of CRISPR loci in *Escherichia coli* and the presence of integrons, plasmids or antibiotic resistance (Touchon et al., 2012). Evaluation of sequence similarities between accessible sequences of plasmids and spacers have seen no spacers matching elements associated in antibiotic resistant gene mobilization, for example, Tn3, intI, ISEcp1 or genes that are antibiotic resistant, for example, production of ESBL, ESBL kind or replicon and sequences of plasmid (Touchon et al., 2012). On examination of *Escherichia coli* it was observed that small numbers of spacers matched the plasmid genes. These spacers were particular for every group of CRISPR containing 15 spacers in a single strain. It is seen that 77% of the strains have a single spacer, 13.2% have double spacers, and 7.5% of them have three or four spacers. Touchon (Touchon et al., 2012) observed that CRISPR have only a tiny effect on the plasmid epidemiology in *Escherichia coli* or genes that are antibiotic resistant. These observations were not similar to those observed by Palmer and Gilmore (Palmer & Gilmore, 2010) on *Enterococci specie*s where an inverse association of CRISPRs with antibiotic resistance was observed. According to these observations it was seen that because of a different history in the evolution of CRISPR-Cas systems they have different antibiotic resistance effects on various species. The generation of CRISPR loci by Cas gene cluster partial or complete deletion and the presence of anti-CRISPR proteins, including the toxin-antitoxin or systems of restriction and modification for plasmid competition may also be the reason behind the CRISPR-Cas system having different antibiotic resistance effects on various species (Pawluk et al., 2016; Touchon & Rocha, 2010).

High numbers of repeats occur in the complete Cas system, intermediate repeats are present when the Cas system has erosion. Large numbers of repeats in *Escherichia coli* related strains, for example, BL21, Bl21-DE3, and B-REL606 occur when detachable system relics are present only. These repeats do not contain Cas genes and *Escherichia coli* strains SMS35 are devoid of CRISPR2 because of sucrose operon insertion (Touchon & Rocha, 2010). The system integrity and function can be indicated by the repeats count. Degradation of DNA plasmid containing chromosomal DNA similar spacers resulting in degradation of chromosomes is triggered by the CRISPR-Cas system. If gene expression interference is performed by the CRISPR-Cas system then there will be a chance of foreign component manipulation by the host. Evolution of bacterial genome mechanism results in CRISPR-Cas system inhibition or escape. So, this results in the utilization of CRISPR as anti-CRISPR in *Escherichia coli* and *Pseudomonas aeruginosa* (Pawluk et al., 2016; Touchon & Rocha, 2010).

11.6 ROLE PLAYED BY THE CRISPR-CAS SYSTEM IN NEUTRALIZING ANTIBIOTIC-RESISTANT GENES

Cas proteins are channelized by partial repeats flanked by RNA-based spacers. These Cas proteins target and cleave DNA encoding protospacers that are a matched CRISPR array. Giving in vivo information helps the CRISPR-Cas system to target and cleave the DNA, targeting of the population of the bacteria performed by the CRISPR array for encoding antibiotic resistance (Bikard et al., 2014; Hale et al., 2012; Yosef et al., 2015).

A bacterial genome sequence targeted accidentally by the CRISPR-Cas system is cytotoxic, leading to cell lysis due to initiation of chromosomal lesions that are irreversible in nature as mentioned in certain research work (Citorik et al., 2014; Vercoe et al., 2013).

Due to the broadly moderated CRISPR-Cas system in microorganisms and archaea, the disengagement, improvement, and advancement of carriers for delivery and vectors of the CRISPR-Cas system will be required for the production of RNA-directed nucleases fit for focusing on additional strains, including multidrug-resistant (MDR) microorganisms and key individuals from endogenous microbiota (Citorik et al., 2014). RNA-directed nucleases lead to the predominance of particular genes including the virulence determinants as present in the wild-type populations and genes that are antibiotic resistant in nature, this occuring depending on the higher organisms' delivery systems (Esvelt et al., 2014). The CRISPR nanocomplex which is polymer based and also contains transmissible bacterial plasmids by the process of conjugation, helps the Cas9 nuclease to target particular bacterial pathogenic DNA sequences and genes that are antibiotic resistant amongst the populations of the microbes (Citorik et al., 2014; Kim et al., 2016).

Gomaa (Gomaa et al., 2014) utilized the typed I-E CRISPR-Cas arrangement of *Escherichia coli*, that encodes six Cas genes that are divided into two operons which are cas ABCDE and Cas3 (Brouns et al., 2008) to explicit and fundamental sequences in the genomes of various sequences. By focusing on different and diverse locations including ftsA, nusB, msbA, and asd all through the genome, and degrees of removal showed by concurrent targeting of numerous locations that are similar to only one of the locations are performed to accomplish elimination (Gomaa et al., 2014). Bikard (Bikard et al., 2014) took advantage of the Cas9 phagemid, that is a plasmid that is intended to be packaged with the phage capsids (Melnikov et al., 1984). It also helps in destroying MRSA strains obtained from a mixed colony of bacteria. A CRISPR-Cas9 plasmid for targeting plasmids that are tetracycline resistant, for example, pUSA01 and pUSA02 has been designed. For treating *Staphylococcus aureus USA300*$^{\Phi}$ isolated clinically the developed Cas9 and crRNA of gene mecA90 which is methicillin resistant in the phagemid (pDB121::mecA) are utilized. Cell lysis was not recorded in any case of Cas9 plasmids targeting plasmids that are tetracycline resistant. Although, 99.99% of the cells were sensitive to tetracycline and there there was a reduction in proportion of *Staphylococcus aureus USA300*$^{\Phi}$ which was 50% before treatment and became 0.4% after pDB121::mecA phagemid treatment. Citorik, in his study applied two methodologies, namely, M12-based phagemid and conjugative plasmid. These two methodologies were utilized for conveying Cas9 nuclease into bacteria that target blaNDM-1 and blasSHV-18. Both of them are utilized for encoding pan-resistant to beta-lactams and also resistant to antibiotics (J K Rasheed et al., 2000; J Kamile Rasheed et al., 2013).

There were 2 to 3 log reductions of cells that were viable when treated with the resistant strains. The CRISPR-Cas system was seen to have the ability to distinguish between susceptible and resistant strains. It was observed that the gyrA phagemid was cytotoxic particularly only for gyrA chromosome causing mutations which in turn causes *Eschereschai coli* to be quinolone resistant and for the isogenic strains having the type gyrA gene not to be cytotoxic (Citorik et al., 2014).

A nonviral method was applied by Kang (Kang et al., 2017) in his study. In this study a nanocomplex of Cas9 utilizes Cas9, sgRA that targets mecA and branched polyethyleneimine(bPEI) which is a polymer of cationic nature. The widely utilized carriers for the delivery of genes (Chung et al., 2011; Patnaik & Gupta, 2013) is the bpEI which was also utilized as a packaging carrier for sgRNA that increases delivery of Cas9 into MRSA strains (Rouet et al., 1994). Cas9 conjugation with bPEI as described by Kang (Kang et al., 2017) has the ability for bacterial uptake which is more compared to the normal mixing of Cas9 with bPEI and native mixture of Cas9 with lipofectamine that does not show any sign of uptake (M. Wang et al., 2016; Zuris et al., 2015). This may be due to the increase in the protein polarity or bpEI polymer having high cationic properties (Kang et al., 2017). One disadvantage of the CRISPR system killing mechanism by plasmids is from conjugation having less frequency utilized for developing efficient phagemid or bacteriophage delivery systems. In any case, CRISPR nucleases distribution by the delivery of conjugative plasmid remains an intriguing choice because of conjugative plasmids that don't need a cell receptor (Pérez-Mendoza & de la Cruz, 2009; Smillie et al., 2010), are impervious to limitation change systems (Oliveira et al., 2014), and have wide host ranges (Jain & Srivastava, 2013). The rate of transfer of conjugative plasmid can be enhanced by increasing the cell to cell contact (Hausner & Wuertz, 1999) by using conjugative plasmids that encode and enhance the formation of biofilm. They are also appropriate as molecular tools used for delivery in CRISPR nucleases for regulating the microbial composition (López-Igual et al., 2019; Peters et al., 2019).

The enterohemorrhagic *Escherichia coli* (EHEC) treatment as shown by Citorik (Citorik et al., 2014) requiring the use of the *eae* phagemid that targets a virulence factor which is encoded by chromosomes of *Escherichia coli O157:H7,* known as intimin that is required for the colonization and pathology of the intestine (Kaper et al., 2004). It gives a result by reducing the viable cell count by 20-fold. The CRISPR-Cas system as suggested by the data has the ability to reduce antibiotic resistance loci that are virulent in nature or bacterial metabolic pathways by not changing bystanders. This is due to the CRISPR-Cas ability to remove specific genome bacteria. The utilization of the CRISPR-Cas system to modify bacterial composition that further helps in developing new drug therapies for modifying probiotic or prebiotic human microflora to reduce the harmful impact of diseases. These therapies are not very functional with proper mechanisms (Sonnenburg & Fischbach, 2011). Disadvantages of phagemids are that production of more phages post infection is not possible by a phagemid. So, there is a large amount of phagemid for treatments which exceeds the required population size. Moreover, phagemids do not have broad host range which is another disadvantage.

CRISPR systems are most likely suited for killing target genes efficiently. Some studies suggest that the most suited for biofilm and planktonic conditions are a mixture of the phage-, conjugative-, and nanoparticles that are polymeric in nature in delivery systems. One of the major problems in controlling infectious diseases through microorganisms modification is the shortage of appropriate tools for bacterial and microbe modification control (Roach et al., 2017). Antimicrobial agents that are sequence specific are CRISPR-based nucleases and they have the ability to decrease the impact of the bacterial infection and thus can be used in future to develop a widely used delivery system.

11.7 REGULATION OF QUORUM SENSING (QS) OF BACTERIAL BIOFILM BY THE CRISPR-CAS SYSTEM

The CRISPR-Cas system of type-I-C with Cas genes in *Myxococcus xanthus* are present within the regulatory system that helps in regulating different quorum-sensing (QS) signals that are intercellular in nature and also signaling cascades that are intracellular in nature. They further check the ascoma formation (Julien et al., 2000; Kaiser et al., 2010). Some studies have shown the significant role played by the CRISPR-Cas system in controlling the virulent nature of the bacteria with the help of endogenous genes. Studies are suggestive of bacterial physiology control by modifying pathogenesis which is done by the Cas9 gene that is present within the type II CRISPR-Cas system. For instance, suppression of the endogenous bacterial gene blp which is a bacterial lipoprotein is carried out by the the functioning of the Cas9 gene present in the type II-b CRISPR-Cas system for the intracellular bacteria *Francisella novicida*, that finally helps in pathogenesis and commensalism (Sampson et al., 2013; Sampson & Weiss, 2013). The attaching and invading of the epithelial cells of the colon and rectum by *Campylobacter jejuni* is mainly carried by the Cas 9 gene present in the CRISPR- Cas system of type II-C (Louwen et al., 2013). The characteristics of a virulent nature are controlled by endogenous genes and the CRISPR-Cas system of type II plays an important part in it. It is seen that *Pelobacter carbinolicus* encodes the type I system of archaea that plays a significant part in gene expression regulation (Aklujkar & Lovley, 2010). *Pseudomonas aeruginosa* contains the Cas3 gene of the type I-F CRISPR-Cas system which helps in preventing biofilm formation by interaction with particular genes present within the chromosomes of prophages (Cady & O'Toole, 2011). Some studies suggest that the bacterial mRNA QS regulator LasR are targeted by *Pseudomonas aeruginosa* containing the Cas3 gene that causes the recognition of receptor 4 (TLR4) and also the pathogen type recognition receptor (PRR) less intensely, thus the host defenses and responses that are pro-inflammatory in nature for both cells and models are destroyed (R. Li et al., 2016). Quorum sensing has the potential for determining and responding to the density of cell population by regulating the complex genes (Nealson et al., 1970). Regulation of particular phenotypic expressions is carried out by quorum sensing (QS) utilized by bacteria and this is performed by autoinducers (AIs) which are signaling molecules that are secreted and that determine their behaviors. Examples of a few quorum sensing (QS) regulated phenotypes involve the formation of biofilm, expression of

TABLE 11.2
Clustered Regularly Interspaced Short Palindromic Repeats and CRISPR-associated (CRISPR-Cas) Applications in the Field of Biotechnology

Application in biotechnology	CRISPR-Cas system and relevant activities	References
Base editing	SpyCas9 fused to cytidine deaminase or adenosine to inosine deaminase; Cas12a fused to cytidine deaminase; Cas13 fused to adenosine deaminase	(Abudayyeh et al., 2017; Gaudelli et al., 2017; Komor et al., 2016)
Gene activation	Fusions of SpyCas9, Cas12a, and CASCADE to transcriptional activation Domains	(Breinig et al., 2019; Perez-Pinera et al., 2013; You et al., 2019)
Gene editing	1. Spy Cas9 (type II-A) and its high-fidelity variants; light and chemically induced Cas9 expression; sgRNA regulation to control Cas9 2. Cas12a (type V-A), sequence specific DNA cleavage producing staggered ends 3. CASCADE and Cas3 for sequence-specific DNA degradation and Cas8- FokI fusion for sequence-specific DNA cleavage 4. Cmr (type III-B), sequence specific interference of transcriptionally active DNA	(Cameron et al., 2019; Cho et al., 2013; Hur et al., 2016; Jinek et al., 2012; Y. Li et al., 2016; Nihongaki et al., 2015; Yan et al., 2019)
Gene knockdown via ssRNA cleavage	Cas9, Cmr, Cas13-based methods	(Abudayyeh et al., 2017; Batra et al., 2017; Zebec et al., 2014)
Nucleic acid detection	Cas12a, Cas13, Csm6	(Gootenberg et al., 2018)
Transcriptional repression	Nuclease inactivated Spy Cas9 and its tag-fusion derivatives, nuclease inactivated Cas12a, CASCADE without Cas3	(Huang et al., 2017; Qi et al., 2013; Zhang et al., 2017)

certain virulence factors, movement and other factors such as fixation of nitrogen, bioluminescence and spore formation for certain bacteria (Pan & Ren, 2009). The impact on the quorum sensing (QS) behaviour depends on the bacterial mode of recognition and transducing of autoinducer (AI) signals. One of the recognized autoinducer (AI) molecules is quorum sensing (QS) molecule autoinducer 2 (AI-2) that is synthesized by the important regulator of quorum sensing (QS), which is LuxS (Lowery et al., 2008). In salmonella (Vendeville et al., 2005) LuxS/AI-2 quorum sensing (QS) is seen to be regulating an unknown operon expression which encodes a LuxS–regulated(lxr) operon which is an ATP-binding cassette transporter of (ABC)-type (M E Taga et al., 2001). The extracellular concentration of autoinducers 2 (AI-2) increases with increase in density of bacterial population until the threshold is reached. The lsr transporter helps to transport the autoinducer 2 (AI-2) into the cell. This lsr transporter consists of Lsr A, LrB, LsrC, and LsrD which encode intracellular phosphorylated kinase, Lsrk and changes made by LsrF and LsrG (Michiko E Taga et al., 2003). The virulent nature of *Salmonella* can be reduced by the downregulation of quorum sensing (QS) by LsrR in excess. This is mainly achieved by causing the avoidance of *Salmonella* by macrophage oxidative killing (Choi et al., 2010). The type three secretion system 1 of *Escherichia coli* which is encoded by an LEE operon is regulated by the quorum sensing (QS) system. It is mainly related to infection at the initial stage where translocation of T3SS effectors throughout the cell membrane of the host causing bacterial invasion of epithelial cells of the host thus causing inflammation of the intestine (Galán, 2001) (Table 11.2).

REFERENCES

Abudayyeh, O. O., Gootenberg, J. S., Essletzbichler, P., Han, S., Joung, J., Belanto, J. J., Verdine, V., Cox, D. B. T., Kellner, M. J., Regev, A., Lander, E. S., Voytas, D. F., Ting, A. Y., and Zhang, F. (2017). RNA targeting with CRISPR-Cas13. *Nature*, *550*(7675), 280–284.

Aklujkar, M., and Lovley, D. R. (2010). Interference with histidyl-tRNA synthetase by a CRISPR spacer sequence as a factor in the evolution of Pelobacter carbinolicus. *BMC Evolutionary Biology*, *10*, 230.

Arslan, Z., Hermanns, V., Wurm, R., Wagner, R., and Pul, Ü. (2014). Detection and characterization of spacer integration intermediates in type I-E CRISPR-Cas system. *Nucleic Acids Research*, *42*(12), 7884–7893.

Babu, M., Beloglazova, N., Flick, R., Graham, C., Skarina, T., Nocek, B., Gagarinova, A., Pogoutse, O., Brown, G., Binkowski, A., Phanse, S., Joachimiak, A., Koonin, E. V, Savchenko, A., Emili, A., Greenblatt, J., Edwards, A. M., and Yakunin, A. F. (2011). A dual function of the CRISPR-Cas system in bacterial antivirus immunity and DNA repair. *Molecular Microbiology*, *79*(2), 484–502.

Barrangou, R., Fremaux, C., Deveau, H., Richards, M., Boyaval, P., Moineau, S., Romero, D. A., and Horvath, P. (2007). CRISPR provides acquired resistance against viruses in prokaryotes. *Science (New York, N.Y.)*, *315*(5819), 1709–1712.

Barrangou, R., and Horvath, P. (2017). A decade of discovery: CRISPR functions and applications. *Nature Microbiology*, *2*(7), 17092.

Batra, R., Nelles, D. A., Pirie, E., Blue, S. M., Marina, R. J., Wang, H., Chaim, I. A., Thomas, J. D., Zhang, N., Nguyen, V., Aigner, S., Markmiller, S., Xia, G., Corbett, K. D., Swanson, M. S., and Yeo, G. W. (2017). Elimination of toxic microsatellite repeat expansion RNA by RNA-targeting Cas9. *Cell*, *170*(5), 899–912.e10.

Bhaya, D., Davison, M., and Barrangou, R. (2011). CRISPR-Cas systems in bacteria and archaea: Versatile small RNAs for adaptive defense and regulation. *Annual Review of Genetics*, *45*, 273–297.

Bikard, D., Euler, C. W., Jiang, W., Nussenzweig, P. M., Goldberg, G. W., Duportet, X., Fischetti, V. A., and Marraffini, L. A. (2014). Exploiting CRISPR-Cas nucleases to produce sequence-specific antimicrobials. *Nature Biotechnology*, *32*(11), 1146–1150.

Bourgogne, A., Garsin, D. A., Qin, X., Singh, K. V, Sillanpaa, J., Yerrapragada, S., Ding, Y., Dugan-Rocha, S., Buhay, C., Shen, H., Chen, G., Williams, G., Muzny, D., Maadani, A., Fox, K. A., Gioia, J., Chen, L., Shang, Y., Arias, C. A., … Weinstock, G. M. (2008). Large scale variation in Enterococcus faecalis illustrated by the genome analysis of strain OG1RF. *Genome Biology*, *9*(7), R110.

Breinig, M., Schweitzer, A. Y., Herianto, A. M., Revia, S., Schaefer, L., Wendler, L., Cobos Galvez, A., and Tschaharganeh, D. F. (2019). Multiplexed orthogonal genome editing and transcriptional activation by Cas12a. *Nature Methods*, *16*(1), 51–54.

Brouns, S. J. J., Jore, M. M., Lundgren, M., Westra, E. R., Slijkhuis, R. J. H., Snijders, A. P. L., Dickman, M. J., Makarova, K. S., Koonin, E. V, and van der Oost, J. (2008). Small CRISPR RNAs guide antiviral defense in prokaryotes. *Science (New York, N.Y.)*, *321*(5891), 960–964.

Burley, K. M., and Sedgley, C. M. (2012). CRISPR-Cas, a prokaryotic adaptive immune system, in endodontic, oral, and multidrug-resistant hospital-acquired Enterococcus faecalis. *Journal of Endodontics*, *38*(11), 1511–1515.

Cady, K. C., and O'Toole, G. A. (2011). Non-identity-mediated CRISPR-bacteriophage interaction mediated via the Csy and Cas3 proteins. *Journal of Bacteriology*, *193*(14), 3433–3445.

Cameron, P., Coons, M. M., Klompe, S. E., Lied, A. M., Smith, S. C., Vidal, B., Donohoue, P. D., Rotstein, T., Kohrs, B. W., Nyer, D. B., Kennedy, R., Banh, L. M., Williams, C., Toh, M. S., Irby, M. J., Edwards, L. S., Lin, C.-H., Owen, A. L. G., Künne, T., … Sternberg, S. H. (2019). Harnessing type I CRISPR–Cas systems for genome engineering in human cells. *Nature Biotechnology*, *37*(12), 1471–1477.

Charpentier, E., Richter, H., Oost, J., and White, M. (2015). Biogenesis pathways of RNA guides in archaeal and bacterial CRISPR-Cas adaptive immunity. *FEMS Microbiology Reviews*, 39, 428–441.

Cho, S. W., Kim, S., Kim, J. M., and Kim, J. S. (2013). Targeted genome engineering in human cells with the Cas9 RNA-guided endonuclease. *Nature Biotechnology*, *31*(3), 230–232.

Choi, J., Park, J., and Ryu, S. (2010). Repressed quorum sensing by overexpressing LsrR Hampers Salmonella evasion from oxidative killing within macrophages. *Journal of Microbiology and Biotechnology*, *20*(12), 1624–1629.

Chung, H. J., Hong, C. A., Lee, S. H., Jo, S. D., and Park, T. G. (2011). Reducible siRNA dimeric conjugates for efficient cellular uptake and gene silencing. *Bioconjugate Chemistry*, *22*(2), 299–306.

Citorik, R. J., Mimee, M., and Lu, T. K. (2014). Sequence-specific antimicrobials using efficiently delivered RNA-guided nucleases. *Nature Biotechnology*, *32*(11), 1141–1145.

Cox, D. B. T., Gootenberg, J. S., Abudayyeh, O. O., Franklin, B., Kellner, M. J., Joung, J., and Zhang, F. (2017). RNA editing with CRISPR-Cas13. *Science (New York, N.Y.)*, *358*(6366), 1019–1027.

Díez-Villaseñor, C, Almendros, C., García-Martínez, J., and Mojica, F. J. M. (2010). Diversity of CRISPR loci in Escherichia coli. *Microbiology (Reading, England)*, *156*(5), 1351–1361.

Díez-Villaseñor, C., Guzmán, N. M., Almendros, C., García-Martínez, J., and Mojica, F. J. M. (2013). CRISPR-spacer integration reporter plasmids reveal distinct genuine acquisition specificities among CRISPR-Cas I-E variants of Escherichia coli. *RNA Biology*, *10*(5), 792–802.

Esvelt, K. M., Smidler, A. L., Catteruccia, F., and Church, G. M. (2014). Concerning RNA-guided gene drives for the alteration of wild populations. *ELife*, *3*, e03401.

Fineran, P. C., and Charpentier, E. (2012). Memory of viral infections by CRISPR-Cas adaptive immune systems: Acquisition of new information. *Virology*, *434*(2), 202–209.

Flannagan, S. E., Chow, J. W., Donabedian, S. M., Brown, W. J., Perri, M. B., Zervos, M. J., Ozawa, Y., and Clewell, D. B. (2003). Plasmid content of a vancomycin-resistant Enterococcus faecalis isolate from a patient also colonized by Staphylococcus aureus with a VanA phenotype. *Antimicrobial Agents and Chemotherapy*, *47*(12), 3954–3959.

Galán, J. E. (2001). Salmonella interactions with host cells: Type III secretion at work. *Annual Review of Cell and Developmental Biology*, *17*, 53–86.

Gaudelli, N. M., Komor, A. C., Rees, H. A., Packer, M. S., Badran, A. H., Bryson, D. I., and Liu, D. R. (2017). Programmable base editing of A•T to G•C in genomic DNA without DNA cleavage. *Nature*, *551*(7681), 464–471.

Gleditzsch, D., Pausch, P., Müller-Esparza, H., Özcan, A., Guo, X., Bange, G., and Randau, L. (2019). PAM identification by CRISPR-Cas effector complexes: Diversified mechanisms and structures. *RNA Biology*, *16*(4), 504–517.

Gomaa, A. A., Klumpe, H. E., Luo, M. L., Selle, K., Barrangou, R., and Beisel, C. L. (2014). Programmable removal of bacterial strains by use of genome-targeting CRISPR-Cas systems. *MBio*, *5*(1), e00928-13.

Gootenberg, J. S., Abudayyeh, O. O., Kellner, M. J., Joung, J., Collins, J. J., and Zhang, F. (2018). Multiplexed and portable nucleic acid detection platform with Cas13, Cas12a, and Csm6. *Science (New York, N.Y.)*, *360*(6387), 439–444.

Grissa, I., Vergnaud, G., and Pourcel, C. (2007). CRISPRFinder: A web tool to identify clustered regularly interspaced short palindromic repeats. *Nucleic Acids Research*, *35*(Web Server issue), W52–7.

Haft, D. H., Selengut, J., Mongodin, E. F., and Nelson, K. E. (2005). A guild of 45 CRISPR-associated (Cas) protein families and multiple CRISPR/Cas subtypes exist in prokaryotic genomes. *PLoS Computational Biology*, *1*(6), e60.

Hale, C. R., Majumdar, S., Elmore, J., Pfister, N., Compton, M., Olson, S., Resch, A. M., Glover, C. V. C. 3rd, Graveley, B. R., Terns, R. M., and Terns, M. P. (2012). Essential features and rational design of CRISPR RNAs that function with the Cas RAMP module complex to cleave RNAs. *Molecular Cell*, *45*(3), 292–302.

Han, P., Niestemski, L. R., Barrick, J. E., and Deem, M. W. (2013). Physical model of the immune response of bacteria against bacteriophage through the adaptive CRISPR-Cas immune system. *Physical Biology*, *10*(2), 25004.

Hausner, M., and Wuertz, S. (1999). High rates of conjugation in bacterial biofilms as determined by quantitative in situ analysis. *Applied and Environmental Microbiology*, *65*(8), 3710–3713.

Heler, R., Samai, P., Modell, J. W., Weiner, C., Goldberg, G. W., Bikard, D., and Marraffini, L. A. (2015). Cas9 specifies functional viral targets during CRISPR–Cas adaptation. *Nature*, *519*(7542), 199–202.

Hille, F., and Charpentier, E. (2016). CRISPR-Cas: Biology, mechanisms and relevance. *Philosophical Transactions of the Royal Society of London. Series B, Biological Sciences*, *371*(1707), 20150496.

Horvath, P., and Barrangou, R. (2010). CRISPR/Cas, the immune system of bacteria and archaea. *Science (New York, N.Y.)*, *327*(5962), 167–170.

Horvath, P., Romero, D. A., Coûté-Monvoisin, A.-C., Richards, M., Deveau, H., Moineau, S., Boyaval, P., Fremaux, C., and Barrangou, R. (2008). Diversity, activity, and evolution of CRISPR loci in Streptococcus thermophilus. *Journal of Bacteriology*, *190*(4), 1401–1412.

Huang, Y.-H., Su, J., Lei, Y., Brunetti, L., Gundry, M. C., Zhang, X., Jeong, M., Li, W., and Goodell, M. A. (2017). DNA epigenome editing using CRISPR-Cas SunTag-directed DNMT3A. *Genome Biology*, *18*(1), 176.

Huo, Y., Nam, K. H., Ding, F., Lee, H., Wu, L., Xiao, Y., Farchione, M. D., Zhou, S., Rajashankar, K., Kurinov, I., Zhang, R., and Ke, A. (2014). Structures of CRISPR Cas3 offer mechanistic insights into cascade-activated DNA unwinding and degradation. *Nature Structural & Molecular Biology*, *21*(9), 771–777.

Hur, J. K., Kim, K., Been, K. W., Baek, G., Ye, S., Hur, J. W., Ryu, S.-M., Lee, Y. S., and Kim, J.-S. (2016). Targeted mutagenesis in mice by electroporation of Cpf1 ribonucleoproteins. *Nature Biotechnology*, *34*(8), 807–808.

Jackson, S. A., McKenzie, R. E., Fagerlund, R. D., Kieper, S. N., Fineran, P. C., and Brouns, S. J. J. (2017). CRISPR-Cas: Adapting to change. *Science (New York, N.Y.)*, *356*(6333), eaal5056.

Jain, A., and Srivastava, P. (2013). Broad host range plasmids. *FEMS Microbiology Letters*, *348*(2), 87–96.

Jansen, R., Embden, J. D. A. van, Gaastra, W., and Schouls, L. M. (2002). Identification of genes that are associated with DNA repeats in prokaryotes. *Molecular Microbiology*, *43*(6), 1565–1575.

Jinek, M., Chylinski, K., Fonfara, I., Hauer, M., Doudna, J. A., and Charpentier, E. (2012). A programmable dual-RNA-guided DNA endonuclease in adaptive bacterial immunity. *Science (New York, N.Y.)*, *337*(6096), 816–821.

Julien, B., Kaiser, A. D., and Garza, A. (2000). Spatial control of cell differentiation in Myxococcus xanthus. *Proceedings of the National Academy of Sciences*, *97*(16), 9098–9103.

Kaiser, D., Robinson, M., and Kroos, L. (2010). Myxobacteria, polarity, and multicellular morphogenesis. *Cold Spring Harbor Perspectives in Biology*, *2*(8), a000380–a000380.

Kang, Y. K., Kwon, K., Ryu, J. S., Lee, H. N., Park, C., and Chung, H. J. (2017). Nonviral genome editing based on a polymer-derivatized CRISPR nanocomplex for targeting bacterial pathogens and antibiotic resistance. *Bioconjugate Chemistry*, *28*(4), 957–967.

Kaper, J. B., Nataro, J. P., and Mobley, H. L. T. (2004). Pathogenic Escherichia coli. *Nature Reviews Microbiology*, *2*(2), 123–140.

Keseler, I. M., Mackie, A., Peralta-Gil, M., Santos-Zavaleta, A., Gama-Castro, S., Bonavides-Martínez, C., Fulcher, C., Huerta, A. M., Kothari, A., Krummenacker, M., Latendresse, M., Muñiz-Rascado, L., Ong, Q., Paley, S., Schröder, I., Shearer, A. G., Subhraveti, P., Travers, M., Weerasinghe, D., … Karp, P. D. (2013). EcoCyc: Fusing model organism databases with systems biology. *Nucleic Acids Research*, *41*(Database issue), D605–12.

Kim, J.-S., Cho, D.-H., Park, M., Chung, W.-J., Shin, D., Ko, K. S., and Kweon, D.-H. (2016). CRISPR/Cas9-mediated re-sensitization of antibiotic-resistant Escherichia coli harboring extended-spectrum β-lactamases. *Journal of Microbiology and Biotechnology*, *26*(2), 394–401.

Komor, A. C., Badran, A. H., and Liu, D. R. (2017). CRISPR-based technologies for the manipulation of eukaryotic genomes. *Cell*, *168*(1–2), 20–36.

Komor, A. C., Kim, Y. B., Packer, M. S., Zuris, J. A., and Liu, D. R. (2016). Programmable editing of a target base in genomic DNA without double-stranded DNA cleavage. *Nature*, *533*(7603), 420–424.

Koonin, E. V. (2018). Hunting for treasure chests in microbial defense islands. *Molecular Cell*, *70*(5), 761–762.

Kunin, V., Sorek, R., and Hugenholtz, P. (2007). Evolutionary conservation of sequence and secondary structures in CRISPR repeats. *Genome Biology*, *8*(4), R61.

Li, R., Fang, L., Tan, S., Yu, M., Li, X., He, S., Wei, Y., Li, G., Jiang, J., and Wu, M. (2016). Type I CRISPR-Cas targets endogenous genes and regulates virulence to evade mammalian host immunity. *Cell Research*, *26*(12), 1273–1287.

Li, Y., Pan, S., Zhang, Y., Ren, M., Feng, M., Peng, N., Chen, L., Liang, Y. X., and She, Q. (2016). Harnessing Type I and Type III CRISPR-Cas systems for genome editing. *Nucleic Acids Research*, *44*(4), e34.

Liu, T., Liu, Z., Ye, Q., Pan, S., Wang, X., Li, Y., Peng, W., Liang, Y., She, Q., and Peng, N. (2017). Coupling transcriptional activation of CRISPR-Cas system and DNA repair genes by Csa3a in Sulfolobus islandicus. *Nucleic Acids Research*, *45*(15), 8978–8992.

López-Igual, R., Bernal-Bayard, J., Rodríguez-Patón, A., Ghigo, J. M., and Mazel, D. (2019). Engineered toxin-intein antimicrobials can selectively target and kill antibiotic-resistant bacteria in mixed populations. *Nature Biotechnology*, *37*(7), 755–760.

Louwen, R., Horst-Kreft, D., de Boer, A. G., van der Graaf, L., de Knegt, G., Hamersma, M., Heikema, A. P., Timms, A. R., Jacobs, B. C., Wagenaar, J. A., Endtz, H. P., van der Oost, J., Wells, J. M., Nieuwenhuis, E. E. S., van Vliet, A. H. M., Willemsen, P. T. J., van Baarlen, P., and van Belkum, A. (2013). A novel link between Campylobacter jejuni bacteriophage defence, virulence and Guillain-Barré syndrome. *European Journal of Clinical Microbiology & Infectious Diseases: Official Publication of the European Society of Clinical Microbiology*, *32*(2), 207–226.

Lowery, C. A., Dickerson, T. J., and Janda, K. D. (2008). Interspecies and interkingdom communication mediated by bacterial quorum sensing. *Chemical Society Reviews*, *37*(7), 1337–1346.

Makarova, K. S., Aravind, L., Grishin, N. V, Rogozin, I. B., and Koonin, E. V. (2002). A DNA repair system specific for thermophilic Archaea and bacteria predicted by genomic context analysis. *Nucleic Acids Research*, *30*(2), 482–496.

Makarova, K. S., Grishin, N. V, Shabalina, S. A., Wolf, Y. I., and Koonin, E. V. (2006). A putative RNA-interference-based immune system in prokaryotes: Computational analysis of the predicted enzymatic machinery, functional analogies with eukaryotic RNAi, and hypothetical mechanisms of action. *Biology Direct*, *1*, 7.

Makarova, K. S., Haft, D. H., Barrangou, R., Brouns, S. J. J., Charpentier, E., Horvath, P., Moineau, S., Mojica, F. J. M., Wolf, Y. I., Yakunin, A. F., van der Oost, J., and Koonin, E. V. (2011). Evolution and classification of the CRISPR–Cas systems. *Nature Reviews Microbiology*, *9*(6), 467–477.

Makarova, K. S., and Koonin, E. V. (2015). Annotation and classification of CRISPR-Cas Systems. *Methods in Molecular Biology (Clifton, N.J.)*, *1311*, 47–75.

Makarova, K. S., Wolf, Y. I., Alkhnbashi, O. S., Costa, F., Shah, S. A., Saunders, S. J., Barrangou, R., Brouns, S. J. J., Charpentier, E., Haft, D. H., Horvath, P., Moineau, S., Mojica, F. J. M., Terns, R. M., Terns, M. P., White, M. F., Yakunin, A. F., Garrett, R. A., van der Oost, J., … Koonin, E. V. (2015). An updated evolutionary classification of CRISPR–Cas systems. *Nature Reviews Microbiology*, *13*(11), 722–736.

Manson, J. M., Hancock, L. E., and Gilmore, M. S. (2010). Mechanism of chromosomal transfer of Enterococcus faecalis pathogenicity island, capsule, antimicrobial resistance, and other traits. *Proceedings of the National Academy of Sciences of the United States of America*, *107*(27), 12269–12274.

Marraffini, L. A., and Sontheimer, E. J. (2008). CRISPR interference limits horizontal gene transfer in staphylococci by targeting DNA. *Science (New York, N.Y.)*, *322*(5909), 1843–1845.

McGinn, J., and Marraffini, L. A. (2019). Molecular mechanisms of CRISPR–Cas spacer acquisition. *Nature Reviews Microbiology*, *17*(1), 7–12.

Melnikov, A. A., Tchernov, A. P., Fodor, I., and Bayev, A. A. (1984). Lambda phagemids and their transducing properties. *Gene*, *28*(1), 29–35.

Mohanraju, P., Makarova, K. S., Zetsche, B., Zhang, F., Koonin, E. V, and van der Oost, J. (2016). Diverse evolutionary roots and mechanistic variations of the CRISPR-Cas systems. *Science (New York, N.Y.)*, *353*(6299), aad5147.

Mojica, F. J. M., Díez-Villaseñor, C., García-Martínez, J., and Almendros, C. (2009). Short motif sequences determine the targets of the prokaryotic CRISPR defence system. *Microbiology (Reading, England)*, *155*(3), 733–740.

Nealson, K. H., Platt, T., and Hastings, J. W. (1970). Cellular control of the synthesis and activity of the bacterial luminescent system. *Journal of Bacteriology*, 104(1): 313–322.

Nihongaki, Y., Kawano, F., Nakajima, T., and Sato, M. (2015). Photoactivatable CRISPR-Cas9 for optogenetic genome editing. *Nature Biotechnology*, *33*(7), 755–760.

Nishimasu, H., and Nureki, O. (2017). Structures and mechanisms of CRISPR RNA-guided effector nucleases. *Current Opinion in Structural Biology*, *43*, 68–78.

Oliveira, P. H., Touchon, M., and Rocha, E. P. C. (2014). The interplay of restriction-modification systems with mobile genetic elements and their prokaryotic hosts. *Nucleic Acids Research*, *42*(16), 10618–10631.

Palmer, K. L., and Gilmore, M. S. (2010). Multidrug-resistant enterococci lack CRISPR-cas. *MBio*, *1*(4), e00227–10.

Pan, J., and Ren, D. (2009). Quorum sensing inhibitors: a patent overview. *Expert Opinion on Therapeutic Patents*, *19*(11), 1581–1601.

Patnaik, S., and Gupta, K. C. (2013). Novel polyethylenimine-derived nanoparticles for in vivo gene delivery. *Expert Opinion on Drug Delivery*, *10*(2), 215–228.

Pawluk, A., Staals, R. H. J., Taylor, C., Watson, B. N. J., Saha, S., Fineran, P. C., Maxwell, K. L., and Davidson, A. R. (2016). Inactivation of CRISPR-Cas systems by anti-CRISPR proteins in diverse bacterial species. *Nature Microbiology*, *1*(8), 16085.

Pérez-Mendoza, D., and de la Cruz, F. (2009). Escherichia coli genes affecting recipient ability in plasmid conjugation: Are there any?. *BMC Genomics*, *10*, 71.

Perez-Pinera, P., Kocak, D. D., Vockley, C. M., Adler, A. F., Kabadi, A. M., Polstein, L. R., Thakore, P. I., Glass, K. A., Ousterout, D. G., Leong, K. W., Guilak, F., Crawford, G. E., Reddy, T. E., and Gersbach, C. A. (2013). RNA-guided gene activation by CRISPR-Cas9-based transcription factors. *Nature Methods*, *10*(10), 973–976.

Peters, J. M., Koo, B.-M., Patino, R., Heussler, G. E., Hearne, C. C., Qu, J., Inclan, Y. F., Hawkins, J. S., Lu, C. H. S., Silvis, M. R., Harden, M. M., Osadnik, H., Peters, J. E., Engel, J. N., Dutton, R. J., Grossman, A. D., Gross, C. A., and Rosenberg, O. S. (2019). Enabling genetic analysis of diverse bacteria with Mobile-CRISPRi. *Nature Microbiology*, *4*(2), 244–250.

Qi, L. S., Larson, M. H., Gilbert, L. A., Doudna, J. A., Weissman, J. S., Arkin, A. P., and Lim, W. A. (2013). Repurposing CRISPR as an RNA-guided platform for sequence-specific control of gene expression. *Cell*, *152*(5), 1173–1183.

Rasheed, J K, Anderson, G. J., Yigit, H., Queenan, A. M., Doménech-Sánchez, A., Swenson, J. M., Biddle, J. W., Ferraro, M. J., Jacoby, G. A., and Tenover, F. C. (2000). Characterization of the extended-spectrum beta-lactamase reference strain, Klebsiella pneumoniae K6 (ATCC 700603), which produces the novel enzyme SHV-18. *Antimicrobial Agents and Chemotherapy*, *44*(9), 2382–2388.

Rasheed, J. K, Kitchel, B., Zhu, W., Anderson, K. F., Clark, N. C., Ferraro, M. J., Savard, P., Humphries, R. M., Kallen, A. J., and Limbago, B. M. (2013). New Delhi metallo-β-lactamase-producing Enterobacteriaceae, United States. *Emerging Infectious Diseases*, *19*(6), 870–878.

Rath, D., Amlinger, L., Rath, A., and Lundgren, M. (2015). The CRISPR-Cas immune system: Biology, mechanisms and applications. *Biochimie*, *117*, 119–128.

Richter, C., Chang, J. T., and Fineran, P. C. (2012). Function and regulation of clustered regularly interspaced short palindromic repeats (CRISPR) / CRISPR associated (Cas) systems. *Viruses*, *4*(10), 2291–2311.

Roach, D. R., Leung, C. Y., Henry, M., Morello, E., Singh, D., Di Santo, J. P., Weitz, J. S., and Debarbieux, L. (2017). Synergy between the host immune system and bacteriophage is essential for successful phage therapy against an acute respiratory pathogen. *Cell Host & Microbe*, *22*(1), 38–47.e4.

Roberts, A. P., and Mullany, P. (2009). A modular master on the move: The Tn916 family of mobile genetic elements. *Trends in Microbiology*, *17*(6), 251–258.

Rouet, P., Smih, F., and Jasin, M. (1994). Expression of a site-specific endonuclease stimulates homologous recombination in mammalian cells. *Proceedings of the National Academy of Sciences of the United States of America*, *91*(13), 6064–6068.

Sampson, T. R., Saroj, S. D., Llewellyn, A. C., Tzeng, Y. L., and Weiss, D. S. (2013). A CRISPR/Cas system mediates bacterial innate immune evasion and virulence. *Nature*, *497*(7448), 254–257.

Sampson, T. R., and Weiss, D. S. (2013). Cas9-dependent endogenous gene regulation is required for bacterial virulence. *Biochemical Society Transactions*, *41*(6), 1407–1411.

Shah, S. A., Erdmann, S., Mojica, F. J. M., and Garrett, R. A. (2013). Protospacer recognition motifs: Mixed identities and functional diversity. *RNA Biology*, *10*(5), 891–899.

Shmakov, S. A., Sitnik, V., Makarova, K. S., Wolf, Y. I., Severinov, K. V, and Koonin, E. V. (2017). The CRISPR spacer space is dominated by sequences from species-specific mobilomes. *MBio*, *8*(5).

Silas, S., Makarova, K. S., Shmakov, S., Páez-Espino, D., Mohr, G., Liu, Y., Davison, M., Roux, S., Krishnamurthy, S. R., Fu, B. X. H., Hansen, L. L., Wang, D., Sullivan, M. B., Millard, A., Clokie, M. R., Bhaya, D., Lambowitz, A. M., Kyrpides, N. C., Koonin, E. V, and Fire, A. Z. (2017). On the origin of reverse transcriptase-using CRISPR-Cas systems and their hyperdiverse, enigmatic spacer repertoires. *MBio*, *8*(4), e00897-17.

Silas, S., Mohr, G., Sidote, D. J., Markham, L. M., Sanchez-Amat, A., Bhaya, D., Lambowitz, A. M., and Fire, A. Z. (2016). Direct CRISPR spacer acquisition from RNA by a natural reverse transcriptase-Cas1 fusion protein. *Science (New York, N.Y.)*, *351*(6276), aad4234.

Smillie, C., Garcillán-Barcia, M. P., Francia, M. V., Rocha, E. P. C., & de la Cruz, F. (2010). Mobility of plasmids. *Microbiology and Molecular Biology Reviews: MMBR*, *74*(3), 434–452.

Sonnenburg, J. L., and Fischbach, M. A. (2011). Community health care: Therapeutic opportunities in the human microbiome. *Science Translational Medicine*, *3*(78), 78ps12.

Sorek, R., Lawrence, C. M., and Wiedenheft, B. (2013). CRISPR-mediated adaptive immune systems in bacteria and archaea. *Annual Review of Biochemistry*, *82*, 237–266.

Stern, A., Keren, L., Wurtzel, O., Amitai, G., and Sorek, R. (2010). Self-targeting by CRISPR: Gene regulation or autoimmunity? *Trends in Genetics: TIG*, *26*(8), 335–340.

Taga, M. E, Semmelhack, J. L., and Bassler, B. L. (2001). The LuxS-dependent autoinducer AI-2 controls the expression of an ABC transporter that functions in AI-2 uptake in Salmonella typhimurium. *Molecular Microbiology*, *42*(3), 777–793.

Taga, M. E, Miller, S. T., and Bassler, B. L. (2003). Lsr-mediated transport and processing of AI-2 in Salmonella typhimurium. *Molecular Microbiology*, *50*(4), 1411–1427.

Touchon, M., Charpentier, S., Pognard, D., Picard, B., Arlet, G., Rocha, E. P. C., Denamur, E., and Branger, C. (2012). Antibiotic resistance plasmids spread among natural isolates of Escherichia coli in spite of CRISPR elements. *Microbiology (Reading, England)*, *158*(12), 2997–3004.

Touchon, M., and Rocha, E. P. C. (2010). The small, slow and specialized CRISPR and anti-CRISPR of Escherichia and Salmonella. *PloS One*, *5*(6), e11126.

Vendeville, A., Winzer, K., Heurlier, K., Tang, C. M., and Hardie, K. R. (2005). Making "sense" of metabolism: Autoinducer-2, LUXS and pathogenic bacteria. *Nature Reviews Microbiology*, *3*(5), 383–396.

Vercoe, R. B., Chang, J. T., Dy, R. L., Taylor, C., Gristwood, T., Clulow, J. S., Richter, C., Przybilski, R., Pitman, A. R., and Fineran, P. C. (2013). Cytotoxic chromosomal targeting by CRISPR/Cas systems can reshape bacterial genomes and expel or remodel pathogenicity islands. *PLOS Genetics*, *9*(4), e1003454.

Vogeleer, P., Tremblay, Y. D. N., Mafu, A. A., Jacques, M., and Harel, J. (2014). Life on the outside: Role of biofilms in environmental persistence of Shiga-toxin producing Escherichia coli. *Frontiers in Microbiology*, *5*, 317.

Wang, J., Li, J., Zhao, H., Sheng, G., Wang, M., Yin, M., and Wang, Y. (2015). Structural and mechanistic basis of PAM-Dependent spacer acquisition in CRISPR-Cas systems. *Cell*, *163*(4), 840–853.

Wang, M., Zuris, J. A., Meng, F., Rees, H., Sun, S., Deng, P., Han, Y., Gao, X., Pouli, D., Wu, Q., Georgakoudi, I., Liu, D. R., and Xu, Q. (2016). Efficient delivery of genome-editing proteins using bioreducible lipid nanoparticles. *Proceedings of the National Academy of Sciences of the United States of America*, *113*(11), 2868–2873.

Wei, Y., Terns, R. M., and Terns, M. P. (2015). Cas9 function and host genome sampling in Type II-A CRISPR-Cas adaptation. *Genes & Development*, *29*(4), 356–361.

Wolfgang, W. (2003). The Enterococci: Pathogenesis, molecular biology, and antimicrobial resistance. *International Journal of Medical Microbiology*, *292*(7), 462.

Wright, A. V, Nuñez, J. K., and Doudna, J. A. (2016). Biology and applications of CRISPR systems: Harnessing nature's toolbox for genome engineering. *Cell*, *164*(1–2), 29–44.

Yan, W. X., Hunnewell, P., Alfonse, L. E., Carte, J. M., Keston-Smith, E., Sothiselvam, S., Garrity, A. J., Chong, S., Makarova, K. S., Koonin, E. V, Cheng, D. R., and Scott, D. A. (2019). Functionally diverse type V CRISPR-Cas systems. *Science (New York, N.Y.)*, *363*(6422), 88–91.

Yosef, I., Manor, M., Kiro, R., and Qimron, U. (2015). Temperate and lytic bacteriophages programmed to sensitize and kill antibiotic-resistant bacteria. *Proceedings of the National Academy of Sciences*, *112*(23), 7267–7272.

You, L., Ma, J., Wang, J., Artamonova, D., Wang, M., Liu, L., Xiang, H., Severinov, K., Zhang, X., and Wang, Y. (2019). Structure studies of the CRISPR-Csm complex reveal mechanism of co-transcriptional interference. *Cell*, *176*(1–2), 239–253.e16.

Zebec, Z., Manica, A., Zhang, J., White, M. F., and Schleper, C. (2014). CRISPR-mediated targeted mRNA degradation in the archaeon Sulfolobus solfataricus. *Nucleic Acids Research*, *42*(8), 5280–5288.

Zhang, X., Wang, J., Cheng, Q., Zheng, X., Zhao, G., & Wang, J. (2017). Multiplex gene regulation by CRISPR-ddCpf1. *Cell Discovery*, *3*(1), 17018.

Zhu, W., Clark, N. C., McDougal, L. K., Hageman, J., McDonald, L. C., and Patel, J. B. (2008). Vancomycin-resistant Staphylococcus aureus isolates associated with Inc18-like vanA plasmids in Michigan. *Antimicrobial Agents and Chemotherapy*, *52*(2), 452–457.

Zuris, J. A., Thompson, D. B., Shu, Y., Guilinger, J. P., Bessen, J. L., Hu, J. H., Maeder, M. L., Joung, J. K., Chen, Z.-Y., and Liu, D. R. (2015). Cationic lipid-mediated delivery of proteins enables efficient protein-based genome editing in vitro and in vivo. *Nature Biotechnology*, *33*(1), 73–80.

12 Small Molecule Associated Inhibition of Metabolic Pathways in Biofilm

Sreejita Ghosh, Moupriya Nag, Dibyajit Lahiri, Debasmita Bhattacharya and Rina Rani Ray

12.1 INTRODUCTION

Biofilms formed by bacteria are a matrix-surrounded, self-made, surface-adhered mass of cells that support the lives of a huge proportion of a drug-resistant population of bacterial cells. Bacterial pathogens residing within the biofilm matrix can withstand treatment with antibacterials by three distinct but inter–related mechanistic pathways: (a) sequestration of antibacterials by the EPS components of biofilms, (b) the external thick biofilm layer, called the exracellular matrix consisting of proteins, polysaccharides, lipids, extracellular DNA and other complex sugars and which can physically hinder antibacterial penetration through the biofilm and (c) creating an internal environmental condition which is hostile because it contains very low amount of oxygen and nutrients, thereby making the bacteria metabolically inactive so that they can withstand antibacterials at any low to medium doses. Since most of the antibacterials are fabricated to disrupt targets like biosynthesis of the cell wall, replication of DNA and biosynthesis of proteins but all o these processes are either completely or partially absent in the sessile cells growing within the biofilm. In contrast to the antibiotics, a majority of the fabricated anti–biofilm compounds must target the supplementary pathways without having any impact on the survival of bacteria thereby exerting low selection pressure for resistant mutant development. In most of the cases, a mixture of these compounds along with the front–line drugs has shown a better therapeutic impact.

Over the past two decades, it has been observed that the sessile form of development (biofilm mediated growth) is the most dominant life style of a majority of the species of bacteria (Flemming and Wuertz, 2019). Inside biofilms, bacterial cells are present as densely packaged microcolonies covered by a shielding bipolymeric matrix. The bacterial capability of forming biofilms is a conventional characteristic, which at the time of evolution has protected the bacterial cells from antimicrobials as well as grazing amoeba. At the time of evolution of the higher microbes, bacteria adapted to forming biofilms at different sites of the host body and nowadays, the intensive uses of medical devices in modernized healthcare sectors has provided several fresh niches for the formation of biofilm by bacteria. Within the biofilm mode of bacterial living state, the bacteria are resistant against currently available antibiotic therapies and also the human immune system (Ciofu et al., 2015). Hence, biofilms are considered as the major causative agents of persistent infectious diseases involving medical implants and other infections such as cystic fibrosis (CF), urinary tract infections (UTIs), chronic wound infections and chronic obstructive pulmonary disease (COPD) (Rybtke et al., 2015). The National Institute of Health (US) has given an estimation that over 80% of all the microbe mediated infections among the developed countries are due to biofilms.

DOI: 10.1201/9781003281238-12

The persistent infections caused due to biofilms have encouraged researchers to study the underlying molecular mechanisms in detail for the formation and dispersal of biofilms. As a result, these studies have led to the development of novel mechanisms and molecules, which can act as potential targets for the small molecule anti–biofilm drugs. Based on this knowledge, formulation of numerous active anti–biofilm small molecule drugs under development. In this article, we will describe the currently available information regarding the various metabolic pathways leading to the development of biofilms by Gram negative opportunistic pathogenic bacteria such as the ESKAPE pathogens, which are the major cause of all persistent and chronic infections. Apart from that, we will also discuss the different small molecule chemicals that have been identified so far as inhibiting the pathways that lead to the formation and dispersal of biofilms.

12.2 METABOLIC PATHWAYS FOR BIOFILM FORMATION

In this portion, we will discuss about how biofilm forms in bacteria such as *Pseudomonas aeruginosa*, *Escherichia coli* and *Acinetobacter baumannii* emphasizing the metabolic pathways that are being targeted by the newly developed small molecule anti-biofilm compounds.

Formation of biofilm needs an extracellular polymeric matrix, which under most conditions is produced by the organism, which forms the biofilm. Studies indicate that different species of bacteria employ the c–di–GMP signalling pathway for regulating the extracellular matrix production and biofilm formation (Jenal et al., 2017). Formation of the c–di–GMP molecule is being catalyzed by diguanylate cyclase enzymes (DGCs) while c–di–GMP degradation is catalyzed by a particular phosphodiesterase enzyme (PDE) in bacteria. Typically, bacteria produce numerous and different types of PDEs and DGCs and the results available indicate that c–di–GMP signal specificity is warranted via physical communications of particular PDE, c–di–GMP and DGC effectors (Sarenko et al., 2017). Apart from their catalytic sites, these enzymes frequently comprise regulatory sites and are considered to control the lifestyle (sessile versus planktonic) of the bacteria as a result of the environmental conditions. An increased c–di–GMP level in cells indicates that the components of the matrix of biofilm are also elevated and induce bacteria into forming biofilms while a reduced level of c–di–GMP inside the cells leads to the down-regulation of the production of the constituents of the biofilm matrix thereby resulting in the dispersal of bacterial cells living within the biofilm so that they can lead a planktonic lifestyle (Christensen et al., 2013). The PDEs and DGCs that break and make the c–di–GMP possess conserved catalytic HD–GYP (His–Asp–Gly–Tyr–Pro)/ EAL (Glu–Ala–Leu) and GGDEF (Gly–Gly–Asp–Glu–Phe) sites respectively. Genomic studies have suggested that the EAL and GGDEF sites are those motifs which are abundantly found in bacteria (Jenal et al., 2017). Hence, c–di–GMP is considered to be the primary regulator of biofilm formation in every Gram negative bacteria found to date and also among the majority of Gram positive bacteria (Jenal et al., 2017). However, c–di–GMP is not secreted by humans or some other mammals.

Quorum sensing (QS) is one of the regulatory processes which takes a significant part in the formation of biofilm amongst a wide variety of species of bacteria and even takes part as an overall controller for virulence factor expression (Juhas et al., 2005). QS is a cellular communication in bacteria that depends on the sensing, response and production of the extracellular signal molecules. QS supports a bacterial group to specifically alter gene expression in a synchronous manner particularly with response to the modifications in the density of their population. Among Gram negative bacteria, most of the QS systems utilize acyl homoserine lactone (AHL) signalling molecules (Ghosh et al., 2022). These systems work using one or multiple AHL synthases, which produce AHL molecules that can pass through the membranes of bacteria and once a threshold concentration is reached, the AHL molecules bind to it, thereby activating the transcriptional activators resulting in the activation of a particular target gene transcription. Signalling molecule concentration is correlated to the bacterial population density thereby activating gene expression regulation depending on the density.

12.2.1 Metabolic Pathways in *Pseudomonas aeruginosa*

Biofilms formed by *P. aeruginosa* are responsible for numerous chronic and persistent infections such as pneumonia, CF, wound infections, COPD-associated infections, bacterial prostatitis, otitis media and infections associated with medical devices and implants mainly related to endotracheal tubes and urinary tract catheters (—Cifou et al., 2015).

The process of the formation of biofilms by *P. aeruginosa* can be activated by the adherence activity of numerous components such as type IV pili, flagella, extracellular DNA (eDNA), cup fimbriae and Psl polysaccharides (Ma et al., 2009). Many of such components are even significant components in the formation of an extracellular matrix at late phases of biofilm development. Psl exopolysaccharide is associated with the surface of the cells and serves as an adhesion in the early phases of biofilm development, but gets shuffled to being a peripheral exopolysaccharide during the late phases of biofilm development (Ma et al., 2009). The adhesion CdrA, which is located at the surfaces, can bind to Psl since it is needed for the conglomeration of the cells of *P. aeruginosa* and this cell aggregation process is mediated by Psl (Borlee et al., 2010). *P. aeruginosa* produces a cationic exopolysaccharide, called Pel and it has been demonstrated to form cross-links amongst the eDNA within the matrix of biofilms (Jennings et al., 2015). Overproduction of an exopolysaccharide, called alginate, facilitates mucoid strains of *P. aeruginosa* to form persistent biofilm-associated infections within the lungs of those patients suffering from CF (Hoiby, 1977). Furthermore, *P. aeruginosa* exhibits small variants in the colonies, which overproduce Pel and Psl exopolysaccharides showing increased persistence within the lungs of CF patients (Starkey et al., 2009). Apart from this, the LecB/LecA lectins as well as Fap, which is a functional amyloid protein, act as components of the matrix of the biofilms of *P. aeruginosa* (Dueholm et al., 2013).

Biosynthesis of components such as Pel, Psl, CdrA, alginate, Cup fimbriae and type IV pili of the matrix in biofilms of *P. aeruginosa*, becomes up-regulated by c–di–GMP (Fazli et al., 2014). In contrast to this, c–di–GMP down-regulates *P. aeruginosa* motility (Simm et al., 2004). The level of c–di–GMP in *P. aeruginosa* is controlled by the action of 17 proteins having a GGDEF site, 9 proteins possessing a HD–GYP/ EAL site and 16 proteins containing an EAL and a GGDEF site. Christensen et al., 2013, designed a recombinant strain of the bacteria *P. aeruginosa*, in which the cellular content of c–di–GMP can be decreased by adding an inducer, which activates PDE gene transcription and indicates that the *in vitro* biofilms of *P. aeruginosa* become dispersed with response to a decrement of the cellular level of c–di–GMP. Additionally, studies suggested that murine implant-related biofilm–associated infections in *P. aeruginosa* can be treated by reducing the level of c–di–GMP in bacteria (Christensen et al., 2013).

The QS system in *P. aeruginosa* is regulated by the LasR/ LasI and RhlR/ RhlI proteins, which produce and respond to the signalling molecules C4–homoserine lactones (C4–HSL) and 3–oxo–C–12–homoserine lactones (HSL). Furthermore, *P. aeruginosa* makes use of a Pqs system, which can produce and respond to 2–heptyl–3–hydroxyl–4–quinolone, also called as PQS. These systems are arranged in a hierarchical manner and LasR regulates the Pqs and Rhl systems. The production of numerous compounds is mediated by QS and all these compounds are required for persistence and formation of biofilms in *P. aeruginosa*. Out of these, one of the factors that is regulated by QS is eDNA, which helps in maintaining the stability of biofilms of *P. aeruginosa* thereby making them tolerant towards antimicrobials. In addition to this, rhamnolipid production is regulated by QS and such amphipathic molecules aid in developing the biofilms thereby making *P. aeruginosa* cells resistant to phagocytosis by the immune cells of the host. Hence, QS inhibition reduces the antimicrobial resistance of biofilms formed by *P. aeruginosa* which in turn decreases the resistivity towards the host immunity.

12.2.2 Metabolic Pathways in *Escherichia coli*

E. coli is the most common cause of infections in urinary tracts like pyelonephritis (infections of the upper portion of urinary tract or kidney infections) and cystitis (infection in the lower portion of the urinary tract or in the bladder). These bacteria can also develop biofilms on walls of the bladder, catheter substances and in epithelial cells of the bladder and this formation of biofilm is a recurrent event when infection relapses.

Formation of biofilms by *E. coli* is mediated by a wide variety of extracellular matrix constituents and adhesins. Flagella even play a significant role in transporting bacteria to the surfaces thereby adhering them to the surfaces. The curli fibers contain protein materials and are the most important components of the biofilm matrix formed by *E. coli* and these fibers are required during the early stages of adherence of the cells of bacteria to the proteins inside the cells of the host (Serra et al., 2013). Curli fibers are made of two types of proteins, namely CsgB and CsgA, out of which, CsgB precedes CsgA polymerization (Hammar et al., 1995). P pili and type I pili surface adhesive structures are significant in infections caused by *E. coli.*

Type I pili are the most important for the irreversible adherence of the bacterial cells to the surfaces. The strains which lack type I pili demonstrated remarkable impairment in the formation of biofilm (Niba et al., 2007). With the help of a mouse model, it was observed that strains of *E. coli* that are uropathogenic and can express type I pili possess an advantage in survival within the bladder and those *E. coli* without the P pili or type I pili are immensely attenuated in causing infections of the urinary tracts (Sivick and Mobley, 2010). Type I pili possesses repetitive subunits of FimA, which are arranged in a rod–coli structure, which is tipped with molecules called FimH adhesins. Lack of FimH decreases adherence both within the mice bladders as well as *in vitro* (Mulvey et al., 1998). FimH is the cause for certain adhesions with residues of mannose over the epithelial cells and hence facilitates infections. P pili have subunits of PapA, which are arranged in a helical structure with PapG adhesion being anchored at the tip. PapG facilitates the cells of *E. coli* in binding with the epithelial cells of the kidney of the host. An autotransporter protein known as antigen43 is associated with the increased intracellular persistence of urinary tract infections (UTIs). Additionally with the protein constituents, the exopolysaccharides such as poly–GlcNAc (PGA), cellulose and colanic acid are significant structural parts of the biofilm matrix of *E. coli.* Furthermore, eDNA can serve as a significant constituent of the matrix in the biofilms of *E. coli.*

Biosynthesis of matrix components such as PGA, cellulose and curli fimbirae in *E. coli* biofilms, is up-regulated by c–di–GMP. In contrast to this, c–di–GMP down-regulates *E. coli* motility. The level of c–di–GMP in *E. coli* is decided through the action of numerous EAL/GGDEF/HD–GYP sites containing proteins including the laboratory strain of *E. coli* K12, which possesses 13PDEs and 12DGCs. Results suggested that there exists a hierarchical arrangement of PDEs and DGCs having a small number of master regulators (Sarenko et al., 2017). The PdeH (previously known as YhjH) is a PDE, which was demonstrated to remove the global impacts of numerous DGCs and hence restrict them from functioning on neighbouring systems (Sarenko et al., 2017). DgcE was found to be the major DGC, which primarily controls level of c–di–GMP in *E. coli* (Sarenko et al., 2017).

A QS system based on AHL is not encoded by *E. coli.* On the contrary, it definitely makes use of a furanosyl borate diester (Autoinducing peptide 2/ AI–2) associated system of QS and results indicated that such a system is responsible for regulating the matrix component biosynthesis in the biofilms (Beloin et al., 2008).

12.2.3 Metabolic Pathways in *Acinetobacter baumannii*

Most of the common infections, which are caused due to *A. baumannii* include meningitis, pneumonia, soft tissue and skin infections, UTIs, bacteremia and wound infections. The enhanced utilization of urinary and central venous catheterization and mechanical ventilation has immensely enhanced the occurrence of infections from *A. baumannii*. The capacity of *A. baumannii* to develop biofilms is a significant contributing factor for the recalcitrance of these infections against treatment with antibiotics.

The metatbolic pathways involved in the *A. baumannii* formation of biofilms, are not studied in comparison to that of *E. coli* and *P. aeruginosa*. Conversely, numerous extracellular constituents and adhesins have been observed to take a significant part in the formation of biofilms by *A. baumannii* (Gaddy et al., 2009). flagella are absent in these bacteria but adhesion to surfaces is supported by an outer membrane protein OmpA and Csu pili. Other than being adhered to abiotic surfaces, OmpA also binds to the epithelial cells thereby mediating the formation of biofilms on biotic surfaces. In addition to this, a biofilm associated protein (BAP) has been found to take an important part in the formation of biofilms by *A. baumannii* (Loehfelm et al., 2008). Studies suggested that Bap is significant in cellular adhesions thereby maintain the mature biofilm structure (Loehfelm et al., 2008). Furthermore, *A. baumannii* helps in the biosynthesis of the exopolysaccharides such as poly–β–1, 6–N–acetylglucosamine (PNAG) and alginate that can work as significant components of the matrix of biofilms. eDNA also is a component of the biofilm matrices formed by *A. baumannii*.

There is a knowledge gap about the function of the c–di–GMP signal pathway in the regulation of the formation of biofilms by *A. baumannii*. On the other hand, a study utilizing an *in-silico* pharmacopore–associated screen for identifying small molecule inhibitors of the DGC enzymes revealed that the hit components can prevent formation of biofilms by *A. baumannii* (Sambanthamoorthy et al., 2014). Csu pili biosynthesis has been found to be regulated by the BfmSR–2–component system within *A. baumannii*.

QS has been found to take a major part in the regulation of the formation of biofilms by *Acinetobacter sp*. (Anbazhagan et al., 2012). *A. baumanii* contains an AHL-associated system of QS having AbaI acting as AbaR the AHL synthase serving as the receptor for AHL. A mutant of *abaI* cannot produce AHL and was found to possess defects during the late phases of the development of biofilms (Niu et al., 2008). Additionally, it was demonstrated that adding AHL to the cells of *A. baumanii* led to the elevated expression of Csu pili thereby stimulating the formation of biofilms (Luo et al., 2015). Furthermore, a reduced Fe^{3+} concentration was observed to stimulate AHL-associated QS among *A. baumanii* thereby promoting robust formation of biofilm (Modarresi et al., 2015). *A. baumanii* can produce a quorum quenching enzyme called AidA and it was observed that AidA activation among *A. baumanii* led to the inhibition of formation of biofilms (Lopez et al., 2017).

12.3 SMALL MOLECULE BIOFILM INHIBITORS

The background of small molecule DCG-inhibitors began during 2006 with a study done by Barraud et al., 2006, who found that the nitric oxide (NO) molecule is capable of dispersing the biofilm formed by *P. aeruginosa*. The researchers demonstrated that a synergistic treatment strategy with NO as the donor and an antimicrobial substance can eliminate the biofilms. They studied different NO acting as donors, particularly sodium nitroprusside (SNP)–17, S–nitroso–N–acetylpenicillamine (SNAP)–19 and S–nitroso–L–glutathione (GSNO)–18 in synergy with antibacterial substances such as hydrogen peroxide, tobramycin and sodium dodecyl sulfate. Out of the already studied NO serving as donors, SNP was found to be the most active biofilm inhibitor amongst *P. aeruginosa*.

Even though the NO donors were found to possess antibiofilm activity, it wasn't until 2009 that their biochemical properties were unraveled (Barraud et al., 2009). It was observed that decreased non-toxic NO levels instigated the activity of PDE among *P. aeruginosa* leading to a total decline of the c–di–GMP causing the dispersal of biofilms. Apart from the reduction in the level of intracellular c–di–GMP, NO–donors even led to the negative regulation of pyoverdine biosynthesis (Kang et al., 2017). Pyoverdine is a siderophore, which is responsible for producing iron required for biofilm formation.

Sulfathiazole contains an aromatic sulfonamide, free aniline and a thiazole ring and all these components are enriched with electrons and are polar. This leads to the formation of a molecular structure having two planar sides bent around the moiety of sulfonamide. Consequently, it was also found that azathioprine 21, which is an anti–inflammatory drug used for treating various autoimmune diseases, can hamper c–di–GMP synthesis in *E. coli*.

When sulfathiazole 20 and azathioprine 21 are compared, it can be seen that they possess similar structural characteristics having two hetero aromatic portions, which are being bent around the polar moiety (sulfonamide and sulfide respectively). Azathioprine 21 and sulfathiazole 20 do not interrupt the biosynthesis of c–di–GMP by inhibiting the action of DCG. Instead, azathioprine 21 and sulfathiazole 20 were found to impede the metabolism of nucleotides thereby resulting in the modification of pools of nucleotides, which will affect the availability of c–di–GMP substrate.

Sambanthamoorthy et al., 2012 screened almost 66 000 compounds for inhibiting the activity of DCG in *Vibrio cholera*. Following numerous optimization and screening rounds, eight compounds were found to posses good antagonistic activity against various DCG enzymes having N–(4–anilinophenyl) benzamide 22 demonstrating remarkable decrease in the formation of biofilm by *P. aeruginosa*.

Sambanthamoorthy et al., 2014 conducted an *in silico* screening of pharmacophore, based on approximately 15 000 small molecules for their DCG inhibition ability thereby controlling development of biofilms. Four compounds: LP 3145 24, LP 3134 23, LP 1062 26 and LP 4010 25 were identified, which remarkably decreased the activity of Wspr in *P. aeruginosa*. More studies found that these compounds remarkably restricted formation of biofilms by *A. baumanii* and *P. aeruginosa* within a continuous flow-system. All of these four molecules were observed to disperse *P. aeruginosa* biofilms thereby inhibiting the development of biofilms in urinary catheters while only one molecule could disperse the biofilm of *A. baumanii*.

Fernicola et al., 2016 adopted a virtual screening approach to report the *in silico* screening of 2.3 $\times 10^7$ compounds from the ZINC database, with the goal of finding active inhibitors of DCG thereby targeting the active site. The active site of DCG PleD in *Caulobacter crescentus* was utilized as a structural template. Seven tested compounds demonstrated remarkable decrease in PleD DCG action with Amb379455 27b and Amb2250085 27a comprising sulfonohydrazide, the catechol moiety and the nitro group were found to be most effective inhibitors of PleD. More extensive studies showed that 27a was increasingly sensitive towards metals like Mg^{2+} that entirely wiped out the inhibitory action of WspR, PleD and YfiN. In contrast to this, 27b was found not to be interrupted by the existence of a divalent metal ion. 27b demonstrated strong inhibitory effect on DCG including WspR and YfiN in *P. aeruginosa* via binding with the DCG active site.

It is important to take note of the structural similarity while comparing Amb379455 27b with the earlier stated LP 3134 23. Both of these inhibitors of DCG contain the N–benzylidenebenzohydrazide moiety, which is thought to interact with the amino group within the active site (Asn335 residue). In addition to this, both of them comprise of the pyrogallol unit, which is a moiety having strong coordination and high polarity characteristics. Polyphenols have the attribute of forming strong bonds with proteins. Also, pyrogallol can go through reversible oxidation for generating reactive oxygen species (ROS) promoting damage of DNA. Studies suggested that the compounds containing polyphenols can hinder QS in bacteria via the blockage of the AHL-mediated communication among bacteria (Huber et al., 2003).

Lierberman et al., 2014 recognized a small molecule called ebselen (Eb) 28a, which acts as an allosteric inhibitor of c–di–GMP to its receptor sites having an RxxD domain, consisting of PelD and WspR in *P. aeruginosa*. It was indicated that Eb 28a covalently altered DCGs through the formation of a bond between the thiol of cystein residue and selenium of Eb 28a within the allosteric inhibitory site. It was also found that the allosteric DCG inhibition led to the inhibition of the activity of DCGs. The oxidation state of selenium is significant for its action since the corresponding ebselen oxide (EbO) 28a demonstrated decreased inhibiton.

Zheng et al., 2016 screened 250 000 compounds for finding a small molecule inhibitor of PDE, which can get bound to the EAL domain of YahA among *E. coli* cells. Numerous binders of PDE were being identified followed by testing for their inhibitory actions on RocR of *P. aeruginosa* having just the derivative of benzoisothiazoline 29a that can effectively inhibit c–di–GMP hydrolysis. 29a was also tested against a variety of PDEs obtained from various bacteria and was found to be increasingly discriminating towards RocR inhibition. The significance of the benzoisothiazolinone moiety was unmistakably demonstrated as 29c and 29b showing a significant reduction in inhibitory action. RocR is the example of a PDE associated with pathogenicity and does not impact the entire level of c–di–GMP within *P. aeruginosa* and hence, RocR inhibition may reduce pathogenicity also. 29a was the first example of a PDE inhibitor that is cell-permeable and can selectively inhibit PDE without impacting the other PDEs. The PDEs, which were tested, were DipA in *P. aeruginosa*, YahA in *E. coli*, PA4108 in *P. aeruginosa*, Pvr in *P. aeruginosa* and RocR in *P. aeruginosa*. In addition to this, they possessed no activity on the DCGs D70E and WspR from *P. aeruginosa*.

Wang et al., 2011 identified two compounds called 925 30b and H19 30a, which could interfere with the binding of c–di–GMP to Alg44 (PilZ domain) of *P. aeruginosa* thereby inhibiting the capacity of alginate production, which is an exopolyssacharide polymer helping in the formation of biofilm. The sulfhydryl group, which can form a disulfide covalent bond with Cys–98 in Alg44 was concluded to be of great significance for action. Apart from this, a wide range of alkyl substituents was tested without any enhancement in inhibitory action.

In a study conducted by Kim et al., 2018, a link was found between c–di–GMP and QS signaling of *P. aeruginosa*. Terrein 31 was observed to inhibit the formation of biofilm via QS receptor blocking and this similar activity is also observed in furanone C–30 32. In contrast to this, terrain 31 was observed to reduce the levels of c–di–GMP whereas furanone 31 32 increased it. Hence, this study indicated that the system of QS regulates the c–di–GMP levels by controlling PDE and DGC activities. The activity of DGC was thought to be controlled by Las–QS through PDE and LasR by Rhl–QS.

12.3.1 Small Molecule QS Inhibitors

This section will present a synopsis about small molecules, which are being identified as potential QS inhibitors (QSIs) to target mainly *A. baumanii* and *P. aeruginosa*. Inhibition of QS is the strategy of attenuating virulence without blocking the processes required for growth of bacteria. Here, we will shed light on those compounds, which target the AI–1 system among *A. baumanii* and *P. aeruginosa* and not on the cover molecules, which target the PQS system that is just utilized by *P. aeruginosa*. So far, *P. aeruginosa* is the most widely studied microbe with relation to AHL–associated QS inhibition, while the number of researches conducted on inhibition of QS in *A. baumanii* is much less.

The search for small molecules, which can decrease the effectiveness of QS system, has achieved immense importance for over two decades resulting in the identification of a wide variety of inhibitors of QS. It has been indicated that a biofilm formed by *P. aeruginosa* that lacks QS can be established either by a mutant strain of *P. aeruginosa* deficient in QS or on treating with QSIs led to a reduced survival chance within a lung infection of mouse model in comparison to the infection caused by *P. aeruginosa* having a fully functional system of QS (Bjarnsholt et al., 2005). Three parameters, which are most probably involved in this modulation of *P. aeruginosa*, are eDNA, rhamnolipids and pyocycanin. All these are completely or partially regulated by QS and are found to give protection against antibiotics and also host immune cells.

12.3.1.1 Small Molecule Inhibiting AHL-associated QS

The usual perception is that there exist three targets for chemically affecting AHL–based systems of QS using small molecules: (i) signal molecules (AHL–ligand), (ii) production of signals (LuxI–type synthase) and (iii) receptors of signals (LuxR type of receptor). In contrast to this, this is a perception with differences and modifications between AHL based systems of QS in bacteria. For example, the AHL–based system of QS in *P. aeruginosa* contains 2 LuxIR–type of systems, RhlIR and LasIR, which are interconnected in contrast to *A. baumanii* QS, which is only mediated by a signle LuxIR system called AbaIR. Furthermore, system of QS in *P. aeruginosa* is composed of extra systems such as PQS and IQS linked to the Rhl and Las system thereby rendering the inhibition of these systems more complicated.

AHL synthase inhibition can serve as a potential strategy for disrupting the cell–cell communication, but presently, there are very few studies which thoroughly investigat the small molecules that can target the production of signals. Studies have been done on the ability of analogues of S–adenosylmethionine (SAM) that is the donor of amino to form the homoserine lactone ring moiety (Parsek et al., 1999). Three analogues sinefungin 47b, S–adenosyl–homocysteine 47a and butyryl SAM have been observed to inhibit RhlI synthase in *P. aeruginosa* thereby hindering production of AHL *in vitro*.

Direct targeting of QS signaling pathways can be done by enzymatic biodegradation of the signal molecules such as AHL–acylase, AHL–lactonase or paraoxonase thereby breaking the AHL molecules through the action of oxidoreductases, which can reduce carbonyl to hydroxyl groups. Numerous AHL–lactonases have been found till now including the MomL enzyme, which can effectively biodegrade numerous kinds of AHLs (Tang et al., 2015). In recent years, MomL has been found to decrease biofilm formation by *A. baumanii* and *P. aeruginosa* and also to increase the susceptibility of the biofilm to being degraded by the action of various antibiotics. In contrast to this, no action was observed within a biofilm model of wounds or within a model of *Caenorhabditis elegans* thereby putting a question mark on the anti-QS property of the MomL *in vivo* (Zhang et al., 2017). The first lactonase that was identified in a soil bacteria *Bacillus sp.* strain 240B1, was AiiA lactonase (Dong et al., 2000). During recent years, this AiiM lactonase served as a potential inhibitor of QS inside *P. aeruginosa* thereby reducing the intensity of the infections caused by *P. aeruginosa* in an acute pneumonia mouse model (Migiyama et al., 2013). Paraoxonases (PON) have been found in numerous types of mammals. They posses strong biodegradation action towards the acetylated AHL chains particularly when the chain is long. The signal molecule 3–oxo–C–12–HSL in *P. aeruginosa* is readily hydrolyzed by PON leading to QS system attenuation (Teiber et al., 2008).

Most studies that are investigating the inhibition of QS metabolic pathways via small molecules have directed their attention towards the identification of those compounds that can interact and block the LuxR–homolog receptor. The use of analogs of AHL, which match with homolog of LuxR, which is a signal–receptor complex results in signal disruption. Numerous studies have identified with success with many such compounds which can act as a wide range of inhibitors of the metabolic pathways involved in QS by incorporating modifications of the acyl chain or lactone moiety within the native signaling molecules. The X-ray crystal structure produced, obtained from a few LuxR–type receptors was complicated, having natural AHL ligands, which can lead to the fabrication of synthetic AHL ligands. The crystalline domain structure of the ligand–binding in LasR has been already solved by Bottomley et al., 2007, while AbaR structure has not been studied till now. Furthermore, the quest for analogs of AHL requires a remarkable effort for identifying small molecules, which are completely different from the original AHL molecules. The methods that are usually employed include random chemical library screening, which consists either of the isolates obtained from natural products or synthetic compounds in combination with bioassay–based natural product fractionation.

12.3.1.2 Small Molecules Inhibiting Pili Formation

In this section, curlicides and pilicides will be described, which lead to the inhibition of the early stages of the formation of biofilms by *E. coli* and its compounds, which disrupt c–di–GMP signaling among *E. coli* and *P. aeruginosa* and also the compounds inhibiting QS among *A. baumanii* and *P. aeruginosa*. A direct differentiation among the compound efficacies is difficult to obtain because of the absence of standardization among the assays utilized in various studies. Also, in studies investigating identical protein modulation, there is frequently a modification in the utilized strains of bacteria, assay and growing media. This provides a quantitative differentiation of the activities (such as IC_{50} values) from different studies which is challenging, and in some instances, also misleading.

Attachment of bacteria is a primary step in most of the infections caused by bacteria and failure to get attached results in the elimination of the pathogenic bacteria. In accordance with this, bacteria have evolved to develop more sophisticated fimbriae and pili systems for attaching to the epithelial surfaces thereby supporting colonization and invasion of the underlying tissues. Type P and type I pili are two of these pili systems, which are involved in invasion and attachment of UPEC strains within the host resulting in UTIs. Particularly, type I pili are held responsible for infection within the lower portion of the urinary tracts resulting in cystitis (bladder infection) whereas P pili are related to pyelonephritis (kidney infection). The type I and type p pili consist of pili rods linked with the flexible end of the tip of fibrillum enabling communications with the targets.

Often, pili are arranged through an immensely conserved process known as the chaperone–usher pathway (CUP) employed by several adhesion organelles among Gram–negative bacteria such as *P. aeruginosa* and *E. coli*. Pili construction via CUP takes place from top to the bottom, namely, the first subunit that needs to be introduced is the adhesion (FimH or PapG), which is succeeded by the incorporation of the remaining fibrillum tip and subunits of the adaptor (PapE, F, and K or FimF and G), then the base of the pili (FimA or PapA) and finally, in the case of the Pap systems, the subunits for termination and anchorage (PapH). Construction of pili via CUP systems comprises of numerous feasible targets for the designing of compounds, which can obstruct this arrangement thereby blocking formation of pili.

Out of all these compounds, mannocides can compete to get bound within the mannose binding domain found in the pilus lectin FimH of type I pili thereby hindering its bond with the receptors enriched with mannose within eukaryotic cells. Until now, compounds such as biphenylmannosides have been found to efficient *in vitro* for preventing formation of biofilm in the UPEC and disrupting the already formed biofilms (Han et al., 2010). Administration of these compounds orally is effective to eradicate chronic UTIs among mice and for elevating the action of antibiotics such as sulfamethoxazole and trimethoprim (Cusumano et al., 2011). Arrangement of pili in bacteria needs chaperones present in the periplasm. Hence, the design of compounds which can bind to these chaperones is gaining importance. By pili assembly inhibition, these compounds will disrupt the attachment of bacteria to the host and hence they can act as good anti-biofilm substances. Such kinds of compounds are known as pilicides.

P pili get assembled with the help of chaperone PapD, which binds with and covers the interactive portions present on each subunit of the pili thereby restricting premature cluster formation at the time of their release inside the periplasmic space. The elevated percentage of structural similarity along with conserved mode of action among different pathogens and CUP proteins create compounds, which use this system as their targets and hence behave as efficient broad spectrum anti-microbial drugs.

Peptides with C-terminal such as PapH and PapG have been demonstrated to become bound with a chaperone leading to potential inhibition of being further bound to the subunits of pilus (Karlsson et al., 1998). This has led to the development of effective probe molecules, which are used to study the intricate mechanisms of the CUPs. On the other hand, using peptides as drug molecules has numerous limitations such as ineffective absorption when administered orally, fast biodegradation and/or being excreted. Hence, attention is given to fabricating small-molecule peptidomimetics as effective drug molecules.

The vast amount of knowledge that is known about the C–terminus of PapG and their interactions with the chaperones has been exploited for the design of the small drug-like molecules based on the structures mimicking the C–terminus of the PapG proteins (Svensson et al., 2001). The scaffolds made of the dihydrothiazolo ring in fusion with 2–pyridone is the most broadly studied structure of pilicides, which target the Lys112/ Arg8 cleft portion of the chaperone. This scaffold contains the carboxylic acid moiety, which is essential for the association of Lys112 and Arg8 residues within the cleft of the chaperone. In addition to this, the scaffold can again be enhanced by installing substituents in different places for introduction of significant hydrophobic motifs (such as Leu311 residue) and/or exploitation of other effective interactions.

The significance of the bicyclic 2-pyridone system was elaborately explained as monocyclic 2-pyridone showing a significantly lowered bioactivity in comparison to the bicyclic compound (Pemberton et al., 2008).

The first generation bicyclic 2-pyridone pilicides lacked the characteristic of being soluble in water thereby restricting their uses. In contrast to this, introduction of a substituent of aminomethylene in the C6 position, which is open within the 2-pyridone scaffold, resolved this limitation leading the 2-pyridone to bind with chaperones within a low millimolar range.

The carboxylic acid groups present on the scaffolds of the pilicides is important for pilicide activity and exchange of these carboxylic acids with other functional groups such as $-CH_2OH$, $-CO_2Me$, -CHO, $-CH_2OMe$ or CH_3 effectively decreased the activity of pilicides. On the other hand, isoesters of carboxylic acids comprised of acyl sulfonamides, tetrazoles and hydroxamic acids, which tolerated and in some instances even led to an increased potency of pilicides.

The functionality of the carboxylic acid was first fabricated for interaction with Lys112 and Arg8 in the cleft of the chaperone. Later on, a study based on NMR, elucidated the binding site of the pilicides revealing that the chaperone cleft is not the only site of binding (Hendenstrom et al., 2005). Instead, this study further indicated that pilicides have an impact on the formation of pilus in *E. coli* either through binding within the cleft of the chaperone or via modulating the arrangement of the flexible F1–G1 loop of the chaperone and both of them are portions of the surfaces, which are involved in interacting with the subunits of the pilus. Another site for binding was shortly indicated by an X-ray crystallographic study of the complex of PapD–plicide (Pinkner et al., 2006).

The impact of modulating the C8 and C7 terminii of the scaffold of 2-pyridone have been studied extensively on the basis of two major objectives; firstly, for improving the peptide mimicking characteristics of the recognized hits and secondly, for establishing the structure-activity relationship (SAR) about the activity of pilicides (Chorell et al., 2012). Generally, the characteristics of the substituents at both C8 and C7 positions were found to strongly impact the bioactivity (Chorell et al., 2012).

Chorell et al., 2012 conducted a comprehensive study on the SAR effects of substituting C7. On the basis of this study, two wide trends of activity were delineated: (i) extended linkers of ether are allowed and in some instances, can also improve action and (ii) hydrophobic sterically demanding substituents of C7 are required whereas, linkers, which incorporate basic amines, sulfonamides and amides and also heteroaryls reduce the capacity to inhibit biofilms. Overall, the most interesting substituents of C7 were identified as naphthoxymethyl, 1–naphthylmethyl, 2, 3–dimethylphenoxymethyl and 3–tolyethyl.

Similarly, hydrophobic, aromatic and large substituents are necessary in the R1 position of 2–pyridone for significant binding with the chaperone. Compounds having indole, phenyl, 3,4–methylenedioxyphenyl or thiophene groups at position C8 were effective whereas the introduction of pyridine and small substituents such as cyclopropyl, methoxy and isopropyl led to development of compounds having decreased potency. Moreover, in addition to this, the substituent R1 impacts 2–pyridone situated at position C2. For instance, 2–pyridone having cyclopropyl at position R1 was more susceptible towards racemization in comparison to 2–pyridone having phenyl at position R1.

12.3.1.3 Small Molecules Inhibiting Second Messenger Molecule Siganling

Aggregation of nucleotide second messenger molecules in presence of specific non–optimal conditions of growth can remodel cellular metabolism thereby translating into visible phenotypic modifications. One of the most well-understood second messenger molecules are guanosine pentaphospahte (pppGpp) and guanosine tetraphosphate (ppGpp) (together called (p)ppGpp), which play a major part among a huge number of biological mechanisms for ensuring survival in the presence of nutrient–limiting situations. Intracellular levels of (p)ppGpp seem to possess a confirmatory association with QS, formation and dispersal of biofilms in several bacteria such as *Enterococcus faecalis*, *Vibrio cholera* and *Bordetella pertussis*, in which ppGpp absence led to decreased formation of biofilms. However, the null mutant strains of ppGpp produce more biofilm in comparison to the wild type strain of *Actinobacillus pleuropneumoniae* and *Francisella novicida* (Valentini and Filloux, 2016). In the case of *E. coli*, a relA spot deletional mutant strain produced more biofilms in LB media but less bioiflms when grown on minimal media. Liu et al., 2017 suggested that (p)ppGpp mediates regulation of biofilm in *Helicobacter pylori* and *Pseudomonas putida* respectively.

Another significant second messenger bis(3',5')–cyclic diguanylic acid (c–di–GMP) is known to regulate various physiological processes with response to external stumuli. Cyclic di–GMP signaling is one of the most evolved positive regulators of the formation of biofilm amongst Gram negative bacteria. c–di–GMP is also known to be a contributing factor in lifestyle transition in most of the bacteria such as *E. coli*, *P. aeruginosa* and *Salmonella enterica* Serovar Typhimurium, in which, usually high and low concentrations of c–di–GMP facilitate sessile and motile states, respectively. In *Mycobacterium smegmatis*, c–di–GMP has been found to play a critical role in QS and formation of biofilms (Ghosh et al., 2020). This nucleotide messenger is biologically synthesized by the action of enzymes such as diguanylate cyclases (DGCs) that can convert two GTP molecules into c–di–GMP and themselves get disintegrated by phosphodiesterase enzymes (PDEs) to GMP and/or 5'–phosphoguanylyl–(3'–5')–guanosine (pGpG). The rate of biosynthesis and biodegradation directs the intracellular level of c–di–GMP contributing to sluggish formation of biofilms, biofilm growth and resistivity against drugs.

c–di–GMP as well as (p)ppGpp pathways of signaling and metabolism directly lead to the development of antimicrobial resistance and thereby can act as a promising therapeutic strategy for the reduction of the population within biofilms resulting in enhanced susceptibility against the standard drugs. Two of the second messenger synthetase enzymes, mainly DGC and Rel are considered to be the most lucid targets for drugs to reduce the intracellular c–di–GMP and (p)ppGpp concentration, respectively. Syal et al., 2017 reported that vitamin C shares similarity in structure to a substrate of the Rel enzyme, guanosine diphosphate (GDP) thereby inhibiting biosynthesis of ppGpp by directly becoming bound with the Rel enzyme within *M. smegmatis* and consequently it inhibits the formation of biofilms. In another study conducted by this research group, it was demonstrated that synthetic (p)ppGpp analogs can directly block biosynthesis of ppGpp thereby serving as an effective inhibitor of biofilms among mycobacteria. During recent years, Dutta et al., 2019, described the strategy of the screening of compounds followed by identification of a lead compound X9, which can particularly target Rel_{Mtb} thereby eradicating the persisters starving due to lack of nutrients.

c–di–GMP is thought to be one of the major regulators of the biofilm phenotype and contributes to the formation of biofilm in several bacteria. A synthetic organic molecule known as terrain, has been observed to decrease c–di–GMP as well as the level of biofilm formation by *P. aeruginosa* (Kim et al., 2018). Glycosylated triterpenoid saponin has been found to serve as an inhibitor of DCG enzyme. Recently, Sambanthanmoorthy et al., 2012, developed a high-throughput screening strategy for the identification of numerous small molecules for DCG inhibition in *P. aeruginosa* and *V. cholerae* subsequently reducing formation of biofilms.

The evolving second messenger cyclic–di–AMP (c–di–AMP) of bacteria plays an important function in the formation of biofilm. Peng et al., 2016 reported that c–di–AMP can bind to the receptors CabPA thereby promoting the formation of biofilm in *Streptococcus mutans*. Diadenylate cyclase (DAC) enzymes can transform two molecules of ATP to c–di–AMP and are regarded as one of the confirmed targets for c–di–AMP reduction. Bromophenol–TH was the first ever found inhibitor of DisA synthetase (enzyme involved in biosynthesis of c–di–AMP) having restricted therapeutic applications (Zheng et al., 2014).

12.4 CONCLUSION

By having a complete understanding of the biology of the biofilms, some of the selected small molecules have produced promising results in biofilm eradication. However, one of the limitations which still exists is the lack of *in vivo* results and in most of the instances, the mode of action of these small molecules is still not clear and thus no complete structure–activity reports are obtained. In some of the clinical cases, by the time any biofilm–associated infection is detected, the bacterial pathogen might have progressed towards a more complex phase of forming biofilms or might have already formed a matured biofilm. In such cases, thorough knowledge of the microbe–specific biology and the metabolic pathways involved in the formation of biofilm, is required to target the metabolic pathways by the use of the developed small molecules. Furthermore, the fact that there are multiple metabolic pathways and numerous number of genes involved in the formation of biofilm for a single microbe, hence it is very difficult to inhibit the formation of biofilms via targeting one such pathway or gene. Hence, a very accurate and succinct strategy will be the use of small–molecule associated chemical strategy for dispersing the biofilms formed *in vivo* irrespective of the type of bacteria forming that biofilm. Research in future must be aimed towards achieving more insights into these interesting small molecules possessing chemical properties like that of a drug and has very low chances of developing a mutant that is resistant to these compounds. A nanoparticle–based strategy can also aid to enhancing drug penetration within the biofilm structure and nanoparticle–small molecule comples approach include 'smart release' and 'on–demand activation' of bioactive small molecules, which will remarkably alter the outcome of the treatment approach. The interdisciplinary approach to developing these anti–biofilm small molecules for clinical research will serve as a significant step forward in the fight against resistance towards antimicrobials.

REFERENCES

Anbazhagan, D., Mansor, M., Yan, G.O., Md Yusof, M. Y., Hassan, H., and Sekaran, S. D. (2012). Detection of quorum sensing signal molecules and identification of an autoinducer synthase gene among biofilm forming clinical isolates of Acinetobacter spp. *PloS one*, *7*(7), e36696.

Barraud, N., Hassett, D. J., Hwang, S. H., Rice, S. A., Kjelleberg, S., and Webb, J. S. (2006). Involvement of nitric oxide in biofilm dispersal of Pseudomonas aeruginosa. *Journal of Bacteriology*, *188*(21), 7344–7353.

Barraud, N., Schleheck, D., Klebensberger, J., Webb, J. S., Hassett, D. J., Rice, S. A., and Kjelleberg, S. (2009). Nitric oxide signaling in Pseudomonas aeruginosa biofilms mediates phosphodiesterase activity, decreased cyclic di-GMP levels, and enhanced dispersal. *Journal of Bacteriology*, *191*(23), 7333–7342.

Beloin, C., Roux, A., and Ghigo, J. M. (2008). Escherichia coli biofilms. *Current Topics in Microbiology and Immunology*, *322*, 249–289.

Bjarnsholt, T., Jensen, P. Ø., Burmølle, M., Hentzer, M., Haagensen, J., Hougen, H. P., Calum, H., Madsen, K. G., Moser, C., Molin, S., Høiby, N., and Givskov, M. (2005). Pseudomonas aeruginosa tolerance to tobramycin, hydrogen peroxide and polymorphonuclear leukocytes is quorum-sensing dependent. *Microbiology (Reading, England)*, *151*(2), 373–383.

Borlee, B. R., Goldman, A. D., Murakami, K., Samudrala, R., Wozniak, D. J., and Parsek, M. R. (2010). Pseudomonas aeruginosa uses a cyclic-di-GMP-regulated adhesin to reinforce the biofilm extracellular matrix. *Molecular Microbiology*, *75*(4), 827–842.

Bottomley, M. J., Muraglia, E., Bazzo, R., and Carfì, A. (2007). Molecular insights into quorum sensing in the human pathogen Pseudomonas aeruginosa from the structure of the virulence regulator LasR bound to its autoinducer. *Journal of Biological Chemistry*, *282*(18), 13592–13600.

Chorell, E., Pinkner, J. S., Bengtsson, C., Banchelin, T. S., Edvinsson, S., Linusson, A., Hultgren, S. J., and Almqvist, F. (2012). Mapping pilicide anti-virulence effect in Escherichia coli, a comprehensive structure-activity study. *Bioorganic and Medicinal Chemistry*, *20*(9), 3128–3142.

Christensen, L. D., van Gennip, M., Rybtke, M. T., Wu, H., Chiang, W. C., Alhede, M., Høiby, N., Nielsen, T. E., Givskov, M., and Tolker-Nielsen, T. (2013). Clearance of Pseudomonas aeruginosa foreign-body biofilm infections through reduction of the cyclic Di-GMP level in the bacteria. *Infection and Immunity*, *81*(8), 2705–2713.

Ciofu, O., Tolker-Nielsen, T., Jensen, P. Ø., Wang, H., and Høiby, N. (2015). Antimicrobial resistance, respiratory tract infections and role of biofilms in lung infections in cystic fibrosis patients. Adv*anced* Drug Deliv*ery* Rev*iews*, *85*, 7–23. DOI: 10.1016/j.addr.2014.11.017

Cusumano, C. K., Pinkner, J. S., Han, Z., Greene, S. E., Ford, B. A., Crowley, J. R., Henderson, J. P., Janetka, J. W., and Hultgren, S. J. (2011). Treatment and prevention of urinary tract infection with orally active FimH inhibitors. *Science Translational Medicine*, 3(109), 109ra115.

Dong, Y. H., Xu, J. L., Li, X. Z., and Zhang, L. H. (2000). AiiA, an enzyme that inactivates the acylhomoserine lactone quorum-sensing signal and attenuates the virulence of Erwinia carotovora. *Proceedings of the National Academy of Sciences of the United States of America*, *97*(7), 3526–3531.

Dueholm, M. S., Søndergaard, M. T., Nilsson, M., Christiansen, G., Stensballe, A., Overgaard, M. T., Givskov, M., Tolker-Nielsen, T., Otzen, D. E., and Nielsen, P. H. (2013). Expression of fap amyloids in Pseudomonas aeruginosa, P. fluorescens, and P. putida results in aggregation and increased biofilm formation. *MicrobiologyOpen*, *2*(3), 365–382.

Dutta, N. K., Klinkenberg, L. G., Vazquez, M. J., Segura-Carro, D., Colmenarejo, G., Ramon, F., Rodriguez-Miquel, B., Mata-Cantero, L., Porras-De Francisco, E., Chuang, Y. M., Rubin, H., Lee, J. J., Eoh, H., Bader, J. S., Perez-Herran, E., Mendoza-Losana, A., and Karakousis, P. C. (2019). Inhibiting the stringent response blocks *Mycobacterium tuberculosis* entry into quiescence and reduces persistence. *Science Advances*, *5*(3), eaav2104.

Fazli, M., Almblad, H., Rybtke, M. L., Givskov, M., Eberl, L., and Tolker-Nielsen, T. (2014). Regulation of biofilm formation in Pseudomonas and Burkholderia species. *Environmental Microbiology*, *16*(7), 1961–1981.

Fernicola, S., Paiardini, A., Giardina, G., Rampioni, G., Leoni, L., Cutruzzolà, F., and Rinaldo, S. (2015). In silico discovery and in vitro validation of catechol-containing sulfonohydrazide compounds as potent inhibitors of the diguanylate cyclase pleD. *Journal of Bacteriology*, *198*(1), 147–156.

Flemming, H. C., and Wuertz, S. (2019). Bacteria and archaea on Earth and their abundance in biofilms. *Nature Reviews. Microbiology*, *17*(4), 247–260.

Gaddy, J. A., Tomaras, A. P., and Actis, L. A. (2009). The Acinetobacter baumannii 19606 OmpA protein plays a role in biofilm formation on abiotic surfaces and in the interaction of this pathogen with eukaryotic cells. *Infection and Immunity*, *77*(8), 3150–3160.

Ghosh, A., Jayaraman, N., and Chatterji, D. (2020). Small-molecule inhibition of bacterial biofilm. *ACS Omega*, *5*(7), 3108–3115.

Ghosh, S., Lahiri, D., Nag, M., Dey, A., Pandit, S., Sarkar, T., Pati, S., Abdul Kari, Z., Ishak, A. R., Edinur, H. A., and Ray, R. R. (2022). Phytocompound mediated blockage of quorum sensing cascade in ESKAPE pathogens. *Antibiotics (Basel, Switzerland)*, *11*(1), 61.

Hammar, M., Arnqvist, A., Bian, Z., Olsén, A., and Normark, S. (1995). Expression of two csg operons is required for production of fibronectin – and congo red-binding curli polymers in *Escherichia coli* K-12. *Molecular Microbiology*, *18*(4), 661–670.

Han, Z., Pinkner, J. S., Ford, B., Obermann, R., Nolan, W., Wildman, S. A., Hobbs, D., Ellenberger, T., Cusumano, C. K., Hultgren, S. J., and Janetka, J. W. (2010). Structure-based drug design and optimization of mannoside bacterial FimH antagonists. *Journal of Medicinal Chemistry*, *53*(12), 4779–4792.

Hedenström, M., Emtenäs, H., Pemberton, N., Aberg, V., Hultgren, S. J., Pinkner, J. S., Tegman, V., Almqvist, F., Sethson, I., and Kihlberg, J. (2005). NMR studies of interactions between periplasmic chaperones from uropathogenic E. coli and pilicides that interfere with chaperone function and pilus assembly. *Organic and Biomolecular Chemistry*, *3*(23), 4193–4200.

Hoiby N. (1977). *Pseudomonas aeruginosa* infection in cystic fibrosis. Diagnostic and prognostic significance of pseudomonas aeruginosa precipitins determined by means of crossed immunoelectrophoresis. A survey. *Acta Pathologica et Microbiologica Scandinavica*, Supplement, *262*, 1–96.

Huber, B., Eberl, L., Feucht, W., and Polster, J. (2003). Influence of polyphenols on bacterial biofilm formation and quorum-sensing. *Zeitschrift fur Naturforschung. C, Journal of Biosciences*, *58*(11–12), 879–884.

Jenal, U., Reinders, A., and Lori, C. (2017). Cyclic di-GMP: Second messenger extraordinaire. *Nature Reviews. Microbiology*, *15*(5), 271–284.

Jennings, L. K., Storek, K. M., Ledvina, H. E., Coulon, C., Marmont, L. S., Sadovskaya, I., Secor, P. R., Tseng, B. S., Scian, M., Filloux, A., Wozniak, D. J., Howell, P. L., and Parsek, M. R. (2015). Pel is a cationic exopolysaccharide that cross-links extracellular DNA in the *Pseudomonas aeruginosa* biofilm matrix. *Proceedings of the National Academy of Sciences of the United States of America, 112*(36), 11353–11358.

Juhas, M., Eberl, L., and Tümmler, B. (2005). Quorum sensing: The power of cooperation in the world of Pseudomonas. *Environmental Microbiology, 7*(4), 459–471.

Kang, D., Turner, K. E., and Kirienko, N. V. (2017). PqsA promotes pyoverdine production via biofilm formation. *Pathogens (Basel, Switzerland), 7*(1), 3.

Karlsson, K. F., Walse, B., Drakenberg, T., Roy, S., Bergquist, K. E., Pinkner, J. S., Hultgren, S. J., and Kihlberg, J. (1998). Binding of peptides in solution by the Escherichia coli chaperone PapD as revealed using an inhibition ELISA and NMR spectroscopy. *Bioorganic and Medicinal Chemistry, 6*(11), 2085–2101.

Kim, B., Park, J. S., Choi, H. Y., Yoon, S. S., and Kim, W. G. (2018). Terrein is an inhibitor of quorum sensing and c-di-GMP in *Pseudomonas aeruginosa*: A connection between quorum sensing and c-di-GMP. *Scientific Reports, 8*(1), 8617.

Lieberman, O. J., Orr, M. W., Wang, Y., and Lee, V. T. (2014). High-throughput screening using the differential radial capillary action of ligand assay identifies ebselen as an inhibitor of diguanylate cyclases. *ACS Chemical Biology, 9*(1), 183–192.

Loehfelm, T. W., Luke, N. R., and Campagnari, A. A. (2008). Identification and characterization of an *Acinetobacter baumannii* biofilm-associated protein. *Journal of Bacteriology, 190*(3), 1036–1044.

López, M., Mayer, C., Fernández-García, L., Blasco, L., Muras, A., Ruiz, F. M., Bou, G., Otero, A., Tomás, M., and GEIH-GEMARA (SEIMC) (2017). Quorum sensing network in clinical strains of *A. baumannii*: AidA is a new quorum quenching enzyme. *PloS one, 12*(3), e0174454.

Luo, L. M., Wu, L. J., Xiao, Y. L., Zhao, D., Chen, Z. X., Kang, M., Zhang, Q., and Xie, Y. (2015). Enhancing pili assembly and biofilm formation in *Acinetobacter baumannii* ATCC19606 using non-native acyl-homoserine lactones. *BMC Microbiology, 15*, 62.

Ma, L., Conover, M., Lu, H., Parsek, M. R., Bayles, K., and Wozniak, D. J. (2009). Assembly and development of the *Pseudomonas aeruginosa* biofilm matrix. *PLoS Pathogens, 5*(3), e1000354.

Migiyama, Y., Kaneko, Y., Yanagihara, K., Morohoshi, T., Morinaga, Y., Nakamura, S., Miyazaki, T., Hasegawa, H., Izumikawa, K., Kakeya, H., Kohrogi, H., and Kohno, S. (2013). Efficacy of AiiM, an N-acylhomoserine lactonase, against Pseudomonas aeruginosa in a mouse model of acute pneumonia. *Antimicrobial Agents and Chemotherapy, 57*(8), 3653–3658.

Modarresi, F., Azizi, O., Shakibaie, M. R., Motamedifar, M., Mosadegh, E., and Mansouri, S. (2015). Iron limitation enhances acyl homoserine lactone (AHL) production and biofilm formation in clinical isolates of Acinetobacter baumannii. *Virulence, 6*(2), 152–161.

Mulvey, M. A., Lopez-Boado, Y. S., Wilson, C. L., Roth, R., Parks, W. C., Heuser, J., and Hultgren, S. J. (1998). Induction and evasion of host defenses by type 1-piliated uropathogenic Escherichia coli. *Science (New York, N.Y.), 282*(5393), 1494–1497.

Niba, E. T., Naka, Y., Nagase, M., Mori, H., and Kitakawa, M. (2007). A genome-wide approach to identify the genes involved in biofilm formation in E. coli. *DNA Research: An International Journal for Rapid Publication of Reports on Genes and Genomes, 14*(6), 237–246.

Niu, C., Clemmer, K. M., Bonomo, R. A., and Rather, P. N. (2008). Isolation and characterization of an autoinducer synthase from Acinetobacter baumannii. *Journal of Bacteriology, 190*(9), 3386–3392.

Parsek, M. R., Val, D. L., Hanzelka, B. L., Cronan, J. E., Jr, and Greenberg, E. P. (1999). Acyl homoserine-lactone quorum-sensing signal generation. *Proceedings of the National Academy of Sciences of the United States of America, 96*(8), 4360–4365.

Pemberton, N., Pinkner, J. S., Edvinsson, S., Hultgren, S. J., and Almqvist, F. (2008). Synthesis and evaluation of dihydroimidazolo and dihydrooxazolo ring-fused 2-pyridones—targeting pilus biogenesis in uropathogenic bacteria. *Tetrahedron*, 64, 9368–9376.

Peng, X., Zhang, Y., Bai, G., Zhou, X., and Wu, H. (2016). Cyclic di-AMP mediates biofilm formation. *Molecular Microbiology, 99*(5), 945–959.

Pinkner, J. S., Remaut, H., Buelens, F., Miller, E., Aberg, V., Pemberton, N., Hedenström, M., Larsson, A., Seed, P., Waksman, G., Hultgren, S. J., and Almqvist, F. (2006). Rationally designed small compounds inhibit pilus biogenesis in uropathogenic bacteria. *Proceedings of the National Academy of Sciences of the United States of America, 103*(47), 17897–17902.

Rybtke, M., Hultqvist, L. D., Givskov, M., and Tolker-Nielsen, T. (2015). *Pseudomonas aeruginosa* biofilm infections: Community structure, antimicrobial tolerance and immune response. *Journal of Molecular Biology*, *427*(23), 3628–3645.

Sambanthamoorthy, K., Luo, C., Pattabiraman, N., Feng, X., Koestler, B., Waters, C. M., and Palys, T. J. (2014). Identification of small molecules inhibiting diguanylate cyclases to control bacterial biofilm development. *Biofouling*, *30*(1), 17–28.

Sambanthamoorthy, K., Luo, C., Pattabiraman, N., Feng, X., Koestler, B., Waters, C. M., and Palys, T. J. (2014). Identification of small molecules inhibiting diguanylate cyclases to control bacterial biofilm development. *Biofouling*, *30*(1), 17–28.

Sambanthamoorthy, K., Sloup, R. E., Parashar, V., Smith, J. M., Kim, E. E., Semmelhack, M. F., Neiditch, M. B., and Waters, C. M. (2012). Identification of small molecules that antagonize diguanylate cyclase enzymes to inhibit biofilm formation. *Antimicrobial Agents and Chemotherapy*, *56*(10), 5202–5211.

Sarenko, O., Klauck, G., Wilke, F. M., Pfiffer, V., Richter, A. M., Herbst, S., Kaever, V., and Hengge, R. (2017). More than enzymes that make or break cyclic Di-GMP-Local Signaling in the interactome of GGDEF/EAL domain proteins of *Escherichia coli*. *mBio*, *8*(5), e01639–17.

Serra, D. O., Richter, A. M., and Hengge, R. (2013). Cellulose as an architectural element in spatially structured Escherichia coli biofilms. *Journal of Bacteriology*, *195*(24), 5540–5554.

Simm, R., Morr, M., Kader, A., Nimtz, M., and Römling, U. (2004). GGDEF and EAL domains inversely regulate cyclic di-GMP levels and transition from sessility to motility. *Molecular Microbiology*, *53*(4), 1123–1134.

Sivick, K. E., and Mobley, H. L. (2010). Waging war against uropathogenic Escherichia coli: Winning back the urinary tract. *Infection and Immunity*, *78*(2), 568–585.

Starkey, M., Hickman, J. H., Ma, L., Zhang, N., De Long, S., Hinz, A., Palacios, S., Manoil, C., Kirisits, M. J., Starner, T. D., Wozniak, D. J., Harwood, C. S., and Parsek, M. R. (2009). Pseudomonas aeruginosa rugose small-colony variants have adaptations that likely promote persistence in the cystic fibrosis lung. *Journal of Bacteriology*, *191*(11), 3492–3503.

Svensson, A., Larsson, A., Emtenäs, H., Hedenström, M., Fex, T., Hultgren, S. J., Pinkner, J. S., Almqvist, F., and Kihlberg, J. (2001). Design and evaluation of pilicides: Potential novel antibacterial agents directed against uropathogenic Escherichia coli. *Chembiochem: A European Journal of Chemical Biology*, *2*(12), 915–918.

Syal, K., Flentie, K., Bhardwaj, N., Maiti, K., Jayaraman, N., Stallings, C. L., and Chatterji, D. (2017). Synthetic (p)ppGpp analogue is an inhibitor of stringent response in mycobacteria. *Antimicrobial Agents and Chemotherapy*, *61*(6), e00443–17.

Tang, K., Su, Y., Brackman, G., Cui, F., Zhang, Y., Shi, X., Coenye, T., and Zhang, X. H. (2015). MomL, a novel marine-derived N-acyl homoserine lactonase from Muricauda olearia. *Applied and Environmental Microbiology*, *81*(2), 774–782.

Teiber, J. F., Horke, S., Haines, D. C., Chowdhary, P. K., Xiao, J., Kramer, G. L., Haley, R. W., and Draganov, D. I. (2008). Dominant role of paraoxonases in inactivation of the Pseudomonas aeruginosa quorum-sensing signal N-(3-oxododecanoyl)-L-homoserine lactone. *Infection and Immunity*, *76*(6), 2512–2519.

Valentini, M., and Filloux, A. (2016). Biofilms and cyclic di-GMP (c-di-GMP) Signaling: Lessons from pseudomonas aeruginosa and other bacteria. *Journal of Biological Chemistry*, *291*(24), 12547–12555.

Wang, J., Zhou, J., Donaldson, G. P., Nakayama, S., Yan, L., Lam, Y. F., Lee, V. T., and Sintim, H. O. (2011). Conservative change to the phosphate moiety of cyclic diguanylic monophosphate remarkably affects its polymorphism and ability to bind DGC, PDE, and PilZ proteins. *Journal of the American Chemical Society*, *133*(24), 9320–9330.

Zhang, Y., Brackman, G., and Coenye, T. (2017). Pitfalls associated with evaluating enzymatic quorum quenching activity: The case of MomL and its effect on *Pseudomonas aeruginosa* and *Acinetobacter baumannii* biofilms. *PeerJ*, *5*, e3251.

Zheng, Y., Tsuji, G., Opoku-Temeng, C., and Sintim, H. O. (2016). Inhibition of *P. aeruginosa* c-di-GMP phosphodiesterase RocR and swarming motility by a benzoisothiazolinone derivative. *Chemical Science*, *7*(9), 6238–6244.

Zheng, Y., Zhou, J., Sayre, D., Sintim, H. (2014). Identification of bromophenolthiohydantoin as an inhibitor of DisA, a c-di-AMP synthase, from a 1000 compound library, using the coralyne assay. *Chemical Communications* 50, 11234–11237.

13 Bacteriophage Associated Inhibition of Biofilm

A Review

Dipro Mukherjee, Sanchari Das, Sayantani Garai, Dibyajit Lahiri, Moupriya Nag and Debasmita Bhattacharya

13.1 INTRODUCTION

Lytic bacteriophages can turn into a new type of agent which will be anti-biofilm. Biofilms which are present on different types of medical devices can also be eradicated with the help of bacteriophages [1]. Different microbes which comprise the microflora of the skin, oral cavity, genitourinary tract, gastrointestinal tract and the external environment of the human body or which can be found in infected tissues can undergo colonization in the indwelling medical devices for the formation of a biofilm [2]. Organisms whose isolation processes are quite simple include *Pseudomonas aeruginosa, Enterococcus faecium, Staphylococcus aureus, coagulase negative staphylococci, Enterococcus faecalis, Escherichia coli, Proteus mirabilis, Klebsiella pneumoniae, Viridans streptococci* and *Acinetobacter species* [3]. Those infections that are in association with different medical devices result in substantial mortality and morbidity and give rise to a large increased cost in patient care [4].

In any natural environment it is less common to find that a single species of microorganism will form a biofilm. A biofilm is more likely to consist of numerous ranges of species, thereby making a natural biofilm a unique and complex environment and ubiquitous to that specific environmental constraint. A biofilm is in a constant state of change and adaptation and biofilm formation is a multistep complex process that is controlled broadly by two separate factors and their combination, genetic and the physic-chemical [5]. The intricate process of the biofilm formation may vary from species to species and different strain to strain and also may vary due to the overruling of the environmental factors depending upon where the biofilm is formed but the overall main process of biofilm formation remains universal and similar in steps. A biofilm is a multi-layered community of the cells of different kinds forming a consortium which remains embedded in hydrated extracellular polymeric substances recreated and formed by the organisms itself. This helps the cells to colonize the living, inert or the boundary surface. This influences the organisms which are living in the extracellular polymeric substance matrix of the biofilm to become virulent and the main cause of this is the antimicrobial resistance formed due to the aforementioned encapsulation. This biofilm matrix is the preferred way for the bacteria to live and hence it is the natural state of its existence [6, 7].

13.2 BACTERIOPHAGES

Bacteriophages are natural biological entities responsible for infecting bacterial cells. Virions, the complete infectious virus particles contain a RNA or DNA core wrapped with a protein coat or

DOI: 10.1201/9781003281238-13

capsid or lipoprotein coat. Phages usually act as obligate parasites since they are unable to procreate and carry out their life cycle without involving a suitable host [8]. The nucleic acid inside bacteriophages provides all the information necessary for its procreation inside the bacterial host. Virulent phages involve a lytic cycle for their multiplication. During this cycle the bacteriophages seize the host cell by attaching to its surface followed by inserting their genetic materials, allowing the genomic materials to replicate thus creating several new phages. Cell lysis and the release of intact bacteriophages mark the end of this lytic cycle [9, 10]. Adsorption marks the opening of the lytic cycle which is assisted by tail fibres that are responsible for primary specific binding of bacteriophages to the surfaces of bacterial host cells. The host range is determined by the specificity of receptors for an individual bacteriophage strain [11]. Some phages are strain-specific, whereas others are broad-spectrum, infecting numerous bacterial cells within a specific species or maybe a couple of related-species. Based upon the burst size of the specific phage strain, an individual virion of bacteriophage infecting a bacterial cell may further lead to the generation of numerous daughter phages during the lytic cycle of the bacteriophage [12]. The eclipse period of the lytic cycle is the intermediate time period which exists between the induction of nucleic acid of bacteriophages into the bacterial cells and the discharge of mature progeny phages from the infected bacterial cells while the latent period is defined as the time taken for a single phage particle to replicate inside an infected bacterial host cell [13]. Lysogeny can be categorized as a process where a bacteriophage infects a bacterial cell by incorporating its nucleic acid into the genome of the host bacterial cell or in some cases the nucleic acid may sometimes form circular self-replicating genomic units often found in the cytoplasm of the infected bacterial host cell. In such a circumstance, if the lysis of the bacterial host cell fails to happen and no offspring of the bacteriophages are created, such phages are termed as temperate phages and the bacterial host cells consisting of such phages are named as lysogenic. Lysogens after several progenies can be transformed to mature lytic phages by initiating the lytic cycle of bacteriophage [14, 15].

13.3 ADVANTAGES OF USING PHAGES FOR BIOFILM CONTROL

Several infectious disorders currently in plants and animals have been treated with phages. In humans, phage therapy has been employed to treat a number of infectious disorders induced by pathogens such as *Staphylococcus, Shigella, Salmonella, P. aeruginosa, Streptococcus, and E. coli.* A new interest in phage therapy has emerged as concerns about antibiotic resistance have intensified in global healthcare centres despite the fact that most of the prior research projects weren't clinical studies with strict negative controls [16]. Antimicrobial resistance has emerged as a result of the wise application of antimicrobial compounds in healthcare and various other sectors [17, 18]. Several works showed evidence of how phages can be utilized for treating appliance-related infections, thus reducing the propagation of antimicrobial-resistant microorganisms.

There are several scientific proofs that demonstrated the capability of bacteriophages to control biofilms. A few studies also showed that T4 bacteriophages may form units having burst sizes ranging from 12–200 plaque determined by the physiology of its host cells of *E. coli.* Even under low-nutrient conditions, a few bacteriophages have been found to be efficiently targeting their host bacterial cells [19, 20]. A strain of *E. coli* was developed at a dilution rate in a chemostat having a minimum amount of glucose in it and was treated with T4 bacteriophages who were found to be effective against the biofilms of *E. coli,* despite the fact that the rate of bacteriophage assembly and synthesis was directly related to the quantity of protein synthesized in the host cell. Greater nutrition levels may lead to increase in the rate of cell development, which further leads to larger burst size and shorter eclipse and latent phases [21].

Biofilm organisms are generally found to be resistant to biocides, disinfectants, and antimicrobial medications, and biofilm reduction or extermination usually demands elevated doses of antimicrobials for extended durations. Comparing with the development of plaque on a host bacterial lawn, some researchers have even documented that T4 phage particles have the ability to move radially across a biofilm, indicating that biofilm-associated infections can be treated applying phages in a single dose. Antimicrobial drugs have the disadvantage of requiring greater antimicrobial doses in order to neutralize or complete eradicate biofilm cells as the maturation of biofilm happens [22, 23]. The susceptibility of *P. aeruginosa* biofilms on Phage F116 were not significantly affected even as they aged. Biofilms that had been cultivated for 20 days prior to treatment were found to be effective against phages [24]. Due to diffusion limitations, the efficiency of antimicrobial organisms might also affect the EPS matrix of biofilm. It was also observed that the EPS matrix of *P. aeruginosa* and alginate gel phages enable phages to diffuse [25].

The EPS matrix of biofilm can be easily degraded by polysaccharide depolymerases produced by phages. Past research also showed evidence which revealed that during plaque formation there is a release of a specific enzyme by bacteria when infected by *K. pneumoniae* phages. This capsular enzyme, named depolymerase usually aims for the capsular polysaccharide linked to the bacterial cell surface which was found to be directly hooked to the phages, facilitating them to infiltrate the capsule and finally reach the cell surface of the host. Certain researchers described phage-induced depolymerase specific for capsular polysaccharides and EPS of stains of *Klebsiella aerogenes*. In various bacteria species, including *Azotobacter agilis, Lactococcus lactis, Bacillus subtilis, P. aeruginosa, and Alcaligenes faecalis*, phage-associated depolymerases significantly influence phage adsorption to the cell surface of the bacteria [26–29]. After specific binding, the enzyme, depolymerase may be either released as a free enzyme or as a free as well as a bound enzyme. Numerous reviews showed evidence that biofilms are extensively influenced by the phage-induced depolymerases [30]. The development of biofilms of several bacterial stains including *Serratia marcescens* and *Enterobacter,* which is now called *Pantoea agglomerans,* even the phage-resistant ones, when treated with phage SF153b, showed considerable reduction [31]. The biofilms of this organism were dramatically reduced by the phage-free enzyme which further affected the phage-resistant biofilms due to the depolymerase produced by the phages [32]. According to certain studies, *P. aeruginosa* alginates and its EPS viscosity were reduced in suspensions containing only depolymerase [33]. Researchers also found out that phages could be employed to eliminate some major barriers in controlling biofilms such as, extracellular matrix present in biofilms, influence of biofilm age and the antimicrobial tolerance [34, 35].

When phages' ability to maintain clinically relevant biofilms is being evaluated, numerous critical characteristics of the phages must be taken into consideration. Each one is examined carefully, and different proposals are set forth to overcome each obstacle. These obstacles are as follows [36].

13.4 NARROW HOST RANGE

The tail fibres of bacteriophages get attached to respective sites with the help of certain molecules present on or directly associated with the cell surface of the host bacterium. Gram-negative and Gram-positive bacteria have distinct bacteriophage receptors. Major interactions between Gram-negative bacteria and its phages are observed with the lipopolysaccharide constituents, transport proteins and porins found in the outer membrane of the Gram-negative bacteria, the capsular polysaccharide also known as K antigens or with the pili or glycocalyx and flagella. However, in the case of Gram-positive bacteria the phages usually target the peptidoglycan components, associated proteins, teichoic acids and lipoteichoic acids [37–39]. The increased levels of specificity might be a disadvantage, particularly in the presence of poly-microbic biofilms found on certain indwelling medical equipment. The host range can be expanded by the multiple adhesin tip of tail fibres of specific bacteriophages. E. coli phage (FK1–5) codes two different tail fibre proteins, K5 lyase and K1 endosialidase. The K1 endosialidase is responsible for cleaving the a-2, 8-linked poly-Nacetylneuraminic acid polymer present in the K1 capsule while on the other hand the tail fibre protein, K5 lyase is responsible for degrading the b-glucuronic acid and a-N-acetylglucosamine contained in a capsular polymer [40]. Biofilms comprising of different bacterial species or strains, create phage cocktails carrying a single phage that have the ability to break down numerous extracellular polymers. A polyvalent *Staphylococcus* phage named bacteriophage K has the ability to lyse a vancomycin-resistant strain of *S. aureus* and numerous methicillin-resistant strains of *S. aureus* which were among the ten distinct *Staphylococcus epidermidis* strains and nine different *Staphylococcus species*. According to several sources, there are several approaches for selecting phages with higher potency infecting several host cells [41, 42]. The phages in a particular population that adhere to and proliferate in the host bacteria are isolated using this procedure [43]. A bacteriophage population with an entirely different host affinity might develop from a repeated cycle of this approach. In an investigation, researchers extracted phages from natural waters by treating them with two distinct bacterial hosts strains. Utilising this approach, researchers developed a phage capable of lysing *Proteus vulgaris, Sphaerotilus natans, Shigella flexneri, P. aeruginosa* (multiple strains), *E. coli* and *Rhodospirillum rubrum*. This demonstrates that the constraints of specificity of the host organism might be addressed by adopting one or more of the procedures mentioned to create phage cocktails having broad spectrum potency [44].

13.5 PHAGE RESISTANCE

Mutations in either the bacterium or phage influence the rate at which phages adhere to their bacterial hosts. The bacterial host cells might undergo mutation and might become resistant to phage adsorption because of the consequence of the following: (i) modifications in receptor site structure (ii) changes in receptor site exposure, (iii) decrement in receptor site density, or (iv) because of a manufacturing failure or due to the incorporation of receptor sites into the cell membrane, the receptor sites are lost completely [45–47]. Bacterial cells that have undergone lysogeny might become susceptible to invasion caused by another phage of the same kind. In any lysogenic bacterial cell, a phage-coded repressor blocks phage gene transcription and also binds to the operator sites of incoming phage DNA, preventing lytic growth. Another recently discovered method is that clustered, regularly interspaced short palindromic repeats (CRISPRs), which are common in the genomes of several bacteria, may impart phage resistance. Phages could also transform to develop lytic phage strains against these mutant bacterial cells. Such host range mutants vary from wild-type phages mainly in two distinct ways. Firstly, they show modified specificity, which allows the phages to bind to the resistant bacteria's surfaces, and, secondly, they show lowered selectivity, which significantly reduces the threshold for irreversible binding to the surface of bacterial cells, allowing the phage's genetic material components to be discharged more conveniently into resistant bacteria. The majority of bacteriophage mutants with specific host ranges can thrive on both resistant and wild-type bacteria. Coevolution of bacteria and lytic phages enables equilibrium to be achieved between them [48–50]. The mutation rate was measured approximately around 8108 for phage resistance by Lenski and Levin in a chemostat when EB and a lytic phage were grown together. Bacterial mutations that are resistant to phages appeared almost 100 hours after the bacteria and phage were treated together, resulting in the predominance of the phage-sensitive strains. Strains that are phage-sensitive, on the other hand, persisted due to their faster development rate. In none of their trials, mutants of the phage T4 (phage mutants having a broad host range that can thrive on both mutant and parent cultures of *E. coli*) host range were discovered [51]. In a chemostat, some researchers treated *E. coli* O157:H7 along with its phage, subsequently collecting samples of bacteria and bacteriophages. At multiple time points, bacterial mutants that are phage-resistant and phage mutants retaining specific host range were obtained from the chemostat [52]. The bacterial count was initially reduced by 3–4 logs after treatment with phage PP01, but this change was reversed, and so was the phage count. Wild-type organisms were immediately flushed out of the system, accompanied by the discovery of different mutant species of bacteriophages and bacteria [53]. These conclusions indicated that, followed by a quick decline in viable count, resistant strains of bacteria start to emerge and continue to thrive. In the process of what's described as a mutual arms race, both bacteria and phage seem to be continually evolving [54].

Mutations that change or prevent the production of certain phage receptor sites confer viral resistance to bacteria. Despite the fact that phage resistance might be lost due to the elimination of these receptors, it may also affect the organism's capacity to exploit its environment and compete with the bacterial strain of wild type [55–58]. The extermination of certain strains of bacteriophages is prevented by this feature of phage resistant bacteria. In spite of the evolution of the resistant strain of bacteriophages, some researchers believe that they have the ability to survive provided bacteria that are phage-sensitive have a competitive edge against their resistant counterparts [59]. Naturally occurring bacterial communities exhibiting virulent phages coexist with resistant and sensitive clones of bacteria [60].

Some research has analysed the significance of biofilms in the evolution of bacteriophage resistance. It was also demonstrated that biofilms of an *E. coli* strain were subjected to two separate lytic phages (OP7061 and OP10081). The established subpopulations possessed a persistent phenotype that was immune to bacteriophage lysis [61, 62]. Each strain formed significantly improved clumps in liquid media, with improved biofilm development, as well as the generation of fimbria-like extracellular structures, that were all missing in the parent strain. This shows that biofilm formation might be a phage resistance mechanism [63].

The capability of bacteriophages to co-evolve with their bacterial hosts is considered to be lower. In addition, the potential of phages to co-evolve with their bacterial hosts is considered to be lower when compared to that of the bacterial host cells [64]. Hence, there are bacterial mutants which exhibit phage resistance despite the lack of corresponding mutants for the host range. The research indicated that this technique can be employed to efficiently minimize initial attachment of bacteria or to significantly diminish (without eliminating) biofilms [65]. Phage cocktails that are created by combining host range mutants with broad-spectrum bacteriophages may be beneficial sometimes. It would be quite essential to identify phage-resistant bacterial mutants regarding their potential to establish biofilms. It is indeed possible that the capacity of the bacterial mutants showing resistance against bacteriophages to adhere to substrata is significantly diminished or boosted [66].

13.6 PHAGE INACTIVATION BY THE PATIENT'S IMMUNE SYSTEM

Furthermore, it is important to consider how the immune system of the patient will respond to bacteriophages when they are introduced therapeutically. As phages are antigenic in nature, serum antibodies as well as cellular immunity will react to them. Human or animal serum has the ability to inactivate a variety of phages. Antibody titers may significantly rise when they are exposed to bacteriophages repeatedly [67–69]. As per certain researches using animal models, the bacteriophages can be eliminated from the blood circulation by the cellular immune system.

Furthermore, Kallings observed that, following phage treatment, around 5–25% of phage activity survived; however, the inactivation rate reached 99–100% only after bacteriophages were incubated with serum for 18 hours. A study also demonstrated that, in 54 of 57 patients, serum phage antibody titers seem to be either undetectable or identified in an extremely low amounts 10 days after starting the onset of phage therapy [70]. In flowing or static systems, bacteriophages might develop into biofilms, where EPS may serve to protect phages from clearance and inactivation by immune cells. Bacterial polysaccharides were also reported to significantly inhibit the phage-neutralizing action of human serum [71].

13.7 A STRATEGY FOR OVERCOMING INACTIVATION BY THE IMMUNE SYSTEM

In mice, mutant phages of E and *Salmonella typhimurium* l and P22 were observed to be resistant to entrapment by the cell's immune system and remained in their circulatory system without a significant drop-in activity. Unless serum antibody titers are present prior to treatment, the antibodies to phages have a very weak response. It was observed that the amount of antibody titers escalated when the bacteriophages were treated repeatedly. A phage introduced into the cellular environment will be inactivated by the immune system [72]. There is also an unproved possibility that bacteriophages might develop into biofilms in order to protect themselves from inactivation. There maybe be a way of designing mutant bacteriophages which can resist clearance by the cell's immune system [73].

13.8 SAFETY OF PHAGE PREPARATIONS FOR USE IN HUMANS

Phage exposure starts at birth in human beings. It has been reported that in human beings several bacteriophages can be recovered from the oral cavity and the gastrointestinal tract. In addition, humans are constantly exposed to bacteriophages in their surroundings. Many such observations, combined with evidence from several research projects into bacteriophage therapy, strongly imply that therapeutic phage treatments should be effectively accepted amongst individuals. Despite this, it is important to consider the safety of bacteriophage preparations for treating infections caused by bacteria [74, 75].

Another uncertainty is whether the phage will have a damaging effect on the immune system of the patient. It was observed in several studies that phage therapy may modify turnover as well as function of the neutrophil, yet this was linked to a favourable outcome for the patient. A separate concern relates to the purity of bacteriophage preparations. Major structural components, lipopolysaccharides, which make up the cell walls of Gram-negative bacteria, are discharged at the time of bacteriophage cell lysis [76].

Lipopolysaccharides (endotoxins) may yield a number of biological consequences, which include endotoxic shock, fever, and hypotension, if they are present in significant amounts. Employing ultrafiltration and chromatography to purify phage preparations can ensure that they are endotoxin-free. A lysis deficient (LyD) strain of the *E. coli* bacteriophage was also discovered, which killed without lysing its bacterial hosts [77]. The amount of endotoxin discharged by the LyD phage was notably lower in treated mice than in control mice. This demonstrates that it would be feasible to develop phages that destroy without lysing their host cells. However, the only drawback of this approach is of not allowing the progeny bacteriophage to infect other biofilm cells [78].

Furthermore, bacteriophages may encode extracellular toxins and proteins involved in host adhesion, invasion of cells, and intracellular survival. Virulent strains of *S. aureus, E. coli, Corynebacterium diphtheriae, P. aeruginosa, Vibrio cholerae, Streptococcus pyogenes, Shigella spp., Clostridium tetani, and Clostridium botulinum* are commonly known instances of phage mediated virulence. Such virulence factors are acquired by host cells of bacteria during lysogeny, without damaging the cell instead resulting in the integration of the bacteriophage genome containing the virulence genes, into the bacterial genome as prophage. To prevent the accidental transmission of genes that encode bacterial toxins and perhaps several other virulence factors, all phages that seem to be potentially suitable for bacteriophage therapies should have their genomes thoroughly sequenced [79, 80] (Table 13.1).

Bacteria have the ability to resist several antibiotics such as minocycline and rifampin by sequestering them onto central venous catheters (CVC) and this approach is being employed currently to prevent biofilm development [81]. It may also be possible to encapsulate bacteriophages into surface coating or sequester them onto a surface in order to inhibit development of biofilm on an intravascular catheter or any other indwelling gadgets [82]. In a laboratory setting, coating hydrogel catheters using *Staphylococcus epidermidis* phage 456 has the ability to suppress the viability

TABLE 13.1
Biofilm Study Therapies and Phage-Mediated Biology Phage Technology for Biofilm Control: A Potential Application

Sl No.	Target	Context	Reviewed	References
1.	*Aggregatibacter actinomycetemcomitans*	Polystyrene microtiter plate	Not Reviewed	[124]
2.	*Escherichia coli*	Modified Robbins' device	Reviewed	[150]
3.	*Klebsiella pneumoniae*	Microtiter plate and Glass	Not Reviewed	[107]
4.	*E. coli* and *P. aeruginosa*	Flow cells	Reviewed	[155]
5.	*Staphylococcus aureus*	Silicone discs	Not Reviewed	[109]
6.	*Staphylococcus aureus*	Rabbit wound model	Not Reviewed	[111]
7.	*Pseudomonas aeruginosa*	Wastewater filtration system	Not Reviewed	[114]
8.	*Staphylococcus aureus*	Rat implant model	Not Reviewed	[112]
9.	*Pseudomonas putida*	Polystyrene peg in 96-well microtiter plate	Not Reviewed	[125]
10.	*Pseudomonas aeruginosa*	Biofilter filtration system	Not Reviewed	[115]
11.	*Pseudomonas aeruginosa*	Rat implant model	Not Reviewed	[113]
12.	*Pseudomonas aeruginosa*	Pulmonary models including epithelial-cell monolayer	Not Reviewed	[116]
13.	*Pseudomonas aeruginosa*	Root canal model	Not Reviewed	[110]
14.	*Escherichia coli*	Microtiter plate	Not Reviewed	[117]
15.	*Pseudomonas aeruginosa*	Microtiter plate	Not Reviewed	[127]
16.	*Escherichia coli*	Stainless steel	Reviewed	[147]
17.	*Staphylococcus aureus*	Microtiter plate	Not Reviewed	[118]
18.	*Campylobacter jejuni*	Glass	Not Reviewed	[129]
19.	*Pseudomonas aeruginosa*	Silicone catheter segment	Not Reviewed	[119]
20.	*Listeria monocytogenes*	Stainless steel	Not Reviewed	[120]
21.	*Escherichia coli*	Calgary biofilm device	Not Reviewed	[121]
22.	*E. coli* and *Pseudomonas aeruginosa*	Silicone rubber disks	Not Reviewed	[126]
23.	*Staphylococcus aureus*	Microtiter plate	Reviewed	[134]
24.	*Pseudomonas aeruginosa*	Microtiter plate	Not Reviewed	[128]
25.	*P. aeruginosa, Acinetobacter johnsonii,* and *Bacillus subtilis*	Ultrafiltration membrane model	Reviewed	[137]
26.	*E. coli* and *Proteus mirabilis*	Hydrogel-coated catheters	Not Reviewed	[130]
27.	*Pseudomonas aeruginosa*	Hydrogel-coated catheters	Reviewed	[131]
28.	*Sphaerotilus natans*	Stainless steel coupons and wire screen	Reviewed	[132]
29.	*Pseudomonas fluorescens* and *Staphylococcus lentus*	Stainless steel	Reviewed	[133]
30.	*Listeria monocytogenes*	Stainless steel	Not Reviewed	[135]
31.	*Klebsiella pneumoniae*	Microtiter plate	Reviewed	[138]
32.	*Klebsiella pneumoniae*	Microtiter plate and glass	Not Reviewed	[136]
33.	*Pseudomonas aeruginosa*	Microtiter plate	Reviewed	[139]
34.	*Pseudomonas fluorescens*	Glass	Reviewed	[140]
35.	*Pseudomonas fluorescens*	Stainless steel	Reviewed	[141]
36.	*Escherichia coli*	Polyvinylchloride coupons	Reviewed	[156]
37.	*Staphylococcus epidermidis*	Microtiter plate	Reviewed	[142]
38.	*Pseudomonas aeruginosa*	Poly(methyl)methacrylate discs	Reviewed	[151]
39.	*Staphylococcus aureus*	Polystyrene microtiter plate	Reviewed	[143]
40.	*Pseudomonas aeruginosa*	Microtiter plate	Not Reviewed	[123]
41.	*Escherichia coli*	Pegs in microtiter plates	Reviewed	[144]
42.	*Staphylococcus epidermidis*	Hydrogel-coated catheters	Reviewed	[145]

(Continued)

TABLE 13.1 (Continued)
Biofilm Study Therapies and Phage-Mediated Biology Phage Technology for Biofilm Control: A Potential Application

Sl No.	Target	Context	Reviewed	References
43.	*Enterobacter agglomerans* and *Serratia marcescens*	Modified Robbins' device	Reviewed	[153]
44.	*Escherichia coli*	3-channel flow chamber	Reviewed	[146]
45.	*Pseudomonas fluorescens*	Inox plate placed in microtiter tray	Reviewed	[148]
46.	*Enterobacter cloace*	Glass	Reviewed	[149]
47.	*Staphylococcus epidermidis*	Catheter	Reviewed	[152]
48.	*Acinetobacter baumannii*	Microtiter plate	Not Reviewed	[122]
49.	*Listeria monocytogenes*	Stainless steel	Reviewed	[154]
50.	*Listeria monocytogenes*	Stainless steel	Not Reviewed	[108]
51.	*Listeria monocytogenes*	Stainless steel	Reviewed	[157]

counts of biofilm by *S. epidermidis* 414 coating catheter surfaces by more than 4 logs when exposed for over a time period of 24 hours. It further implies that using phage-covering surface coatings on catheters or other medical instruments might diminish biofilm development by *S. epidermidis* and perhaps other clinically important bacteria [83]. The selection of phages is considered as a significant topic in this approach (namely, how to select a phage cocktail which would be responsible for inhibiting biofilm development with the assistance of numerous bacterial strains that may form colonies on device surfaces). Since most phages have got a restricted host range, designing a coating that might be able to inhibit the development of colonies by the whole range of organisms recognized to form colonies in the central venous catheters would be impossible. To prevent biofilm formation by strains of *S. aureus*, for example, a phage coating could be developed that could target *S. aureus* and might prevent or significantly reduce biofilm formation by this pathogenic strain [84, 85]. Although multiple studies have indicated that for a small fraction of strains isolated clinically that are incapable of being typed or lysed using each and every one of the 21 phages in the sequence, a phages cocktails employed for pre-treatment of the catheter might contain the whole international bacteriophage lysing set for *S. aureus*. Numerous different phage collections, with staphylococcal phages (like phage K) that have a broad range against their host cells, or phages acquired from different environmental sources could be used to expand the host range [86].

Other major problems include the ability of coatings of matrix in order to adsorb the bacteriophage. Besides the coating matrix's ability to bind phages, other important questions relate to the phage's infectivity and the stability once it is incorporated, as well as the efficacy of the phage in the mere existence of serum proteins such as fibrin. Another important question is whether the patient will tolerate the coating or not [87–89]. In contrast to antimicrobial-treated catheters, which work primarily by releasing a controlled amount of an antimicrobe, phages could easily bind to surfaces and attach to bacteria thus infecting it. Several research works have documented Phago-Bio-Derm, described as a unique poly (ester amide) polymer having a biodegradable property that was impregnated using multiple phage strains along with an antibiotic to have a good ability to heal several chronic wounds [90].

Several indwelling medical gadgets or catheters could be treated with phages in order to inhibit the development of bacterial biofilm. Before evaluating in vivo conditions, phage cocktails must be carefully selected, the coating matrix must be optimized successfully, and animal models involved in in-vitro must be validated as well [91, 92].

13.9 CONCLUSION

Bacteriophages are used for the treatment of different types of infections in humans. Bacteriophages also possess effectiveness against biofilms of different bacteria, possessing clinical relevance and leading to the fact that bacteriophages cannot give rise to a brand-new strategy for the control of biofilms on different types of medical devices [93]. The nature of biofilms which are in association with different medical devices proves that the ways of treatment that are most effective will prevent rather than control the existence of biofilms and will affect the components of the extracellular matrix of the biofilm and will ignore too much use of antimicrobial agents [94]. Bacteriophages have the capability to meet these requirements, and thus it is possible for the development of treatments which can be tolerated by patients [95]. These do not undergo inactivation by the immune system of the patients, and they do not introduce bacterial toxin genes or virulence. The bacteriophages' formulations can also be made artificially for avoiding or minimizing bacteriophage resistance [96]. Newer bacteriophage strains can be easily isolated from the environment and bacteriophages can be engineered genetically for an increased efficacy against different types of biofilms. The continuity of different types of infections associated with healthcare, amongst which many are in association with medical devices and the increase in antimicrobial resistance which is still thriving demands for novel strategies for treatment of those target organisms in association with biofilms [97–99]. These novel anti-biofilm agents and these newer strategies of treatments can possess bacteriophages which undergo lytic lifecycles [100].

REFERENCES

1. Donlan, R.M., and Costerton, J.W. (2002) Biofilms: Survival mechanisms of clinically relevant microorganisms. *Clin. Microbiol. Rev*. 15: 167–193.
2. Donlan, R.M., (2008) Biofilms on central venous catheters: Is eradication possible? In: Romeo, T., ed., *Bacterial Biofilms. Current Topics in Microbiology and Immunology*, pp. 133–161, Springer-Verlag.
3. Donlan, R.M. (2005) New approaches for the characterization of prosthetic joint biofilms. *Clin. Orthop. Relat. Res*. 437: 12–19.
4. Stewart, P.S., et al. (2004) Biofilm antimicrobial resistance. In: Ghannoum, M. and O'Toole, G.A., eds., *Microbial Biofilms*, pp. 250–268, ASM Press.
5. Marr, K.A., et al. (1997) Catheter-related bacteremia and outcome of attempted catheter salvage in patients undergoing hemodialysis. *Ann. Intern. Med*. 127: 275–280.
6. Carmen, J.C., et al. (2005) Treatment of biofilm infections on implants with low-frequency ultrasound and antibiotics. *Am. J. Infect. Control*. 33: 78–82.
7. Johnson, L.L., et al. (1998) Treatment of bacterial biofilms on polymeric biomaterials using antibiotics and ultrasound. *J. Biomater. Sci. Polym. Ed*. 9: 1177–1185.
8. Rasmussen, T.B., et al. (2005) Screening for quorum-sensing inhibitors (QSI) by use of a novel genetic system, the QSI Selector. *J. Bacteriol*. 187: 1799–1814.
9. Balaban, N., et al. (2005) Prevention of staphylococcal biofilmassociated infections by the quorum sensing inhibitor RIP. *Clin. Orthop. Relat. Res*. 437: 48–54.
10. Alkawash, M.A., et al. (2006) Alginate lyase enhances antibiotic killing of mucoid Pseudomonas aeruginosa in biofilms. *APMIS*. 114: 131–138.
11. Fox, J. (2000) Phage treatments yield healthier tomato, pepper plants. *ASM News*. 66: 455–456.
12. Barrow, P., et al. (1998) Use of lytic bacteriophage for control of experimental Escherichia coli septicemia and meningitis in chickens and calves. *Clin. Diagn. Lab. Immunol*. 5: 294–298.
13. Sulakvelidze, A., et al. (2001) Bacteriophage therapy. *Antimicrob. Agents Chemother*. 45: 649–659.
14. Tenover, F.C., et al. (2001) Development and spread of bacterial resistance to antimicrobial agents: An overview. *Clin. Infect. Dis*. 33 (Suppl 3): S108–S115.
15. Schwalbe, R.S., et al. (1987) Emergence of vancomycin resistance in coagulase-negative staphylococci. *N. Engl. J. Med*. 316: 927–931.

16. Kaplan, A.H., et al. (1988) Recovery of resistant enterococci during vancomycin prophylaxis. *J. Clin. Microbiol.* 26: 1216–1218.
17. Sieradzki, K., et al. (2003) Evolution of a vancomycin-intermediate staphylococcus aureus strain in vivo: Multiple changes in antibiotic resistance phenotypes of a single lineage of methicillin-resistant S. aureus under the impact of antibiotics administered for chemotherapy. *J. Clin. Microbiol.* 41: 1687–1693.
18. Archibald, L., et al. (1997) Antimicrobial resistance in isolates from inpatients and outpatients in the United States: Increasing importance of the intensive care unit. *Clin. Infect. Dis.* 24: 211–215.
19. Hughes, K.A., et al. (1998) Biofilm susceptibility to bacteriophage attack: The role of phage-borne polysaccharide depolymerase. *Microbiology*. 144: 3039–3047.
20. Doolittle, M.M., et al. (1995) Lytic infection of *Esherichia coli* biofilms by bacteriophage T4. *Can. J. Microbiol.* 41: 12–18.
21. Doolittle, M.M., et al. (1996) Tracing the interaction of bacteriophage with bacterial biofilms using fluorescent and chromogenic probes. *J. Ind. Microbiol.* 16: 331–341.
22. Roy, B., et al. (1993) Biological inactivation of adhering Listeria monocytogenes by Listeriaphages and a quaternary ammonium compound. *Appl. Environ. Microbiol.* 59: 2914–2917.
23. Hadas, H., et al. (1997) Bacteriophage T4 development depends on the physiology of its host *Escherichia coli*. *Microbiology*. 143: 179–185.
24. Corbin, B.D., et al. (2001) Bacteriophage T4 multiplication in a glucoselimited Escherichia coli biofilm. *Can. J. Microbiol.* 47: 680–684.
25. Ceri, H., et al. (1999) The Calgary biofilm Device: New technology for rapid determination of antibiotic susceptibilities of bacterial biofilms. *J. Clin. Microbiol.* 37: 1771–1776.
26. Anwar, H., et al. (1992) Dynamic interactions of biofilms of mucoid Pseudomonas aeruginosa with tobramycin and piperacillin. *Antimicrob. Agents Chemother.* 36: 1208–1214.
27. Amorena, B., et al. (1999) Antibiotic susceptibility assay for Staphylococcus aureus in biofilms developed in vitro. *J. Antimicrob. Chemother.* 44: 43–55.
28. Hanlon, G.W., et al. (2001) Reduction of exopolysaccharide viscosity as an aid to bacteriophage penetration through Pseudomonas aeruginosa biofilms. *Appl. Environ. Microbiol.* 67: 2746–2753.
29. Briandet, R., et al. (2008) Fluorescence correlation spectroscopy to study diffusion and reaction of bacteriophages inside biofilms. *Appl. Environ. Microbiol.* 74: 2135–2143.
30. Adams, M.H., and Park, B.H. (1956) An enzyme produced by a phagehost cell system II. The properties of the polysaccharide depolymerase. *Virology*. 2: 719–736.
31. Sutherland, I.W. (1967) Phage-induced fucosidases hydrolyzing the exopolysaccharide of Klebsiella aerogenes Type 54. *Biochem. J.* 104: 278–285.
32. Reese, J.F., et al. (1974) Factors influencing the adsorption of Bacteriophage 2 to cells of Pseudomonas aeruginosa. *J. Virol.* 13: 22–27.
33. Bartell, P.F., et al. (1966) Polysaccharide depolymerase associated with bacteriophage infection. *J. Bacteriol.* 92: 56–62.
34. Deveau, H., et al. (2002) Effect of exopolysaccharides on phage-host interactions in Lactococcus lactis. *Appl. Environ. Microbiol.* 68: 4364–4369.
35. Mare, I.J. and Smit, J.A. (1969) A capsule-depolymerizing enzyme from Alcaligenes faecalis infected with Bacteriophage A6. *J. Gen. Virol.* 5: 551–552.
36. Eklund, C., and Wyss, O. (1962) Enzyme associated with bacteriophage infection. *J. Bacteriol.* 84: 1209–1215.
37. Kimura, K., and Itoh, Y. (2003) Characterization of poly-g-glutamate hydrolase encoded by a bacteriophage genome: Possible role in phage infection of Bacillus subtilis encapsulated with poly-g-glutamate. *Appl. Environ. Microbiol.* 69: 2491–2497.
38. Kutter, E., et al. (2005) Molecular mechanisms of phage infection. In: Kutter, E. and Sulakvelidze, A., eds., *Bacteriophages: Biology and Application*, pp. 165–222, CRC Press.
39. Blair, J.E. and Williams, E.O. (1961) Phage typing of staphylococci. *Bull. World Health Organ.* 24: 771–784.
40. O'Flaherty, S., et al. (2005) Potential of the polyvalent anti /Staphylococcus Bacteriophage K for control of antibiotic-resistant staphylococci from hospitals. *Appl. Environ. Microbiol.* 71: 1836–1842.
41. Scholl, D., et al. (2001) Bacteriophage K1-5 encodes two different tail fiber proteins, allowing it to infect and replicate on both K1 and K5 strains of *Escherichia coli*. *J. Virol.* 75: 2509–2515.

42. Cerca, N., et al. (2007) Susceptibility of Staphlococcus epidermidis planktonic cells and biofilms to the lytic action of Staphylococcus Bacteriophage K. *Lett. Appl. Microbiol.* 45: 313–317.
43. Stewart, G., et al. Amersham International PLC. Selective virus culture. World Patent Number WO9523848A1.
44. Jensen, E.C., et al. (1998) Prevalence of broad-host-range lytic bacteriophages of Sphaerotilus natans, Escherichia coli, and Pseudomonas aeruginosa. *Appl. Environ. Microbiol.* 64: 575–580.
45. Lenski, R.E. (1988) Dynamics of interactions between bacteria and virulent bacteriophage. In: Marshall, K.C., ed., *Advances in Microbial Ecology*, pp. 1–44, Plenum Press.
46. Sorek, R., et al. (2008) CRISPR – A widespread system that provides acquired resistance against phages in bacteria and archeae. *Nat. Rev. Microbiol.* 6: 181–186.
47. Lenski, R.E. and Levin, B.R. (1985) Constraints on the coevolution of bacteria and virulent phage: A model, some experiments, and predictions for natural communities. *Am. Nat.* 125: 585–602.
48. Mizoguchi, K., et al. (2003) Coevolution of Bacteriophage PP01 and Escherichia coli O157:H7 in continuous culture. *Appl. Environ. Microbiol.* 69: 170–176.
49. Lacqua, A., et al. (2006) Emergence of biofilm-forming subpopulations upon exposure of Escherichia coli to environmental bacteriophages. *Appl. Environ. Microbiol.* 72: 956–959.
50. Kallings, L.O. (1961) Studies on the inactivation of bacterial viruses by normal human serum: 1. Kinetics of the Coli T2 Phage inactivation at various serum concentrations. *Acta Pathol. Microbiol. Scand.* 51: 389–40251 Kallings, L.O. (1961) Studies on the inactivation of bacterial viruses by normal human serum. *Acta Pathol. Microbiol. Scand.* 52: 273–286.
52. Jerne, N.K. and Avegno, P. (1956) The development of the phageinactivating properties of serum during the course of specific immunization of an animal: Reversible and irreversible inactivation. *J. Immunol.* 76: 200–208.
53. Guttman, B., et al. (2005) Basic phage biology. In: Kutter, E. and Sulakvelidze, A., eds., *Bacteriophages: Biology and Application*, pp. 165–222, CRC Press.
54. Inchley, C.J., (1969) The activity of mouse Kupffer cells following intravenous injection of T4 bacteriophage. *Clin. Exp. Immunol.* 5: 173–187.
55. Geier, M.R., et al. (1973) Fate of Bacteriophage Lambda in non-immune germ-free mice. *Nature.* 246, 221–223.
56. Uhr, J.W. and Weissmann, G. (1965) Intracellular distribution and degradation of bacteriophage in mammalian tissues. *J. Immunol.* 94: 544–550.
57. Kucharewicz-Krukowska, A. and Slopek, S. (1987) Immunogenic effect of bacteriophage in patients subjected to phage therapy. *Arch. Immunol. Ther. Exp. (Warsz.).* 35, 553–561
58. Landy, M. et al. (1958) Influence by endotoxic polysaccharides with inactivation of coliphage by human serum. *Br. J. Exp. Pathol.* 39: 618–630.
59. Merril, C.R. et al. (1996) Long-circulating bacteriophage as antibacterial agents. *Proc. Natl. Acad. Sci. U. S. A.* 93: 3188–3192.
60. Gorski, A. and Weber-Dabrowska, B. (2005) The potential role of endogenous bacteriophages in controlling invading pathogens. *Cell. Mol. Life Sci.* 62: 511–519.
61. Sulakvelidze, A. and Kutter, E. (2005) Bacteriophage therapy in humans. In: Kutter, E. and Sulakvelidze, A., eds., *Bacteriophages: Biology and Applications*, pp. 381–436, CRC Press.
62. Weber-Dabrowska, B., et al. (2002) Effect of phage therapy on the turnover and function of peripheral neutrophils. *FEMS Immunol. Med. Microbiol.* 34: 135–138.
63. Rietschel, E.T.H., et al. (1982) Bacterial endotoxins: Chemical structure, biological activity and role in septicaemia. *Scand. J. Infect. Dis. Suppl.* 31: 8–21.
64. Boratynski, J., et al. (2004) Preparation of endotoxin-free bacteriophages. *Cell. Mol. Biol. Lett.* 9: 253–259.
65. Matsuda, T., et al. (2005) Lysis-deficient bacteriophage therapy decreases endotoxin and inflammatory mediator release and improves survival in a murine peritonitis model. *Surgery.* 137: 639–646.
66. Boyd, E.F., (2005) Bacteriophages and bacterial virulence. In: Kutter, E., ed., *Bacteriophages: Biology and Applications*, pp. 223–265. CRC Press.
67. Raad, I., et al. (1995) Antibiotics and prevention of microbial colonization of catheters. *Antimicrob. Agents Chemother.* 39: 2397–2400.
68. Curtin, J.J. and Donlan, R.M. (2006) Using bacteriophages to reduce formation of catheter-associated biofilms by Staphylococcus epidermidis. *Antimicrob. Agents Chemother.* 50: 1268–1275.

69. Anaissie, E., et al. (1995) Role of catheter colonization and infrequent hematogenous seeding in catheter-related infections. *Eur. J. Clin. Microbiol. Infect. Dis.* 14: 134–137.
70. Tenover, F.C., et al. (1994) Comparison of traditional and molecular methods of typing isolates of Staphylococcus aureus. *J. Clin. Microbiol.* 32: 407–415.
71. Schaefler, S., et al. (1984) Methicillin-resistant Staphylococcus aureus strains in New York City hospitals: Inter-hospital spread of resistant strains of type 88. *J. Clin. Microbiol.* 20: 536–538.
72. de Gialluly, C., et al. (2003) Identification of new phages to type Staphylococcus aureus strains and comparison with a genotypic method. *J. Hosp. Infect.* 55: 61–67.
73. Vintov, J., et al. (2003) Association between phage types and antimicrobial resistance among bovine Staphylcoccus aureus from 10 countries. *Vet. Microbiol.* 95: 133–147.
74. Markoishvili, K., et al. (2002) A novel sustained-release matrix based on biodegradable poly(ester amide)s and impregnated with bacteriophages and an antibiotic shows promise in management of infected venous stasis ulcers and other poorly healing wounds. *Int. J. Dermatol.* 41: 453–458.
75. Sharma, M., et al. (2005) Inactivation of Esherichia coli O157:H7 in biofilm on stainless steel by treatment with alkaline cleaner and a bacteriophage. *J. Appl. Microbiol.* 99: 449–459.
76. Lu, T.K., and Collins, J.J. (2007) Dispersing biofilms with engineered enzymatic bacteriophage. *Proc. Natl. Acad. Sci. U. S. A.* 104: 11197–11202.
77. Sillankorva, S. et al. (2004) Bacteriophage F SI infection of Pseudomonas fluorescens planktonic cells versus biofilms. *Biofouling*. 20: 133–138.
78. Hall, M.R., McGillicuddy, E., and Kaplan, L.J. (2014) Biofilm: Basic principles, pathophysiology, and implications for clinicians. *Surg Infect (Larchmt)*. 15: 1–7.
79. Flemming, H.C, and Wingender, J. (2010) The biofilm matrix. *Nat Rev Microbiol.* 8: 623–33.
80. Krisch, H.M. (2003) The view from Les Treilles on the origins, evolution and diversity of viruses. *Res Microbiol.* 154: 227–9.
81. Brüssow, H., Canchaya, C., and Hardt, W.D. (2004) Phages and the evolution of bacterial pathogens: From genomic rearrangements to lysogenic conversion. *Microbiol Mol Biol Rev*. 68: 560–602.
82. Pal, C., Macia, M.D., Oliver, A., Schachar, I., and Buckling, A. (2007) Coevolution with viruses drives the evolution of bacterial mutation rates. *Nature*. 450: 1079–81.
83. Galán, J.C. (2008) Phage-shaping evolution of bacterial pathogenicity and resistance. In: Nombela, C., Cassell GH, Gutierrez-fuentes JA., *Evolutionary Biology of Bacterial and Fungal Pathogens*. ASM Press, 167–84.
84. Casas, V., and Maloy, S. (2011) Role of bacteriophage-encoded exotoxins in the evolution of bacterial pathogens. *Future Microbiol* 6: 146173.
85. Weinbauer, M.G., Agis, M., Bonilla-Findji, O., Malits, A., and Winter, C. (2007) Bacteriophage in the environment. In: Mc Grath, S. and van Sinderen, D., *Bacteriophage: Genetics and Molecular Biology*. Academic Press; 61–92.
86. Danovaro, R.,, Corinaldesi, C., Dell'Anno, A., et al. (2011) Marine viruses and global climate change FEMS. *Microbiol. Rev.* 35: 993–1034.
87. Suttle, C.A. (2007) Marine viruses – major players in the global ecosystem. *Nat. Rev. Microbiol.* 5: 801–12.
88. Breitbart, M. (2012) Marine viruses: Truth or dare. *Ann. Rev. Mar. Sci.* 4: 425–48.
89. Hyman, P., and Abedon, S.T. (2012) *Bacteriophages in Health and Disease*. CABI Press.
90. Abedon, S.T. (2008) *Bacteriophage Ecology: Population Growth, Evolution, and Impact of Bacterial Viruses*. Cambridge University Press.
91. Abedon, S.T. (2011) *Bacteriophages and Biofilms: Ecology, Phage Therapy, Plaques Hauppauge*Nova Science Publishers.
92. Vu, B., , Chen, M., Crawford, R.J., and Ivanova, E.P. (2009) Bacterial extracellular polysaccharides involved in biofilm formation. *Molecules*. 14: 2535–54.
93. Lei, Y.,, Oshima, T., Ogasawara, N., and Ishikawa, S. (2013) Functional analysis of the protein veg, which stimulates biofilm formation in Bacillus subtilis. *J Bacteriol.* 195: 1697–705.
94. Sutherland, I. (2001) Biofilm exopolysaccharides: A strong and sticky framework. *Microbiology*. 147: 3–9
95. Cai, W., De La Fuente, L., and Arias, C.R. (2013) Biofilm formation by the fish pathogen *Flavobacterium columnare*: Development and parameters affecting surface attachment. *Appl. Environ. Microbiol.* 79(18): 5633–42.

96. Worthington, R.J., Richards, J.J., and Melander, C. (2012) Small molecule control of bacterial biofilms. *Org. Biomol. Chem.* 10: 7457–74.
97. Mosig, G., and Eiserling, F. (2006) T4 and related phages: Structure and development. In: Calendar, R. and Abedon, S.T., *The Bacteriophages*. Oxford, Oxford University Press.
98. Sutherland, I.W., Hughes, K.A., Skillman, L.C., and Tait, K. (2004) The interaction of phage and biofilms. *FEMS Microbiol Lett.* 232: 1–6
99. Scholl, D., and Merril, C. (2005) Polysaccharide-degrading phages. In: Waldor M.K., Friedman D.I., and Adhya S.L., *Phages: Their Role in Bacterial Pathogenesis and Biotechnology*. ASM Press, 400–14.
100. Drulis-Kawa, Z., Majkowska-Skrobek, G., Maciejewska, B., Delattre, A.S., and Lavigne, R. (2012) Learning from bacteriophages – Advantages and limitations of phage and phage-encoded protein applications. *Curr. Protein Pept. Sci.* 13: 699–722.
101. Shen, Y., Mitchell, M.S., Donovan, D.M., and Nelson, D.C. (2012) Phage-based enzybiotics. In: Hyman, P. and Abedon, S. T., *Bacteriophages in Health and Disease*. CABI Press, 217–39.
102. Sulakvelidze, A., and Kutter, E. (2005) Bacteriophage therapy in humans.In: Kutter, E. and Sulakvelidze, A., *Bacteriophages: Biology and Application*. Boca Raton, Florida, CRC Press, 381–436.
103. Kutter, E. (2008) Phage therapy: Bacteriophages as naturally occurring antimicrobials. In: Goldman, E. and Green, L.H., *Practical Handbook of Microbiology*. CRC Press, 71330.
104. Chanishvili, N., Malkhazova, I., and Khurtsia, N. (2009) Phage therapy against intestinal infections In: Chanishvili, N. and Sharp, R., *A Literature Review of the Practical Application of Bacteriophage Research*. Eliava Institute, 33–58.
105. Housby, J.N., and Mann, N.H. (2009) Phage therapy. *Drug Discov. Today*. 14: 536–40.
106. Letkiewicz, S., , Miedzybrodzki, R., Klak, M., Jonczyk, E., WeberDabrowska, B., and Górski, A. (2010) The perspectives of the application of phage therapy in chronic bacterial prostatitis. *FEMS Immunol. Med. Microbiol.* 60: 99–112.
107. Miedzybrodzki, R., Borysowski, J., Weber-Dabrowska, B., et al. (2012) Clinical aspects of phage therapy. *Adv. Virus Res.* 83: 73–121.
108. Loc-Carrillo, C., and Abedon, S.T. (2011) Pros and cons of phage therapy. *Bacteriophage.* 1: 111–4.
109. Christie, G.E., Allison, H.A., Kuzio, J., McShan, M., Waldor, M.K., and Kropinski, A.M. (2012) Prophage-induced changes in cellular cytochemistry and virulence. In: Hyman, P. and Abedon, S.T., *Bacteriophages in Health and Disease*. CABI Press, 33–60.
110. Kuhl, S., Hyman, P., and Abedon, S.T. (2012) Diseases caused by phages. In: Hyman P. and Abedon S.T., *Bacteriophages in Health and Disease*. CABI Press, 21–32.
111. Gill, J.J., and Hyman, P. (2010) Phage choice, isolation and preparation for phage therapy. *Curr. Pharm. Biotechnol.* 11: 2–14.
112. Chan, B.K., and Abedon, S.T. (2012) Bacteriophage adaptation, with particular attention to issues of phage host range In: Quiberoni, A. and Reinheimer, J., *Bacteriophages in Dairy Processing*. Nova Science Publishers, 25–52.
113. Chan, B.K.,, Abedon, S.T., and Loc-Carrillo, C. (2013) Phage cocktails and the future of phage therapy. *Future Microbiol*. 8: 769–83.
114. Ormala, A.M., and Jalasvuori, M. (2013) Phage therapy: Should bacterial resistance to phages be a concern, even in the long run? *Bacteriophage*. 3: e24219.
115. Bush, K.,, Courvalin, P., Dantas, G., et al. (2011) Tackling antibiotic resistance. *Nat. Rev. Microbiol.* 9: 894–6.
116. Hughes, J.M. (2011) Preserving the lifesaving power of antimicrobial agents. *J. Am. Med. Assoc.* 305: 1027–8.
117. Lungren, M.P., Christensen, D., Kankotia, R., Falk, I., Paxton, B.E., and Kim, C.Y., et al (2013) Bacteriophage K for reduction of biofilm on central venous catheter material. *Bacteriophage*. 3: e26825
118. Singh, R., Ray, P., Das, A., and Sharma, M. (2010) Penetration of antibiotics through *Staphylococcus aureus* and *Staphylococcus epidermidis* biofilms. *J. Antimicrob. Chemother*. 65: 1955–8.
119. Tseng, B.S., Zhang, W., Harrison, J.J., et al. (2013) The extracellular matrix protects *Pseudomonas aeruginosa* biofilms by limiting the penetration of tobramycin. *Environ. Microbiol.* 15: 2865–78.
120. Sertic, V. (1929) Origine de la lysine d'une race du bacttriophage. *Compt. Rend. Soc. Biol.* 100: 477.
121. Biziulevicius, G.A., Biziuleviciene, G., and Kazlauskaite, J. (2008) A list of enzyme preparations covered by the term enzybiotics should not be restricted to bacteriophage-encoded peptidoglycan hydrolases (lysins). *J. Pharm. Pharmacol.* 60: 531–2.

122. Skillman, L.C., Sutherland, I.W., and Jones, M.V. (1999) The role of exopolysaccharides in dual species biofilm development. *J. Appl. Microbiol.* 85: 13S–8S.
123. Mushtaq, N., , Redpath, M.B., Luzio, J.P., and Taylor, P.W. (2004) Prevention and cure of systemic *Escherichia coli* K1 infection by modification of the bacterial phenotype. *Antimicrob. Agents Chemother.* 48: 1503–8.
124. Mushtaq, N., , Redpath, M.B., Luzio, J.P., and Taylor, P.W. (2005) Treatment of experimental *Escherichia* coli infection with recombinant bacteriophage-derived capsule depolymerase. *J. Antimicrob. Chemother.* 56: 160–5.
125. Scorpio, A., Tobery, S.A., Ribot, W.J., and Friedlander, A.M. (2008) Treatment of experimental anthrax with recombinant capsule depolymerase. *Antimicrob. Agents Chemother.* 52: 1014–20.
126. Scorpio, A., Chabot, D.J., Day, W.A., et al. (2007) Poly-glutamate capsuledegrading enzyme treatment enhances phagocytosis and killing of encapsulated *Bacillus anthracis. Antimicrob. Agents Chemother.* 51: 215–22.
127. Avery, O.T., and Dubos, R. (1931) The protective action of a specific enzyme against type III pneumococcus infection in mice. *J Exp Med.* 54: 73–89.
128. Goodner, K.R., Dubos, R., and Avery, O.T. (1932) The action of a specific enzyme upon the dermal infection of rabbits with type III pneumococcus. *J. Exp. Med.* 55: 393–404.
129. Waseh, S., Hanifi-Moghaddam, P., Coleman, R., et al. (2010) Orally administered P22 phage tailspike protein reduces salmonella colonization in chickens: Prospects of a novel therapy against bacterial infections. *PLoS One.* 5: e13904
130. Prevelige, P.E.Jr. (2006) Bacteriophage P22. In: Calendar, R. and Abedon, S.T., *The Bacteriophages.* Oxford, Oxford University Press, 457–68.
131. Steinbacher, S., Miller, S., Baxa, U., et al. (1997) Phage P22 Tailspike Protein: Crystal Structure of the Head-Binding Domain at 2.3 Å, fully refined structure of the endorhamnosidase at 1.56 Å resolution, and the molecular basis of O-antigen recognition and cleavage. *J. Mol. Biol.* 267(4): 865–80.
132. Yilmaz, C., Colak, M., Yilmaz, B.C., Ersoz, G., Kutateladze, M., and Gozlugol, M. (2013) Bacteriophage therapy in implant-related infections: An experimental study. *J. Bone Joint Surg. Am.* 95: 117–25.
133. Zhang, Y., Hunt, H.K., and Hu, Z. (2013) Application of bacteriophages to selectively remove *Pseudomonas aeruginosa* in water and wastewater filtration systems. *Water Res.* 47: 4507–18.
134. Zhang, Y., and Hu, Z. (2013) Combined treatment of *Pseudomonas aeruginosa* biofilms with bacteriophages and chlorine. *Biotechnol. Bioeng.* 110: 286–95.
135. Alemayehu, D., Casey, P.G., McAuliffe, O., et al. (2012) Bacteriophages MR299-2 and NH-4 can eliminate *Pseudomonas aeruginosa* in the murine lung and on cystic fibrosis lung airway cells. *MBio.* 3: e00029–12.
136. Chibeu, A., Lingohr, E.J., Masson, L., et al. (2012) Bacteriophages with the ability to degrade uropathogenic *Escherichia coli* biofilms. *Viruses.* 4: 471–87.
137. Kelly, D., McAuliffe, O., Ross, R.P., and Coffey, A. (2012) Prevention of *Staphylococcus aureus* biofilm formation and reduction in established biofilm density using a combination of phage K and modified derivatives. *Lett. Appl. Microbiol.* 54: 286–91.
138. Liao, K.S., Lehman, S.M., Tweardy, D.J., Donlan, R.M., and Trautner, B.W. (2012) Bacteriophages are synergistic with bacterial interference for the prevention of *Pseudomonas aeruginosa* biofilm formation on urinary catheters. *J. Appl. Microbiol.* 113: 1530–9.
139. Montanez-Izquierdo, V.Y., Salas-Vazquez, D.I., and Rodriguez-Jerez, J.J. (2012) Use of epifluorescence microscopy to assess the effectiveness of phage P100 in controlling *Listeria monocytogenes* biofilms on stainless steel surfaces. *Food Control.* 23: 470–7.
140. Ryan, E.M., Alkawareek, M.Y., Donnelly, R.F., and Gilmore, B.F. (2012) Synergistic phage-antibiotic combinations for the control of *Escherichia coli* biofilms *in vitro. FEMS Immunol. Med. Microbiol.* 65:395–8.
141. Thawal, N.D., Yele, A.B., Sahu, P.K., and Chopade, B.A. (2012) Effect of a novel podophage AB7-IBB2 on *Acinetobacter baumannii. Biofilm Curr. Microbiol.* 65: 66–72.
142. Ahiwale, S., Tamboli, N., Thorat, K., Kulkarni, R., Ackermann, H. and Kapadnis, B. (2011) *In vitro* management of hospital *Pseudomonas aeruginosa* biofilm using indigenous T7-like lytic phage. *Curr Microbiol.* 62: 335–40.

143. Castillo-Ruiz, M., Vines, E.D., Montt, C., et al. (2011) Isolation of a novel *Aggregatibacter actinomycetemcomitans* serotype b bacteriophage capable of lysing bacteria within a biofilm. *Appl. Environ. Microbiol.* 77: 3157–9.
144. Cornelissen, A., Ceyssens, P.J., T'Syen, J., et al. (2011) The T7-related *Pseudomonas putida* phage phi15 displays virion-associated biofilm degradation properties. *PLoS One*. 6: e18597.
145. Knezevic, P., Obreht, D., Curcin, S., et al. (2011) Phages of *Pseudomonas aeruginosa*: Response to environmental factors and *in vitro* ability to inhibit bacterial growth and biofilm formation. *J. Appl. Microbiol.* 111: 245–54.
146. Pires, D., Sillankorva, S., Faustino, A., and Azeredo, J. (2011) Use of newly isolated phages for control of *Pseudomonas aeruginosa* PAO1 and ATCC 10145 biofilms. *Res. Microbiol.* 162: 798–806.
147. Siringan, P., Connerton, P.L., Payne, R.J., and Connerton, I.F. (2011) Bacteriophagemediated dispersal of *Campylobacter jejuni* biofilms. *Appl. Environ. Microbiol.* 77: 3320–6.
148. Carson, L., Gorman, S.P., and Gilmore, B.F. (2010) The use of lytic bacteriophages in the prevention and eradication of biofilms of *Proteus mirabilis* and *Escherichia coli*. *FEMS Immunol. Med. Microbiol.* 59: 447–55.
149. Fu, W., Forster, T., Mayer, O., Curtin, J.J., Lehman, S.M., and Donlan, R.M. (2010) Bacteriophage cocktail for the prevention of biofilm formation by *Pseudomonas aeruginosa* on catheters in an *in vitro* model system. *Antimicrob. Agents Chemother.* 54: 397–404.
150. Gino, E., Starosvetsky, J., Kurzbaum, E., and Armon, R. (2010) Combined chemical-biological treatment for prevention/rehabilitation of clogged wells by an iron-oxidizing bacterium. *Environ. Sci. Technol.* 44: 3123–9.
151. Sillankorva, S., Neubauer, P., and Azeredo, J. (2010) Phage control of dual species biofilms of *Pseudomonas fluorescens* and *Staphylococcus lentus*. *Biofouling*. 26: 567–75.
152. Son, J.S., Lee, S.J., Jun, S.Y., et al. (2010) Antibacterial and biofilm removal activity of a podoviridae *Staphylococcus aureus* bacteriophage SAP-2 and a derived recombinant cell-wall-degrading enzyme. *Appl. Microbiol. Biotechnol.* 86: 1439–49.

14 Nanoparticle Associated Inhibition of Biofilm

Suhasini Mallick, Dibyajit Lahiri, Moupriya Nag, Tanmay Sarkar and Rina Rani Ray

14.1 INTRODUCTION

Biofilms are the root cause of many diseases. The list of infections that offshoot from biofilms is numerous, namely: osteomyelitis, native valve endocarditis, biliary tract infections, necrotizing fasciitis, cystic fibrosis, pneumonia and many more [1]. They are different from their planktonic forms on the basis of morphology as well as functioning. The planktonic or free-living singular bacterial cells lack external and variety of defense systems that can be registered while advancing as a blooming microbial community. It is advantageous for the overall fitness of the microbe, drafting maximum survival at a cheaper cost. The biofilms can be rephrased as the "city of microbe"[2], while the extracellular polymeric substances (EPS) as the "house of the biofilm cells."[3]. They have evolved over a span of 3.2–3.4 billion years and have crafted themselves to defend against harmful conditions [4]. Apart from posing a significant threat to health and economy, they are difficult to identify and remove. The EPS delivers a habitable and hospitable environment to the microbes residing inside it. Apart from acting as a mechanical barrier, it also serves to cushion the persister cells in the biofilm community. These cells are capable of withstanding high dosages of antibiotic administration and regrow themselves to their former glory once the antibiotic is withdrawn [5]. Hence, delivering plasticity and tenacity to the biofilm.

Estimates suggest that 65% of the overall infections are results of biofilm formation [6]. The plausible reason that biofilms are ulterior infection portals is due to the microbial diversity they encompass and the unhindered transfer of horizontal gene transfers (HGT) among the different species within them [6–9]. Transforming biofilms as a moat of resistance to the microbes inside. Keeping that in mind researchers are scanning potential antibiofilm agents, among which nanoparticles (NP) furnish high and speedy outcomes. In general, nanotechnology is a fast-growing and emerging domain that focuses on the nanoscale implementation of science. It encompasses intricacy since the scale of working is very refined in nature. The amalgamation of biotechnology further takes it a notch higher than other applications as it is mostly associated with living applications. The other sectors of nanotechnical applications include drug delivery, pollution abatement, cosmetics, biomedical prospects, infrastructures, informatics and many more [10–14]. Since it renders a vast diversity, it extends a larger dimension of applications due to the differential response of nanocomponents with respect to its bulk contemporaries. This exhibition of significantly different properties is due to a higher surface area-volume ratio, greater molecular interactions, adaptability of different morphologies and improved mechanical abilities [15–19]. Furthermore, as these modifications are achieved at the nanoscale, it is imperative to achieve greater antibiofilm potential using the NPs. The most appealing aspect of NPs is that not only can they be modified as per use but also use similar patterns to cutdown a divergent range of biofilms composed of a single or multiple species. Furthermore,

 DOI: 10.1201/9781003281238-14

crafting and testing new molecules takes decades and might end up being inefficient, meanwhile, microbes evolve at a faster rate making alternative antimicrobial strategies a universal necessity [20,21]. Although this chapter primarily focuses on the NP-mediated eradication of biofilm, it is important to understand and estimate the relative benefits of it over its contemporaries.

14.2 BACKGROUND OF NANOTECHNOLOGY

The prefix, nano, originates from Greek and Latin word roots, nanos and nanus respectively, translating as a dwarf that is one billionth of a unit [22]. The virtue of nanotechnology was given by an American physicist, Richard Feynman in 1959. His lecture, "There's Plenty of Room at the Bottom" at the annual meeting of the American Physical Society held at the California Institute of Technology was the most pivotal moment in the history of nanotechnology. He made the visionary hypothesis of scaling down machines to create molecular nuances [23]. Later on, nanotechnology gained momentum during the early 1970s when Japanese Professor Norio Taniguchi coined the term 'Nanotechnology' in 1974 and was the first to define it as: "Nano-technology mainly consists of the processing of separation, consolidation, and deformation of materials by one atom or one molecule"[22]. Following that, the term eventually trickled down and progressed in the first book on nanotechnology, named, "Engines of creation: The Coming Era of Nanotechnology" by K. Eric Drexler in 1986 [24]. As a result, the prospects of molecular engineering described in the book became popular [24].

Nanoparticles are strictly gauged as particles with at least one dimension under the 100 nm range [25]. They can be categorized into four different sets as 0D, 1D, 2D and 3D based on the overall shape of the nanomaterial [26]. Another distinguishing feature of NPs is the presence of layers generated during their nucleation. The three general wraps are 1) outer surface layer, often attached to functional additives such as surfactants, metal ions, polymers, or small molecules, 2) middle shell layer, mostly comprising of chemically different material than the inner layer and 3) the inner core layer, which is the NP entity in itself [27]. Due to the involvement of a circuitous NP installation, it sets them apart from the rest of the antibiofilm agents. As the former encompasses more ingredients to rip apart the biofilms. On that note, the blueprint or the type of ingredients contributing to coherent NP fabrication is discussed below.

14.3 MODULES OF NANOFABRICATION

Nanofabrication is the adeptness involved in increasing the overall efficiency of the NPs. It is easy to be lost in the labyrinth of multidisciplinary ways to categorize the various NPs. Before diving into the possible mechanisms of NPs towards biofilms, it is necessary to give a summary of the subject.

14.4 APPROACH OF SYNTHESIS

Specificity is the predecessor of effectivity; the superior impact of the NP will rely on how it was built. The keynotes to approaching the architecture of an NP are to both stipulate and simulate its creation as per requirements for a broad spectrum of categories and processes. Based on that, the terms: top-down and bottom-up approach in reference to nanotechnology were initially cropped up in 1989 by the Foresight Institute.

Top-down approach

The process revolves around breaking larger particles into smaller subunits [28]. The established processes include grinding/milling, lithography, chemical vapor deposition, physical vapor deposition and other decomposition techniques [29,30]. The disadvantages of such an approach are expense, low output, time-consuming and lags in curating self-assembling properties [31].

Bottom-up approach

The process revolves around the creation of the nano system itself. It uses different sub-components linked repeatedly in specified proportions and timing to build the nano system from scratch. The process is intricate. The molecules used in the framework either self-assemble themselves or rely on positional metrics for assembly. Examples of this approach are green synthesis, spinning, sol-gel and biochemical synthesis [29]. The flaws arising from this approach are not scalable, expensive equipment, difficulty in controlling synthesis and hazardous [31].

14.5 TYPES OF SYNTHESIS

The techniques of synthesis can be illustrated as two major subplots, non-biological and biological methods. The non-biological methods involve physical and chemical routes of synthesis. Further, they are heterogeneous in approach as both top-down and bottom-up approaches are involved. The common physical routes are high-energy ball milling, inert gas condensation, laser pyrolysis, pulse vapor deposition, flash spray pyrolysis and electro-spraying. Meanwhile, the common chemical routes are sol-gel, co-precipitation, polyol synthesis, chemical vapor deposition, hydrothermal synthesis, and alkaline precipitation. The type of route employed for synthesis depends on the context of the application. Chemical routes create liabilities in the form of toxic spills or unwarranted impacts while physical routes mostly incur higher costs. Therefore, there is a rise in the generation of frugal, biodegradable, and biocompatible NPs via biological routes of synthesis [32].

The biological methods are speedily being explored due to the presence of numerous biogenic resources. The resources spread across single-celled organisms, multicell organisms, excretes and extracts derived from microorganisms, algae, fungi, yeast, plants, animals, and biological particles [33]. Since there are multitudinous modes of synthesizing nanoparticles via chemical, physical and biological ways, there is an abundance of hassle-free conspicuous generation of NPs (Table 14.1).

14.6 CURRENT ANTI-BIOFILM STRATEGIES

Treatment of biofilms may be subject to numerous existing strategies. The most popular of them is the antibiotic-mediated strategy. While its effectiveness and longevity of effectiveness vary tremendously, it can obstruct pre-mature biofilm growth. This in turn offers the challenge of biofilm detection which is a very scrupulous diagnostic task. Compounding that, antibiotics are also dependent on dosage, timing, duration, and selective sensitivity in terms of application. Another breach in antibiotic efficiency is the development of drug resistance. The global community is under grave threat in the form of drug resistance as the UN Ad hoc Interagency Coordinating Group on Antimicrobial Resistance estimates 10 million deaths a year by 2050 due to it. Among other non-invasive strategies is the use of molecular inhibitors such as enzymes, polyphenols, or peptides that may serve well under ideal settings but are highly prone to degradation on entering active host systems [45]. A pitfall of potential antibiofilm chemicals is to be validated extensively among in vivo models to authenticate their pharmacokinetics and biocompatibility while upholding the antibiofilm efficacy [46].Antibiofilm bioweapons such as bacteriophages are used to discard biofilms by triggering their dispersion but there has been evidence of resistance developed towards this [47].

On the other hand, environmental turbulence such as sonication, photodynamic therapy (PTD) and use of nanoparticles demonstrate controllable intensity with little/no impedance arising from the microbial community. Sonication treatment dislodges the biofilms from their site of attachment,

TABLE 14.1
Antibiofilm Nanoparticles

Synthesis	Technique		Nanoparticle	Size Shape	Target Microbe	Reference
PHYSICAL	MICROWAVE IRRADIATION		Biopolymer capped silver	50–100 nm Polygon	*Staphylococcus aureus* MTCC 740 and *Escherichia coli* MTCC 9492	[34]
CHEMICAL	HYDROTHERMAL		Carbon quantum dot	16.5 nm and 15.1 nm -	*Porphyromonasgingivalis*	[35]
	COPRECIPITATION		Glycerol iron oxide	4.7 ± 0.2 nm Spherical	*P. aeruginosa* 1397 and *E. faecalis ATCC 29212*	[36]
	SOL-GEL		Copper	55 ± 2 nm Spherical	*Pseudomonas aeruginosa*	[37]
MIXED	COPRECIPITATION CUM LASER ABLATION TECHNIQUE		$Mn_{0.5}Zn_{0.5}Fe_2O_4$	~12 nm Spherical	*Bacillus subtilis*	[38]
BIOLOGICAL	**BIOGENIC SOURCE**					
	LEAF	*Euphrasia officinalis*	Gold and silver	40.3–AgNPs and 49.72–AuNPs Quasi-spherical	*S. aureus* and *P. aeruginosa*	[39]
		Psidium guajava	Selenium	30–50 nm spherical	*Enterococcus faecalis*	[40]
	FLOWER	*Rosa f loribunda charisma*	Magnesium oxide	35.25–55.14 nm polyhedral	*Staphylococcus epidermidis*, *Streptococcus pyogenes*, and *Pseudomonas aeruginosa*	[41]
	MICROORGANISM	*Pseudomonas deceptionensis*	Silver	10–30 nm spherical	*S. aureus* and *P. aeruginosa*	[42]
		Aspergillus tubingensis		35 ± 10 nm -	*Bacillus subtilis*	[43]
	SEAWEED	*Ulva lactuca*	Zinc oxide	10–50 nm Triangles, hexagons, rods and rectangles	*B. licheniformis*, *B. pumilus*, *E. coli* and *P. vulgaris*	[44]

thereby implementing debridement of the biofilm. While that has a lucrative advantage over its chemical competitors [48], it does not promise to kill the biofilm [49], making it liable to reframe itself at another interface. Moreover, sonication remains inconsistent with delivering uniform repeatable antibiofilm effects. Meanwhile, PTD is a promising approach to biofilm removal. This uses a targeted generation of reactive oxygen species (ROS) to terminate biofilms with the help of a photosensitizer molecule. The limitation arises as mostly it is best suited for localized, easily accessible, and upfront areas where light penetration faces no obstruction [50,51]. Additionally, use of invasive devices to enhance photonic adequacy requires technical skills, making this prone to error as well. Therefore, not only the aforementioned strategies are lagging in certain aspects but also, risk collateral damage to the surrounding tissues.

Nanoparticles (NP) offer a sophisticated and balanced approach to biofilm remediation. This not only impeaches antimicrobial resistance (AMR) [52]but is also capable of extending its effectivity by delivering antimicrobial agents [53]. Moreover, as antimicrobial agents, NPs gain superiority over traditional antibiotics due to multiple killing formats, weakening the relative emergence of resistance [54]. Interestingly, the NP can increase the accumulation of antibiotics within bacterial cells by blocking their efflux pumps. Localized or site-specific response strategies for the unloading of antibiofilm agents are a major advantage offered by nanoparticles. They respond to internal/external stimuli such as target cells, pH, concentration gradient or enzymatic responses to pay off their deliverable or inherent antimicrobial impact. Keeping that aside, more evidence comes to light regarding the increment/abatement of horizontal gene transfer (HGT) compelled by the influence of NPs [55]. This strictly depends on the nature and type of the NP because some metal NPs increase HGTs while metal oxide NPs dribble in both the upregulation and downregulation of HGTs. Generation of reactive oxygen species (ROS) is one of the artillery of NPs against biofilm inhibition. While that is an advantage, the overproduction of ROS results in the increase of HGTs leaching AMR into the bacteria [56,57]. This can be checked upon by suitably configuring the production of ROS with the help of a ROS scavenger. Here enters nanobiotechnical engineering to process the effectivity of the NP against biofilm with minimal liability.

14.7 MECHANISM OF INHIBITION OF BIOFILM

The biofilm matrix has a microenvironment within and has water-filled channels that act as efflux pumps to eradicate irritants. The mechanical imbalance spurred by NPs paves the way to biofilm disruption. It is the first stage of attack since the state of the bacterial cells in the shielded biofilm differs depending upon the nature, tenure of growth and accessibility to resources. The killing module of the NP initially acts by causing membrane disintegration of the cells. Hence, even persister cells fall prey to the mechanism [58]. Amphiphilic balance among NPs can manipulate the hydrophobic and electrostatic interactions within the EPS matrix, thereby maximizing uptake across the layers of biofilm [54,59].Preliminarily based on the charge and type of NP it interacts with the EPS. If the NP is positively charged then it produces a Fenton reaction, causing free radicle production that degrades the EPS and vanquishes the bacteria [60]. Further, it disintegrates the lipid bilayer and the lipopolysaccharide membrane of the bacteria. Thus, causing membrane fluidization and modifying its hydrophobicity. Upon disruption, the NPs, and ions leach across the cytosol, consequently, causing leakage of cytoplasmic content [61]. Additionally, the ions also create reactive oxygen species (ROS) in the form of hydroxyl radical (OH), hypochlorous acid (HOCI), superoxide anion (O_2^-), hydrogen peroxide (H_2O_2), and singlet oxygen radicals that create oxidative stress on the cell and change its permeability [62]. The ions from NPs bind to the thiol groups of cell membrane proteins resulting in protein deactivation [63,64]. Apart from binding to the protein, it can also delocalize the cell membrane protein, leading to loss of structural integrity [65]. The high ion concentrations also hinder the formation of surface-associated amyloids by countering the growth of fibrillar configuration thereby restricting surface colonization by the bacteria [66–70].

These NPs alter the surface dynamics by changing the zeta potential. The NPs with positive zeta potential more easily interact with the bacterial cells than NPs with negative zeta potential [71,72]. Similarly, the NPs with higher hydrophobicity rapidly infiltrate biofilms and kill the bacterial cells [73]. Proteomic studies suggest the involvement of NPs in the inhibition of proteins from nitrogen metabolism, chemotaxis, flagellar assembly, ribosomes, oxidative phosphorylation pathway, and the

TABLE 14.2
Downregulation of Biofilm-associated Genes by Nanoparticles

Nanoparticle	Microorganism	Downregulation Of Gene	Reference
ZnO	*Pseudomonas aeruginosa*	QS genes (*LasR, rhlI and pqsR*), adhesion factors–lectins (*Lec*A), biofilm formation protein (*PelA*), host defensive (*exoS*) and virulence (*lasA)*	[88]
Lectin conjugated-SiO_2		QS regulatory *(rhlR)* gene	[89]
Chitosan-propolis	*Staphylococcus epidermidis*	Intercellular adhesion genes (*IcaABCD, embp*),a metalloprotease (*sepA*), autolysin E (*atlE*), *rsbU*—regulatory gene (*rsbU*) *genes*	[72]
Carbon quantum dot drugs	*Porphyromonasgingivalis*	Adhesion *(FimA, RgpA, RgpB* and *KGP)* genes	[35]
Silver	*Acinetobacter baumannii*	Virulence *(kpsMII, afa/draBC,bap, OmpA,* and*csuA/B)*genes	[90]
Mycofabricated silver	*Pseudomonas aeruginosa*	QS-virulence (*lasA, lasB, rhlA, lasI, lasR, rhlI, rhlR, phzA1, and fabH2)* and house-keeping *(proC)* genes	[91]
Alpha-mangostin (AMG) loaded polymeric NPs	*Staphylococcus aureus*	Adhesion *(icaC)*, adhesion surface proteins (*clfB*), Fibronectin binding protein A (*fnbA*) genes and strain specific (*ebpS* and *fnbB*) genes.	[92]

electron transport chain [74]. It can inhibit the enzymatic activities involved in the peptidoglycan synthesis of biofilm formation, thereby causing failure of biofilm growth in the initial stages [75]. It also causes impairment in the uptake and release of phosphate ions, which leads to an imbalance in cellular transport, leading to spillage of ATP, enzymes, reducing sugars, ions, and proteins [76,77]. It can disrupt the sugar metabolism pathway by interacting with associated enzymes, thereby switching off the energy circuit in the bacterial cell [78]. The underlying mechanism of disrupting enzymes and respiratory pathways is the NP-mediated interference with the sulfhydryl groups, thereby disrupting its enzymes, membrane integrity and respiratory pathways. Furthermore, it also dampens bacterial cell proliferation and stirs instability in the biofilm matrix [79–81].

Once within the bacterial cell, they rip apart the genetic machinery of the cell by interacting with the functional groups of proteins and nucleic acids, such as the carboxyl (–COOH), amino (-NH), and mercapto (-SH) groups [82]. It happens because these NPs are known to bind and tear DNA bases by using Van der Waals forces and short-range hydrophobic forces [83–85]. Upon investigation, it revealed that free ions of the NP do not link to the phosphate group of nucleic acids but link to just the nucleosides, where it disintegrates the double helical structure by breaking H-bonds [63]. Studies also show that the NPs are also capable of linking with the tRNA of the bacterial cell, causing obstruction in ribosomal synthesis, and resulting in a reduction of biofilm formation [86]. Meanwhile, if the NP is target-specific in nature it can also directly inhibit the production of certain necessary steps involved in the adhesion and maturation of the biofilm [87]. Further, there are numerous ways by which the NPs can interact and cause turbulence in the genetic makeup of the bacterial cells (Table 14.2).

14.8 IMPACT OF NP CHARACTERISTICS ON BIOFILM ERADICATION

The characteristics of NPs are an important aspect as they fine-tune the recognition, diagnosis, and therapeutics simultaneously and carefully. Since there are numerous types of NPs and interactive ways to extirpate biofilms, the following two broad subthemes are described.

14.9 TYPE OF NPS

14.9.1 Connatural Properties

The NPs capable of projecting their antibiofilm dynamics upon the biofilm matrix are termed as connatural properties of NPs. The mechanism of NP action over the biofilm community has been discussed earlier. With reference to the scheme of working, most NPs show unique inherent native agility for suppressing flourishing biofilms. These NPs are often the most direct and simplest form of restricting biofilms. These properties can either come from solo organic or inorganic NPs or a combination of both. The most studied among them is the use of metallic NPs. Metals, even in their bulk forms possess heating, conducting, magnetic, antimicrobial, and antibiofilm capabilities [93]. Such qualities escalate in their nanoform and can be maneuvered as per requirements.

The research community finds metals to be a comprehensive solution to taming biofilms due to their flexibility in action and ability to adapt/conjugate with novel nano systems and are also used in a variety of biomedical processes [94]. Among them, silver NPs have a massive range of antimicrobial operations in different formulations. One of the studies on the interaction between silver NPs and *Acinetobacter baumannii* delivered an antibiofilm effect while repressing the expression of genes related to biofilm formation and its associated virulence [90]. Meanwhile, other metallic NPs can also be modulated via external stimuli to bring about an effect on a stipulated span and duration. Magnetic stimulation of NPs is given as a non-invasive technique to extinguish biofilms since magnetic fields can pass through body tissues [95]. Upon exposing Fe_2O_3 NPs to a magnetic field, there is an incidental rise in their ambient temperature. The thermal exposure enunciates static friction upon biofilms, thereby dispersing the bacterial cells [96]. Since planktonic bacterial cells have less defense compared to biofilm-embedded cells, this technique has the advantage of discarding infectious bacteria by rejigging the bacterial arrangement. Moreover, since the thermal response strictly relies upon magnetization, it shows targeted activity on biofilms with no subsidiary impacts on healthy cells [97].

Metal oxide NPs such asZrO_2, MgO, Al_2O_3, CeO_2, Fe_2O_3, SnO, TiO_2 and ZnO offer higher versatility in their functioning [98]. ZnO NPs can deliver inhibitory results against *Streptococcus pneumoniae* biofilms even at sub-MIC doses. It ensures such activity by blocking the adhesion of bacterial cells thereby promoting them to stay in a nomadic planktonic condition [99]. Carbon quantum dots were hydrothermally synthesized from metronidazole (MCDs) and tinidazole (TCDs). These quantum dots successfully inhibited *P. nigrescens* selectively while having no inhibitory impact on *E. coli* and *S. aureus.* Thus, implying that selective inhibition of microbes is an advantage in natural settings where beneficial microbes of the natural microflora are necessary. Meanwhile, MCDs were superior in biofilm inhibition compared to TCDs [35]. Hence concluding that the innate properties of NPs can render changes to the antibiofilm efficiency.

On the other hand, supplementary measures are also available in the form of delivering other antibiofilm agents with the help of nano transportation of the same. These nano cargos preliminarily act as a vehicle for antimicrobial molecules, creating sustenance for both, the molecules, and the NP, thereby ensuring prolonged and adequate efficiency in eradicating biofilms. The next section gives an overview of the molecules loaded on nano systems and the facets of their antibiofilm proficiency.

14.9.2 NPs as Nano Vehicles

An ideal transport system essentially must entail a nontoxic vehicle that defends therapeutic integrity while remaining penetrative inside biofilm matrices [100]. Conveniently, NPs are seen to qualify for such parameters by tailoring their formations [101]. Among different ways of delivery, lipid-coated NPs shift a huge advantage to themselves since they render a charged surface over the NP, thereby preventing the degradation or inactivation of antimicrobial molecules upon entering a biofilm microenvironment. In a study conducted by Li et al., the lipid-coated NP is capable of overcoming dual barriers of gastric mucus and *Helicobacter pylori* biofilm matrix. The lipid layer held rhamnolipids and clarithromycin on the surface over a chitosan NP. The formulated NP could remove EPS, disrupt the biofilm architecture, and obstruct the reformation of the biofilms by halting bacterial adhesions [96]. Another example would be that of stimuli mediated hybrid micelles, responding to pH and lipase. The hybrid model holds an anionic surface that restricts nonspecific interactions and enhances accumulation of the NP at the required site. Upon encountering bacterial lipases and an acidic microenvironment of *P. aeruginosa* biofilms, the micelles shrink in size and reverse their charges, releasing d-tyrosine to disperse the biofilm matrix and azithromycin to destruct bacterial cells [102]. Interestingly the NPs can also incur a cationic charge as seen in the case of clindamycin-loaded poly (lactic-coglycolic acid) PLGA-PEI (polyethyleneimine) NPs. The delivery console was used to release clindamycin to impair bacterial burden among methicillin resistant *S. aureus* infections [103].

Apart from antibiotics, NPs employs bacteriophages and enzymes as adjuvant to create a higher anti-biofilm effect. As in the case of chitosan-coated Fe_3O_4 colloidal nanoparticle clusters (CS-Fe_3O_4), that conjugated along with the polyvalent bacteriophage PEL1, to penetrate across *P. aeruginosa* and *E. coli* biofilms. The subunits worked synchronously, the NP clusters engaged in cell lysis alongside paving the entry of bacteriophage into the biofilm, that further terminated the bacterial cells [104]. Similarly, deoxyribonuclease-decorated gold nanoclusters (DNase–AuNCs) hampered biofilm by utilizing the DNase to inhibit, and degrade the bacterial extracellular matrix, and to give easy access to the Au NPs to enter and act on the bacterial cell resulting in biofilm dispersion. Meanwhile, upon combining the NP with fluorescent DNase and using phototherapy, not only enabled the killing of the biofilms but also the detection of the biofilms upon simultaneous interaction of the pathogens with DNase [105].

There is evidence of elite nanocarriers that can switch their surface charge as per microenvironment to effectively deliver the antimicrobial agents that react with the biofilms upon receiving external energy. A supramolecular nanocarrier (α-CD-Ce6-NO-DA), comprised of α-cyclodextrin-conjugated nitric oxide (NO) prodrug along with Ce6 prodrug into PEG block polypeptide copolymer to act upon biofilms. In the acidic ambience of the biofilm, the nanocarriers take up a positive charge while remaining negatively charged at a physiological pH of 7.4. The former environment facilitates effective penetration through the biofilm and latches to the negatively charged bacterial surfaces. Following this it then releases NO molecules. NO worked in two ways, one by dramatically reducing the GSH concentration in biofilm to enhance the PDT efficiency, and the other by reacting with light-triggered ROS, producing the RNS (ONOO–) for stronger bactericidal ability [106].

14.9.3 Morphology

Size, shape, and composition will act as a mandate for the antibiofilm behaviour of the NP. The size is probably the most important parameter in terms of biofilm eradication. The reason is that a smaller size enables speedy and better penetration into the biofilm matrix. This rapid diffusion has been studied in *P. fluorescens* biofilm and is related to the increased radius of the NP [107]. This can be well explained with the help of a studied analysis. Upon evaluating the efficacy of varying sizes and shapes of NO-releasing Si NPs against *P. aeruginosa* and *S. aureus* biofilms, insights were drawn on the significance of particle size and shape have over biofilms. Particle sizes of 14, 50, and 150 nm with identical release of total NO release were examined. Out of them, the smallest NP had the highest antibacterial influence. Similarly, upon varying the aspect ratio as an indicator of NP shape, increasing the aspect ratio had a higher antimicrobial effect. Thus, rod-shaped NPs were more effective in biofilm hindrance than spherical NPs [108]. Likewise, the antibacterial effect of Au nano-stars and nanoflowers was higher compared to Au nanospheres due to the augmentation of increased surface area, generating an elevated probability of interactions with components of biofilm and bacterial cells [109].

14.10 ANTIBIOFILM ACTIVITIES

14.10.1 Nanoparticles (NPs)

14.10.1.1 Biogenic NPs

Biogenic synthesis of NPs originates from the idea of utilizing biospheric resources or biological processes to generate the NP. It can be achieved using viruses, bacteria, fungi, algae, seaweed, plankton, plants, and the like [110]. For instance, the reduction activity of crude plant extracts on biofilms is inferior to the NPs formulated with the same extract [44]. Hence, apart from being non-toxic, it is also more effective than ethnobiological antimicrobials. Similarly, microbial synthesis of NPs is another approach to stabilize the nanoforms and hinder biofilm formation [111].

The key to biogenic synthesis is that the resources can be developed seamlessly and are easy to develop. As seen in the case of mycofabricated Ag NPs, they were crafted using the metabolites of *Rhizopus arrhizus BRS-07* to combat *P. aeruginosa* biofilm, showcased impact on different sections of the biofilm, including an inhibitory effect on several virulence factors by contracting the production of pyochelin, rhamnolipid, pyocyanin, pyoverdine, alginate, LasA protease and LasB elastase. In addition to that, it also prevented the AHL production from blocking the QS machinery of the bacteria [91]. The Au NP produced via the mycelium of *Laccaria* stabilized the NPs which caused the reduction in pyocyanin production from *P. aeruginosa* [112]. Similarly, the fruit extract of *Averrhoa bilimbi* was used as a reducing and stabilizing agent for ZnO NP synthesis. Due to this mode of generation, the NP contained greater organic compounds bound to its surface compared to its commercial form (probably not produced by biogenic modules). On comparing the two against *E. coli* biofilms, the biogenic NP was effective in biofilm inhibition but not better than the commercial one. Interestingly, the biogenic ZnO NP lost its potency with an increase in concentration while the opposite was documented for commercial ZnO NPs [113].

14.10.1.2 Non-biogenic NPs

The basis of non-biogenic sources is to create a nano system that fits according to the required application, thereby giving liberty to eliminate any chance events upon usage alongside giving detailed and uniform features to the NP. For instance, the creation of gentamicin-loaded phosphatidylcholine (PA) -decorated Au NPs. Here, the PA was acquired via thin-film rehydration along with an aqueous solution of $HAuCl_4$. Then the NPs were loaded with gentamicin by mixing with gentamicin sulfate solution. The framed NP was an effective antibiofilm agent against both Gram-positive and Gram-negative bacteria strains [114]. Even though the NPs are of non-biogenic origin, they can also be conjugated with biological extracts to reduce the load on the NP as an antibiofilm agent. In fact, the biological resource can be extracted from the pathogen to be treated itself. In a study, four NPs of different types were prepared via laser ablation. It comprised a comparative assessment of platinum (Pt), gold (Au), silica oxide (SiO_2), and tin oxide (SnO_2) NPs against *P. aeruginosa* biofilm. The twist of this experiment was the conjugation of lectin extracted from *P. aeruginosa* upon NPs. Pt, Au and SnO_2 could inhibit biofilm formation but SiO_2 showed the highest antibiofilm potential and could also repress the QS regulatory gene-*rhlR* [89]. CuO NPs inhibited aquaculture pathogens of *Vibrio parahaemolyticus* (ATCC 17802), *Vibrio alginolyticus* (ATCC 17749) and *Aeromonas hydrophila* (ATCC 7966) in both *in vivo* and *in vitro* analysis. The one-pot synthesis of these NPs delivered a reduction of the cell surface hydrophobicity by 95% and of EPS production by 85% of *V. alginolyticus* biofilm [115].

14.10.2 Nanomaterials

The elementary translation of nanomaterials is the materials that are framed with the help of nanoparticles to uplift their potency. There are other well-stated definitions given by the International Organization for Standardization (ISO) as "material with any external nanoscale dimension or having internal nanoscale surface structure" and the European Union Commission (EU) as "a manufactured or natural material that possesses unbound, aggregated or agglomerated particles where external dimensions are in the1–100 nm size range". In order to halt the free roaming of toxic detached NPs and also retain sustenance in the antibiofilm effect, nanomaterials consist of NPs embedded in them for various applications. Ag NPs an anchored to an amino-silanized glass surface as an antibiofilm medium against *Staphylococcus epidermidis* RP62A. Apart from that the modified biomaterial retained stability in an aqueous medium with extended release of the NP, successfully maintaining a high localized NP concentration that consequently resulted in a strong antibiofilm effect [116]. Nano-decorated textiles offer a handy solution to avoid biofilms, especially in nosocomial settings. The one-pot synthesis of N-heterocyclic molecule-coated gold NPs is readily fabricated onto textiles via sonication. This modified textile possesses antibiofilm properties against *Staphylococcus aureus* (MRSA) and multi-drug resistant *Escherichia coli* [117].

14.10.3 Nanocomposites

The British Standards Institution defines a nanocomposite as a multiphase structure with at least one phase on the nanoscale dimension [118]. Since biofilm is a complex and dynamic system that can either contain mono-species or poly-species matrix within itself, inhibitory tactics must remain general yet effective. Especially in clinical settings, identifying species to treat them can eventually result in morbidity. Hence the idea of nanocomposite (NC) shines through to quench ghoulish multi-species biofilms. One such example would be that of ZnO/Ag NC embellished surfaces. The NC could establish antibiofilm activity against monospecies of *S. aureus* and *E. coli* in a dose-dependent setting. In the meantime, on casting ZnO NPs in oligotrophic conditions against *C. albicans* biofilm, it induced biofilm increment. Such scenarios can be avoided upon recruiting ZnO/Ag NC since Ag counteracted on the biofilm [119]. Thus, it is of significant interest to form NC that can inhibit the biofilms in a mixed microbial environment. Such derivatives are most functional towards frequently touched surfaces of hospital settings to halt the spread of nosocomial infections. The nanocomposites can also be of metallic-non-metallic composition. For example, NCs of reduced graphene oxide (RGO) with metal oxides of zinc, silver and nickel were tested as an antibiofilm agent. Among them, RGO-NiO NC exhibited highest antibiofilm activity with an inhibition percentage of 98.03% for *C. neoformans*, 94.60% for *B. subtilis* and 91.74% for *P. aeruginosa* [120].

14.11 APPLICATIONS OF NP-ASSISTED BIOFILM INHIBITION

14.11.1 Pathogenic Inhibitions

Pathogenesis is almost synonymous with biofilm formation. Biofilms are auxiliary to the persistence of chronic wounds as it is prevalent among at least half of such wounds [121]. 80% of all chronic infections arise due to untreated and undiscovered biofilms [122]. Microbes are shrewd enough to take cues from their immediate environment and make the most out of wound scenarios. It has been noted that Gram-positive bacteria, *S. aureus* and *S. pyogens* infect the wound site and prepare the area to be inhabitable for Gram-negative bacteria, *E. coli* and *P. aeruginosa.* Such interplay among the bacteria consistently delays wound healing [123]. In an *in vivo* mouse wound study, polyethylene glycol (PEG) coated AgNPs were embedded in hydrogels to fight the prevalent MRSA infection. In addition to promoting healing of the wound, it also had higher antibacterial efficacy over the commercial silver-based cream. Hence, not only do NPs terminate the infection but are better than established wound remedies [124].

Biofilm cause numerous infections of which 55.52% were of UTIs followed by 35.78% in ventilator-associated pulmonary infection and 11.52% of catheter-related infection [125]. The hematogenous spread of infections from colonized intravenous devices gives rise to other complicated health issues [126]. Additionally, host inflammatory molecules may also facilitate the adhesion of biofilms to the surface of the devices [127]. Orthopedic implants suffer from the repercussion of periprosthetic joint infection (PJI),which is largely caused by biofilm infection whereby it is prevalent among 80% of the PJI cases [128–130]. Reinfection is an inevitable occurrence. Such scenarios must be dealt with through invasive and painful surgical procedures to discard infected tissues [130]. While this is not feasible in most scenarios, a factor of uncertainty prevails since complete biofilm inhibition is not conclusive based on exemptive methods. Compounding the problem is the ability of microbes to escape via the blood stream and initiate biofilm growth elsewhere. Bone and bacteria targeting NPs have been examined against orthopedic implant-related infections. The mesoporous silica NPs utilized D6 and $UBI_{29\text{-}41}$ peptides to reach the target sites and then eventually release the vancomycin to interact with the MRSA infection [131].

Denture-related diseases [132] can be avoided by utilizing NPs. Incorporation of CuO NP on soft denture liners resulted in the inhibition of bacterial strains, *Streptococcus mutans* (ATCC35668), *Streptococcus salivarius* (ATCC9222) and *Streptococcus sobrinus* (ATCC27607) and fungal strains, *Candida albicans* (CBS 10261, 1905, 1912, 1949, 2730). Similarly, CuO NP also has agricultural applications in inhibiting plant pathogenic *Ralstonia solanacearum* biofilm [133].

14.11.2 Non-pathogenic Inhibitions

Biofilms encase many enzymes that help them defend themselves but in turn, these enzymes are the reason behind biofouling [134]. Biofouling due to biofilms in different industrial units increases the overall operational and maintenance costs [135]. It can also pose a threat to public health by acting as a reservoir of pathogens in drinking water distribution systems. Ag NPs have been used as a preventive measure to abstain from the growth of microfouling biofilms. These NPs could prevent several microbes namely *Myroides odoratimimus, Micrococcus luteus, Halomonas aquamarine, Proteus mirabilis, Exiguobacterium aurantiacum, Exiguobacterium arabatum, Jeotgalibacillus alimentarius, Bacillus megaterium, Bacillus pumilus, Halotalea alkalilenta* and *Arthrobacter mysorens* [136].

Pathogenic biofilms in food processing units can go unchecked leading to post-processing contamination [137]. Hence it is important to have a potent external packaging to prevent incoming microbes. A possible solution can arise from the use of nanotechnology. On framing a Ag-ZnO-RGO-polyethylene glycol (PEG) NC as a potential food packaging material, it demonstrated antibiofilm activity against *S. aureus* and *P. aeruginosa*. Further on varying the silver concentration to 0.05 M in the NC, a steep 95% biofilm inhibition was observed in addition to the removal of preformed biofilms. Meanwhile, on transforming it into an agar-based nanosheet, it retained its antimicrobial properties for 90 days, thus delivering the aspect of longevity in terms of application as a nano packing material [138].

14.11.3 Biocompatibility

Most NP-oriented studies must undergo scrupulous biocompatibility tests. For instance, under the worst-case scenarios, surface-ZnO NPs were exposed to human keratinocyte cells directly. This caused the viability of the cells since zinc is a wound-healing agent by itself [119,139]. Similarly, doping of the metal oxide can reduce its toxicity but retain it as a pro-oxidative and pro-inflammatory agent [140]. Further doping of ZnO NPs with Mn and Fe ions does not impact its overall antibacterial effects. Studies have shown that it rendered higher antibacterial activity on *S. aureus, P. aeruginosa, E. coli, S. typhi, Proteus mirabilis, K. pneumoniae* and *B. subtilis* as compared to undoped NP [141,142]. Additionally, it is also identified as a GRAS (generally recognized as safe) by the US FDA [143]. Moreover, the addition of non-toxic NPs such as MgO NP to form a ZnMgO NC, is also a viable approach to reducing toxicity.

On the other hand, toxicity depends on several factors such as type, interactions, concentration, and so forth. AgNPs have low toxicity in ecosystems [144,145]. Meanwhile, other studies have discovered that the toxicity of the NP is due to the leaching of solid Ag particles and soluble Ag [146,147]. Therefore, it is necessary to bind them and condition them to prospective release. There are other non-toxic alternatives within the nano-community itself. Meanwhile, there are alternatives such as poly (lactic-coglycolic acid) PLGA, a synthetic biocompatible biopolymer used for numerous biomedical applications [148], a biological technique to determine toxicity is the testing of NPs against green algae. The algae are sensitive to changes in chemical parameters and is often chosen as an indicator to assess the bioactivity of industrial wastes. Unlike CuO, ZnO, TiO_2 NPs, and SiO_2NPsdo not restrain the growth of the algae, thereby qualifying as non-toxic. Meanwhile, it enhances algal growth which can have ecological implications to some degree [149]. Another counter technique to reduce problems of NP toxicity is to encourage the biogenic synthesis of NPs. Especially because chemically synthesized NPs retain traces of toxic elements that are difficult to identify as well as difficult to track [150].

Macrophages are essential regulators of the human defense system [151]. They can respond to changes in the local microenvironment, differentiating into pro-inflammatory and anti-inflammatory states. This event is described as macrophage polarization [152]. Nanomaterials can stimulate and initiate macrophagic polarization to promote the host defense system [153]. Thus, rendering beneficial properties by regulating the NPs in different fields that require biocompatibility.

CONCLUSION AND FUTURE PROSPECTS

The future of antibiofilm NPs is directed towards collaborative domains of intelligent and responsive nanomaterials, such as chemo-responsive, magneto-responsive, bio-responsive, pH-responsive, and photodynamic and photothermal nanotechnology. Additionally, Artificial intelligence (AI) assisted/modified/optimized NPs and machine learning coordinated antibiofilm methods, strategies and predictive models are rapidly being undertaken to advance the inhibitory/regulatory measures against biofilms. The monitoring of biofilm inhibition wirelessly via the help of advanced NPs and nanomaterials is also helping researchers understand the pathogenicity of different microbes.

The nanoparticles crafted to inhibit biofilm must render focus on its physiochemical aspects and ecological significance. There is a pressing urgency to derive knowledge on the chronological impact of NPs released untreated in the environment, inviting unprecedented pathogenic risks. Additionally, there are numerous pitfalls in the regulatory aspects of NPs' synthesis and application that seek uniformity globally. Moreover, there is a substantial gap between the laboratory-scale studies of NPs on biofilm inhibition and its translative impact in long-term clinical settings. This is mainly due to the complex processes occurring simultaneously, in an actively altering biological setting render unpredictability of results over the variety of NP-triggered/altered responses. Similarly, there is a need for case-specific NPs to minimize harmful implications. A major scope of utilizing NPs in inhibition of biofilm is in developing personalized medical treatments across different bacterial strains. Even though there is a grave need of improvement of existing NPs, it has delivered greater efficiency than other antibiofilm agents and has room for enhancements. Clinical biofilm is a repetitive problem, to which NPs can deliver lasting solutions when assisted with continued research.

REFERENCES

1. Costerton, W., Veeh, R., Shirtliff, M., Pasmore, M., Post, C., Ehrlich, G. The application of biofilm science to the study and control of chronic bacterial infections. *J. Clin. Invest.* **2003**, *112*, 1466–1477.
2. Watnick, P., Kolter, R. Biofilm, city of microbes. *J. Bacteriol.* **2000**, *182*, 2675–2679.
3. Yang, L., Barken, K.B., Skindersoe, M.E., Christensen, A.B., Givskov, M., Tolker-Nielsen, T. Effects of iron on DNA release and biofilm development by Pseudomonas aeruginosa. *Microbiology*. **2007**, *153*, 1318–1328.

4. de Carvalho, C.C.C.R. Biofilms: Microbial Strategies for Surviving UV Exposure. *Adv. Exp. Med. Biol.* **2017**, *996*, 233–239.
5. Brauner, A., Fridman, O., Gefen, O., Balaban, N.Q. Distinguishing between resistance, tolerance and persistence to antibiotic treatment. *Nat. Rev. Microbiol.* **2016**, *14*, 320–330.
6. Lewis, K. Riddle of biofilm resistance. *Antimicrob. Agents Chemother.* **2001**, *45*, 999–1007.
7. Burmølle, M., Webb, J.S., Rao, D.,Hansen, L.H., Sørensen, S.J., Kjelleberg, S. Enhanced biofilm formation and increased resistance to antimicrobial agents and bacterial invasion are caused by synergistic interactions in multispecies biofilms. *Appl. Environ. Microbiol.* **2006**, *72*, 3916–3923.
8. Madsen, J.S., Burmølle, M., Hansen, L.H., Sørensen, S.J. The interconnection between biofilm formation and horizontal gene transfer. *FEMS Immunol. Med. Microbiol.* **2012**, *65*, 183–195.
9. Wolcott, R., Costerton, J.W., Raoult, D., Cutler, S.J. The polymicrobial nature of biofilm infection. *Clin. Microbiol. Infect. Off. Publ. Eur. Soc. Clin. Microbiol. Infect. Dis.* **2013**, *19*, 107–112.
10. Kruis, F.E., Fissan, H., Peled, A. Synthesis of nanoparticles in the gas phase for electronic, optical and magnetic applications—a review. *J. Aerosol Sci.* **1998**, *29*, 511–535.
11. Stark, W.J., Stoessel, P.R., Wohlleben, W., Hafner, A. Industrial applications of nanoparticles. *Chem. Soc. Rev.* **2015**, *44*, 5793–5805.
12. Singh, R., Nalwa, H.S. Medical applications of nanoparticles in biological imaging, cell labeling, antimicrobial agents, and anticancer nanodrugs. *J. Biomed. Nanotechnol.* **2011**, *7*, 489.
13. Tsuzuki, T. Commercial scale production of inorganic nanoparticles. *Int. J. Nanotechnol.* **2009**, *6*, 567–578.
14. Hassan, S.M., Ashour, M., Soliman, A.A.F., Hassanien, H.A., Alsanie, W.F., Gaber, A., Elshobary, M.E. The potential of a new commercial seaweed extract in stimulating morpho-agronomic and bioactive properties of Eruca vesicaria (L.) Cav. *Sustain.* 2021, 13, 4485
15. Hajipour, M.J., Fromm, K.M., Akbar Ashkarran, A., Jimenez de Aberasturi, D., Larramendi, I.R. de, Rojo, T., Serpooshan, V., Parak, W.J., Mahmoudi, M. Antibacterial properties of nanoparticles. *Trends Biotechnol.* **2012**, *30*, 499–511.
16. Thanh, N.T.K., Maclean, N., Mahiddine, S. Mechanisms of nucleation and growth of nanoparticles in solution. *Chem. Rev.* **2014**, *114*, 7610–7630.
17. Biju, V., Itoh, T., Anas, A., Sujith, A., Ishikawa, M. Semiconductor quantum dots and metal nanoparticles: Syntheses, optical properties, and biological applications. *Anal. Bioanal. Chem.* **2008**, *391*, 2469–2495.
18. Guo, D., Xie, G., Luo, J. Mechanical properties of nanoparticles: Basics and applications. *J. Phys. D. Appl. Phys.* **2013**, *47*, 13001.
19. Jain, P.K., Huang, X., El-Sayed, I.H., El-Sayed, M.A. Review of some interesting surface Plasmon Resonance-enhanced properties of noble metal nanoparticles and their applications to biosystems. *Plasmonics.* **2007**, *2*, 107–118.
20. Arias, C.A., Murray, B.E. Antibiotic-resistant bugs in the 21st century—a clinical super-challenge. *N. Engl. J. Med.* **2009**, *360*, 439–443.
21. Arias, L.R., Yang, L. Inactivation of bacterial pathogens by carbon nanotubes in suspensions. *Langmuir.* **2009**, *25*, 3003–3012.
22. Bayda, S., Adeel, M., Tuccinardi, T., Cordani, M., Rizzolio, F. The history of nanoscience and nanotechnology: From chemical-physical applications to nanomedicine. *Molecules.* **2019**, *25*, 112.
23. Feynman, R. There's plenty of room at the bottom. *Eng. Sci.* **1960**, *23*, 22–36.
24. Drexler, E. *Engines of Creation: The Coming Era of Nanotechnology*, Anchor Library of Science, 1987.
25. Philippot, L., Andersson, S.G.E., Battin, T.J., Prosser, J.I., Schimel, J.P., Whitman, W.B., Hallin, S. The ecological coherence of high bacterial taxonomic ranks. *Nat. Rev. Microbiol.* 2010, *8*, 523–529.
26. Tiwari, J.N., Tiwari, R.N., Kim, K.S. Zero-dimensional, one-dimensional, two-dimensional and three-dimensional nanostructured materials for advanced electrochemical energy devices. *Prog. Mater. Sci.* **2012**, *57*, 724–803
27. Shin, W.K., Cho, J., Kannan, A.G., Lee, Y.S., Kim, D.W. Cross-linked composite gel polymer electrolyte using Mesoporous Methacrylate-Functionalized SiO2 nanoparticles for Lithium-Ion polymer batteries. *Sci. Rep.* **2016**, *6*, 26332.
28. Wang, Y., Xia, Y. Bottom-up and top-down approaches to the synthesis of monodispersed spherical colloids of low melting-point metals. *Nano Lett.* **2004**, *4*, 2047–2050.
29. Iravani, S. Green synthesis of metal nanoparticles using plants. *Green Chem.* **2011**, *13*, 2638–2650.

30. Jahid, I.K., Ha, S.-D. A review of microbial biofilms of produce: Future challenge to food safety. *Food Sci. Biotechnol.* **2012**, *21*, 299–316.
31. Khan, I., Saeed, K., Khan, I. Nanoparticles: Properties, applications and toxicities. *Arab. J. Chem.* **2019**, *12*, 908–931.
32. Bartolucci, C., Antonacci, A., Arduini, F., Moscone, D., Fraceto, L., Campos, E., Attaallah, R., Amine, A., Zanardi, C., Cubillana-Aguilera, L.M., et al. Green nanomaterials fostering agrifood sustainability. *TrAC Trends Anal. Chem.* **2020**, *125*, 115840.
33. Chaudhari, A. Biogenic synthesis of nanoparticles and potential applications an eco-friendly approach. *J. Nanomed. Nanotechnol.* **2013**, *4*, 165.
34. Velusamy, P., Su, C.-H., Venkat Kumar, G., Adhikary, S., Pandian, K., Gopinath, S.C.B., Chen, Y., Anbu, P. Biopolymers regulate silver nanoparticle under microwave irradiation for effective antibacterial and antibiofilm activities. *PLoS One*. **2016**, *11*, e0157612–e0157612.
35. Liang, G., Shi, H., Qi, Y., Li, J., Jing, A., Liu, Q., Feng, W.,Li, G.,Gao, S. Specific anti-biofilm activity of carbon quantum dots by destroying *P. gingivalis* biofilm related genes. *Int. J. Nanomedicine*. **2020**, *15*, 5473–5489.
36. Iconaru, S.L., Prodan, A., Le Coustumer, P., Predoi, D. Synthesis and antibacterial and antibiofilm activity of iron oxide glycerol nanoparticles obtained by coprecipitation method. *J. Chem.* **2013**, *2013*.
37. LewisOscar, F., Mubarak Ali, D., Nithya, C., Priyanka, R., Gopinath, V., Alharbi, N.S., Thajuddin, N. One pot synthesis and anti-biofilm potential of copper nanoparticles (CuNPs) against clinical strains of *Pseudomonas aeruginosa*. *Biofouling*. **2015**, *31*, 379–391.
38. Shahid, S.A., Anwar, F., Shahid, M., Majeed, N., Azam, A., Bashir, M., Amin, M., Mahmood, Z., Shakir, I. Laser-assisted synthesis of Mn0.50$Zn_{0.50}Fe_2O_4$ Nanomaterial: Characterization and *in Vitro* Inhibition activity towards *Bacillus subtilis* biofilm. *J. Nanomater.* **2015**, *2015*, 896185.
39. Singh, H., Du, J., Singh, P., Yi, T.H. Ecofriendly synthesis of silver and gold nanoparticles by *Euphrasia officinalis* leaf extract and its biomedical applications. *Artif. cells, nanomedicine, Biotechnol.* **2018**, *46*, 1163–1170.
40. Miglani, S., Tani-Ishii, N. Biosynthesized selenium nanoparticles: characterization, antimicrobial, and antibiofilm activity against *Enterococcus faecalis*. *PeerJ*. **2021**, *9*, e11653.
41. Younis, I.Y., El-Hawary, S.S., Eldahshan, O.A., Abdel-Aziz, M.M., Ali, Z.Y. Green synthesis of magnesium nanoparticles mediated from *Rosa floribunda* charisma extract and its antioxidant, antiaging and antibiofilm activities. *Sci. Rep.* **2021**, *11*, 16868.
42. Jo, J.H., Singh, P., Kim, Y.J., Wang, C., Mathiyalagan, R., Jin, C.G., Yang, D.C. *Pseudomonas deceptionensis* DC5-mediated synthesis of extracellular silver nanoparticles. *Artif. cells, Nanomedicine, Biotechnol.* **2016**, *44*, 1576–1581.
43. Rodrigues, A.G., Ruiz, R.de.C., Selari, P.J.R.G., de Araújo, W.L., de Souza, A.O. Anti-biofilm action of biological silver nanoparticles produced by *Aspergillus tubingensis* and antimicrobial activity of fabrics carrying it. *Biointerface Res. Appl. Chem.* **2021**, *11*, 14764–14774.
44. Ishwarya, R., Vaseeharan, B., Kalyani, S., Banumathi, B., Govindarajan, M., Alharbi, N.S., Kadaikunnan, S., Al-Anbr, M.N., Khaled, J.M., Benelli, G. Facile green synthesis of zinc oxide nanoparticles using Ulva lactuca seaweed extract and evaluation of their photocatalytic, antibiofilm and insecticidal activity. *J. Photochem. Photobiol. B*. **2018**, *178*, 249–258.
45. Dürig, A., Kouskoumvekaki, I., Vejborg, R.M., Klemm, P. Chemoinformatics-assisted development of new anti-biofilm compounds. *Appl. Microbiol. Biotechnol.* **2010**, *87*, 309–317.
46. Beloin, C., Renard, S., Ghigo, J.-M., Lebeaux, D. Novel approaches to combat bacterial biofilms. *Curr. Opin. Pharmacol.* **2014**, *18*, 61–68.
47. Siringan, P., Connerton, P.L., Payne, R.J.H., Connerton, I.F. Bacteriophage-mediated dispersal of *Campylobacter jejuni* biofilms. *Appl. Environ. Microbiol.* **2011**, *77*, 3320–3326.
48. Karbysheva, S., Di Luca, M., Butini, M.E., Winkler, T., Schütz, M., Trampuz, A. Comparison of sonication with chemical biofilm dislodgement methods using chelating and reducing agents: Implications for the microbiological diagnosis of implant associated infection. *PLoS One*. **2020**, *15*, e0231389.
49. Sandbakken, E.T., Witsø, E., Sporsheim, B., Egeberg, K.W., Foss, O.A., Hoang, L., Bjerkan, G., Løseth, K., Bergh, K. Highly variable effect of sonication to dislodge biofilm-embedded *Staphylococcus epidermidis* directly quantified by epifluorescence microscopy: An in vitro model study. *J. Orthop. Surg. Res.* **2020**, *15*, 522.

50. Wainwright, M., Maisch, T., Nonell, S., Plaetzer, K., Almeida, A., Tegos, G.P., Hamblin, M.R. Photoantimicrobials-are we afraid of the light?. *Lancet. Infect. Dis.* **2017**, *17*, e49–e55.
51. Wainwright, M. The problem with dyes in infection control. *Dye. Pigment.* **2017**, *146*,402–407.
52. Kzar, A.J. Antimicrobial and biofilm inhibitory activity of nanoparticles against clinical isolates from urinary tract infection. *Indian J. Forensic Med. &amp, Toxicol.* **2020**, 14,194–199.
53. Mihai, M.M., Dima, M.B., Dima, B., Holban, A.M. Nanomaterials for wound healing and infection control. *Mater. (Basel, Switzerland).* **2019**, *12*, 2176.
54. Pelgrift, R.Y., Friedman, A.J. Nanotechnology as a therapeutic tool to combat microbial resistance. *Adv. Drug Deliv. Rev.* **2013**, *65*, 1803–1815
55. Otinov, G.D., Lokteva, A.V, Petrova, A.D., Zinchenko, I.V, Isaeva, M.V, Kovtunov, E.A., Koshel, E.I. Positive and negative effects of metal oxide nanoparticles on antibiotic resistance genes transfer. *Antibiot. (Basel, Switzerland).* **2020**, *9*, 742.
56. Lu, J., Wang, Y., Jin, M., Yuan, Z., Bond, P., Guo, J. Both silver ions and silver nanoparticles facilitate the horizontal transfer of plasmid-mediated antibiotic resistance genes. *Water Res.* **2020**, *169*, 115229.
57. Zhang, S., Wang, Y., Song, H., Lu, J., Yuan, Z., Guo, J. Copper nanoparticles and copper ions promote horizontal transfer of plasmid-mediated multi-antibiotic resistance genes across bacterial genera. *Environ. Int.* **2019**, *129*, 478.
58. Hurdle, J.G., O'Neill, A.J., Chopra, I., Lee, R.E. Targeting bacterial membrane function: An underexploited mechanism for treating persistent infections. *Nat. Rev. Microbiol.* **2011**, *9*, 62–75.
59. Gupta, A., Landis, R.F., Rotello, V.M. Nanoparticle-based antimicrobials: Surface functionality is critical. *F1000Research.* **2016**, *5*, F1000–364.
60. Chaurasia, A.K., Thorat, N.D., Tandon, A., Kim, J.H., Park, S.H., Kim, K.K. Coupling of radiofrequency with magnetic nanoparticles treatment as an alternative physical antibacterial strategy against multiple drug resistant bacteria. *Sci. Rep.* **2016**, *6*, 33662.
61. Ikuma, K., Decho, A.W., Lau, B.L.T. When nanoparticles meet biofilms-interactions guiding the environmental fate and accumulation of nanoparticles. *Front. Microbiol.* **2015**, *6*, 591.
62. Hwan, S., Lee, H.S., Ryu, D.S., Choi, S.J., Lee, D.S. Antibacterial activity of silver-nanoparticles against Staphylococcus aureus and Escherichia coli. *Microbiol. Biotechnol. Lett.* **2010**, 39, 1106.
63. Klueh, U., Wagner, V., Kelly, S., Johnson, A., Bryers, J.D. Efficacy of silver-coated fabric to prevent bacterial colonization and subsequent device-based biofilm formation. *J. Biomed. Mater. Res.* **2000**, *53*, 621–631.
64. Rai, M.K., Deshmukh, S.D., Ingle, A.P., Gade, A.K. Silver nanoparticles: The powerful nanoweapon against multidrug-resistant bacteria. *J. Appl. Microbiol.* **2012**, *112*, 841–852.
65. Jena, P., Bhattacharya, M., Bhattacharjee, G., Satpati, B., Mukherjee, P., Senapati, D., Srinivasan, R. Bimetallic gold–silver nanoparticles mediate bacterial killing by disrupting the actin cytoskeleton MreB. *Nanoscale.* **2020**, *12*, 3731–3749.
66. Schwartz, K., Syed, A.K., Stephenson, R.E., Rickard, A.H., Boles, B.R. Functional amyloids composed of phenol soluble modulins stabilize *Staphylococcus aureus* biofilms. *PLoS Pathog.* **2012**, *8*, e1002744.
67. Tõugu, V., Karafin, A., Zovo, K., Chung, R.S., Howells, C., West, A.K., Palumaa, P. Zn(II)–and Cu(II)-induced non-fibrillar aggregates of amyloid-beta (1-42) peptide are transformed to amyloid fibrils, both spontaneously and under the influence of metal chelators. *J. Neurochem.* **2009**, *110*, 1784–1795.
68. Abelein, A., Gräslund, A., Danielsson, J. Zinc as chaperone-mimicking agent for retardation of amyloid β peptide fibril formation. *Proc. Natl. Acad. Sci. U. S. A.* **2015**, *112*, 5407–5412.
69. Ban, D.K., Paul, S. Nano zinc oxide inhibits fibrillar growth and suppresses cellular toxicity of Lysozyme Amyloid. *ACS Appl. Mater. Interfaces.* **2016**, *8*, 31587–31601.
70. Dutta, A., Bhattacharyya, S., Kundu, A., Dutta, D., Das, A.K. Macroscopic amyloid fiber formation by staphylococcal biofilm associated SuhB protein. *Biophys. Chem.* **2016**, *217*, 32–41.
71. Arakha, M., Saleem, M., Mallick, B.C., Jha, S. The effects of interfacial potential on antimicrobial propensity of ZnO nanoparticle. *Sci. Rep.* **2015**, *5*, 9578.
72. Ong, T.H., Chitra, E., Ramamurthy, S., Ling, C.C.S., Ambu, S.P., Davamani, F. Cationic chitosan-propolis nanoparticles alter the zeta potential of *S. epidermidis*, inhibit biofilm formation by modulating gene expression and exhibit synergism with antibiotics. *PLoS One.* **2019**, *14*, e0213079.
73. Shebl, R.I., Farouk, F., Azzazy, H.M.E.S. Effect of surface charge and hydrophobicity modulation on the antibacterial and antibiofilm potential of magnetic iron nanoparticles. *J. Nanomater.* **2017**, *2017*.

74. Zhang, Y., Pan, X., Liao, S., Jiang, C., Wang, L., Tang, Y., Wu, G., Dai, G., Chen, L. Quantitative proteomics reveals the mechanism of silver nanoparticles against multidrug-resistant *Pseudomonas aeruginosa* biofilms. *J. Proteome Res.* **2020**, *19*, 3109–3122.
75. Khan, S.T., Ahamed, M., Musarrat, J., Al-Khedhairy, A.A. Anti-biofilm and antibacterial activities of zinc oxide nanoparticles against the oral opportunistic pathogens *Rothia dentocariosa* and *Rothia mucilaginosa*. *Eur. J. Oral Sci.* **2014**, *122*, 397–403.
76. Schreurs, W.J., Rosenberg, H. Effect of silver ions on transport and retention of phosphate by *Escherichia coli*. *J. Bacteriol.* **1982**, *152*, 7–13.
77. Lok, C.N., Ho, C.M., Chen, R., He, Q.Y., Yu, W.Y., Sun, H., Tam, P.K.H., Chiu, J.F., Che, C.M. Proteomic analysis of the mode of antibacterial action of silver nanoparticles. *J. Proteome Res.* **2006**, *5*, 916–924.
78. Bhattacharya, R., Mukherjee, P. Biological properties of "naked" metal nanoparticles. *Adv. Drug Deliv. Rev.* **2008**, *60*, 1289–1306.
79. Lansdown, A.B.G. Silver. I: Its antibacterial properties and mechanism of action. *J. Wound Care.* **2002**, *11*, 125–130.
80. Chaw, K.C., Manimaran, M., Tay, F.E.H. Role of silver ions in destabilization of intermolecular adhesion forces measured by atomic force microscopy in *Staphylococcus epidermidis* biofilms. *Antimicrob. Agents Chemother.* **2005**, *49*, 4853–4859.
81. Jung, W.K., Koo, H.C., Kim, K.W., Shin, S., Kim, S.H., Park, Y.H. Antibacterial activity and mechanism of action of the silver ion in *Staphylococcus aureus* and *Escherichia coli*. *Appl. Environ. Microbiol.* **2008**, *74*, 2171–2178.
82. Yu, J., Zhang, W., Li, Y., Wang, G., Yang, L., Jin, J., Chen, Q., Huang, M. Synthesis, characterization, antimicrobial activity and mechanism of a novel hydroxyapatite whisker/nano zinc oxide biomaterial. *Biomed. Mater.* **2014**, *10*, 15001.
83. Carnerero, J.M., Jimenez-Ruiz, A., Castillo, P.M., Prado-Gotor, R. Covalent and non-Covalent DNA-Gold-Nanoparticle interactions: New avenues of research. *Chemphyschem.* **2017**, *18*, 17–33.
84. Jiang, WY., Ran, S.Y. Two-stage DNA compaction induced by silver ions suggests a cooperative binding mechanism. *J. Chem. Phys.* **2018**, *148*, 205102.
85. Koo, K.M., Sina, A.A.I., Carrascosa, L.G., Shiddiky, M.J.A., Trau, M. DNA–bare gold affinity interactions: Mechanism and applications in biosensing. *Anal. Methods.* **2015**, *7*, 7042–7054, doi:10.1039/C5AY01479D
86. Cui, Y., Zhao, Y., Tian, Y., Zhang, W., Lü, X., Jiang, X. The molecular mechanism of action of bactericidal gold nanoparticles on *Escherichia coli*. *Biomaterials.* **2012**, *33*, 2327–2333.
87. Johansson, E.M.V, Crusz, S.A., Kolomiets, E., Buts, L., Kadam, R.U., Cacciarini, M., Bartels, K.M., Diggle, S.P., Cámara, M., Williams, P., et al. Inhibition and dispersion of Pseudomonas aeruginosa biofilms by glycopeptide dendrimers targeting the fucose-specific lectin LecB. *Chem. Biol.* **2008**, *15*, 1249–1257.
88. Abdelraheem, W.M., Mohamed, E.S. The effect of zinc oxide nanoparticles on *Pseudomonas aeruginosa* biofilm formation and virulence genes expression. *J. Infect. Dev. Ctries.* **2021**, *15*, 826–832.
89. Nsayef Muslim, S., Mohammed Ali, A.N., Auda, I.G. Anti-biofilm and anti-virulence effects of silica oxide nanoparticle-conjugation of lectin purified from *Pseudomonas aeruginosa*. *IET Nanobiotechnology.* **2021**, *15*, 318–328.
90. Hetta, H.F., Al-Kadmy, I.M.S., Khazaal, S.S., Abbas, S., Suhail, A., El-Mokhtar, M.A., Ellah, N.H.A., Ahmed, E.A., Abd-ellatief, R.B., El-Masry, E.A., et al. Antibiofilm and antivirulence potential of silver nanoparticles against multidrug-resistant *Acinetobacter baumannii*. *Sci. Rep.* **2021**, *11*, 10751.
91. Singh, B.R., Singh, B.N., Singh, A., Khan, W., Naqvi, A.H., Singh, H.B. Mycofabricated biosilver nanoparticles interrupt *Pseudomonas aeruginosa* quorum sensing systems. *Sci. Rep.* **2015**, *5*, 13719.
92. Nguyen, P.T.M., Nguyen, M.T.H., Bolhuis, A. Inhibition of biofilm formation by alpha-mangostin loaded nanoparticles against *Staphylococcus aureus*. *Saudi J. Biol. Sci.* **2021**, *28*, 1615–1621.
93. Ghanati, F., Abdolmaleki, P., Vaezzadeh, M., Rajabbeigi, E., Yazdani, M. Application of magnetic field and iron in order to change medicinal products of *Ocimum basilicum*. *Environmentalist.* **2007**, *27*, 429–434.

94. Burduşel, A.C., Gherasim, O., Grumezescu, A.M., Mogoantă, L., Ficai, A., Andronescu, E. Biomedical applications of silver nanoparticles: An up-to-date overview. *Nanomater. (Basel, Switzerland)*. **2018**, *8*, 681.
95. Yang, H.Y., Li, Y., Lee, D.S. Multifunctional and stimuli-responsive magnetic nanoparticle-based delivery systems for biomedical applications. *Adv. Ther.* **2018**, *1*, 1800011.
96. Li, J., Nickel, R., Wu, J., Lin, F., van Lierop, J., Liu, S. A new tool to attack biofilms: driving magnetic iron-oxide nanoparticles to disrupt the matrix. *Nanoscale*. **2019**, *11*, 6905–6915.
97. Mahmoudi, M., Serpooshan, V. Silver-coated engineered magnetic nanoparticles are promising for the success in the fight against antibacterial resistance threat. *ACS Nano*. **2012**, *6*, 2656–2664.
98. Fernández-García, M., Rodriguez, J.A., Encyclopedia of Inorganic and Bioinorganic Chemistry 2007.
99. Bhattacharyya, P., Agarwal, B., Goswami, M., Maiti, D., Baruah, S., Tribedi, P. Zinc oxide nanoparticle inhibits the biofilm formation of *Streptococcus pneumoniae*. *Antonie Van Leeuwenhoek*. **2018**, *111*, 89–99.
100. Whittam, A.J., Maan, Z.N., Duscher, D., Wong, V.W.,Barrera, J.A.,Januszyk, M., Gurtner, G.C. Challenges and opportunities in drug delivery for wound healing. *Adv. Wound Care*. **2016**, *5*, 79–88.
101. Mallick, S., Nag, M., Lahiri, D., Pandit, S., Sarkar, T., Pati, S., Nirmal, N.P., Edinur, H.A., Kari, Z.A., Ahmad Mohd Zain, M.R., et al. Engineered nanotechnology: An effective therapeutic platform for the chronic cutaneous wound. *Nanomater. (Basel, Switzerland)*. **2022**, *12*, 778.
102. Chen, M.,Wei, J.,Xie, S.,Tao, X.,Zhang, Z.,Ran, P.,Li, X. Bacterial biofilm destruction by size/surface charge-adaptive micelles. *Nanoscale*. **2019**, *11*, 1410–1422.
103. Hasan, N., Cao, J., Lee, J., Hlaing, S.P., Oshi, M.A., Naeem, M., Ki, M.H., Lee, B.L., Jung, Y., Yoo, J.W. Bacteria-targeted clindamycin loaded polymeric nanoparticles: Effect of surface charge on nanoparticle adhesion to MRSA, antibacterial activity, and wound healing. *Pharmaceutics*. **2019**, *11*.
104. Li, L.L., Yu, P., Wang, X., Yu, S.S., Mathieu, J., Yu, H.Q., Alvarez, P.J.J. Enhanced biofilm penetration for microbial control by polyvalent phages conjugated with magnetic colloidal nanoparticle clusters (CNCs). *Environ. Sci. Nano*. **2017**, *4*, 1817–1826.
105. Xie, Y., Zheng, W., Jiang, X. Near-infrared light-activated phototherapy by gold nanoclusters for dispersing biofilms. *ACS Appl. Mater. Interfaces*. **2020**, *12*, 9041–9049.
106. Hu, D., Deng, Y., Jia, F., Jin, Q., Ji, J. Surface charge switchable supramolecular nanocarriers for nitric oxide synergistic photodynamic eradication of biofilms. *ACS Nano*. **2020**, *14*, 347–359.
107. Peulen, T.O., Wilkinson, K.J. Diffusion of nanoparticles in a biofilm. *Environ. Sci. Technol.* **2011**, *45*, 3367–3373.
108. Slomberg, D.L., Lu, Y., Broadnax, A.D., Hunter, R.A., Carpenter, A.W., Schoenfisch, M.H. Role of size and shape on biofilm eradication for nitric oxide-releasing silica nanoparticles. *ACS Appl. Mater. Interfaces*. **2013**, *5*, 9322–9329.
109. Penders, J., Stolzoff, M., Hickey, D.J., Andersson, M., Webster, T.J. Shape-dependent antibacterial effects of non-cytotoxic gold nanoparticles. *Int. J. Nanomedicine*. **2017**, *12*, 2457–2468.
110. Sinha, S., Pan, I., Chanda, P., Sen, S.K. Nanoparticles fabrication using ambient biological resources. *J Appl Biosci*. **2009**, *19*, 1113–1130.
111. Lahiri, D., Nag, M., Sheikh, H.I., Sarkar, T., Edinur, H.A, Siddhartha, P., Rani, R.R. Microbiologically synthesized nanoparticles and their role in silencing the biofilm signaling cascade. *Front. Microbiol.* **2021**, *12*, 636588.
112. Samanta, S., Singh, B.R., Adholeya, A. Intracellular synthesis of gold nanoparticles using an ectomycorrhizal strain EM-1083 of *Laccaria fraterna* and its nanoanti-quorum sensing potential against *Pseudomonas aeruginosa*. *Indian J. Microbiol.* **2017**, *57*, 448–460.
113. Rahayu, E., Wonoputri, V., Samadhi, T.W. Plant extract-assisted biosynthesis of zinc oxide nanoparticles and their antibacterial application. *IOP Conf. Ser. Mater. Sci. Eng.* **2020**, *823*, 12036.
114. Mu, H., Tang, J., Liu, Q., Sun, C., Wang, T., Duan, J. Potent antibacterial nanoparticles against biofilm and intracellular bacteria. *Sci. Rep.* **2016**, *6*, 18877.
115. Chari, N., Felix, L., Davoodbasha, M., Sulaiman Ali, A., Nooruddin, T. In vitro and in vivo antibiofilm effect of copper nanoparticles against aquaculture pathogens. *Biocatal. Agric. Biotechnol.* **2017**, *10*, 336–341.
116. Taglietti, A., Arciola, C.R., D'Agostino, A., Dacarro, G., Montanaro, L., Campoccia, D., Cucca, L., Vercellino, M., Poggi, A., Pallavicini, P., et al. Antibiofilm activity of a monolayer of silver nanoparticles anchored to an amino-silanized glass surface. *Biomaterials*. **2014**, *35*, 1779–1788.

117. Wang, L., Natan, M., Zheng, W., Zheng, W., Liu, S., Jacobi, G., Perelshtein, I., Gedanken, A., Banin, E., Jiang, X. Small molecule-decorated gold nanoparticles for preparing antibiofilm fabrics. *Nanoscale Adv.* **2020**, *2*, 2293–2302.
118. Jeevanandam, J., Barhoum, A., Chan, Y.S., Dufresne, A., Danquah, M.K. Review on nanoparticles and nanostructured materials: History, sources, toxicity and regulations. *Beilstein J. Nanotechnol.* **2018**, *9*, 1050–1074.
119. Rosenberg, M., Visnapuu, M., Vija, H., Kisand, V., Kasemets, K., Kahru, A., Ivask, A. Selective antibiofilm properties and biocompatibility of nano-ZnO and nano-ZnO/Ag coated surfaces. *Sci. Rep.* **2020**, *10*, 13478.
120. Elbasuney, S., El-Sayyad, G.S., Tantawy, H., Hashem, A.H. Promising antimicrobial and antibiofilm activities of reduced graphene oxide-metal oxide (RGO-NiO{,} RGO-AgO{,} and RGO-ZnO) nanocomposites. *RSC Adv.* **2021**, *11*, 25961–25975.
121. James, G.A., Swogger, E., Wolcott, R., Pulcini, E. deLancey, Secor, P., Sestrich, J., Costerton, J.W., Stewart, P.S. Biofilms in chronic wounds. *Wound repair Regen. Off. Publ. Wound Heal. Soc. [and] Eur. Tissue Repair Soc.* **2008**, *16*, 37–44.
122. Qayyum, S., Khan, A.U. Nanoparticles vs. biofilms: A battle against another paradigm of antibiotic resistance. *Medchemcomm.* **2016**, *7*, 1479–1498.
123. Harper, D.R., Parracho, H.M.R.T., Walker, J., Sharp, R., Hughes, G., Werthé, M. Bacteriophages and biofilms. *Antibiotics.* **2014**, *3*, 270–284.
124. Negut, I., Grumezescu, V., Grumezescu, A.M. Treatment Strategies for Infected Wounds. *Molecules.* **2018**, *23*, 2392.
125. Madhurima, P., Sateesh, G., Saritha, Y., Padma, M. The microbiological profile of nosocomial infections in an intensive care unit. *BMC Infect. Dis.* **2014**, *14*, P73–P73.
126. Varma, R.S. Greener approach to nanomaterials and their sustainable applications. *Curr. Opin. Chem. Eng.* **2012**, *1*, 123–128.
127. Hall-Stoodley, L., Costerton, J.W., Stoodley, P. Bacterial biofilms: From the natural environment to infectious diseases. *Nat. Rev. Microbiol.* **2004**, *2*, 95–108.
128. Tzeng, A., Tzeng, T.H., Vasdev, S., Korth, K., Healey, T., Parvizi, J., Saleh, K.J. Treating periprosthetic joint infections as biofilms: Key diagnosis and management strategies. *Diagn. Microbiol. Infect. Dis.* **2015**, *81*, 192–200
129. Gbejuade, H.O., Lovering, A.M., Webb, J.C. The role of microbial biofilms in prosthetic joint infections. *Acta Orthop.* **2015**, *86*, 147–158.
130. Shoji, M.M., Chen, A.F. Biofilms in periprosthetic joint infections: A review of diagnostic modalities, current treatments, and future directions. *J. Knee Surg.* **2020**, *33*, 119–131.
131. Nie, B., Huo, S., Qu, X., Guo, J., Liu, X., Hong, Q., Wang, Y.,Yang, J., Yue, B. Bone infection site targeting nanoparticle-antibiotics delivery vehicle to enhance treatment efficacy of orthopedic implant related infection. *Bioact. Mater.* **2022**, *16*, 134–148
132. Pachava, K.R., Nadendla, L.K., Alluri, L.S.C., Tahseen, H., Sajja, N.P. Invitro antifungal evaluation of denture soft liner incorporated with tea tree oil: A new therapeutic approach towards denture stomatitis. *J. Clin. Diagn. Res.* **2015**, *9*, ZC62–ZC64.
133. Chen, J., Mao, S., Xu, Z., Ding, W. Various antibacterial mechanisms of biosynthesized copper oxide nanoparticles against soilborne *Ralstonia solanacearum*. *RSC Adv.* **2019**, *9*, 3788–3799.
134. Kassinger, S.J., vanHoek, M.L. Biofilm architecture: An emerging synthetic biology target. *Synth. Syst. Biotechnol.* **2020**, *5*, 1–10
135. Eguía, E., Trueba, A., Río-Calonge, B., Girón, A., Bielva, C. Biofilm control in tubular heat exchangers refrigerated by seawater using flow inversion physical treatment. *Int. Biodeterior. Biodegradation.* **2008**, *62*, 79–87.
136. Inbakandan, D., Kumar, C., Abraham, L.S., Kirubagaran, R., Venkatesan, R., Khan, S.A. Silver nanoparticles with anti microfouling effect: a study against marine biofilm forming bacteria. *Colloids Surf. B. Biointerfaces.* **2013**, *111*, 636–643.
137. Cheng, G., Zhang, Z., Chen, S., Bryers, J.D., Jiang, S. Inhibition of bacterial adhesion and biofilm formation on zwitterionic surfaces. *Biomaterials.* **2007**, *28*, 4192–4199
138. Naskar, A., Khan, H., Sarkar, R., Kumar, S., Halder, D., Jana, S. Anti-biofilm activity and food packaging application of room temperature solution process based polyethylene glycol capped Ag-ZnO-graphene nanocomposite. *Mater. Sci. Eng. C.* **2018**, *91*, 743–753.

139. Lin, P.H., Sermersheim, M., Li, H., Lee, P.H.U., Steinberg, S.M., Ma, J. Zinc in Wound Healing Modulation. *Nutrients*. **2017**, *10*, 16.
140. Xia, T., Zhao, Y., Sager, T., George, S., Pokhrel, S., Li, N., Schoenfeld, D., Meng, H., Lin, S., Wang, X., et al. Decreased dissolution of ZnO by iron doping yields nanoparticles with reduced toxicity in the rodent lung and zebrafish embryos. *ACS Nano*. **2011**, *5*, 1223–1235.
141. Bahari, A., Roeinfard, M., Ramzannezhad, A., Khodabakhshi, M., Mohseni, M. Nanostructured features and antimicrobial properties of Fe_3O_4/ZnO nanocomposites. *Natl. Acad. Sci. Lett.* **2019**, *42*, 9–12.
142. Sharma, N., Jandaik, S., Kumar, S., Chitkara, M., Sandhu, I.S. Synthesis, characterisation and antimicrobial activity of manganese–and iron-doped zinc oxide nanoparticles. *J. Exp. Nanosci.* **2016**, *11*, 54–71.
143. Zou, F., Zhou, J., Zhang, J., Li, J., Tang, B., Chen, W., Wang, J., Wang, X. Functionalization of silk with In-Situ synthesized platinum nanoparticles. *Mater. (Basel, Switzerland)*. **2018**, *11*, 1929.
144. Neethu, S., Midhun, S.J., Radhakrishnan, E.K., Jyothis, M. Green synthesized silver nanoparticles by marine endophytic fungus *Penicillium polonicum* and its antibacterial efficacy against biofilm forming, multidrug-resistant *Acinetobacter baumanii. Microb. Pathog.* **2018**, *116*, 263–272.
145. Abo-Shama, U.H., El-Gendy, H., Mousa, W.S., Hamouda, R.A., Yousuf, W.E., Hetta, H.F., Abdeen, E.E. Synergistic and antagonistic effects of metal nanoparticles in combination with antibiotics against some reference strains of pathogenic microorganisms. *Infect. Drug Resist.* **2020**, *13*, 351–362.
146. Faiz, M.B., Amal, R., Marquis, C.P., Harry, E.J., Sotiriou, G.A., Rice, S.A., Gunawan, C. Nanosilver and the microbiological activity of the particulate solids versus the leached soluble silver. *Nanotoxicology*. **2018**, *12*, 263–273.
147. Sotiriou, G.A., Meyer, A., Knijnenburg, J.T.N., Panke, S., Pratsinis, S.E. Quantifying the origin of released Ag+ ions from nanosilver. *Langmuir* **2012**, *28*, 15929–15936.
148. Liu, H., Slamovich, E.B., Webster, T.J. Less harmful acidic degradation of poly(lacticco-glycolic acid) bone tissue engineering scaffolds through titania nanoparticle addition. *Int. J. Nanomedicine*. **2006**, *1*, 541–545.
149. San, N.O., Kurşungöz, C., Tümtaş, Y., Yaşa, Ö., Ortaç, B., Tekinay, T. Novel one-step synthesis of silica nanoparticles from sugarbeet bagasse by laser ablation and their effects on the growth of freshwater algae culture. *Particuology*. **2014**, *17*, 29–35
150. Römling, U., Gomelsky, M., Galperin, M.Y. C-di-GMP: The dawning of a novel bacterial signalling system. *Mol. Microbiol.* **2005**, *57*, 629–639.
151. Taylor, P.R., Martinez-Pomares, L., Stacey, M., Lin, H.H., Brown, G.D., Gordon, S. Macrophage receptors and immune recognition. *Annu. Rev. Immunol.* **2005**, *23*, 901–944.
152. Mills, C.D., Kincaid, K., Alt, J.M., Heilman, M.J., Hill, A.M. M-1/M-2 macrophages and the Th1/Th2 paradigm. *J. Immunol.***2000**, *164*, 6166–6173.
153. Hong, Q., Huo, S., Tang, H., Qu, X., Yue, B. Smart nanomaterials for treatment of biofilm in orthopedic implants. *Front. Bioeng. Biotechnol.* **2021**, *9*, 694635.

15 Biofilms

Formation, Regulation and Detection

Kanchanlata Tungare, Sneha P. Nair, Sharanya D., Ashish Cyril, Santhosh Jathar, Mustansir Bhori, Thankamani Marar and Renitta Jobby

15.1 INTRODUCTION TO BIOFILMS

The imprecise misconception about microbial life is that microbes exist in a planktonic state in a liquid medium and as individual colonies on solid surfaces. On the contrary, microorganisms have always been seen naturally accumulating on a wide variety of surfaces (Donlan & Rodney, 2002). This accumulation is caused due to adherence of the bacterial community, which is enclosed in an extracellular matrix (Costerton et al., 1995). This was first observed by the father of microbiology, Antonie Van Leeuwenhoek in the plaque he collected from his own teeth using a primitive microscope constructed by him. He described microorganisms that attach to and grow universally on exposed surfaces. He described his observations as "animalcules," which led to studies revealing surface-associated microorganisms that were later known as biofilm (Costerton et al., 1999). The term 'film', which refers to bacterial adhesion, aggregation, and multiplication on surfaces, was used in marine microbiology to distinguish adhering (sessile) bacteria from free-swimming planktonic organisms from 1933 to 1935. Even Louis Pasteur (1822–1895) observed and sketched aggregates of bacteria as the cause of wine becoming acetic, but for the next century or so, biofilm growing microorganisms were not of interest and unknown to medical microbiologists. But now, due to the medical and ecological significance of biofilm, a lot of research has been conducted and the biofilm is defined as a complex microbial ecosystem formed by one or diverse species of microorganisms such as bacteria and fungi. This ecosystem leads to the assemblage of a surface-associated syntrophic consortium which is enveloped by an extracellular polymeric substance (EPS) matrix and results in the formation and maintenance of structured multicellular communities (Watnick & Kolter, 2000; López et al., 2010). Owing to the presence of this slimy extracellular matrix, the cells become adherent to the surface and to each other. They keep growing as densely packed thin films, one above the other, as layers, unlike their planktonic counterparts, which grow as individual colonies. The difference between the biofilm-associated cells and their respective planktonic counterparts is clearly identifiable due to the exhibition of a distinct phenotype with respect to the transcription of genes, resulting in the generation of an EPS matrix and a reduced growth rate. This distinct phenotype aids the formation of biofilms on a variety of surfaces, which include natural and artificial aquatic systems, industrial or potable water systems, living plant and animal tissues, biomaterials like contact lenses, and indwelling or implanted medical devices.

DOI: 10.1201/9781003281238-15

Microorganisms live in biofilms, and functions such as substrate exchange, metabolite distribution, and hazardous end product eviction are supported by both identical species and variety, demonstrating synergy among the biofilm-associated group of microorganisms. This synergy in the biofilm forming communities protects the bacteria present within them not only from the shear forces of the environment or the immune system of the host, but also from the attacks of several antimicrobial agents. Although biofilms start as invisible monolayers of microbes, with time they may become thick enough to be visible to the naked eye (Davey et al., 2000). This chapter extends insight about the importance and significance of biofilms with a focus on their formation and detection techniques. In addition to this, a brief literature to quorum sensing and its impact on metabolomic regulation is also provided.

15.2 STAGES OF FORMATION OF BIOFILM

The formation of biofilm can be divided into four successive stages:

(a) Reversible attachment**:** The extracellular DNA (eDNA) initiates the process of matrix production in the microbe, which is eventually taken up by structural proteins and polysaccharides (Whitchurch et al., 2002). The initial reversible attachment, which is the first step of the biofilm formation, will be facilitated by the bacteria coming into contact with the surface where they can get attached and attract the other microorganisms for attachment (Bose and Ghosh, 2011). Though short-lived, it serves as the most important event that led to the initiation of biofilm formation (Heilmann et al., 1997). A thin layer of cells becomes accumulated on the substratum, which still needs to undergo a differentiation process. While some adherent cells may attain differentiation, most of the cells may not get committed to the biofilm surface and resume their planktonic state. The adherent cells are transformed due to the differentiation into the biofilm composed of the EPS matrix. The attachment of the adherent cells at this stage is reversible due to the fact that, though the differentiation has been attained, the cells are only capable of producing a trace amount of EPS (Stoodley et al., 2002).

(b) Irreversible attachment: During this second stage, the bacteria start to form a monolayer by producing an extracellular matrix, which is referred to as "slime," basically for their protection. The interactions at this stage are weak and an irreversible attachment is evident, with a striking difference being the free-living planktonic bacteria in the new arrangement (Garrett et al., 2008). This stage can be subdivided into two parts: weak irreversible attachment and strong irreversible attachment. The event progresses from the overall weak interaction of the cell with the substratum where it is attached. It may continue to have weak interactions with the substratum and may leave it to resume planktonic life again (Liu et al., 2004). Irreversible attachment undergoes a transition change from weak interactions to permanent bonding between the cells and the substratum. The bonding is then followed by the maturation of the cells in the matrix (Stoodley et al., 2002).

(c) Biofilm maturation**:** Gradually, microcolonies are formed with the growing layers of EPS and cell-to-cell communication, called the quorum sensing (QS) mechanism begins. The initiation of QS marks the beginning of the third stage, where the irreversible attachment becomes strong due to cell-to-cell communication (Donlan, 2001). Although, at this point, the rate of cell division in the bacteria is not very high, it becomes resistant to the cleaning actions of the detergent. Furthermore, the bacterial cells become mature and provide a three-dimensional growth of biofilm by recruiting new planktonic cells and integrating them into the film. The three-dimensional growth or the thickening of layers of the biofilm will now not be affected by changing the surface tension of water by detergent alone but will also require mechanical stress (Sutherland, 2001). While the structure is maintained by the EPS layer, the maturation of the biofilm is supplemented by rapid cell division and accompanied by the growth and development of the cells. This leads to the generation of complex architecture through the formation of channels and pores. The newly formed architecture aids in the redistribution of bacteria within the substratum and away from it (Davey and O'Toole, 2000). The main event of maturation is the occurrence of the radically different protein profile that is found in these cells as compared to their planktonic counterparts. Most of these proteins are structural proteins and can be characterized into classes like metabolism, phospholipid and LPS-biosynthesis, membrane transport and secretion, adaptation and protective mechanisms. With this protein profile, mature biofilms can sustain the strong shear forces of nature and high concentrations of antibiotics (Lawrence et al., 1991).

(d) Detachment and dispersal: In the fourth stage, some mature cells detach from the biofilm as planktonic cells and disperse to potentially undertake the process of biofilm formation in a new environment (Davey et al., 2000). Dispersal is the fundamental behaviour of living organisms, and they must disperse in order to be able to colonize new environments by means of disruptive propagules (Donlan and Costerton 2002). Members of the biofilms have been observed to follow the same mechanism. The majority of the bacteria will disperse passively as victims of the shear forces of nature, but they will also undergo active dispersal periodically. The detachment of cells from the biofilm is influenced by the production of matrix breakdown enzymes, surfactants that interfere with surface tension. Other processes that are also regarded to be important for the detachment are the QS mechanism, intracellular di-cyclic GMP level, and changes in the availability of nutrients in the microenvironment. Therefore, detachment and dispersal is a complex and dynamic process that depends on multiple genetic determinants and regulatory processes (Webb, 2009).

15.3 MECHANISM OF PATHOGENICITY

Although the process by which biofilm-associated organisms elicit a response in the host system is crudely understood, numerous mechanisms have been hypothesized that explain why pathogens in biofilms are more likely to cause disease than pathogens in planktonic form: (a) Pathogens in biofilms could cause disease through the seeding and dispersal of a large number of cells that later lead to infection. Biofilms are dynamic in nature and during their life cycle consist of a stage where mature cells are dispersed (Hall-Stoodley and Stoodley, 2005). (b) Phenotypic heterogeneity is likely to be present within pathogens in biofilm. This might increase the chance of the virulent phenotype surviving and growing in the biofilm, which has the ability to infect the host cell (Hall-Stoodley and Stoodley, 2005). (c) In the biofilm, the QS networks are responsible for the virulence mechanism, and it has been observed that there is a correlation between these QS networks and the high cell densities that regulate them (Purevdorj, B. et al. 2002). (d) Biofilm-related organisms are capable of producing a large quantity of toxins and can evade host's immune responses such as phagocytosis (Donlan, 2002). (e) The formation of biofilms also protects the microorganisms inside them from the antimicrobial agents. These are a few of the many suggested mechanisms that make pathogens in biofilms more deadly than their planktonic counterparts.

15.4 ECOLOGICAL SIGNIFICANCE OF BIOFILMS

Biofilms play a vital role in the degradation and production of organic matter to keep up the natural elemental cycles of phosphorus, nitrogen, sulfur and many metals. Once the primary objective of attachment to a surface is achieved, microorganisms undergo many changes, such as shifts in life activities, cell morphology, motility, and resistance to toxic substances (Flemming, 1993). One of the biggest contributions of the biofilm is to plant life, where bacteria like *Bacillus subtilis* form biofilms that can lead to the promotion of plant growth and development and protect plants from the accumulation of plant pathogens (Arkhipova et al., 2005; Choudhary et al., 2009). Some of the ecological significance are as follows:

a. Biofilms play an important role in phytoremediation, where microbes are employed to precipitate, sequester, and modify the oxidation state of heavy metals, thereby cleansing and breaking down a broad range of toxic contaminants into less hazardous compounds.
b. The emergence of micro-consortia symbiosis, for example, in lichens, stromatolites, and other organisms.
c. It promotes plant growth and protects it from the accumulation of plant pathogens (Arkhipova et al., 2005; Choudhary et al., 2009).
d. Formation of ecological niches, for example, anaerobic zones under the biofilms that grow in aerobic environments. (Flemming, 1993).
e. It promotes growth in the desired niche while also improving soil fertility.
f. Environmentally friendly microbial biofilms help in controlling soil and water pollution.
g. It facilitates the preservation of genetic information thus creating opportunities for conjugation and close contact of the microorganisms in the biofilm along with extended retention time.
h. In streams, it acts as a microbial skin that regulates the processing and export of nutrients and organic matter from catchments and influences the dispersal of microorganisms (Battin et al., 2006).
i. It provides sustenance by scavenging and trapping the enrichment of nutrients obtained from the flowing water phase into the gel matrix.

15.5 QUORUM SENSING AND METABOLOMIC REGULATION

In the case of living organisms, not all the genes are expressed constantly. Most of them are expressed according to environmental signals. Cell-cell signaling occurs in a critical environment and helps in maintaining the density or the quorum of the bacteria by repression or induction of the QS controlled gene expression (Shrout et al., 2011). This mechanism of QS enables the bacteria to sense cell density and coordinate behavior in response to nutrient availability, toxic compounds, host–immune response, and defense (Miller Bassler et al., 2001). Nearly 40 years after the first reported discovery of QS in *Vibrio fischeri*, a luminous marine bacteria, the QS mechanism has been extensively understood (Nealson et al., 1979). QS modulates cellular functions, population density-based pathogenesis, nutrient acquisition, transfer of genetic material between the cells, motility, and synthesis of secondary metabolites (Veronica G. Preda, Oana Săndulescu, et.al 2019). To understand the mutual association of biofilm formation and its developmental process, understanding the metabolites involved plays a key role in all the cellular regulatory systems. They get accumulated in the microenvironments of the biofilms in conjunction with the population growth. The optimal threshold concentration maintains the biofilm while the critical threshold concentration triggers a signal transduction cascade which leads to a change in specific target genes that modifies the bacterial phenotypes (Balestrino et al., 2005). The proposed mechanisms for biofilm formation encompass σ factors and two-component systems (TCSs) or quorum sensing. Various σ factors control the expression of various genes, including virulence factors and global regulators (Bo Zhang et al., 2012). This becomes activated when the bacteria is exposed to stressful environmental conditions, for example, nitrogen limitation, starvation, and the like (Gill SR, Fouts DE, Archer GL, et al.2005). The two-component system regulates processes such as the bacterial cell cycle, virulence factor production, and cell-to-cell communication. The histidine kinase sensor protein undergoes autophosphorylation at a conserved histidine residue when a ligand stimulates it. The activation or repression of the genes takes place when the phosphoryl group is transferred to the cognate response regulator. In Gram-negative bacteria, N-acyl homoserine lactones (autoinducer-1 [AI-1]) have been identified as the signal molecules (Jayaraman A, Wood TK, et al. 2008). In Gram-positive bacteria, autoinducer peptides (with no conserved sequence) have been identified as a signal molecule that involves a two-component signal-transduction system (Sturme MH et.al).

15.6 BIOFILM DETECTION METHODS

Biofilms cause life-threatening diseases such as cystic fibrosis (CF), otitis media, periodontitis, infective endocarditis (IE), chronic wounds, and osteomyelitis (Southey-Pillig et al., 2005; Akyildiz et al., 2013; Masters et al., 2019). With the increase in the incidence of biofilms along with the challenges and threats they impose; modern screening methods have gained popularity to address the situation. A new approach was required, taking into consideration that the organism under study was the same but in a different form of association called the biofilm, which was due to the production of slime. Hence, the approach was primarily focused on studying the common phenotype of the organisms that displayed the characteristic features of slime production and secondarily studying the genotype that led to its production. There are both qualitative as well as quantitative methods for the detection of biofilm formation.

15.6.1 Direct Methods

15.6.1.1 Light Microscopy

This is the cheapest, convenient and baseline imaging technique used for observing the morphology of microorganisms adhered to the surface and it requires very simple sample preparation. It correlates with the biofilm cell mass and total biofilm mass. Compared to other microscopy techniques, its resolution is not high enough to determine inter-cellular relationships and morphotype differentiation (Azeredo J et.al. 2017).

15.6.1.2 Transmission Electron Microscopy (TEM)

It is conducted by measuring the elastic and inelastic interactions of an electron beam that is transmitted through a specimen (Kim B.H et.al. 2018). Sample preparation is a time consuming process. The liquid biofilm cells should be sealed properly for successful imaging. To minimize the electron beam scattering, the sample used is thin thereby ensuring a high resolution observation. The internal cross sectional detail of the biofilm can be studied. TEM is capable of providing a practical resolution of up to 1 nm if the sample is prepared according to nanoplast preparations, and 3 nm if the sample is prepared according to epoxy preparations (Lawrence J.et.al.2003). Trained laboratory workers are required, which is due to the necessity of a very thin sample and a vacuum environment. Not much used as it requires a drying step that might cause the destruction.

15.6.1.3 Scanning Electron Microscopy (SEM)

SEM uses an electron beam to produce a magnified image. The two types generally used are conventional and variable-pressure SEM (VP-SEM). Conventional SEM method is considered the best for biofilm visualization because it gets high magnification and high resolution images so as to accurately describe the morphology. It has a combined ability to image with a wide range of magnifications (20–30,000×) coupled with high resolutions (from 50–100 nm). One of the major drawbacks this technique can cause is the destruction of the exopolysaccharide layer and the shrinkage of biofilm during the dehydration and coating with conductive material (Alhede et.al. 2012). The limitation can be overcome by using highly tailored protocols or using alternative SEM. On the other hand, VP-SEM is a high-resolution imaging technique that visualizes fully hydrated cells with very little sample preparation. As compared to conventional SEM, this does not require any dehydration and drying steps that could cause the destruction of the exopolysaccharide layer. When coupled with a fixation step followed by heavy metal staining, it enhances contrast and resolution, which can be used for biofilm characterization. Fixatives used are mostly aldehydes and osmium tetroxide that binds specifically to the lipid layer, thereby preserving the fine features (Bozzola J et.al 1999). SEM is not capable of providing a resolution higher than 50 nm, since the required wavelength of electrons to achieve this results in damaging the biofilm sample (Merino L et.al 2019).

15.6.1.4 Confocal Length Scanning Microscopy (CLSM)

Due to the complexity of the biofilm phenomenon, it is more vital to observe and analyze the structure and dimensions of the colonies generated than the overall number of cells, which is why observation of the sample using confocal laser scanning microscopy (CLSM) becomes critical and this is most widely used microscopy method for biofilms (Neu, Thomas R et.al. 2014). CLSM is different from traditional microscopy since it does not only produce sharp images with high resolution but also creates a 3D image of the sample. Due to the ability of the confocal optics to focus on a very small volume in the sample, this can be done by excluding the light arriving from other locations. At first high-resolution, 2D slices of the sample are produced along the selected axis in the ascending or descending direction. Later, all the slices are arranged one above the other to create a realistic 3D image of the sample. Also, along with the 3D view of the sample volume, different areas of the sample can be selectively marked with fluorescent markers for sequential or simultaneous observation by single or multiple excitation lasers. Although being the most effective way of observing biofilms, it is also among the most expensive techniques, as it requires trained and experienced handlers to obtain accurate observations for analysis (Lawrence & Neu, 1999).

15.6.1.5 Fluorescent In-situ Hybridization (FISH)

Biofilm community is heterogeneous, so in order to identify specific microorganisms, we can employ the FISH method. The technique is based on hybridizing fluorescently labeled probes to ribosomal rRNA in permeabilized whole microbial cells. These probe sequences are complementary to the 16s rRNA sequence of microbes (Volpi EV, Bridger JM, et.al 2008). Rhodamine, FITC and even enzymes such as Horseradish peroxidase can be used as dyes. Analysis begins with fixing the biological material by using 4% Paraformaldehyde and dehydrating the sample, followed by the hybridization of sample with probe. Identification and quantification can be done using a microscope. The biggest advantage is that it has higher sensitivity than the quantitative polymerase chain reaction (PCR) and is not prone to false results, for example, due to contamination with small amounts of DNA (B. Brehm-Stecher, E. Johnson et.al 2006).

15.6.2 Indirect Methods

15.6.2.1 Tube Method

This is one of the easiest and oldest ways to qualitatively understand and observe biofilm growth due to the common phenotype of slime that adheres to surfaces. The tube method, or more popularly known as the Christensen tube method, relies on the ability of bacteria to adhere to smooth surfaces like the surfaces of glass and plastic culture tubes, which are used for experimentation in the laboratory. Cell lining present on the inner surface of the tube is considered to be indicative of adherent growth. On the other hand, the formation of a ring at the junction of the liquid-air interface was not taken into account for the adherence properties of the cells (Christensen et al. 1982). The earliest and most important addition to the standard protocol was the use of stains, which aided in the better visualization of adherent cells. The presence of adherent cells on the surface walls was considered to be evidence of slime formation by cells. This evidence was supported by a transmission electron micrograph (TEM) examination of the tubes, which revealed that this method successfully detected the presence of an EPS in which the bacteria were enmeshed with the tubes rather than cellular adhesion (Christensen et al. 1982). The Christensen tube method with the modification of the addition of the stain to the standard protocol helps to better visually observe the adherent cells in the EPS matrix. Based on the depth of the stain retained by different bacteria in the tubes after following the standard protocol of staining a comparative visual discrimination can be observed which directly corresponds to the strength of the biofilm formed. Judging by the extent of slime formed, darker tubes are indicative of a stronger biofilm compared to the lighter tubes. Although this method only serves to be qualitative, discrimination of biofilm exhibition can be done on a comparative basis (Hassan et al. 2011).

15.6.2.2 Congo Red Agar

A few years later, after the first ever biofilm detection test, Freeman and his colleagues found a viable method of qualitative detection of biofilm on a solid media surface. The Congo red agar (CRA) gets its name due to the direct use of a blend of dyes called congo red in the brain heart infusion agar (Freeman et al. 1989). The inoculated colonies, which initially grew pink but underwent a transformation of color and appeared black, dry, and crystalline, indicated a positive test result for slime producers, which is a characteristic feature of biofilm. On the other hand, the colonies that did not undergo any change in color and remained pink indicated a negative test result and were marked as non-slime producers. (Freeman et al. 1989). The precise mechanism by which color change occurs in CRA agar is unknown, but the fact that it can be associated with the involvement of a secondary product of a microbe is coherent as the observed change is in the latter phase of incubation (Freeman et al. 1989). Although the method is less sensitive due to the fact that the microbes have varying strengths of biofilm formation, it is highly specific for biofilm forming bacteria. (Hassan et al., 2011). Apart from being fast and reproducible, the CRA method has the biggest advantage that the colonies remain viable on the surface medium and can be used for further analysis. The easy-to-use CRA method possesses the limitation of being only a qualitative detection of biofilm formation, although comparative studies of the strength of biofilm formed by the streaked bacteria can be done by observing variations in color intensity produced by the different streaked colonies (Kaiser et al. 2013). As the bacterial biofilm can be classified on the basis of ranging color intensities, Arciola and her colleagues standardized a colorimetric scale of six different kinds of gradation, starting from very red, red, bordeaux, almost black, very black, and black. The categorization of bacteria could be done with the amount of slime produced, which reflected the strength of the biofilm. Strong biofilm producing bacteria were showing very black and these colonies were considered to produce normal quantities of slime, while less darker colonies were considered to indicate less slime production, leading to a weak biofilm. Lastly, very red to bordeaux coloured colonies were considered a failure to produce slime, meaning no biofilm production at all. This classification appears to be promising in terms of providing a faster and more accurate assessment of the phenotypic state of bacteria, owing to a more accurate interpretation of the CRA streaking (Arciola et al., 2001).

15.6.2.3 Microtiter Plate Assay (MtP assay)

Right from the discovery of biofilms and their impact on humans, the need to understand the various aspects of the biofilm life of the organism became very important. Quantitative detection became the need of the hour and the first ever study of quantitative detection of biofilm was described as early as 1970 on the surface of plastic tissue culture plates. Curiosity led to the introduction of 96-well tissue culture and MtP for quantification in the 1980s. The assay started with a basic understanding of the adherence of bacterial culture to the MtP under optimal conditions of nutrition and temperature. In 1996, a rapid screening method by Genevaux and his colleagues was published on the surface adhesion of *Escherichia coli* observed on MtP. Two years later, in 1998, O'Toole and his colleagues adopted the assay for *Pseudomonas aeruginosa* to reinforce his studies conducted on the genetics of biofilm formation. Later on, the protocol became famous and is still widely used for the study of biofilm formation, and *P. aeruginosa* became the model bacteria for the study. It is a batch system in which the biofilm is grown in a small volume and then assessed using a variety of methods. Most assessment methods rely on the use of dyes (for example, crystal violet and resazurin) to stain the biofilm or some of its components, colony forming unit (CFU) counts are also used (Azeredo et.al 2017). The simplicity, requirement of basic laboratory equipment, flexibility of sample size, and variety of samples that could be tested in one single assay were the reasons why the protocol became the first choice of the experimenters (Coffey et al. 2014).

15.7 CONCLUSION

The ability of microorganisms to congregate in the form of biofilms has played a quintessential role in their survival. This allows it to divide the labor and communicate via quorum sensing, which is responsible for the adapting when there is change in stimuli of any kind. As much as these biofilms cause menace by being more pathogenic than their planktonic counterparts, their positive ecological role should not be overlooked. The recent advances in detection methods help identify the biofilms both qualitatively and quantitatively quite efficiently.

REFERENCES

Akyildiz, I., Take, G., Uygur, K., Kizil, Y., Aydil, U. Bacterial biofilm formation in the middle-ear mucosa of chronic otitis media patients. *Indian J. Otolaryngol. Head Neck Surg.* 2013; 65:557–561.

Alhede, M., Qvortrup, K., Liebrechts, R., Høiby, N., Givskov, M., Bjarnsholt, T. Combination of microscopic techniques reveals a comprehensive visual impression of biofilm structure and composition. *FEMS Immunol. Med. Microbiol.* 2012; 65:335–342.

Arciola, C. R., Baldassarri, L., Montanaro, L. Presence of *ica*A and *ica*D genes and slime production in a collection of staphylococcal strains from catheter-associated infections. *J. Clin. Microbiol.* 2001; 39(6):2151–2156.

Arkhipova, T.N., Veselov, S.U., Melentiev, A.I., Martynenko, E.V., Kudoyarova, G.R. Ability of bacterium *Bacillus subtilis* to produce cytokinins and to influence the growth and endogenous hormone content of lettuce plants. *Plant Soil.* 2005; 272:201–209.

Azeredo, J., Azevedo, N.F., Briandet, R., Cerca, N., Coenye, T., Costa, A.R., Desvaux, M., Bonaventura, G.D., Hébraud, M., Jaglic, Z., Kačániová, M., Knøchel, S., Lourenço, A., Mergulhão, F., Meyer, R.L., Nychas, G., Simões, M., Tresse, O., Sternberg, C. Critical review on biofilm methods. *Crit. Rev. Microbiol.* 2017; 43(3):313–351.

Balestrino, D., Haagensen, J.A., Rich, C., Forestier, C. Characterization of type 2 quorum sensing in Klebsiella pneumoniae and relationship with biofilm formation. *J. Bacteriol.* 2005 Apr; 187(8):2870–80.

Battin, T.J., Besemer, K., Bengtsson, M.M., Romani, A.M., Packmann, A.I. The ecology and biogeochemistry of stream biofilms. *Nat. Rev. Microbiol.* 2016 Apr; 14(4):251–63.

Bose, S., Ghosh, A.K. Biofilms: a challenge to medical science. *J. Clin. Diagn. Res.* 2011; 5(1):127–130.

Bozzola, J.J., Russell, L.D. *Electron. Microscopy: Principles and Techniques for Biologists.* Jones & Bartlett Learning; Burlington, MA, USA, 1999.

Brehm-Stecher, B.F., Johnson, E.A. Single-cell microbiology: tools, technologies, and applications. *Microbiol. Mol. Biol. Rev.* 2004 Sep; 68(3):538–59.

Chandki, R., Banthia, P., Banthia, R. Biofilms: A microbial home. *J. Indian Soc. Periodontol.* 2011 Apr; 15(2):111–4.

Choudhary, D.K., Johri, B.N. Interactions of Bacillus spp. and plants—with special reference to induced systemic resistance (ISR). *Microbiol. Res.* 2009; 164(5):493–513.

Christensen, G.D., Simpson, W.A., Bisno, A.L, Beachey, E.H. Adherence of slime-producing strains of Staphylococcus epidermidis to smooth surfaces. *Infect. Immun.* 1982 Jul; 37(1):318–26.

Coffey, B.M., Anderson, G.G. Biofilm formation in the 96-well microtiter plate. *Meth. Mol. Biol. (Clifton, N.J.)* 2014; 1149: 631–41.

Costerton, J.W., Lewandowski, Z., Caldwell, D.E., Korber, D.R., Lappin-Scott, H.M. Microbial biofilms. *Annu. Rev. Microbiol.* 1995; 49:711–45.

Davey, M.E., O'Toole, G.A. Microbial biofilms: from ecology to molecular genetics. *Microbiol. Mol. Biol. Rev.* 2000 Dec; 64(4):847–67.

Donlan, R.M. Biofilms and device-associated infections. *Emerg. Infect. Dis.* 2001 Mar–Apr; 7(2):277–81.

Donlan, R.M. Biofilms: microbial life on surfaces. *Emerg. Infect. Dis.* 2002; 8(9):881–890.

Flemming, H.C. Biofilms and environmental protection. *Water Sci. Technol.* 1993; 27(7–8):1–10.

Freeman, D.J., Falkiner, F.R, Keane, C.T. New method for detecting slime production by coagulase negative staphylococci. *J. Clin. Pathol.* 1989; 42(8):872–4.

Gill, S.R., Fouts, D.E., Archer, G.L., Mongodin, E.F., DeBoy, R.T., Ravel, J., Paulsen, I.T., Kolonay, J.F., Brinkac, L., Beanan, M., Dodson, R.J., Daugherty, S.C., Madupu, R., Angiuoli, S.V., Durkin, A.S., Haft, D.H., Vamathevan, J., Khouri, H., Utterback, T., Lee, C., Dimitrov, G., Jiang, L., Qin, H., Weidman, J., Tran, K., Kang, K., Hance, I.R., Nelson, K.E., Fraser, C.M. Insights on evolution of virulence and resistance from the complete genome analysis of an early methicillin-resistant Staphylococcus aureus strain and a biofilm-producing methicillin-resistant Staphylococcus epidermidis strain. *J. Bacteriol.* 2005; 187(7):2426–38.

Hall-Stoodley, L., Stoodley, P. Biofilm formation and dispersal and the transmission of human pathogens. Trends Microbiol. 2005; 13(1):7–10.

Hassan, A., Usman, J., Kaleem, F., Omair, M., Khalid, A., Iqbal, M. Evaluation of different detection methods of biofilm formation in the clinical isolates. *Braz. J. Infect. Dis.* 2011 Jul-Aug; 15(4):305–11.

Heilmann, C., Hussain, M., Peters, G., Götz, F. Evidence for autolysin-mediated primary attachment of Staphylococcus epidermidis to a polystyrene surface. *Mol. Microbiol.* 1997 Jun; 24(5):1013–24.

Jayaraman, A., Wood, T.K. Bacterial quorum sensing: signals, circuits, and implications for biofilms and disease. *Annu. Rev. Biomed. Eng.* 2008; 10:145–167.

Kamatkar, N.G., Shrout, J.D. Surface hardness impairment of quorum sensing and swarming for Pseudomonas aeruginosa. *PloS One.* 2011; 6(6):e20888.

Kim, B.H., Yang, J., Lee, D., Choi, B.K., Hyeon, T., Park, J. Liquid-phase transmission electron microscopy for studying colloidal inorganic nanoparticles. *Adv. Mater. (Deerfield Beach, Fla.).* 2018; 30(4):10.1002/adma.201703316.

Kirov, S.M., Castrisios, M., Shaw, J.G. Aeromonas flagella (polar and lateral) are enterocyte adhesins that contribute to biofilm formation on surfaces. *Infect. Immun.* 2004 Apr; 72(4):1939–45.

Lawrence, J.R., Neu, T.R. Confocal laser scanning microscopy for analysis of microbial biofilms. *Meth. Enzymol.* 1999; 310:131–144.

Lawrence, J.R., Neu, T.R. Microscale analyses of the formation and nature of microbial biofilm communities in river systems. *Rev. Environ. Sci. Biotechnol.* 2003; 2:85–97.

Lawrence J.R., Swerhone G.D.W., Leppard G.G., Araki T., Zhang X., West M.M., Hitchcock A.P. Scanning transmission X-ray, laser scanning, and transmission electron microscopy mapping of the exopolymeric matrix of microbial films. *Appl. Environ. Microbiol.* 2003; 69:5543–5554.

Lemos Kaiser, T.D., Pereira, E.M., dos Santos, K.R.N., Noia Maciel, E.L., Schuenck, R.P., Ferreira Nunes, A.P. Modification of the congo red agar method to detect biofilm production by Staphylococcus epidermidis, diagnostic. *Microbiol. Infect. Dis.* 2013; 75(3):235–239, ISSN 0732-8893.

Liu, Y, Yang, S.F., Li, Y., Xu, H., Qin, L, Tay, J.H. The influence of cell and substratum surface hydrophobicities on microbial attachment. *J. Biotechnol.* 2004 Jun 10; 110(3):251–6.

López, D., Vlamakis, H., Kolter, R. Biofilms. *Cold Spring Harb. Perspect. Biol.* 2010 Jul; 2(7):a000398.

Masters, E.A., Trombetta, R.P., De Mesy Bentley, K.L., Boyce, B.F., Gill, A.L., Gill, S.R., Nishitani, K., Ishikawa, M., Morita, Y., Ito, H., Bello-Irizarry, S.N., Ninomiya, M., Brodell, J.D., Lee, C.C., Hao, S.P., Oh, I., Xie, C., Awad, H.A., Daiss, J.L., Owen, J.R., Kates, S.L., Schwarz, E.M., Muthukrishnan, G. Evolving concepts in bone infection: redefining "biofilm", "acute vs. chronic osteomyelitis", "the immune proteome" and "local antibiotic therapy". *Bone Res.* 2019; 7:20.

Merino, L., Procura, F., Trejo, F.M., Bueno, D. J., Golowczyc, M.A. Biofilm formation by Salmonella sp. in de poultry industry: detection, control and eradication strategies. *Food Res. Int.* 2019; 119:530–540.

Miller, M.B., Bassler, B.L. Quorum sensing in bacteria. *Annu. Rev. Microbiol.* 2001; 55:165–99.

Muhammad, M.H., Idris, A.L., Fan, X., Guo, Y., Yu, Y., Jin, X., Qiu, J., Guan, X., Huang, T. Beyond risk: bacterial biofilms and their regulating approaches. *Front. Microbiol.* 2020; 11:928.

Nealson, K.H., Hastings, J.W. Bacterial bioluminescence: its control and ecological significance. *Microbiol. Rev.* 1979; 43:496–518.

Neu, T.R., Lawrence, J.R. Investigation of microbial biofilm structure by laser scanning microscopy. *Adv. Biochem. Eng./Biotechnol.* 2014; 146:1–51.

Preda, V.G., Săndulescu, O. Communication is the key: biofilms, quorum sensing, formation and prevention. *Discoveries (Craiova, Romania).* 2019; 7(3):e100.

Preda, V.G., Săndulescu, O. Communication is the key: biofilms, quorum sensing, formation and prevention. *Discoveries.* 2019 Jul-Sep; 7(3):e100.

Purevdorj, B., Costerton, J.W., Stoodley, P. Influence of hydrodynamics and cell signaling on the structure and behavior of Pseudomonas aeruginosa biofilms. *Appl Environ Microbiol.* 2002; 68(9):4457–4464.

Rana, K.L., Kour, D., Yadav, A.N., Yadav, N., Saxena, A.K. Chapter 16–Agriculturally important microbial biofilms: biodiversity, ecological significances, and biotechnological applications. In *New and Future Developments in Microbial Biotechnology and Bioengineering: Microbial Biofilms*. Elsevier, 2020.

Rice, S., Tan, C., Mikkelsen, P., Kung, V., Woo, J., Tay, M., Hauser, A., McDougald, D., Webb, J.S., Kjelleberg, S. The biofilm life cycle and virulence of *Pseudomonas aeruginosa* are dependent on a filamentous prophage. *ISME J.* 2009; 3:271–282.

Sharma, P. Role and significance of biofilm-forming microbes in phytoremediation: A review. *Environ. Technol. Innov.* 2022; 25:102182, ISSN 2352-1864.

Southey-Pillig, C.J., Davies, D.G., Sauer, K. Characterization of temporal protein production in *Pseudomonas aeruginosa* biofilms. *J. Bacteriol.* 2005; 187:8114–8126.

Stewart, P.S. Mechanisms of antibiotic resistance in bacterial biofilms. *Int. J. Med. Microbiol.* 2002 Jul; 292(2):107–13.

Stock, A.M., Robinson, V.L., Goudreau, P.N. Two component signal transduction. *Ann. Rev. Biochem.* 2000 July; 69:183–215.

Sturme, M.H., Kleerebezem, M., Nakayama, J., Akkermans, A.D., Vaugha, E.E., De Vos, W.M. Cell to cell communication by autoinducing peptides in gram-positive bacteria. *Antonie Van Leeuwenhoek.* 2002; 81(1–4):233–243.

Sutherland, I. Biofilm exopolysaccharides: A strong and sticky framework. *Microbiology (Reading).* 2001 Jan; 147(1):3–9.

Volpi, E.V., Bridger, J.M. FISH glossary: An overview of the fluorescence in situ hybridization technique. *Biotechniques.* 2008 Oct; 45(4):385–6, 388, 390 passim.

Watnick, P., Kolter, R. Biofilm, city of microbes. *J Bacteriol.* 2000 May;182(10):2675–9.

Whitchurch, C.B., Tolker-Nielsen, T., Ragas, P.C., Mattick, J.S. Extracellular DNA required for bacterial biofilm formation. *Science.* 2002 Feb 22; 295(5559):1487.

Yadav, M.K. Role of biofilms in environment pollution and control. In: Patra J., Vishnuprasad C., Das G. (eds) *Microbial Biotechnology*. Springer, Singapore. 2017.

Zhang, B., Powers, V. Analysis of bacterial biofilms using NMR-based metabolomics. *Future Med Chem* 2012; 4(10):1273–306.